PLEISTOCENE GEOLOGY AND BIOLOGY

with especial reference to the British Isles

R. G. WEST F.R.S.

LONGMAN
London and New York

SECOND EDITION
LONGMAN GROUP LIMITED
London

Associated companies, branches and representatives
throughout the world

Published in the United States of America
by Longman Inc., New York

First published 1968
Second impression, with revisions,
published as a 'Longman Text' 1972
Second edition 1977
Second impression 1979

Library of Congress Cataloging in Publication Data
West, R. G.
Pleistocene geology and biology, with especial
reference to the British Isles.
Includes bibliographies and index.
1. Glacial epoch. 2. Glacial epoch—Great Britain.
3. Geology, Stratigraphic—Pleistocene. I. Title.
QE696.W47 1977 551.7'92'0941 76-28353
ISBN 0-582-44620-1 pbk.

Printed in Great Britain by
Whitstable Litho Ltd., Whitstable, Kent

CONTENTS

ACKNOWLEDGEMENTS

We are grateful to the following for permission to reproduce copyright material:

Association of American Geographers for diagram by Mackay from *Annals*, Vol. 62 (1972); the author J. Imbrie *et al.* and Academic Press Inc for diagram from *Quaternary Research*, Vol. 3 (1973); the author C. Sancetta *et al.* and Academic Press Inc. for a diagram from *Quaternary Research*, Vol. 3 (1973); Blackwell Scientific Publication Ltd for a diagram by A. G. Smith and J. R. Pilcher from *New Phytologist*, Vol. 72 (1973); Cambridge University Press for a diagram by H. Godwin from *History of the British Flora*, 2nd edn. (1975); Institute of British Geographers for a diagram (after J. B. Sissons *et al.*) from *Trans. Institute British Geographers*, Vol. 39 (1966); the author W. H. Zagwijn and Seel Press House Ltd for a diagram from *Ice Ages: Ancient and Modern*, Ed. A. E. Wright and F. Moseley (1975); Societas Scientiarum Fennica for a diagram by J. J. Donner from *Soc. Scient. Fennica Comm. Phys.-Math.*, Vol. 36; Ulster Journal of Archaeology for a diagram by A. G. Smith and E. H. Willis from *Ulster Journal of Archaeology*, Vol. 24–5 (1962).

We are also grateful to the following:

The Alberta Society of Petroleum Geologists for fig. 8.4c; *American Journal of Science*: fig. 9.6; *Biuletyn Periglacjalny*: fig. 5.17; Blackwell Scientific Publications Ltd: figs. 7.8, 13.11; British Association for the Advancement of Science: figs. 13.23, 13.24; Cambridge University Press: fig. 8.11; J. W. Edwards (Publishers), Inc: fig. 5.5; Ferd. Dümmler Verlag, Bonn: fig. 5.20; The Geological Association: fig. 11.5; The Geological Society: fig. 3.4; The Geological Society of America: figs. 3.5, 5.4, 13.17; Geologische Dienst: fig. 8.7b; Geologie en Mijnbouw: fig. 13.2; Geologisch-Mineralogisch Institut, Leiden: fig. 5.17; Geological Survey of Finland: plates 3, 4a, 4b; Macmillan & Co. Ltd: fig. 3.9; Dr. Sissons and Oliver & Boyd Ltd: fig. 12.22; The Royal Geographical Society: fig. 10.13; The Royal Meteorological Society: fig. 10.16; The Royal Scottish Geographical Society: fig. 3.14; The Royal Society: figs. 7.4, 13.13, 13.14, 13.15, 13.18; Smithsonian Press: fig. 9.4; Sveriges Geologiska Undersökning: fig. 5.1; John Wiley & Sons, Inc: figs. 8.4a, 8.10.

PREFACE TO THE FIRST EDITION

PLEISTOCENE studies are concerned with the reconstruction of conditions of environment and life in the immediate past, and a number of sciences are thereby involved. The geologist studies the chronology, stratigraphy and sediments of the Pleistocene, the palaeontologist and ecologist the record of fossil plants and animals. The geographer is concerned with the history of landforms in terms of the physical events of the Pleistocene, and the archaeologist with the history of man. The climatologist is interested in the climatic changes which are characteristic of the Pleistocene and which are still proceeding at present, and the glaciologist is interested in problems of ice advance and retreat.

It is important that the student in any one of these fields of interest should have a working knowledge of the principles involved in other fields, and in writing this book I have had the aim of providing a general account of various aspects of Pleistocene studies, suitable for use by third-year undergraduates, and by more advanced students, in the subjects of geology, biology, geography and archaeology. As well as the theoretical material, I have tried to include methods of investigation used in Pleistocene studies, as an understanding of field and laboratory work is very necessary in the study of the Pleistocene.

The area I have considered is restricted mainly to the glaciated and periglacial parts of northwest Europe, and the two final chapters deal specifically with the Pleistocene of the British Isles. In these chapters I have aimed to outline the present state of knowledge and the problems which await solution.

The reader will note that limited reference is made to Pleistocene stratigraphy outside the British Isles, to archaeology, and to various aspects of geology, such as marine and cave geology. These are subjects which are already well served by textbooks, and I have not intended this book to be encyclopaedic. Rather, I hope it will be a useful handbook to those interested in the Pleistocene of the British Isles.

Finally, I must thank my colleagues for their invaluable help and advice while writing the book. In particular, those who have helped me on subjects outside my direct experience—Dr J. M. Coles, Dr

B. M. Funnell, Dr K. A. Joysey, Dr H. H. Lamb, Dr V. R. Switsur and Dr E. H. Willis. I am also indebted to Dr F. A. Hibbert for much help with the manuscript, to Dr J. K. St Joseph for air photographs from the Cambridge University Collection and to Mr B. W. Sparks and Professor F. W. Shotton for photographs.

PREFACE TO THE SECOND EDITION

SINCE the first edition was published in 1968 our knowledge of the Pleistocene has been greatly expanded, both by the publication of many original researches in the diverse fields involved in Pleistocene studies and by the publication of several comprehensive textbooks dealing with particular aspects of the subject. These advances reflect the increasing interest which is being taken in the history of the environment in the immediate (geologically speaking) past, an interest fostered by uncertainties of environmental changes in the future.

The aim of this edition remains as with the first edition—to provide a general account of various aspects of Pleistocene studies, with especial reference to the British Isles, suitable for use by undergraduates, and by more advanced students, in the fields of geology, biology, geography and archaeology. The book is not intended to be encyclopaedic, or to cover the Pleistocene of the World, but for the more advanced students, references for further specialised reading are given. The arrangement of the book has remained the same. The chapter headings reflect particular fields of the subject likely to be points of interest for students of different disciplines. The matter has been brought up to date as far as possible. Since glaciology, glacial geology and geomorphology are well served by specialised textbooks, chapters in these aspects of the subject have not been greatly changed or enlarged in this edition. The remaining chapters have been revised and extended, but with a conscious effort to keep the length within reasonable limits. The emphasis on field aspects has been retained and strengthened, since useful contributions to the subject depend on competent fieldwork. In this way, I again hope to provide in this edition a useful and up-to-date handbook for those interested in the study of the Pleistocene of the British Isles, whatever their own field of training may be.

PLATES

between pp. 180–1

CHAPTER 1

THE PLEISTOCENE

THE Pleistocene covers the most recent part of geological time, the last two million years or so. The term was coined by Charles Lyell in 1839 to include recent strata containing molluscan faunas with more than 70 per cent of living species. A few years later, in 1846, Edward Forbes modified the use of the term to make it equivalent to the Glacial Epoch. This definition has remained generally in use. It recognises that the distinctive character of the Pleistocene is the occurrence of repeated extensions of the world's ice fields, on a scale which is recorded only a few times before in the earth's geological history.

Nevertheless, it is now realised that glaciation, though geologically the most obvious result of Pleistocene climatic change, was accompanied by equally important shifts in environmental conditions in non-glaciated regions and in the oceans. Moreover, it is also now evident that the build-up of glaciers started in some areas in the Late Tertiary. Perhaps, then, it is better to say that the Pleistocene is characterised by repeated climatic fluctuations, the most obvious result of which is glaciation.

The Pleistocene is of the greatest interest to the geologist and biologist. It is young enough for the sequences of rocks, faunas and floras to be preserved in great detail. Because ice sheets and valley glaciers still exist, we are able to interpret the phenomena of Pleistocene glaciation in terms of observations of present glacial phenomena. Similarly with the plants and animals; many of the Pleistocene species are still living, and we are able to interpret past conditions of vegetation and fauna, and so climate and environment, in terms of the properties of living species. We can obtain a record of the origin of the present patterns of distribution of living species, including man, whose evolution in the Pleistocene is well documented by the record in Pleistocene deposits of flint tools, and, less frequently, bones.

The climatic changes of the Pleistocene are largely responsible for the appearance of our present landscape. In the highlands erosion by glaciers has moulded the landforms, while in the lowlands a mantle of glacial and periglacial deposits covers much of the country. In the river valleys, terraces have been formed in response to changes of sea-level and to changes of velocity and load caused by climatic and vegetational changes.

There is no reason to suppose that the climatic conditions which lead to repeated glaciations in the Pleistocene have ceased and it is indeed very likely that we are now living in a temperate interglacial period. The cause of the climatic changes remains very much unknown though there are a number of hypotheses. A great deal of information is available about the results of climatic changes in geological and biological terms, but much remains to be discovered of the facts of the Pleistocene before any of the hypotheses can be considered proven.

From the economic point of view, the wide distribution of Pleistocene deposits makes them important in matters of land use, water supply, and as a prime source for sand and gravel for the constructional industries. With expanding requirements for construction materials and with increasing necessity for the rational use of land resources, the need for an appreciation of Pleistocene geology, in terms of sediment and stratigraphy, is obvious. From an environmental point of view, the climatic changes of the immediate past revealed by Pleistocene studies are fundamental for studies of climatic change and its causes. The economic importance of climatic change can hardly be overstressed, since the productivity of the agricultural and fisheries industries is directly dependent on climatic conditions. But even this economic importance has not been conducive to detailed mapping studies of the Pleistocene in Britain, and up to recently, apart from the nineteenth-century pioneers such as Prestwich and Clement Reid, there as been no great tradition of Pleistocene geological research as there has been in the other heavily glaciated countries of northern Europe.

'Drift' and the origin of the glacial theory[1,2]

A distinction made by geologists in the beginning of the nineteenth century was between underlying solid rocks, usually of regular and orderly arrangement, and the irregular beds of such unconsolidated deposits as sand and gravel, which lay scattered about the surface. These superficial deposits were frequently found to contain rock fragments (erratics) foreign to the local underlying solid rocks, and they became known as 'drift', with the implication that they had drifted into their present position by the action of flood-waters.

The widespread occurrence of the drift was taken by many to be the result of the Biblical flood or deluge. This 'Diluvial Theory' of the origin of the drift was brought forward by such geologists as William Buckland and Adam Sedgwick, during the 1820s. At this time Mantell divided the drift into an older part, named the Diluvium, formed by processes no longer operative, and a younger part, the Alluvium, formed by processes still operative (e.g. aggradation in rivers). These terms still persist, especially in Germany, and in general signify Pleistocene and Holocene respectively.

With the increase in polar exploration in the early nineteenth century, it became easier to visualise the agency of floating ice as a conveyor of drift, and the Diluvial Theory was to some extent replaced by a theory postulating the carriage of the drift to its resting place by floating ice. Charles Lyell and Roderick Murchison were amongst the early supporters of this theory. Once this view was accepted, the discovery of drift at high level, for example on Moel Tryfan in north Wales at 430 m, implied a time of considerable submergence, during which floating ice deposited the drift. Such a submergence, then known as the Great Submergence, was accepted by some protagonists of the Floating Ice Theory.

At about the same time, in the 1820s and 1830s, several continental geologists had been trying to explain the presence of large erratic boulders in Switzerland. J. P. Perraudin, J. Venetz, J. de Charpentier and L. Agassiz are associated with the attribution of the movement of these erratic blocks to glacier ice, and so with the origin of the Glacial Theory of the origin of the drift. It was supposed that the former extent of glaciers was far greater and during this great expansion of the glaciers the erratic blocks had been carried to the areas where, on the melting of the ice, they were deposited.

Agassiz introduced his ideas on the transport of the drift by glacier ice to Britain in 1840. He soon convinced Buckland and others of the ability of the Glacial Theory to explain the origin of the drift, but general acceptance of the theory was slow, and it was not until the 1860s and 1870s that the majority of geologists came to regard land-ice, not floating ice, as the agency of transport of the drift.

Stratigraphical studies of the drift started in the 1850s, and revealed evidence for more than one great expansion of former ice sheets and glaciers. The idea of multiple glaciation in the Pleistocene became accepted by most geologists in the latter half of the nineteenth century. In 1877, four glaciations were recognised by Geikie in East Anglia, and in 1909, four by Penck and Brückner in the Alps, and the same number of glaciations was subsequently recognised in other parts of the world. But now evidence has been obtained for six or more cold periods, each separated from one another by temperate stages. Much of this evidence for climatic change has come from biological studies of Pleistocene deposits, as well as from geological evidence.

Use of the term 'Pleistocene'

In this book I have used Pleistocene in the original sense of Lyell, making it include a segment of time up to the present day, and thus avoiding the terms Holocene, Recent or Post-glacial. The reasons for so doing are given in Chapter Eleven, on the subdivision of the Pleistocene.

Pleistocene environments: deposition and erosion

We can usefully distinguish in the Pleistocene of our area three major divisions of environment—the glacial environment, the periglacial environment and the non-glacial environment. The first is the environment which produced the thick glacial deposits which characterise the Pleistocene over much of northwest Europe. The periglacial environment prevailed during the glaciations in non-glacierised areas. If, as appears likely, the total of time covered by the cold stages exceeds that covered by the temperate stages, then this must have been for the biota the most persistent environment of the Pleistocene in northwest Europe. As far as can be seen, the periglacial environment is tundra-like, with vegetation of herbs and shrubs and with low mean temperatures, though the details vary according to climatic variation with time and with latitude and longitude. If the record of the last glaciation is similar to that of the earlier glaciations, then at some times and in some places frozen ground may develop with indications of considerable continentality of climate; at other times and places more oceanic conditions may ameliorate the climate.

The non-glacial environment shows the development of more temperate floras in the intervals between glacial advances. Physical evidence suggests a mean temperature difference of some 5°–7°C between the cold and temperate extremes. During these temperate intervals of the Pleistocene we may envisage a landscape similar to our landscape at present, but largely forested. The main depositional processes were as now: the filling of river valleys, especially seaward, by alluvium (peat, silt and clay), downcutting by rivers in upland areas, the accumulation of beaches and other marine sediments, the filling of lakes with lake sediments, the adjustment of unstable slopes by downwash, and the growth of peat on raised and blanket bogs. The deposits ('non-glacial') formed by these processes in former temperate intervals can now be seen buried beneath glacial deposits of subsequent glaciations.

During the glaciations a greatly different system of erosion and deposition is found. The advance of the ice sheets scoured and eroded the uplands and parts of the adjacent lowlands. Where the ice became slow-moving or ceased movement, it deposited the load of rock debris picked up further upstream, and so extensive sheets of glacial deposits are found in the distal areas of the ice advance. Outside the areas of glaciation, in the periglacial areas, the cold climate caused the formation of solifluction deposits, of loess, and of types of patterned ground now characteristic of arctic regions.

A third category of features includes the river terrace systems, formed by fluctuating conditions of erosion and aggradation, consequent on changes of sea-level, and on climatic and vegetational changes. Their formation continued throughout the periglacial and temperate periods.

The landscapes formed by all these processes are the background against which we must consider the geology and biology of the Pleistocene.

REFERENCE*

FLINT, R. F. 1971. *Glacial and Quaternary geology*. New York: Wiley.

SPARKS, B. W. and WEST, R. G. 1972 *The Ice Age in Britain*. London: Methuen.

WRIGHT, A. E. and MOSELEY, F. (d.). 1975. *Ice Ages: ancient and modern*. Liverpool: Seel House Press.

WRIGHT, W. B. 1937. *The Quaternary Ice Age*. 2nd edn. London: Macmillan.

1 HANSEN, B. 1970. 'The early history of glacial theory in British geology' *J. Glaciol.*, **9**, 135–41.

2 NORTH, F. J. 1943. 'Centenary of the glacial theory', *Proc. Geol. Ass. London*, **54**, 1–28.

* In the bibliography to each chapter, general references precede the numbered specific references.

CHAPTER 2

ICE AND GLACIERS

Glaciers

GLACIAL ice is found above surface in the form of glaciers and ice sheets and is distinct from the ground ice which forms below surface in regions of frozen ground. Glacial ice develops where accumulation of snow exceeds loss by wastage, and at present it covers some 15 million km^2, 10 per cent of the world's land surface. At the time of maximum glaciation in the Pleistocene some 47 million km^2 were covered.

The development of the gross form taken by glacial ice (fig. 2.1) depends on the climate and the landforms. In a mountain region whose altitude exceeds the regional snow line (i.e. the limit to which perpetual snow descends), glaciers will tend to form in corries and valleys (plate 1). At the onset of glaciation in northern Britain the snowline must have been lowered by several hundred metres from its present altitude, in regional terms, of about 1500 m. If the glacier system is extensive there will be main and tributary glaciers. If the main valley glaciers flow down on to the lower foreland regions they will form radiating piedmont glaciers.

If the climate is sufficiently favourable for glacier formation, the bulk of ice grows, becomes independent of the landforms, and buries them. Thus an ice cap may form and grow to become an ice sheet (plate 2). Examples of ice sheets are the great continental ice sheets of Greenland and Antarctica. Such ice sheets may marginally impinge on a mountain system and have outlets through their valleys, so that these outlet glaciers resemble valley glaciers.

Formation of ice

Snow after falling suffers compaction, crushing, melting and re-crystallisation, and these processes result in loss of porosity and permeability and increase in grain size, density and hardness till ice is formed from the snow. Thus new snow flakes, density about 0·1, after a few days become what is known as old snow, which has a granular structure and greater density. This becomes converted to firn with a density of 0·3 and greater. Further compaction and recrystallisation leads to the production of ice, impermeable and with a density greater than 0·8. This ice may start flowing under gravity and by plastic deformation becomes glacier ice.

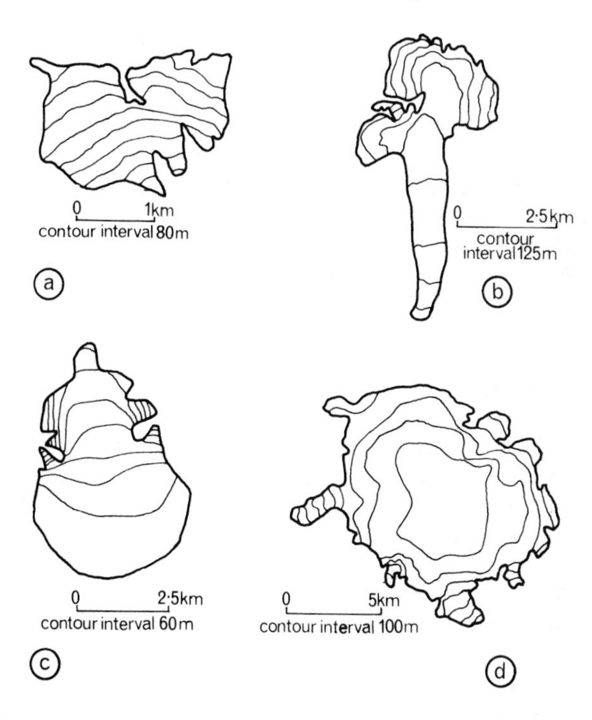

Fig. 2.1 Surface contours and outline of ice of various types of glacier (after Ahlmann).

a. Corrie glacier. Sonntag Kees, the Alps (lat. 47°07′ N, long. 12°18′ E).
b. Valley glacier. Vedretto del Forno, the Alps (lat. 46°19′ N, long. 9°43′ E).
c. Piedmont glacier. Murray Glacier, Spitsbergen (lat. 78°44′ N, long. 11° E). Note the expanded lobate terminal area.
d. Glacier cap or small ice sheet. Hardanger, Norway (lat. 60°32′ N, long. 7°25′ E).
e. (*overleaf*) The Vatnajökull ice sheet in Iceland. There is a central domed plateau with steeper draining glaciers of various shape at the margin.

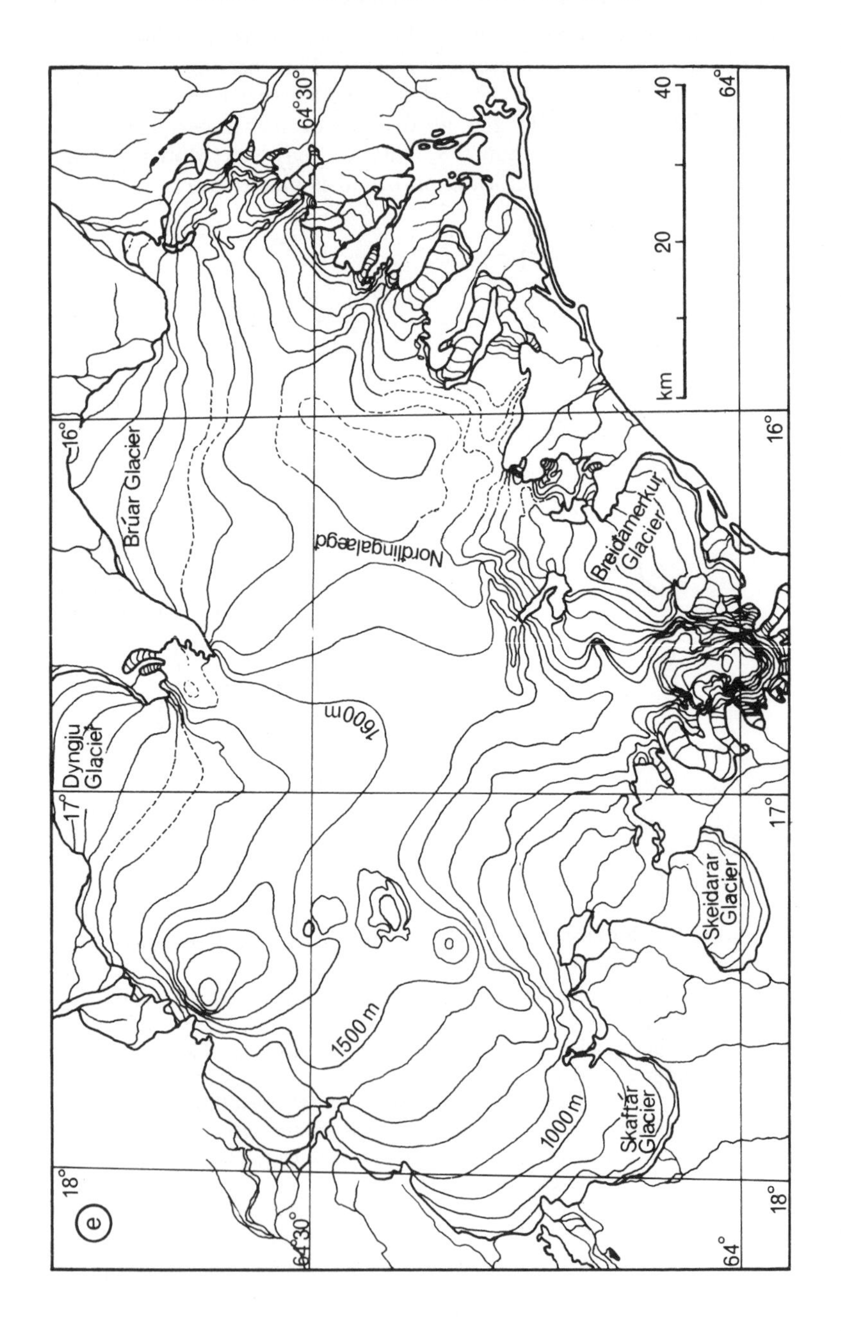
e
Brúar Glacier
Dyngju Glacier
Nordlingalaegd
Breidamerkur Glacier
Skeidarar Glacier
Skaftár Glacier
1600 m
1500 m
1000 m
16°
17°
18°
64°
64°30'
km
20
40

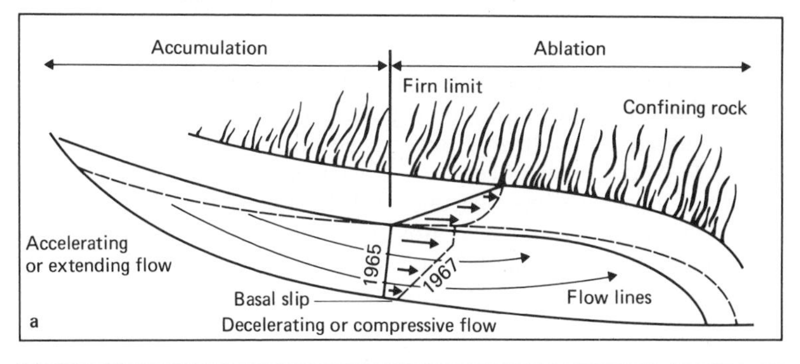

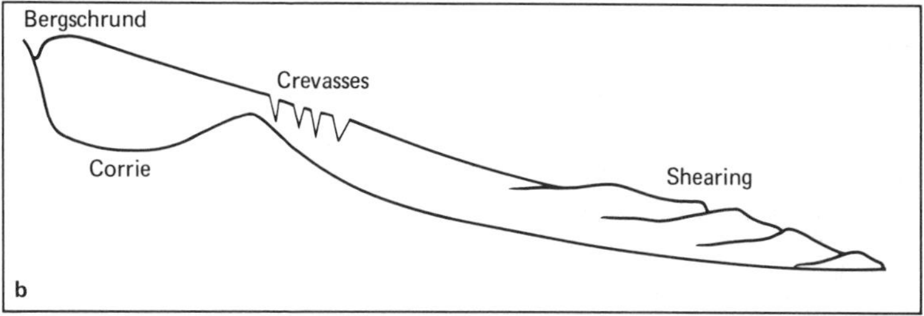

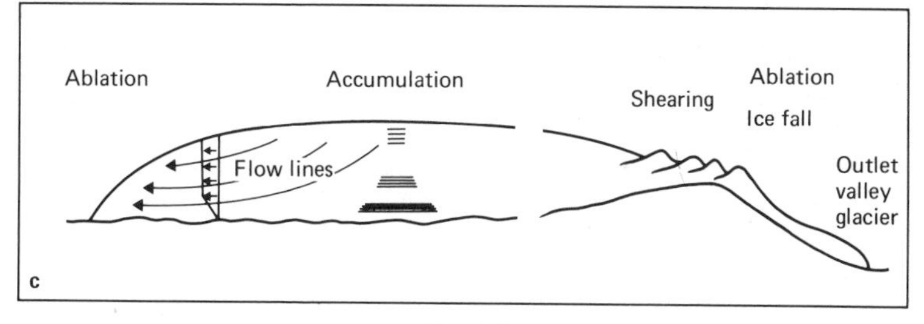

Fig. 2.2

a. Section down centre line of a valley glacier, showing regions of accumulation and ablation, and relative rates of flow in vertical and lateral section over a period of two years. The two vertical lined areas show surplus of accumulation on left and wastage of ice on right.

b. Section down centre line of a valley glacier, showing relation of crevassing to bedrock. Concentration of drainage may occur in bergschrund and crevasse areas, giving kames on down-melting. Shearplane moraines may form at the terminus, on down-melting or push-moraines if readvance occurs.

c. Section through an ice sheet. To the left the ice terminus is in a relatively horizontal area; to the right the ice drains through a col in a ridge and takes the form of a valley glacier. Snow accumulates in annual layers, which thin by deformation with increasing depth. The small arrows indicate relative rates of flow (partly after Dansgaard *et al.*, 1971).

Morphology of glaciers

The firn limit (limit of annual snow blanket at the glacier surface at the end of a melting season) on a glacier divides off the upper zone of accumulation of ice from a lower zone of ablation (wastage) of ice (fig. 2.2). At the headward end of a glacier there is usually a deep, near vertical, crevice, the bergschrund, dividing the ice from the confining rock. The surface of a glacier down its length is very variable and is dependent on the rate of flow and the slope of the subglacial floor. If the subglacial floor is stepped an ice fall will be formed. Crevasses are produced in the ice when stretching, resulting from differential flow rates, overcomes the breaking strength of the ice. Longitudinal stretching, resulting from an increase in slope of the floor, forms transverse crevasses; lateral stretching, from a widening of the ice stream, causes longitudinal crevasses; oblique crevasses are due to shear against side walls, and radial crevasses the result of the type of glacier expansion seen in piedmont glaciers. Crevasses are important to the glacial geologist for they concentrate drainage and thus localise deposition of sediment by the drained meltwaters.

The terminus of a glacier may be sloping or steep. A gentle slope results from considerable ablation and consequent thawing of the ice. In a rapidly advancing glacier the edge may be much steeper, and one ending in standing water, the edge may form a vertical cliff which calves icebergs. The end- and side-moraines associated with glaciers are considered in the next chapter.

Within the ice itself, deformation and flow result in the formation of features important to the glacial geologist. Thrust planes may be formed near the snout of the glacier, resulting from movement of active ice over basal, immobile, or slower moving ice. Folding may be present and seems to be especially characteristic of areas of compressive flow where the ice becomes thicker. Two characters of banding which appear less important to the glacial geologist are foliation of ice, apparent from differences in the size and arrangement of ice crystals, and ogives, curved alternating light and dark bands or swalls or swells appearing on the surfaces of or within glaciers below ice falls.

Cold and warm glaciers

Two types of glacier have been distinguished, cold (or polar) and warm (or temperate) (fig. 2.3). They are not clear-cut divisions for one part of a glacier may be of one type and another part the other.

In cold glaciers, negative temperatures prevail throughout the ice. The base may be frozen or at melting point. There is little melting and melt-water occurs only on the surface. Such glaciers occur in high latitudes. At the prevalent low temperatures, they do not flow as easily as warm glaciers.

In warm glaciers, the ice at a particular depth is at or within 1°C of its pressure melting point at that depth. The surface layers in winter may have negative temperatures. Ablation results in melting and

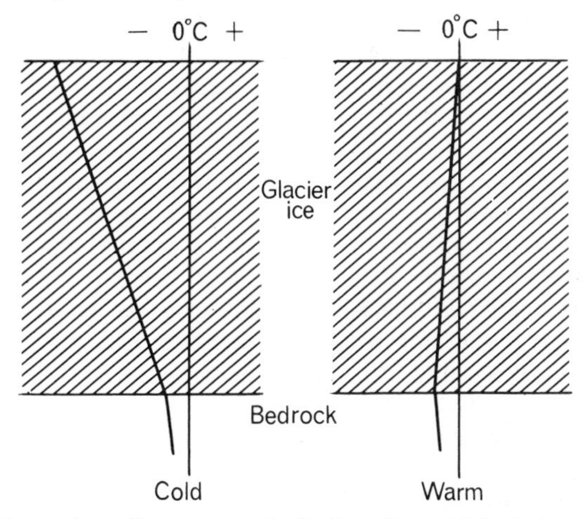

Fig. 2.3 Thermal gradients through the ice of a cold (polar) and a warm (temperate) glacier. At the base of the warm glacier melting will occur because the earth's heat cannot be conducted upwards on the reversed gradient established by the pressure melting temperature (after Sharp).

melt-water is present at the base and at other levels of the glacier. The ice of a warm glacier can be deformed more easily, and flow is relatively faster than in cold glaciers. Warm glaciers are those of highlands in middle and low latitudes, especially in maritime regions.

Because of differences in flow and melt-water formation, cold and warm glaciers probably differ in the nature and degree of erosion and drift transport they cause. Though the differences may not yet be distinguishable from studies of the deposits and landforms left by Pleistocene glaciers, they should be borne in mind when comparing ice work at the present day by cold and warm glaciers. For example, the cold marginal parts of an ice sheet may be effective in picking up drift by a basal freezing-in process,[6] and raising it to the ice surface, later perhaps to form flow till, while a temperate glacier may tend to form lodgement till.[1] In all probability, the Pleistocene ice sheets were predominantly cold except for their southern regions or those parts exposed to a maritime oceanic climate.

Movement of ice

The movement of ice increases with increase of gradient, of thickness and of positive economy (budget). The maximum rate of flow is found near the firn limit because here the quantity of ice passing at a given cross-section is greater, by the increase of accumulation, than that passing at a given point further upstream, and, because of ablation, more than that passing at a given point downstream.

In a valley glacier the horizontal and vertical velocity profiles are generally like that shown in fig. 2.2a. Speeds of greater than 30 m a day are known from outlet glaciers of the Greenland ice sheet, but many glaciers only move at the rate of a few metres per day. The most active glaciers are usually found where winter temperatures are low, with high precipitation, and summer temperatures high, with heavy ablation, conditions found in the montane areas of middle latitudes (the Alps, New Zealand). Besides these regular flows, irregular surges occur where the flow velocity increases by as much as a hundred times over a short period.[3, 8] Such surges have been recorded from valley glaciers and ice sheets. The cause of such surges is not fully understood. They are clearly due to some type of instability associated with glacier flow. Such instability could, for example, result from basal warming of a previously cold glacier base, which could initiate rapid sliding of the glacier ice. Such readjustments and surges are very probably of considerable importance in glacial geology as they may lead to rapid and successive ice advances in differing directions.

In general, two modes of glacier flow occur. The first is extending flow, in which the glacier surface is stretching along the flow line, partly to compensate for surface accumulation and to adjust to a decreasing thickness downstream. The second mode is compressive flow, which is the converse of the above and is normally due to the effects of ablation and increasing thickness downstream. These two modes of flow will produce different modes of erosion and deposition. Thus compressive flow may be associated with the shear planes seen carrying morainic debris to the surface of the ice near the snouts of certain glaciers.

Various processes have been put forward to explain ice flow. Mass movement of ice over the subglacial floor takes place by basal sliding and by internal deformation (creep) or shear throughout the ice mass. Basal sliding appears to be confined mainly to situations where the bedrock contact is at the pressure melting point. Internal shear takes place at all temperatures, and increases towards the bed of the glacier. Such internal deformation within the ice is explained by intragranular adjustments within each ice crystal and by progressive recrystallisation within the ice mass. Regelation of ice is important in the process of basal sliding.

Accumulation and ablation

Glaciers gain ice by accumulation (or nourishment) and lose it by ablation (or wastage).

Accumulation results from snow, frost, hail and freezing rain falling above the snow line in the area of the glacier's nourishment. Drifted and avalanched snow may add to the quantity. Much of the precipitation is in the form of snow, mainly caused by ascent of warm moist air into cooling temperatures over mountain regions. Because of the nature of the nourishment, in most regions winter temperatures are of prime importance in determining the rate of accumulation (fig. 2.4a).

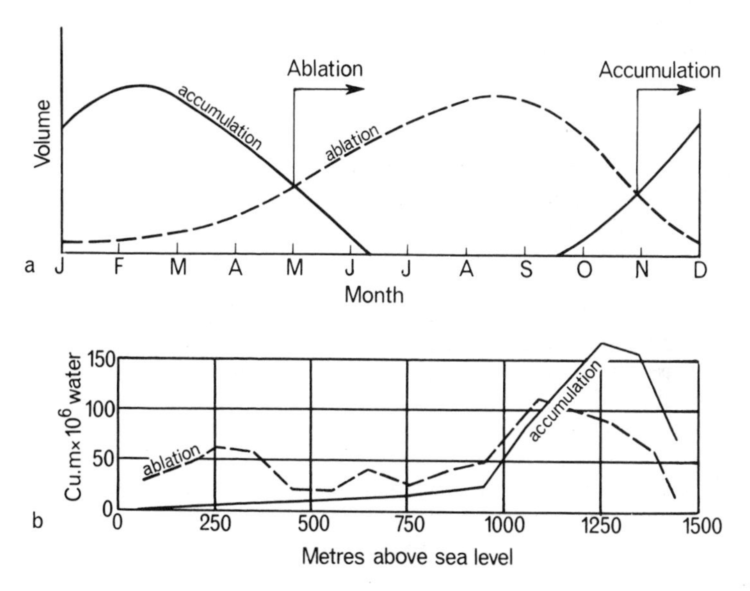

Fig. 2.4

a. The yearly course of accumulation and ablation in a glacier. If the glacier has a positive economy the area enclosed by the accumulation curve will be greater than that enclosed by the ablation curve, and vice-versa if there is a negative economy (after Sharp).

b. Economy of the Hoffell glacier at the east side of the Vatnajökull ice sheet in Iceland. The curves show average accumulation and ablation at successive heights over the period 1935–8 (after Ahlmann).

The area of ablation extends from the glacier terminus upwards to the firn limit of the glacier. Ablation by melting occurs partly through direct radiation of the sun. This effect is reduced by reflection from the snow's surface; it is important in high latitude cold glaciers. Evaporation and deflation of snow are also active as ablation agents in cold glaciers. Conduction, condensation and convection of heat from the atmosphere are important in causing ablation by melting of temperate glaciers in lower latitudes, especially in maritime climates. They result in greater ablation than that caused by radiation. Heat is lost to the ice from warm moist air rising over the glacier's surface. Cooling of the air in contact with the ice results in condensation and the heat liberated by condensation adds to the degree of ablation. If the ice is covered by a thin drift layer ablation may be increased because more heat is absorbed by radiation, but if the drift cover is thick, heat conduction will be slow and reduced ablation will result.

In addition to loss of ice by melting, evaporation and deflation, mass wastage may occur through calving of icebergs from an ice front meeting standing water. Calving results both from buoyant ice breaking off from submerged parts of the ice cliff and from ice masses falling downslope from above the waterline.

The processes of ablation are largely summer activities (fig. 2.4a), so that the degree of ablation in most regions is chiefly determined by summer temperatures, except where calving is dominant.

The balance between accumulation and ablation

The position and movement of the ice margins reflect the ratio of accumulation to ablation over an appropriate period. If the volume of ice gained by accumulation is greater than that lost by ablation (giving a positive economy), the ice front will advance. If ablation is greater than accumulation (giving a negative economy), the ice will thin and as a result there will be a recession of the ice front (fig. 2.4a). Flow may continue as recession occurs, except where the accumulation drops to such a value that flow halts and mass stagnation occurs. If accumulation balances ablation a standstill of the ice front occurs, but owing to short- and long-term climatic fluctuation such a situation is probably never realised for any length of time. Thus the net budget may determine the movement of the ice margin and variation in ice thickness, while total budget will affect the amount of total activity shown by the ice.

The term regimen has been used to describe the activity of a glacier taken as a unit, including the balance between accumulation and ablation (economy), the amount of accumulation and ablation (and thus the effective climatic factors) (fig. 2.4b) and the rate of flow.

Cold glaciers in regions with a continental climate at high latitudes (e.g. the glaciers of Antarctica and northern Greenland) characteristically have low rates of accumulation and low ablation. They have an inactive regimen, with relatively slow rates of flow. The warm glaciers of highlands in middle and low latitudes (e.g. those of Norway) have much higher rates of accumulation and ablation and are said to have an active regimen (fig. 2.4b). Warm glaciers are sensitive to changes of summer temperature, which cause changes in the ablation rate and are thus important in determining whether a glacier's economy will be a positive or negative one. The Greenland ice sheet varies between these two. In the southerly part near the coast, precipitation is higher, and accumulation greater than in the interior. In the southern coastal areas the regimen is much more active, with considerable rates of flow, than in the interior and the north, where flow may be very small. The differing regimens are of interest to the glacial geologist, for the activities of ice resulting in geological features depend on rates of flow and the amount of meltwater.[7]

Continental ice sheets

The characters of valley glaciers, their development, rapidity of flow and nourishment have been considered already in the course of the chapter. Continental ice sheets are of equal interest for mostly it is the Pleistocene equivalents of these that left the landforms and deposits which cover large areas of northwest Europe. The two existing ice sheets, those of Greenland (1·7 million km^2) and Antarctica (12·5 km^2) provide models for the study of Pleistocene ice sheets.[5] They were, in fact, rivalled in size during the Pleistocene by the great Laurentide ice sheet of North America (13·8 million km^2) and the north European or Scandinavian ice sheet (6·7 million km^2).

Recent developments of new survey methods for the study of ice sheets, such as radio echo and seismic profiling and ice drilling, have provided much new evidence of the morphology of ice sheets, ice movement and subglacial topography. The best-known ice sheet is that of Greenland and many of the following observations are based on evidence from it. Apart from emergent mountains (nunataks) an ice sheet completely buries the landscape so no trace of the original relief is left. Presumably ice sheets develop from valley glaciers which extend to form piedmont glaciers which themselves grow and coalesce to form the sheet.

The development of the Scandinavian ice sheet is shown in fig. 2.5.

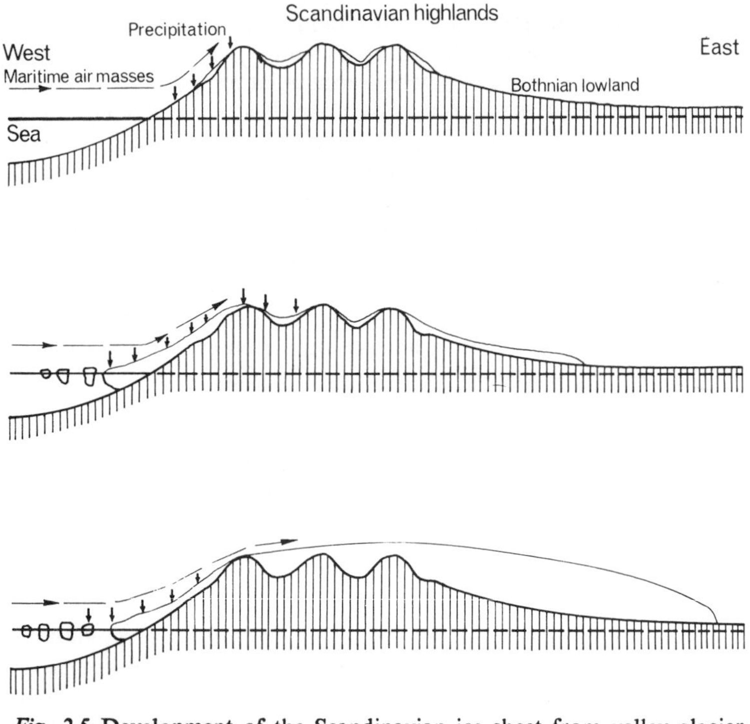

Fig. 2.5 Development of the Scandinavian ice sheet from valley glacier stage to piedmont stage to ice sheet (after Flint).

The result in section is an ice dome with low surface gradient. Ice sheets have not been so well studied as valley glaciers and the distribution of movement within one is not well known. It probably follows the pattern shown in fig. 2.2c, with a slowly subsiding or very slow flowing centre, with flow increasing laterally. This is because nourishment tends to be greatest on the flanks where precipitation is greatest, compared with the drier conditions at higher altitudes near the centre. The low surface gradient is associated with greater ice thickness. Differential flow rates, perhaps related to regions of different accumulation rate and to differences in subglacial gradient, occur within the ice sheet. The greatest rates of flow will be near the firn limit, as already mentioned, and if this is near the margin as with the Greenland ice sheet, maximum flow and thus ice work will occur near the margin. As the ice sheet grows this area of maximum ice work will move outwards. As with valley glaciers, shearplanes will tend to develop at the margin, and perhaps further upstream if the relief is unfavourable to the advance of the ice.

The surface contours of Pleistocene ice sheets have been reconstructed on the basis of profiles of present-day ice sheets, which show a certain regularity of profile in spite of differences in accumulation rate; such regularity has in fact been predicted from studies of the properties of ice.[4] Nevertheless, reconstruction of surface profiles of some ice lobes of the Pleistocene Laurentide ice sheet of North America, based on geological and geomorphological evidence, indicates that surface gradients may have been much lower than expected and a problem arises of explaining ice movement under such conditions of low gradient.[2] Examples of surface profiles are shown in fig. 2.6, and an example of a reconstruction of the Scandinavian ice sheet in Late Weichselian times is shown in fig. 2.7.

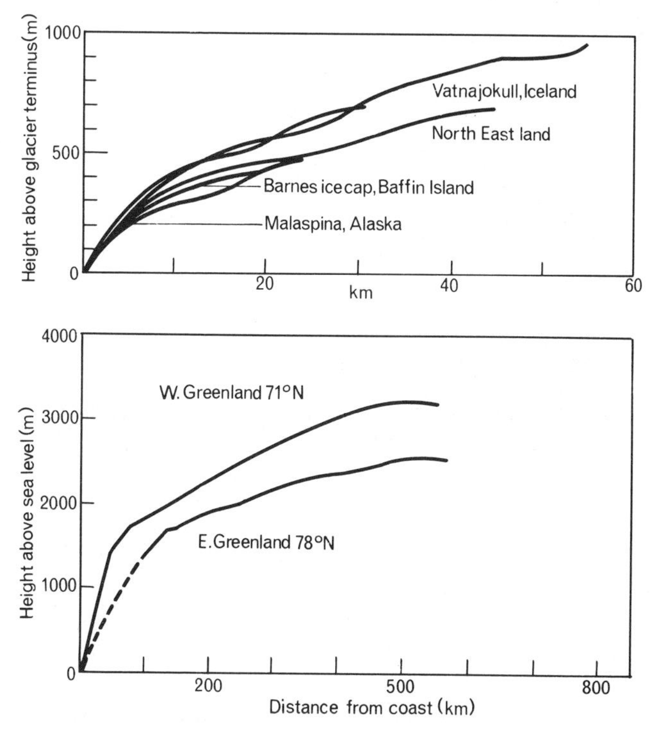

Fig. 2.6 Surface profiles of various small ice caps, in sections where ice rests on an approximately horizontal bed, and at two places at the edge of the Greenland ice sheet, at sections where the ice lies generally within ±500 m of sea-level (after Robin).

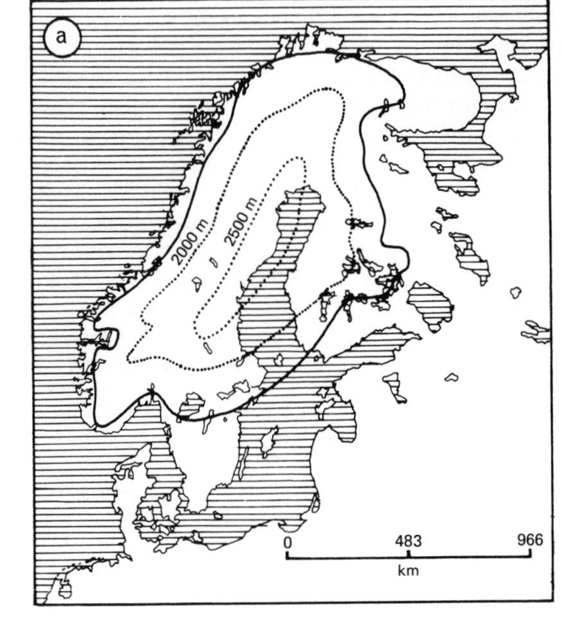

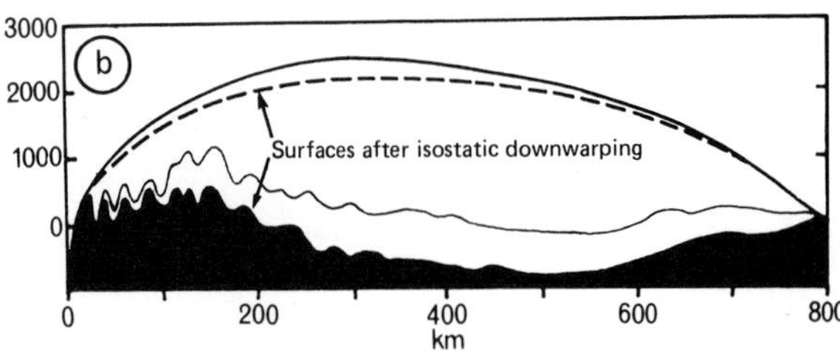

Fig. 2.7

a. Estimated surface contours of the Scandinavian ice sheet in Late Weichselian time (zone III) (after Robin).
b. Estimated profile across the Scandinavian ice sheet in Late Weichselian time (zone III). Estimated positions of ice and bedrock surfaces allowing for isostatic downwarping are also shown (after Robin).

It is unlikely that the balance between accumulation and ablation allowed very long standstills at the margin of an ice sheet. The economy would be very sensitive to climatic change particularly near the southern margins. Shrinkage of an ice sheet takes place primarily by recession through thinning and slower flow, though stagnation *in situ* is likely to occur later. During the course of shrinkage and disappearance small ice caps or valley glaciers will be left on highlands if net accumulation remains in these areas. These may readvance if the accumulation balance increases again. If there is complete stagnation the ice sheet thins and masses of ice become separated by the emergent subglacial landforms, and the ice may survive longest in the basins and valleys.

REFERENCES

AHLMANN, H. W. 1948. *Glaciological research on the north Atlantic coasts.* Royal Geographical Society Research Series, No. 1.

BUDD, W. F. and RADOK, U. 1971. 'Glaciers and other large ice masses', *Reports on Progress in Physics*, **34,** 1–70.

DANSGAARD, W., JOHNSEN, S. J., CLAUSEN, H. B. and LANGWAY, C. C. 1971. 'Climatic record revealed by the Camp Century ice core', in *Late Cenozoic glacial ages*, ed. K. K. Turekian, 37–56. New Haven: Yale University Press.

EMBLETON, C. and KING, C. A. M. 1968. *Glacial and periglacial geomorphology.* London: Arnold.

FLINT, R. F. 1971. *Glacial and Quaternary Geology.* New York: Wiley.

HATTERSLEY-SMITH, G. F. 1972. 'Present Arctic glaciation', *Defence Research Establishment Ottawa (Ontario) Technical Note*, **72,** 1–48.

PATERSON, W. S. B. 1969. *The physics of glaciers.* Pergamon Press.

SHARP, R. P. 1960. *Glaciers.* Eugene: Oregon State System of Higher Education.

SHUMSKII, P. A. 1964. *Principles of structural glaciology.* New York: Dover.

1 BOULTON, G. S. 1972. 'Modern Arctic glaciers as depositional models for former ice sheets', *J. Geol. Soc. Lond.*, **128,** 361–93.

2 MATHEWS, W. H. 1974. 'Surface profiles of the Laurentide ice sheet in its marginal areas', *J. Glaciol.*, **13,** 37–43.

3 NYE, J. F. 1971. 'Causes and mechanics of glacier surges: discussion', *Canadian J. Earth Sci.*, **8,** 306–7.

4 ROBIN, G. DE Q. 1964. 'Glaciology', *Endeavour*, **23,** 102–7.

5 ROBIN, G. DE Q. 1972. 'Polar ice sheets: a review', *Polar Record*, **16,** 5–22.

6 WEERTMAN, J. 1961. 'Mechanism for the formation of inner moraines found near the edge of cold ice caps and ice sheets', *J. Glaciol*, **3,** 965–78.

7 WEERTMAN, J. 1972. 'General theory of water flow at the base of a glacier or ice sheet', *Review Geophys. Space Phys.*, **10,** 287–333.

8 WRIGHT, H. E. 1973. 'Tunnel valleys, glacial surges, and subglacial hydrology of the Superior Lobe, Minnesota', *Geol. Soc. America, Memoir* **136,** 251–76.

CHAPTER 3

GLACIAL GEOLOGY

Glacial erosion and deposition

Distinct zones of erosional and depositional activity by ice in advance and in retreat may be recognised. First, an area of erosion, nearest to the centre of dispersion of the ice; second, an area of deposition of the drift where the load of debris picked up earlier by the eroding ice is deposited; then a zone of end-moraine formation which marks the termination of the ice, and finally a zone of proglacial deposition associated with melt-waters. Figure 3.1 shows the zones of activity and types of deposition.

a

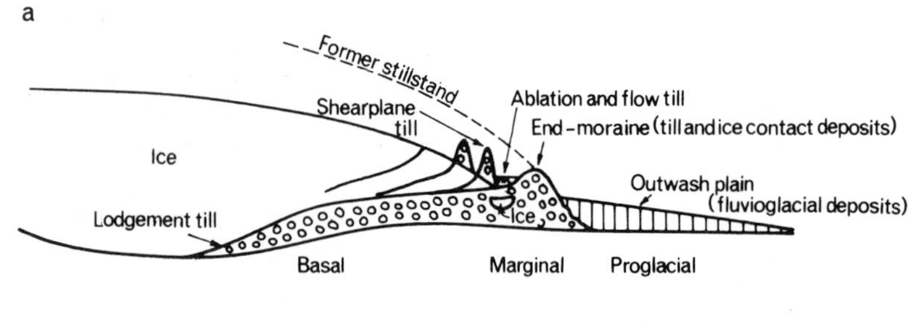

b

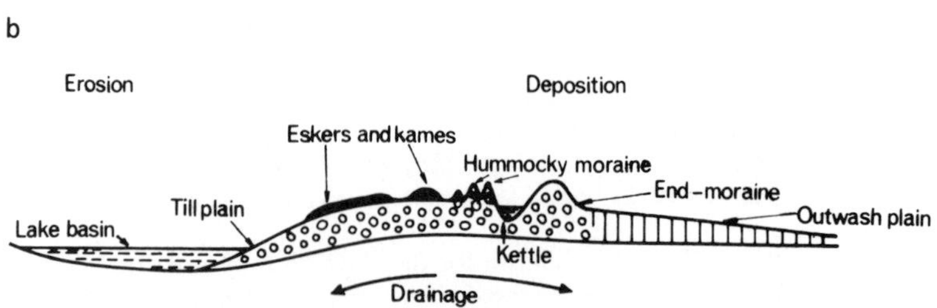

Fig. 3.1 Features associated with ice standstill and retreat.

a. Ice margin near maximum. Zones of basal, marginal and proglacial deposition are distinguished.
b. After retreat. Hummocky moraines of till and ice-contact deposits have been formed.

Thus in a region which has recently suffered glaciation, the inner zone may be recognised by a bare bedrock surface, with a mantle of drift absent or very thin, the next zone by the presence of large sheets of till in the form of ground-moraine, and the outer zones by the presence of end-moraine and ice-contact features running into outwash.

In such a setting we can proceed to describe the various glacial deposits, the form taken by them, and then features resulting from glacial erosion.

Glacial deposits

The history of glaciation is largely based on studies of glacial deposits and of the landforms resulting from accumulation of these deposits and from erosion by glacial ice. The deposits and their landforms are very variable and in order to be able to interpret them in terms of glacial history a good understanding of their origin is necessary. Glacial deposits may be classified according to their constituent sediments or according to the form these deposits take. For the classification of the sediments of the drift the degree of stratification, i.e. the part which water has played in their deposition, is the most important criterion. Thus we have the extremes of till—an unsorted and unstratified deposit formed directly from the ice without the intervention of water-sorting, and of the outwash sediments—those carried forward from the ice by melt-water and then deposited with sorting and stratification. Between these two extremes we have ice-contact stratified deposits, formed next to or near to the ice, but washed to some degree by melt-water. These subdivisions are convenient for describing glacial deposits, but, of course, all grades are possible between them and they cannot be considered as clear-cut entities.

The same may be said of the categories depending on the landforms taken by the glacial deposits. Here again we have three main subdivisions. First are those constructional landforms built up by the direct action of the ice, the moraines, usually composed of till. The moraine may be ground-moraine (that forming the generally low-relief landscape to the interior of the ice margin), or lateral moraine or end-moraine, according to its position in relation to the ice front. The second group is of ice-contact features which consist predominantly of ice-contact deposits, but which may also contain some till. Such features are usually formed by deposition from melt-water in close contact with ice. The last group is of features associated with proglacial sediments; for example, outwash plains and valley trains.

The classification may be summarised as follows:

Sediment 1. Unstratified; e.g. till.
2. Stratified. a. Ice-contact deposits.
b. Proglacial deposits; e.g. outwash sand and gravel, varved clays.

Landform 1. Moraines; e.g. ground-moraine, end-moraine.
2. Ice-contact features; e.g. kames, eskers.
3. Proglacial features; e.g. outwash plains, valley trains.

SEDIMENTS

Till[14]

Eroding ice picks up all shapes and sizes of rocks as it advances over bedrock, as well as the rock flour resulting from glacial abrasion. The deposits laid down directly by ice, known as till, have hardly suffered much sorting action by water and consequently they have a characteristically unsorted and unstratified appearance, all size grades from clay to boulders being mixed together in the deposit, with the softer erratics (e.g. chalk) being clearly striated. A mechanical analysis of till is shown in fig. 5.2. Different size grades may predominate in different tills; for example, the matrix is very largely clayey in the kind of till known as boulder clay.

Various types of till have been described, the classification based on the site of deposition in relation to the ice.[5] Lodgement till is deposited beneath moving ice as the load of debris increases and pressure-melting occurs. Melt (or melt-out) tills are formed subglacially or supraglacially by the melting of stagnant ice. Original englacial structures may be preserved in the sediment so-formed. In the course of melting near the margins of glacial ice, these types of till, especially when silty or clayey, may become heavily charged with water and flow, forming flow or slumped tills[21] by solifluction. Such flow tills are commonly seen in melting end-moraine areas. They may have the appearance of till, but will have been subject to solifluction.

The so-called ablation till may include both melt and flow tills formed at the surface or within ice by ablation of debris-rich ice, and, in valley glaciers, by the accumulation of surface material (supraglacial till) derived in part from the surrounding land surfaces. In ablation till the fine sediment grades are sometimes absent, having been removed by surface melt-water streams.

The differentiation of these till types is important in glacial stratigraphy because it will enable the processes involved in glaciation (ice advance, ice melting, etc.) to be identified and complex till sequences to be broken down into units perhaps to be related to particular processes of ice advance and retreat.[4] However, differentiation may be difficult. More studies are needed on tills being formed by ice sheets and valley glaciers at present.

Tills are characterised by the provenance of their contained erratic stones and matrix, and by their fabric, e.g. orientation of the stones. If the source of a particular rock-type found in the till is known, the direction of the movement which carried the erratic from its source will be apparent. In a region which has undergone multiple glaciation this direct interpretation may not apply, because erratics of an earlier glaciation may be redirected by a later one.

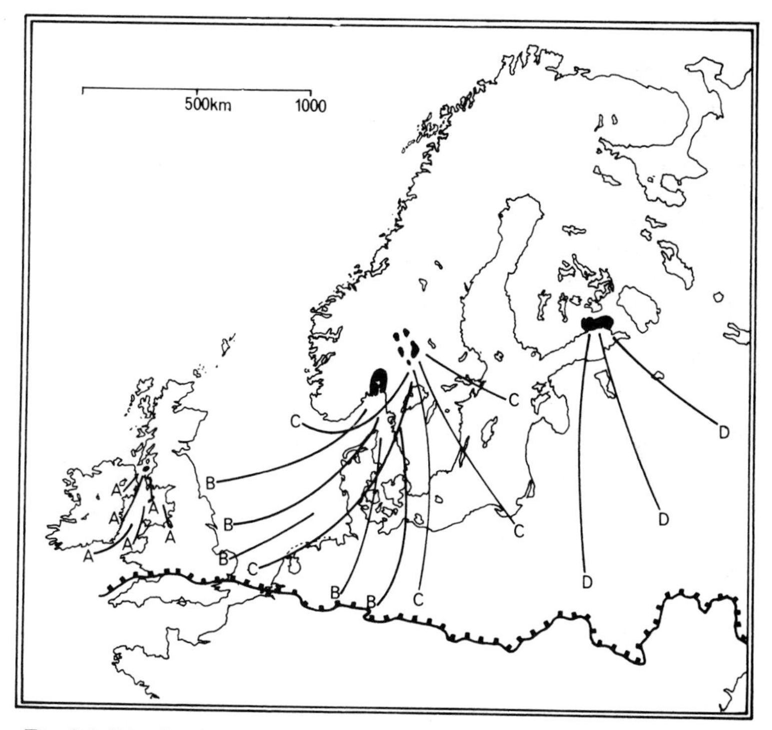

Fig. 3.2 Distribution of some erratics by ice in northwest Europe.

A. Ailsa Craig riebeckite-eurite.
B. Oslo rhomb-porphyry.
C. Dala porphyries.
D. Viipuri Rapakivi granites.
southern ice limit.

Where a rock source is sufficiently well known, it may be possible to map a boulder train, a spread of a particular erratic fanning out from the source rock (fig. 3.2), e.g. the boulder train of Ailsa Craig in the Irish Sea area.

The numerical analysis of erratic types in a till is sometimes used to characterise particular tills. For example, in Denmark the ratio between flints and igneous rocks has been used to characterise different till sheets.[19]

The matrix of a till also gives evidence of the direction of ice movement. Ice moving east in East Anglia picked up the dark blue Mesozoic clays of the Midlands and it is this clay which forms an important part of the till matrix and gives a blue colour to some of the chalky boulder clays of this area. Likewise, the matrix of a till deposited from ice which passes over wide outcrops of sandstone will be sandy.

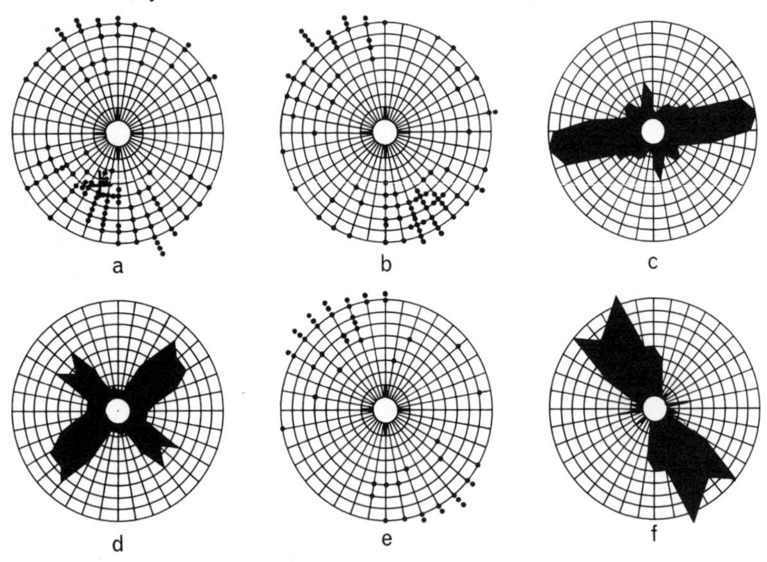

Fig. 3.3 Stone orientation diagrams. a, b, e. Dip diagrams. Each dot represents one stone; the direction of the long axis is shown by the direction of the radius on which the dot lies (north at the top), and dip by the distance of the dot from the outermost circle of the diagram, the concentric circles indicating 10° intervals of dip, 0° on the outermost circle, 80° on the innermost.

a. In talus of a till slope at Sporle, Norfolk, the cross marking the angle and direction of slope of the talus surface.
b. In the parent Gipping Till at the same section.

e. In Norwich Brickearth at Scratby, Norfolk.

c, d, f. Rose diagrams showing the directions in 10° classes of the long axes of stones in a till count. The number of stones in each class is shown by the distance of each peak from the innermost circle, each circle representing one stone in c and d, half a stone in f, with half the number of stones in each class being placed on opposite sides of the centre of the diagram.

c. In Lowestoft till at Scratby, Norfolk. The same cliff section as e and f.
d. In till band in ice, Brageneset, Nordaustlandet, Spitsbergen.
f. In Norwich Brickearth at Scratby, Norfolk; the rose diagram corresponding to the dip diagram e.

The orientation of stones within a till may also give important evidence of direction of ice movement. The striae on an erratic from till are often parallel to the long axis of the erratic, indicating that ice movement is in a direction parallel to the long axis. Such long-axis orientation has also been observed in tills in ice, and it results from the tendency of elongated stones to lodge in till with their long axes parallel to the direction of ice flow.[8, 13] The shape of the stone also affects the mode of orientation which results from the depositional processes.[9] If we take a sample of a hundred stones, for convenience not smaller than 2 cm size, from a till and measure the orientation of their long axes, the preferred orientation will give the direction of movement of the ice which deposited the till.[4, 8, 20, 22, 23, 24, 34] The measurements are simply made with a compass as the stones are excavated from a limited area of the exposure of the till, and the results are plotted on a rose diagram, as shown in fig. 3.3. The dip of the long axes can be measured at the same time.

There is some evidence that elongate stones tend to dip up-glacier. If there is a strongly preferred dip, it is possible that deposition has resulted from solifluction (fig. 3.3a). A random directional orientation may result from a multidirectional slumping associated with the formation of flow till.[29]

Care should be taken in fabric studies of till that the exposures are unaffected by weathering and undeformed by cryoturbation or by subsequent ice movement.[25] The exposures should be in areas of relatively flat landscape; if there is much bedrock relief, ice flow directions may be locally varied by it. Any interpretation of regional ice flow directions must be based on a large number of measurements over the whole region. Interpretations of the ice flow of different glaciations by measurements of a number of successive till sheets is shown in fig. 3.4.

Thin-section studies of tills show that there may be a preferred

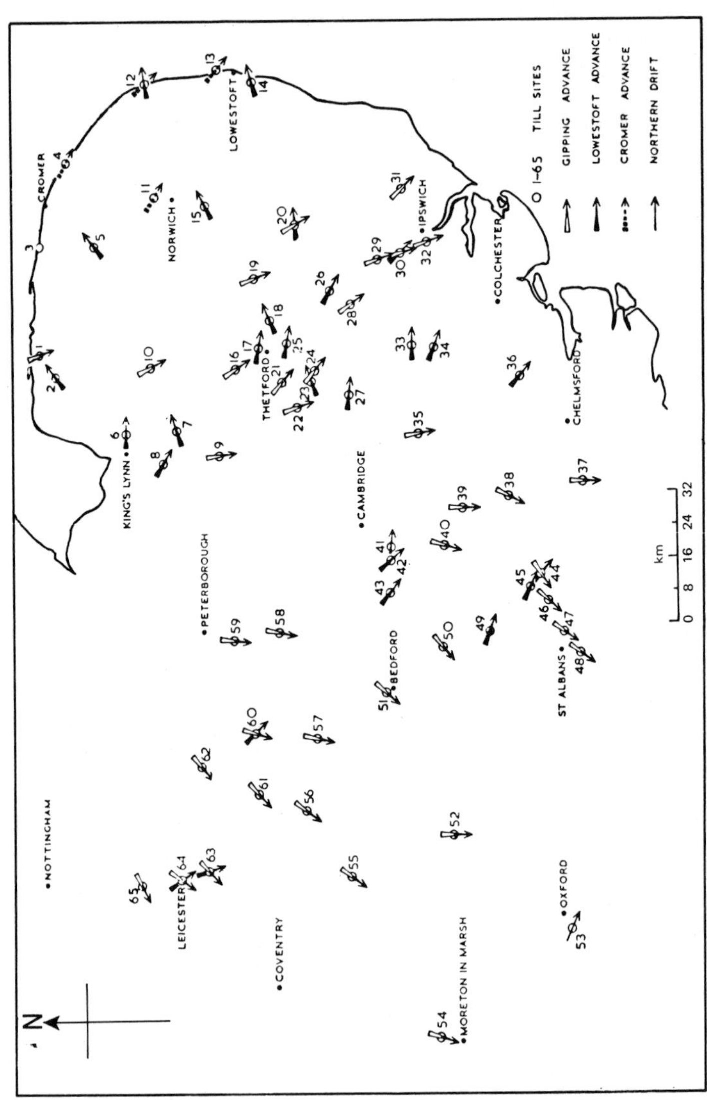

Fig. 3.4 Map showing directions of preferred orientations in East Anglia. Where there are two tills in one section the lower passes beneath the circle marking the site. The advances are based on ice movement directions. Since the orientation measurements were made (1954), increased knowledge of till stratigraphy has made revision necessary.

orientation of small fragments, and the fabric pattern resulting has been found in many examples to be similar to the macrofabric pattern. The microfabric is studied in the laboratory using thin-sections from small blocks of oriented till obtained in the field.[27]

Palaeomagnetic studies have been used to characterise tills in terms of magnetic susceptibility (dependent on the mineralogy of the till)[18] and of magnetic fabric, giving geomagnetic pole positions.[32] The presence of detrital remanent magnetisation in some tills suggests that a water or slurry layer may have occurred at the base of continental glaciers at the time of till deposition.[17] This may explain the known rapid advance of some Pleistocene ice sheet lobes.

Though till is usually structureless to the naked eye, it sometimes shows banding, lamination or contortion (plate 15b). These may be the result of structures present in the till-rich ice which deposited the till. Banding may be caused by alternate layers of coarse and fine-grained till, and lamination by deposition of till from shear-

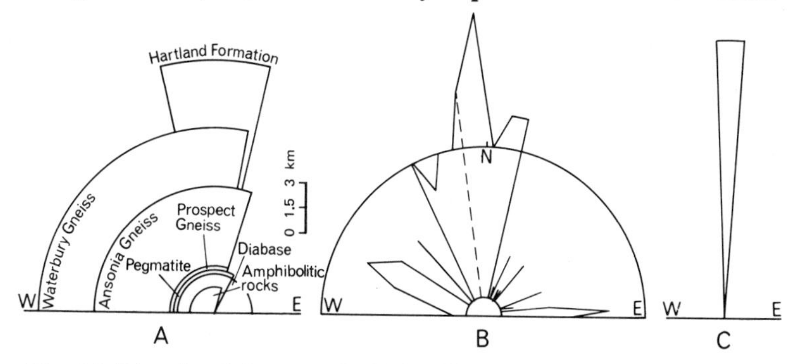

Fig. 3.5 Directional data on the Lake Chamberlain glaciation, southern Connecticut; azimuths of three parameters (Flint, 1961).

A. Probable directions of transport of rock types in the Lake Chamberlain Till. Data derived from 103 of 108 stones collected from the till at the type locality (5 were eliminated as not diagnostic of direction of transport). In plotting the diagram only the area north of the collection locality was considered. Radial lines drawn from locality of collection embrace limits of outcrop area of each rock type; length of line indicates distance to nearest outcrop, regardless of size.

B. Orientation diagram, Lake Chamberlain Till, type locality. Of 107 stones collected, 20 were first eliminated by inspection, because they are essentially equidimensional or have vertical long axes. Dashed line shows median orientation of the principal mode.

C. Directions of striations on bedrock immediately underlying Lake Chamberlain Till at the type locality. All lie within a range of 6°. Lengths of lines have no significance.

planes in ice. Such shearplanes may contain till of rather different constitution, say more chalky, and the result, on the melting of the ice, will be the formation of chalk-rich till laminations in the till. Contortions in till may result from melting-out of contorted ice such as is nowadays seen at the margins of advancing glaciers, especially in push-moraines, from shoving or drag by overriding ice after deposition, by slumping of water-rich till after deposition (flow till), and by diapiric response to post-depositional loading. These possibilities should be borne in mind when studying till, especially till fabric, as they will affect stone orientation. In particular, in shear-plane or narrow-band tills, preferred orientations may be transverse (fig. 3.3d).[13]

The relation of till to underlying beds is various. The contact may be a plane of erosion, or the till may rest on lower sediments with no sign of disturbance, or there may be folding and faulting of the lower sediments. Deformation results from the action of drag and push on weak incompetent beds below. If, on the other hand, deposition is dominant, the lower beds may not show deformation.

As examples of the scope of till studies and interpretation, the papers of Flint,[11] Dreimanis and Reavely,[10] and Krüger[23] are cited. Figure 3.5 shows the properties of a till in Connecticut studied by Flint.[11]

Ice-contact deposits

The characters of ice-contact stratified drift, intermediate between the unsorted till and well-stratified and sorted outwash, arise from the rapid changes in the processes of sedimentation which occur near melting ice. There are frequent marked and sudden changes in the grain size and sorting of the sediments, boulders being interbedded with sand, for example, and there may commonly be intercalations of till (plate 14b). The sediments are often built up against walls of ice, and when this melts, they will be deformed by slumping and faulting at the place of ice-contact. Deformation may also be caused by ice-push as well as ice-collapse.

Proglacial deposits

Melt-waters carry outwash from the ice front. The sediment deposited from the melt-waters depends on the marginal landforms. If it slopes steeply away, no outwash may be deposited as the velocity of the melt-water is sufficient to carry the load and perhaps cause erosion of melt-water channels as well. If the slope is gentle, sediments may be deposited above the local standing water level to form sheets or bands of outwash. Such glacio-fluvial deposits are supra-aqueous. Or the outwash may be discharged into an ice-marginal lake to form glacio-lacustrine deposits or into the sea to form glacio-marine deposits. In all these instances sorting occurs and the stratification of the sediments is far more regular than with the ice-contact sediments.

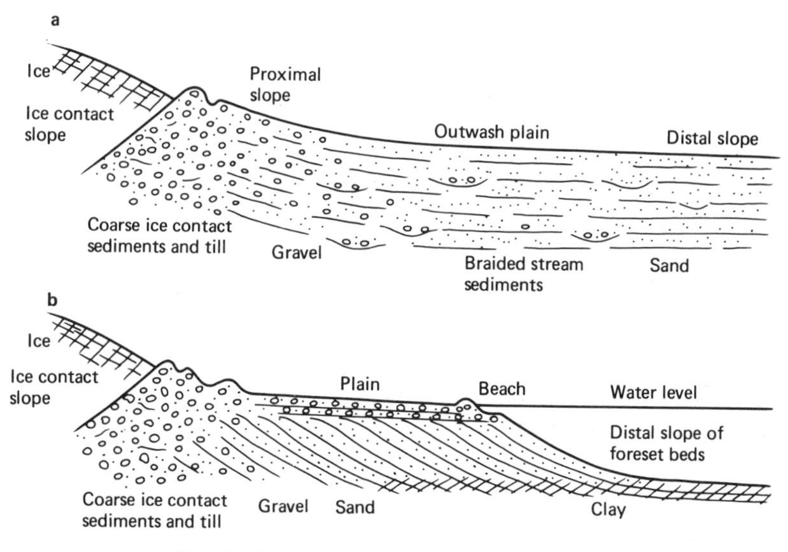

Fig. 3.6 Proglacial land forms and deposits.
a. Section through an outwash plain.
b. Section through a marginal delta.

The grain size deposited at any one place depends initially on the outwash load, whether it is rich in gravel, sand, silt or clay, and secondly on the distance of the place from the immediate melt-water source. Thus the coarsest grades are deposited near the source, the finest further downstream. In the case of the supra-aqueous deposits (fig. 3.6a), the coarser materials build up fans with a characteristic channelling such as would be formed by deposition from streams, while the finest grades are carried downstream and may be deposited only when they reach standing water, perhaps the sea. With sub-aqueous deposition (fig. 3.6b), deltas showing more regular sedimentation are built up by the coarser parts of the load shed by the melt-water, and the finer grades pass into suspension and are deposited distally to the deltas. Rafting of coarse material contained in ice into lake or sea may, however, introduce coarse unsorted or sorted grades into the finer sediments, and the grounding of icebergs may result in deformation of sediments. The wave action in such lakes will re-sort the marginal deposits to produce coarse beach sediments and finer offshore sediments. The finest sediments will be deposited in the deeper water.

It is in proglacial standing water that the laminated clays known as varved clays are formed (plate 3, fig. 3.7).[30] During the summer melt the finer grades are delivered into proglacial lakes, the particles here precipitating in order of their size, the coarsest first. At the onset of winter melting may cease and during this season the finest particles precipitate. The increase of melting in the next summer again introduces abruptly the coarsest grades into the sediment. The result is a sharp winter-to-summer transition and a gradual summer-to-winter one, the sediment between the sharp dividing lines represents one year's sedimentation, known as a varve (plate 3). Varves in freshwater have the clear-cut distinction just described and they are called diatactic varves (plate 4a). Where the water is saline, however, the dividing line between the varves is not so sharp, because flocculation of the finer particles caused by the presence of electrolytes enables these finer particles to precipitate soon after their arrival in the water body, and thus the distinction at the winter to summer change is not so sharp (plate 4b). Such a varve is named symmict. The annual deposition of varves, either diatactic or symmict, enables them to be used as valuable indicators of the passing of time.

The detailed study of the sediments and sedimentary structures of proglacial (and ice-contact) deposits allows the reconstruction of flow regimes and directions and water depth and distribution. Such studies are very important for the reconstruction of glaciation history in ice-marginal areas, especially where the ice-marginal landforms and drainage ways are complex.[1, 12]

Glacitectonic structures[2]

Small-scale *in situ* deformation by ice-push and shear results in folds and faults in drift and bedrock. These structures can be used to determine ice movement direction. Large-scale glacitectonic deformation will result from the pressure of ice acting on valley sides and scarps, as well as through the squeezing of drift and bedrock between advancing ice lobes. In addition to such compressional effects, it has also been suggested that tensional forces may act to displace rafts of bedrock. Some of the coastal sections in northwest Europe are well known for their display of glacitectonic structures, such as the thrust and fold structures of the Danish islands,[19] and the Chalk rafts of the north Norfolk coast.[2] These indicate the magnitude of the forces ice exerted to displace and carry forward masses of drift and bedrock.

The north Norfolk coast also shows sections of contorted drift (plate 15b) and large sand basins. It is thought that diapiric movement of glacial sediments followed loading by thick outwash, with domes of contorted drift forming, separated by basins of outwash. In contrast to the other types of deformation described above, these diapiric structures show no regional preferred orientation.

The quantities of till and stratified drift

The amount of different glacial sediments produced by an ice sheet or valley glacier varies greatly. Thickness of deposit bears a close relation to mode of advance and retreat; time is a subordinate negligible factor. A great thickness of till can be deposited in a short time if ice is carrying a large load of debris, or conversely, an ice advance may leave little or no till if the ice is clean. The quantity of stratified drift produced will vary according to the mode of melting of the ice. A down-melting *in situ* may produce a small amount of stratified drift, but a large amount of till,[16] much of which may be flow till. Near an ice margin reaching far south, ablation may be rapid and may be balanced by ice advance leading to prolonged production of melt-water and stratified deposits. Here the amount of stratified drift may greatly exceed the amount of till, which again may include much flow till.

LANDFORMS

Morainic forms

Moraine is the term for constructional forms built up by the direct action of the glacial ice. Deposits of lodgement till formed in the belt of deposition behind end-moraines of the ice front may take the form of gently undulating or flat plains, known as till plains or ground-moraine flats. The morainic forms of the interior may be of greater relief than this, and many irregularities can be introduced during the disintegration of the ice well behind the ice front.[28] Such disintegration will produce forms, mainly of till but with some stratified drift, which are either linear and directionally related to the ice movement, representing fillings of crevasses or accumulations of till in thrust planes in the ice (the so-called washboard forms), or are distributed more irregularly and form hummocky moraine areas and kettles (plate 5a). These last forms are deposited on the final down-melting of the ice and are not related to the original oriented features of the ice sheet, as are the former.

A special depositional form found in the lodgement zone behind the end-moraines is the drumlin (fig. 3.12, plate 5b). Ideally, drumlins are streamlined, oval, asymmetrical hills, varying greatly in size and shape, from about 100 to 2000 m long, 50 to 500 m wide, and 5 to 70 m high. They have their long axes parallel to the general direction of the ice flow in the region, and with steepest (stoss) sides upstream and gentlest (lee) downstream. Drumlins usually occur not singly but in fields of hundreds and thousands. Their appearance in the mass gives rise to the 'basket of eggs' landscape.

Lodgement till is the main constituent of drumlins, though sorted and stratified drift are sometimes found in them. The till usually shows a preferred orientation of long axes of stones parallel to the long axis of the drumlin. The predominance of till, together with the fact that drumlins are found in the zone of deposition immediately behind end-moraines, indicates that they were formed by actively advancing ice. The production of drumlins, instead of a till plain, is generally thought to be caused by increased lodgement of till on to obstacles, such as rock bosses (often found as cores in drumlins), immovable patches of basal drift in the ice, or overridden moraines, the result being streamlined hillocks with long axes parallel to the flow of the ice.

End-moraines

Where the ice edge has stayed stationary for any length of time a series of deposits, mainly till and ice-contact sediments, is accumulated at the edge. The presence of an end-moraine therefore indicates a balance between the rate of flow of ice forwards and the rate of ablation. The limit of a glaciation may not be marked by a well-developed end-moraine if de-glaciation immediately started when the advance finished. The hummocky form of end-moraines may distinguish them from the more subdued forms found inside and outside the end-moraine belt, but there is often difficulty in separating true end-moraine forms from those hummocky features left by disintegration of ice towards the interior of an ice sheet. In both types of area, hummocky forms may result from compressive flow of ice and intense shearing producing irregularities in deposition.

End-moraines are very diverse in form (fig. 3.7); this diversity results from the position of the ice front in relation to the landforms of the periglacial region, from the relative proportions of till and stratified drift produced, and from the length of time the ice front was stationary. A series of ice lobes may produce lobate end-moraines, with interlobate end-moraines formed where the lobes meet laterally. The morainic material may be piled into a series of ridges, parallel to the ice edge and made discontinuous by melt-water streams, or it may take the form of irregular hummocky topography full of small hillocks and enclosed lake basins. If the ice has pushed up a ridge of sediment in front of it a push-moraine is formed. Here the sediments will be contorted by the ice-push.

If the ice ends in a lake or in the sea the end-moraines are predominantly of stratified drift (fig. 3.6b) and take the form of delta deposits with a steep ice-contact slope and a slope of foreset beds distally to the ice. The deltas may be confluent to produce a long ridge parallel to the ice front or may be isolated.

The so-called recessional end-moraines mark short halts or re-advances in a general recession of an ice front. Their presence implies continued ice flow with much ablation, in contrast to the regional disintegration of an ice sheet resulting from stoppage of flow and increase of ablation. Annual moraines result from pressure ridges forced up by an ice front showing an annual oscillation. It may be difficult to separate recessional moraine landforms from general disintegration landforms, but the former characteristically show a linear arrangement corresponding to the position of the ice margin, absent in the latter.

The end-moraines of corrie and valley glaciers are often formed by girdles of fragmentary rubble, derived in part from the rock debris falling on the surface of the glacier from valley sides, and forming a block-moraine. These end-moraines often extend to the valley sides to link up with lateral moraines, formed as banks at the ice edge next to the confining valley slope (plate 1).

Ice-contact forms

The two characteristic forms associated with ice-contact sediments are the ice-contact slope, marking the place where such sediments were built up against ice, and the kettle or kettle-hole (plate 5a), a depression resulting from the collapse of sediments by the melting of a block of ice buried in them or representing the position of an ice pillar, roof pendant or ice block which prevented deposition of subglacial stratified sediments.[6] In the case of kettles formed by the melting of ice blocks, these may be blocks isolated from the main ice by areas of stagnation, or they may be bergs calved off the ice front; they become buried by ice-contact drift or outwash deposited as the stagnation proceeds. Kettle-holes are very variable in size; most have a diameter from 20 to 500 m and a depth of 5 to 30 m. Most frequently they form lakes in enclosed hollows, but they may be dry or linked with an outside drainage system. Although kettle-

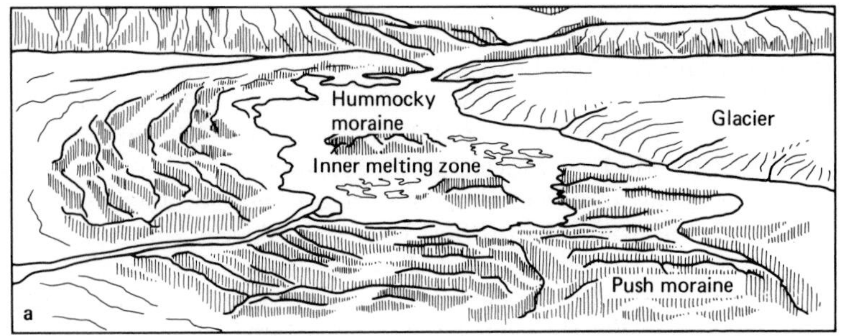
Hummocky moraine
Glacier
Inner melting zone
Push moraine
a

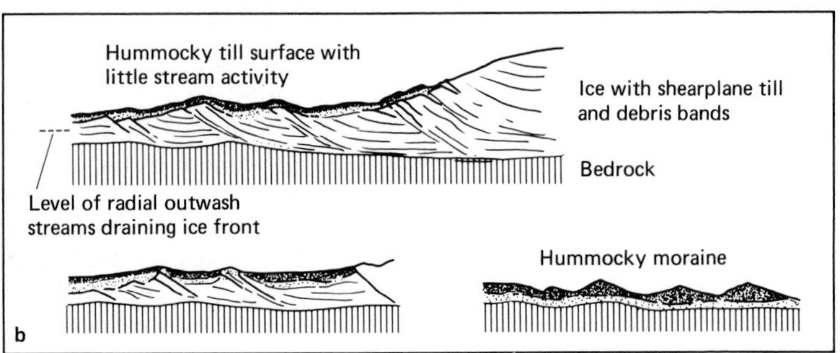
Hummocky till surface with little stream activity
Ice with shearplane till and debris bands
Bedrock
Level of radial outwash streams draining ice front
Hummocky moraine
b

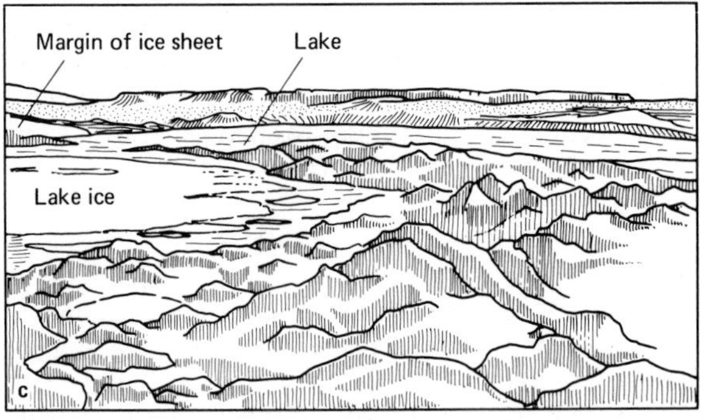
Margin of ice sheet
Lake
Lake ice
c

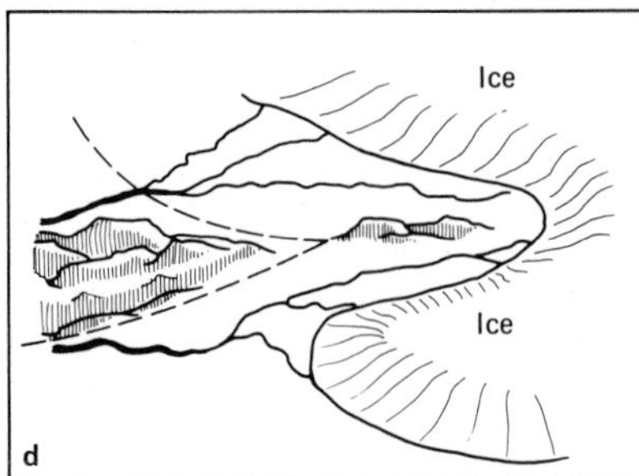
Ice
Ice
d

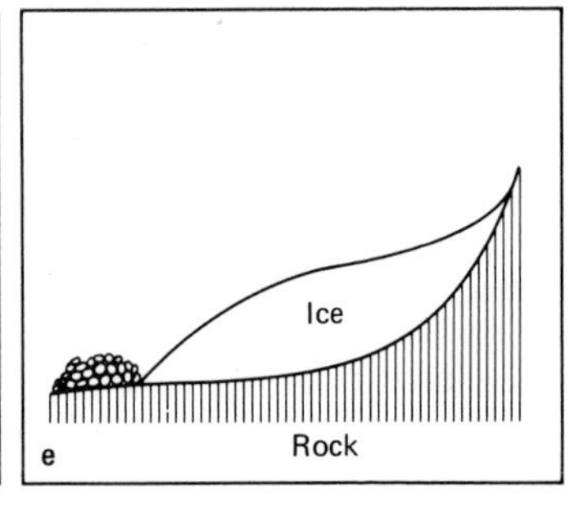
Ice
Rock
e

holes are characteristically found in ice-contact sediments, they have been reported occasionally in till.

The most important constructional features formed against ice by ice-contact stratified sediments are eskers, kames and kame terraces (fig. 3.8, plate 6a). Deposits in former melt-water streams confined by ice appear on the melting of the ice as the long sinuous ridges known as eskers (fig. 3.9). These ridges may range in height from 1 to 30 m and may, with discontinuities, extend up to several miles. They are usually symmetrical in cross-section and the general direction of stream flow is indicated by the arrangement of the irregular stratification of the ice-contact sediment. It is thought that the streams in which eskers form are usually subglacial, but other theories of esker formation contend that in part the streams may be supra- or englacial, and that when the ice melts, the deposits are let down to form the ridges. Probably eskers of all these possible origins occur but the subglacial type appear to be more common. Some eskers are beaded; that is to say, they show regular expansions at intervals. It is supposed that the expansions are deltas formed in an extra-marginal lake at the edge of the ice by the emergent glacial streams, each delta marking a former standstill of the ice front, or are small kames formed by the issuing water at stages of retreat.

Eskers are generally found to be roughly parallel to the last direction of flow of the local ice. Thus in a valley lobe, they may follow the course of the valley. Or in a larger ice lobe which mantles a subglacial topography of some relief they may course uphill and

---

Fig. 3.7 Various types of end-moraine.

a. Sketch of terminus of Usher Glacier, Spitsbergen, 1927 (after Gripp), about 4 kilometres wide, showing push-moraines.
b. Development of hummocky moraine as a result of supraglacial deposition (after Boulton, 1972b).
c. Sketch of hummocky end-moraine at the west margin of the Vestfonna ice sheet, Nordaustlandet, Spitsbergen, 1955. Much slumping of saturated till occurs here.
d. Interlobate moraine. In the re-entrant between two ice lobes a string of hummocky moraine and kames occurs, formed at a time when the lobes were more extensive (dashed line). Kames are shown formed at that time in the area of drainage concentration between the lobes.
e. Block-moraine. Section through terminus of a small valley glacier. Origin of part of the morainic debris may be by the sliding over ice of blocks of rock moved from the upper bed rock slopes by nivation. If the marginal feature is entirely of this origin then it is known as a protalus rampart.

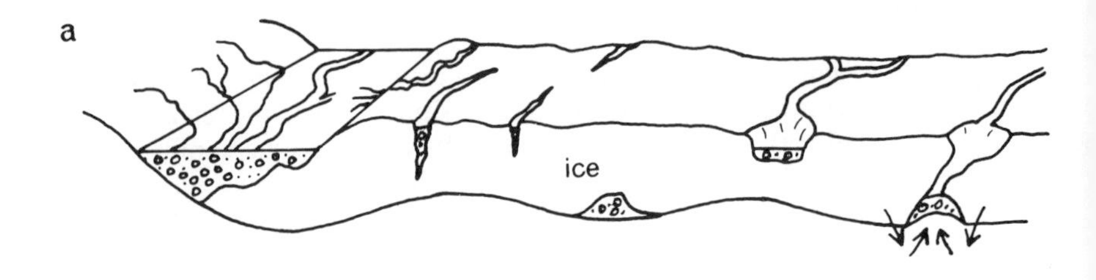

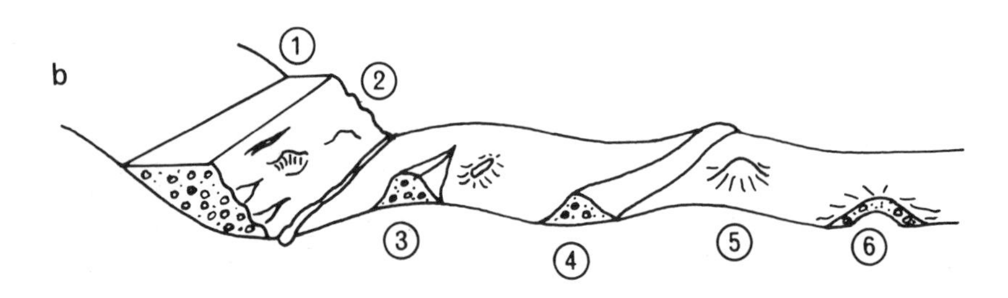

Fig. 3.8 Origin of various landforms built of ice-contact stratified sediments.

1. Kame terrace.
2. Ice-contact slope of terrace, with kettle-hole.
3. Crevasse-fillings.
4. Esker.
5. Kame.
6. Moulin kame with nucleus of till pushed up by ice pressure (marked by arrows).

over divides, as the flow of the subglacial water is dependent not on the subglacial topography but on hydrostatic pressures within the ice sheet.

It is evident from their mode of formation that eskers cannot occur during the rapid advance of ice and they are therefore a sign of ice stagnation, being formed during the thinning and stagnation of the terminal zone of the ice, when melt-water was abundant.

Kame terraces are formed laterally to ice by streams running between the ice and a confining slope. On the melting of the ice, the terrace margin nearest the ice wall forms an ice-contact slope, marking the previous position of the ice. The terrace and its ice-contact slope may be pitted by kettle-holes. The regularity and length of kame terraces depend on the form of drainage of the stagnant ice. Usually kame terraces are not so regular or extended as normal

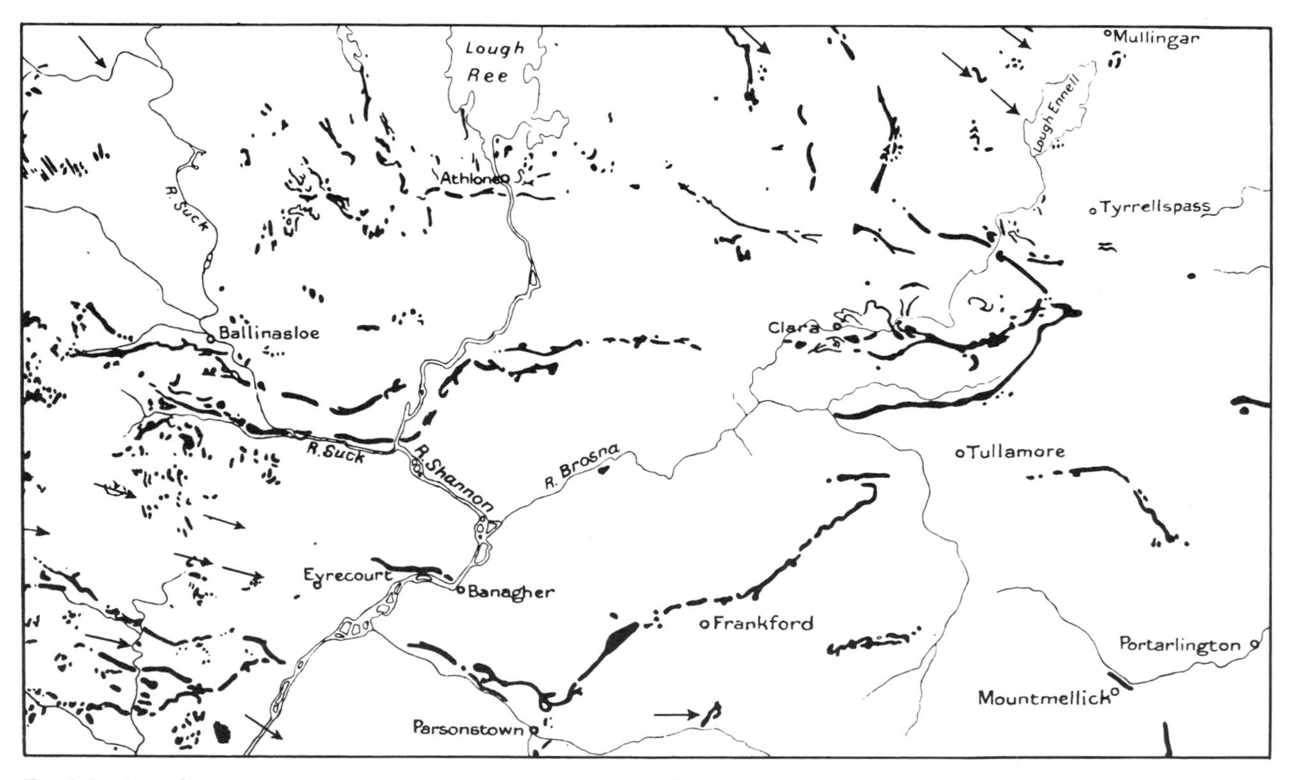

Fig. 3.9 Map of portion of the central plain of Ireland, embracing an area about 80 × 48 km, and showing the distribution and trend of the eskers. The arrows show the direction of the ice motion where this is indicated by glacial striae, drumlins, or the transport of material (Wright, 1937).

terraces formed by incision of rivers in flood plains, but it may be difficult to separate the two types.

Kames are low isolated hills, usually with steep sides, formed of ice-contact stratified drift. Often they are in the form of short ridges. Kames are formed in crevasses, perforations in the ice (moulin kames) or depressions in stagnant ice, or they may be deposited at an ice edge by emergent melt-water streams; in all these instances, the melting of the ice leaves isolated mounds of ice-contact stratified drift. Sometimes the mounds tend to occur on the tops of rises in the subglacial floor, where the ice is thinner during down-melting, and where drainage evidently became concentrated. If a group of elongated kames forms a pattern related to possible crevasse patterns near the margin of an ice sheet, they are usually considered as crevasse-fillings. The surface of a crevasse-filling (fig. 3.10) will be graded to the surface or sub-surface melt-water drainage system, while in a subglacially formed esker, the surface is usually parallel

---

Fig. 3.10 The 'esker' at Blakeney, Norfolk.

a. Plan of 'esker', with approximate contours.
b. Long profile of 'esker'.
c. Cross-profiles of 'esker'.
d. Sketch of relation between gravel surface of 'esker' and the till (pecked line) on which it rests.

The general slope of the 'esker' is towards the west, and its broad-topped east and west ends are almost level. If the ridge were a true subglacial esker it would be expected to show a more irregular crest roughly parallel to the subglacial floor (d), whereas the flat surfaces at the ends point to a more complex form in which crevasse filling played an important part. At the eastern end, where the feature is most marked, the 'esker' is parallel to nearby elongated kames which appear to be crevasse fillings, and the sharp angular bends in the ridge are consistent with its interpretation as a form due to melt-water drainage through a system of crevasses. The overall surface gradient indicates a final northwestwards flow (back into stagnating ice). The bedding in the gravels indicates the same direction of drainage.

The development of the ridge is interpreted as follows. A crevasse, formed where stagnating ice was under tension on the brow of a ridge, was filled from the east by a westward-flowing stream. Fluvioglacial material built up until it formed the flat eastern end of the ridge and overtopped the divide to the west. Some erosion here is indicated by sections which show the gravel to rest in a channel cut in the marly boulder clay. Also the gravel is thin compared with the much greater thicknesses at either end, as is shown in the generalised profile (d). Over the divide the gravel built up on the down-slope in another crevasse system to form the considerable thickness shown at the northwest end.

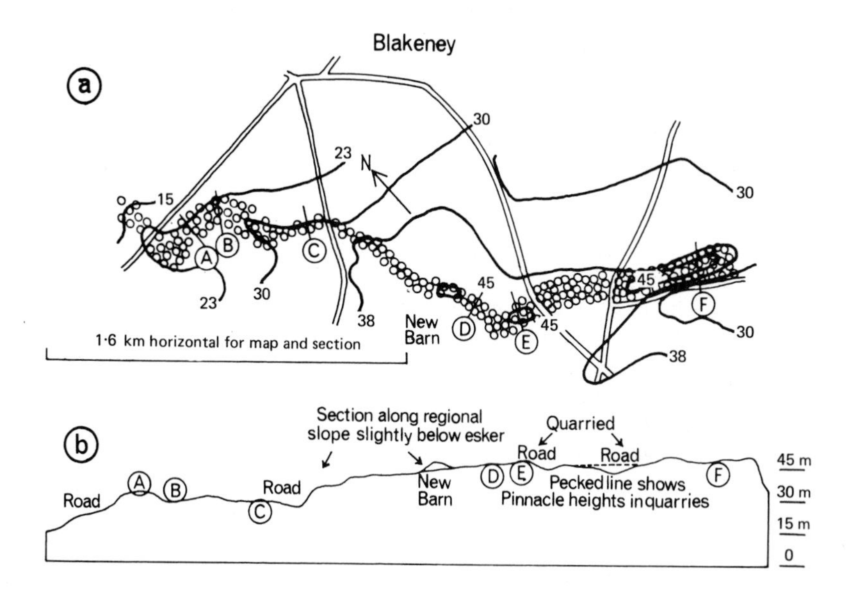
Blakeney
a
N
15
23
30
38
45
A
B
C
D
E
F
New Barn
1·6 km horizontal for map and section
b
Section along regional slope slightly below esker
Quarried
Road
Road
Road
Road
New Barn
Pecked line shows Pinnacle heights in quarries
45 m
30 m
15 m
0

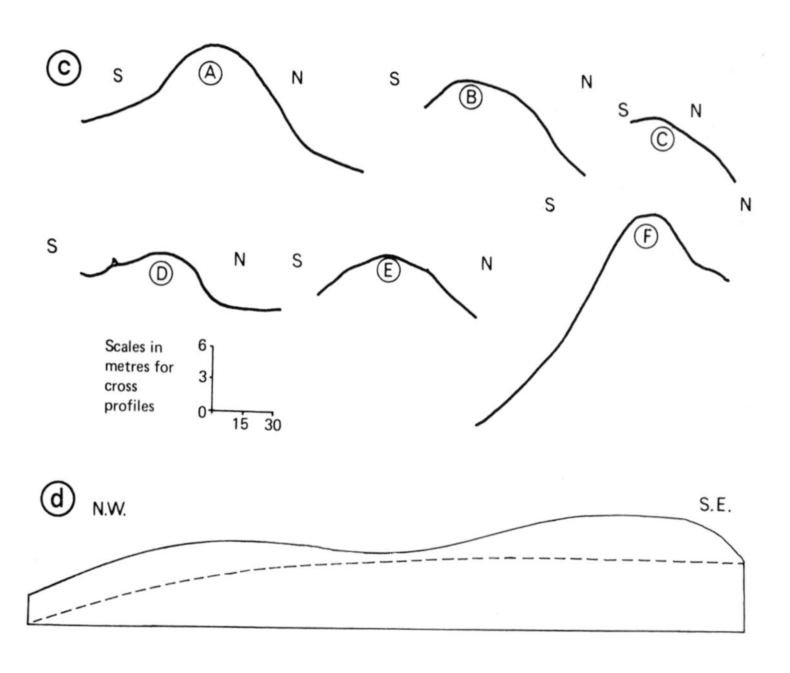
c
S
N
A
B
C
D
E
F
Scales in metres for cross profiles
6
3
0
15
30
d
N.W.
S.E.

to the subglacial floor. A stream depositing sediment in a crevasse may drain subglacially, so that an esker may be confluent with a crevasse-filling. Crevasse-fillings may also be partly formed from basal lodgement deposits squeezed up into the crevasse by ice pressure.

Proglacial forms

Outwash from an ice front may deposit a supra-aqueous outwash plain or fan, or, if there is a marginal lake or sea, a sub-aqueous delta. These, of course, merge with the end-moraine forms already described. Each of the two categories, the supra-aqueous and sub-aqueous, has a characteristic form (fig. 3.6). The outwash plain has a steep ice-contact slope marking the position of the ice (the proximal margin). This slope may be pitted by kettle-holes. From this proximal edge, which is naturally the highest part of the plain, the surface slopes away gradually in the direction of drainage (the distal slope). Outwash plains vary in size, depending on the size of the streams contributing the outwash, the load of the outwash and the length of time of formation. A series of streams close together may form a series of joined outwash fans, which may coalesce to give a single large outwash plain, perhaps up to 4 or 6 km long and a km or more wide. Some outwash plains are pitted by kettle-holes. This indicates that an ice recession left blocks of ice later covered by outwash. The melting of the ice blocks led to the pitting. Outwash plains formed over areas not ice-covered shortly before will not be pitted.

The outwash deltas are of quite different form. They have a flatter surface formed from the topset beds, making a contrast with the steeper distal slope resulting from the inclination of the forest beds. There may be a wave-cut beach at the junction of the flat and the slope. As with the outwash plains the proximal side may be marked by an ice-contact slope, which may be pitted by kettle-holes.

Where outwash is confined to a valley, a valley train with rather a steeper gradient, e.g. 5 m per 1500 m, may be built up by aggradation of the load within a valley. Terraces may be later cut in the infilling, depending on later changes of the base-level of the proglacial stream. Difficulties of separating such terraces from kame-terraces have already been considered.

Under the heading of proglacial landforms there are also the beaches, terraces, bars and lake-clay flats associated with the filling by sediment of ice-marginal lakes or seas.

Summary of the relation of deposition to ice advance and retreat

The surface scheme of glacial deposits and forms shown in fig. 3.1a gives their arrangements at only one stage in a glaciation. The ice has advanced to that point and will recede from it. During the advance lodgement till is the chief deposit. During the standstill, till, ice-contact stratified drift and outwash are formed and till may be deposited under the ice by pressure-melting. During the recession and down-melting, till is deposited from the basal debris-rich ice, and ice-contact stratified drift and outwash are deposited over the previous deposits. Similarly, the till plain results from the advance, the end-moraine and outwash plains and deltas from the standstill and the kames and eskers appear during the recession and down-melting. The processes associated with advance, standstill, recession and down-melting will result in the formation of the sequence of glacial deposits and the distribution of landforms now present. Each process may be of different magnitude in different glaciations, so that much diversity is to be expected in landforms and sequences of deposits.

EROSION

Both ice and melt-water action give rise to erosional features associated with glaciation. We shall first consider the processes concerned in erosion and then describe the effects on the landforms.

Processes

Chemical forms of erosion are of little significance because of the prevailing low temperatures, though chemical erosion by alkaline melt-water has been observed in modern glaciers. Mechancial processes, however, are important. Removal of bedrock by plucking or quarrying will result from pressures such as friction and drag exerted by the ice during flow, especially in well-jointed rocks. In temperate glaciers (and possibly in continental ice sheets developing in temperate regions during the Pleistocene) the ice is at or very near the pressure melting-point at any depth, and water is present throughout. Any release of pressure will bring about freezing, so that water in cracks in the subjacent bedrock may freeze and thaw. This will assist in the quarrying action of the ice. Another possible process is that a release of pressure on parts of the bedrock will allow release of strains inherent in the rock with the result that cracks occur and fragments are removed. It is also possible that large fragments of unconsolidated subglacial sediment or rock may become frozen to the base of the ice and are thus incorporated in the load.

These processes result in fairly large-scale removal of bedrock. At the other end of the scale there is the important process of abrasion of the bedrock by the load in the ice. Abrasion is mechanical erosion on a small scale and leads to the production of 'rock flour' which becomes incorporated in the load together with the larger fragments derived by quarrying. Abrasion is the principal form of erosion of frozen, and therefore unjointed, ground.

The process of nivation is also important in erosion. Nivation is the breaking-up of near-surface rocks by freeze and thaw action and the redistribution of the rubble down slope by such processes as solifluction and melt-water flow.

Small-scale effects

These are the results of abrasion inscribed on the surfaces of the subglacial floor and on rocks within the load. The inscriptions include grooves, striae, and various sets of markings like chattermarks, crescentic gouges and crescentic fractures.

The size of the linear forms depends on the size of the abrading fragment held in the ice. Large grooves may be formed in soft rocks, but in harder, but not too hard, rocks the fine lines of striae, up to about 30 cm long, are more common. They are usually best developed on the upstream (stoss) sides of rock bosses beneath the ice. Other processes involving flow and friction, e.g. grounding icebergs, solifluction, may produce striae, but the regional setting of the particular occurrence should make the origin clear. Polishing of rock surfaces is produced by multitudes of very fine striae etched by fine particles held in the ice.

Linear sole marks have been observed at the base of tills, marking the position of grooves and striations made in the underlying sediment by rock fragments projecting from the base of the ice.[35] Such sole marks are useful for determining the direction of ice movement.

Sets of marks and gouges (fig. 3.11) are often found on the glaciated surfaces of hard brittle rocks. The sets are drawn out along the line of ice advance and may extend 30 cm or so, each mark being up to a few centimetres across. They may result from a series of impacts of stones in the ice load, giving chattermarks, or they may be crescentic in shape. Friction cracks or crescentic fractures are concave downstream, and are supposed to be formed when pressure forward of a stone in the ice load was relieved by cracking of the rock surface. A series of such cracks implies the action was repeated a number of times. Gouges concave downstream and upstream are

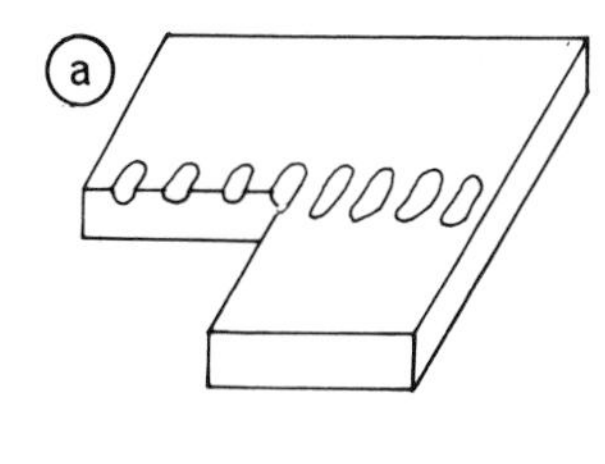

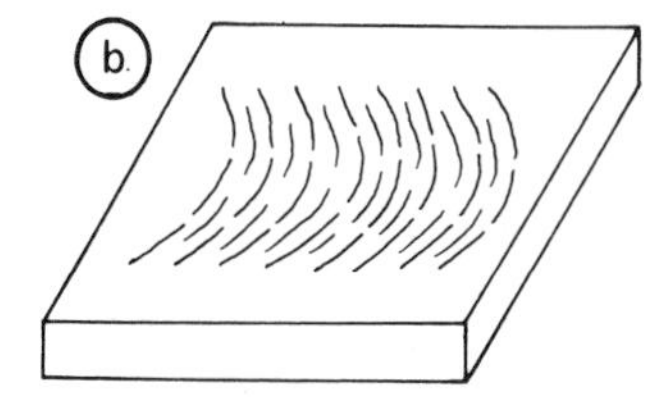

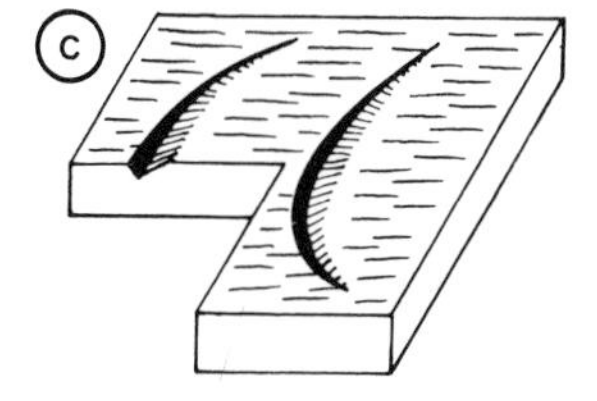

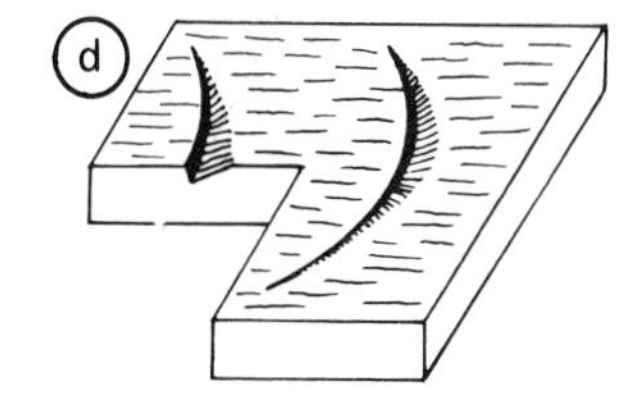

Fig. 3.11 Various small-scale abrasion marks.
a. Chattermarks.
b. Crescentic fractures or friction cracks.
c. Crescentic gouge.
d. Lunate fracture.

also known. They appear to result from pressure producing cones of fracture. The exact method of production of these crescentic features is not clear. The orientation may depend on whether the fragment causing the effect is sliding or rolling.

The orientation of all these small-scale features gives two possible directions of ice flow, but it seems that there are no fool-proof criteria available yet for deciding which is the correct direction. Other evidence of ice movement must be taken into account. Usually the regional setting of their occurrence and the character of the till gives a decisive answer.

Scouring by melt-water load is responsible for pothole formation, which may be on a small or large scale. Potholes cut in Triassic sandstone have been observed beneath last glaciation till in the West Midlands.[26]

Landforms associated with glacial erosion

The characteristic forms resulting from mountain glaciation are the corrie (cirque, cwm), U-shaped valleys, hanging valleys, and rock steps in valleys. Continental glaciation by ice sheets produces regional basins of erosion, and a smoothed relatively driftless aspect to the land. There is of course no clear dividing line to the landforms produced by each, for valley glaciers may develop to piedmont glaciers, and to lowland ice sheets, and the ice of continental ice sheets may reach valley systems and so modify them.

Corries (plate 6b) are depressions, usually semi-circular or oblong, excavated in mountain slopes.[33] They have steep back walls and often contain lakes. They arise by nivation in sheltered places where perennial snow (firn) accumulates, and are therefore most likely to form on north- and east-facing slopes where the snow suffers least ablation. Their distribution is also affected in some degree by rock structure, nivation being more rapid where the rocks are well jointed, and conditions suitable for the deposition of snow drift. Once the firn has settled in a chance hollow, nivation starts as a result of the daily melt and nightly freeze, and the hollow enlarges. If the firn is converted to ice, the ice may accumulate until flow starts, and then ice flow will increase the size of the corrie by abrasion and quarrying. Quarrying will be particularly active in the bergschrund next to the head wall of the corrie, where blocks will be loosened by freeze and thaw. Two types of corries have been distinguished, those produced by nivation, with a floor of debris, and those cleaned out to bedrock by flowing ice.

The lower limit of the accumulation of a bank of firn is the lower limit of the orographic (local) snow-line. Other conditions allowing, corries will therefore tend to develop close to the firn limit, and thus their altitude is an important clue to past changes of the snow-line.

Excavation of corries on mountain areas leads to the characteristic alpine landforms with sharp arrêtes, steep peaks and closely spaced deep corries.

Hanging tributary valleys are also characteristic of mountain glaciation. They form as a result of the greater deepening and widening of the main valley by the larger main glacier compared with the smaller amount of erosion by the minor tributary glaciers. On the melting of the ice, the tributary valleys are left hanging, and become the sites of waterfalls.

The glaciated valley (plate 6b) is U-shaped, because the erosion by the glacier is spread to all parts of the valley floor and side submerged by ice. This contrasts with the V-shaped valley of river erosion, where the point of erosion is at the bottom, and where the slopes are subject to mass-wastage. Erosion by valley glaciers will

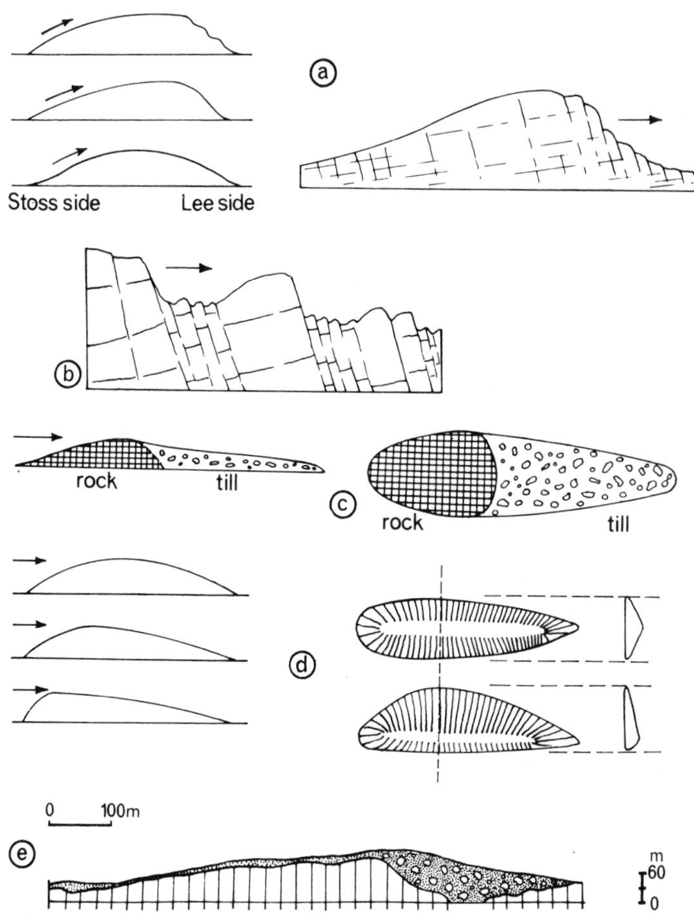

Fig. 3.12 Roches moutonnées, crag and tail, and drumlins. Directions of ice movement shown by arrows (after various authors).

a. Outlines of roches moutonnées (in long-section). The uppermost of the group of three has lee-side plucked, and is enlarged on the right, showing the relation of plucking to jointing. Where the rock is less well jointed, on the stoss-side of the enlarged section or in the two lower examples on the left, the rock surface is moulded by abrasion.
b. Rock steps resulting from variation in jointing of bedrock.
c. Crag and tail. Long-section and plan.
d. Drumlins. Long-section outlines on the left, plans and cross-section outlines on the right.
e. Transverse section through drumlin partially built of bedrock.

also truncate spurs. The height reached by the ice up the valley sides may be marked by the trimline. Below this the rocks are smoothed and polished. Above it nivation has produced an irregular surface. The drowning of valleys deeply excavated by ice is one cause of the formation of fjords.

Rock steps (fig. 3.12) are also characteristic of the floors of glaciated valleys. Changes in gradient, and thus in the erosive power and quarrying action of the ice, and changes in prevalence of jointing of the bedrock, will produce steps in the rock floor.

Roches moutonnées (fig. 3.12) are asymmetrical bosses of rock characteristic of areas showing glaciated bedrock. The upstream side (stoss) is smoothed by abrasion, and the lee side is made irregular and steep by quarrying. If drift is deposited in the lee the form becomes one of crag and tail and the outline may approximate to that of a drumlin. It is only a step away for deposition to replace erosion and for the crag to become the core of a drumlin.

A feature related to erosion by continental ice sheets is the occurrence of large basins between the zones of accumulation and deposition of drift. Such basins may coincide with outcrops of softer rocks. Examples are the Great Lakes of North America and the Baltic in Northern Europe (fig. 3.13). But such basin formation may also be associated with earth movements occurring near the edges of continental shields.

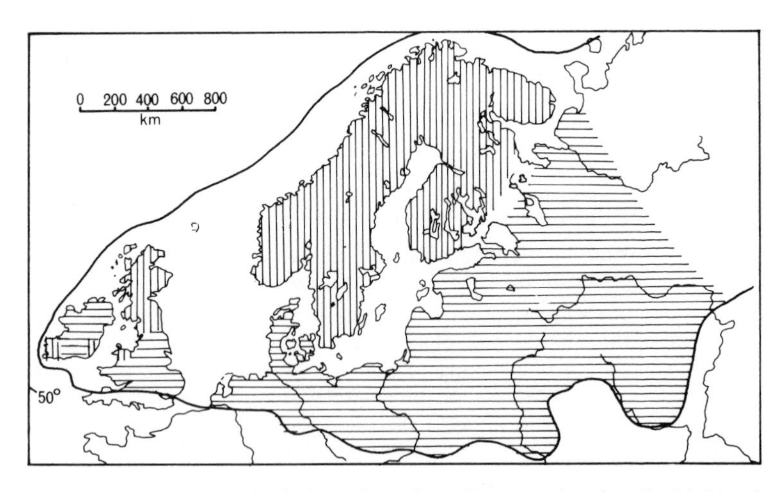

Fig. 3.13 Pleistocene glaciation of northern Europe, showing the highland zone (|||), the marginal drift covered zone (≡) and the intervening lowland area now the Baltic.

Submarine erosional features have been observed around the British Isles. Side-scan sonar pictures have revealed iceberg plough marks in the Norwegian Trough area and off the western coasts,[3] and tunnel valleys associated with the last glaciation have been reported off the Northumberland coast.[7]

Landforms associated with melt-waters

Melt-water carrying drift is an important erosive agent. Beneath the ice, drainage waters and their load may erode potholes (moulins) and substantial subglacial channels (tunnel valleys). These last may also partly be formed from erosion by glacier tongues. Near the margin of the ice, especially on valley slopes, networks of channels (fig. 3.14) indicate the position of marginal, sub-marginal or sub-glacial drainage streams during deglaciation. Drainage of melt-water from marginal lakes leads to the formation of overflow channels. The margins of the later Scandinavian ice sheets in Europe are marked by wide drainage ways (Urstromtäler), often associated with dunes (fig. 5.3); they carried the melt-waters from the outwash plains west to the North Sea area.

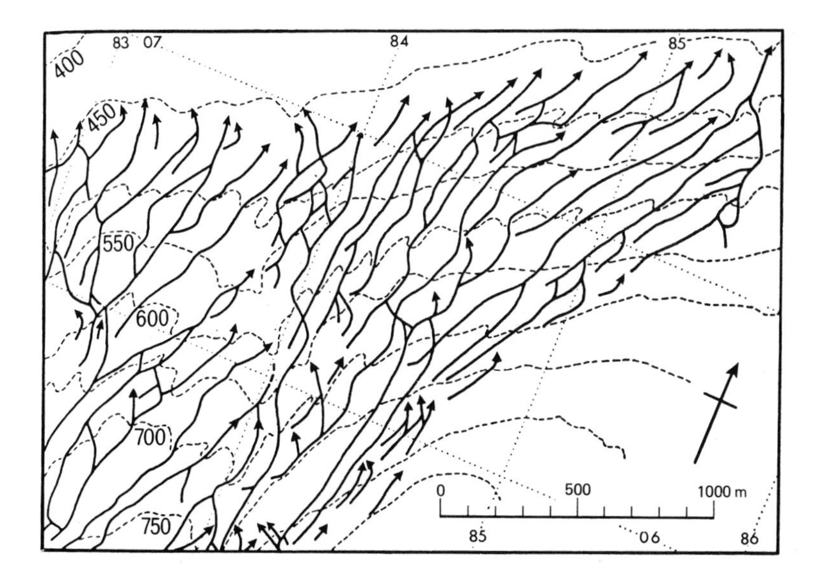

Fig. 3.14 Melt-water channels on the southern side of Strath Allan, Perthshire. Dotted lines are kilometre grid squares (Sissons[31]).

Erosion and transport of drift

Various factors will affect the rate of erosion by ice and thus the size of the load of debris carried by the ice. The rate of movement of the ice, its weight and thickness, are important. So is the load carried by the ice. The debris is concentrated at the base of the ice, for this is where it is obtained from the bedrock, and therefore its abundance and character, whether hard or soft, will help to determine the degree of erosion. Most of the debris is acquired by quarrying, and the degree of hardness and jointing of the bedrock as well as the landforms beneath the ice will therefore also determine the rate of erosion.

With valley glaciers, erosion is much more spectacular than with ice sheets. Deep valleys are gouged out by quarrying and abrasion, the former being more important. The load carried by the ice is both basal, derived by erosion caused by the flowing ice, and superglacial. The superglacial drift is acquired by rock falls on to the surface of the glacier; the surface moraines so formed are drawn out as lines down the surface of the glacier, parallel to the direction of movement. They form the lateral moraines characteristic of valley glacier systems. The lateral moraines from tributary glaciers become medial moraines of a main glacier (fig. 3.15) and these medial moraines will extend down into the body of the main glacier, reflecting their origin in the lateral and basal load of the tributary.

With ice sheets the process of erosion and transport of drift[15] are much more difficult to observe than with valley glaciers. Movement of the ice is most rapid nearer the edge of an ice sheet, while at the centre movement will be at a minimum. In the central parts of the area covered by an ice sheet, little erosion may occur and the pre-glacial weathering rind may be preserved. Erosion will be at a maximum nearer the margin and as the ice sheet grows the region of erosion will also move outwards. Erosion requires time, compared with the rapid deposition of till near the ice margins. The debris load in the ice is principally basal, though englacial and supraglacial drift may be derived by erosion from the rocks emerging through ice sheets and forming nunataks. The zonal arrangement of erosion, deposition, outwash referred to at the beginning of the chapter will be affected by the hardness of the bedrock and the region of erosion is also determined by this. Thus, the trough of erosion forming the Baltic Sea marks, roughly speaking, the boundary between hard rocks of Scandinavia and the younger soft rocks of the north German plain.

Though many individual rock types forming boulder trains are obviously transported great distances by ice, either by a single or successive ice advances, the bulk of the drift is transported less far, perhaps as little as a few kilometres. The evidence for this is the correspondence of the contact of the drift and the bedrock. Thus till may be deposited from the load soon after the debris has been acquired by the ice, depending on the rate of movement of the ice and the size of the load. Besides being carried forward at the base of the ice, the load may also be thrust up along shearplanes in the marginal zone of ice fractures, especially if the ice terminates against a slope. Drift may thus be carried up high above its place of origin. An example is the shelly drift carried up the slopes of north Wales by the Irish Sea ice to a height of 430 m O.D. at Moel Tryfan.

The vast quantities of glacial drift indicate that erosion by ice

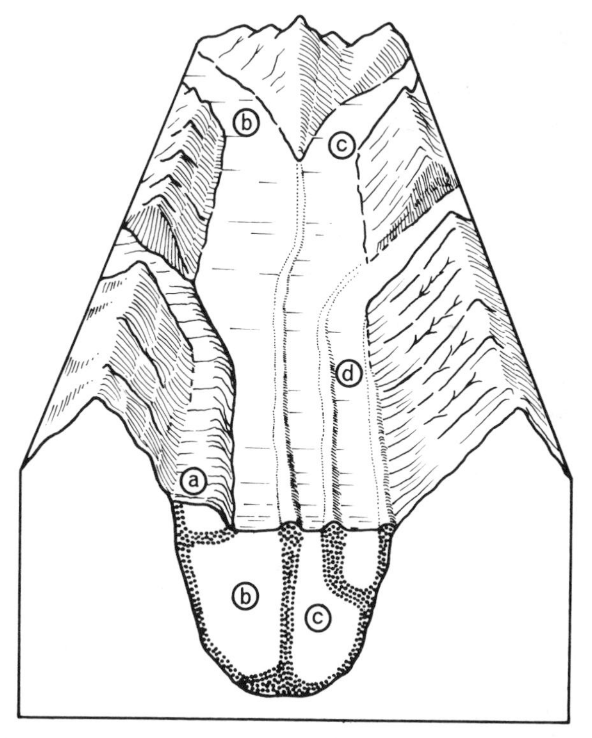

Fig. 3.15 Moraine emplacement in a valley glacier with tributary glaciers. The ice of glacier a with its basal moraine is superimposed on the ice of glacier b. Glaciers b and c are side by side, and d is inset into c. Medial and lateral moraines are present (after Sharp).

was very active. No rough calculations have been made as to the extent of erosion in Britain, but it has been estimated that if all the drift from Scandinavia found in Northern Europe were replaced, it would fill the Baltic and increase the height of Scandinavia's surface by around 25 m.

REFERENCES

CHARLESWORTH, J. K. 1957. *The Quaternary Era*. London: Arnold.

EMBLETON, C. and KING, C. A. M. 1968. *Glacial and periglacial geomorphology*, London: Arnold.

FLINT, R. F. 1971. *Glacial and Quaternary geology*. New York: Wiley.

PRICE, R. J. 1973. *Glacial and fluvioglacial landforms*. Edinburgh: Oliver and Boyd.

THWAITES, F. T. 1959. *Outline of glacial geology*. Ann Arbor.

WRIGHT, W. B. 1937. *The Quaternary Ice Age*. 2nd edn. London: Macmillan.

1 AARIO, R. 1972. 'Associations of bed forms and palaeocurrent patterns in an esker delta, Haapajärvi, Finland', *Ann. Acad. Sci. Fennicae*, A III, No. 111.

2 BANHAM, P. H. 1975. 'Glacitectonic structures: a general discussion with particular reference to the contorted drift of Norfolk'. In *Ice Ages: ancient and modern*, ed. A. E. Wright and F. Moseley, 69-94. Liverpool: Seel House Press.

3 BELDERSON, R. H. and WILSON, J. B. 1973. 'Iceberg plough marks in the vicinity of the Norwegian trough', *Norsk. Geol. Tidsskr.*, **53,** 323–8.

4 BOULTON, G. S. 1972a. 'Till genesis and fabric in Svalbard, Spitsbergen'. In *Till—a symposium*, ed. R. P. Goldthwait, 41–72. Columbus: Ohio State University Press.

5 BOULTON, G. S. 1972b. 'Modern Arctic glaciers as depositional models for former ice sheets', *J. Geol. Soc. Lond.*, **128,** 361–93.

6 CLARK, R. P. K. 1969. 'Kettle holes', J. Glaciol., **8,** 485–6.

7 DINGLE, R. V. 1971. 'Buried tunnel valleys off the Northumberland coast, western North Sea', *Geol. en Mijnbouw*, **50,** 679–86.

8 DONNER, J. J. and WEST, R. G. 1956. 'The Quaternary geology of Brageneset, Nordaustlandet, Spitsbergen', *Norsk Polarinstitutt Skrifter*, No. 109.

9 DRAKE, L. D. 1974. 'Till fabric control by clast shape', *Geol. Soc. America Bull.*, **85,** 247–50.

10 DREIMANIS, A. and REAVELY, G. H. 1953. 'Differentiation of the lower and the upper till along the north shore of Lake Erie', *J. Sedim. Petrology*, **23,** 238–59.

11 FLINT, R. F. 1961. 'Two tills in southern Connecticut', *Bull. Geol. Soc. Am.*, **72,** 1687–92.

12 GIBBARD, P. L. 1976. 'Pleistocene history of the Vale of St. Albans',

Phil. Trans. R. Soc. Lond., B (in press).

13 GLEN, J. W., DONNER, J. J. and WEST, R. G. 1957. 'On the mechanism by which stones in till become oriented', *Am. J. Sci.*, **255,** 194–205.

14 GOLDTHWAIT, R. P. (ed.) 1972. *Till—a Symposium.* Columbus: Ohio State University Press.

15 GOLDTHWAIT, R. P. 1974. 'Till deposition versus glacial erosion'. In *Research in polar and alpine geomorphology*, ed. B. D. Fahey and R. D. 159–66. Norwich: Geoabstects Ltd.

16. GRAVENOR, C. P. and KUPSCH, W. O. 1959. 'Ice-disintegration features in Western Canada', *J. Geol.*, **67,** 48–64.

17 GRAVENOR, C. P., STUPAVSKY, M. and SYMONS, D. T. A. 1973. 'Paleomagnetism and its relationship to till deposition', *Canadian J. Earth Sci.*, **10,** 1068–78.

18 HAAR, S. P. VONDER and JOHNSON, W. H. 1973. 'Mean magnetic susceptibility: a useful parameter for stratigraphic studies of glacial till', *J. Sedim. Petrol.*, **43,** 1148–51.

19 HANSEN, S. 1965. 'The Quaternary of Denmark', in *The Quaternary*, vol. 1, ed. K. Rankama, 1–90. New York: Interscience Publishers.

20 HARRISON, P. W. 1957. 'New technique for three-dimensional fabric analysis of till and englacial debris containing particles from 3 to 40 mm in size', *J. Geol.*, **65,** 98–105.

21 HARTSHORN, J. H. 1958. 'Flowtill in southeastern Massachusetts', *Geol. Soc. America Bull.*, **69,** 477–82.

22 HOLMES, C. D. 1941. 'Till fabric', *Bull. Geol. Soc. Am.*, **52,** 1299–1354.

23 KRÜGER, J. 1970. 'Till fabric in relation to direction of ice movement', *Geografisk Tidsskr.*, **69,** 133–70.

24 LUNDQUIST, G. 1948. 'Blockens orientering i olika Jordarter', *Sver. Geol. Unders.*, Ser. C, No. 492 (Årsbok 42, No. 6).

25 MACCLINTOCK, P. and DREIMANIS, A. 1964. 'Reorientation of till fabric by overriding glacier in the St. Lawrence Valley', *Am. J. Science*, **262,** 133–42.

26 MORGAN, A. V. 1970. 'Late Weichselian potholes near Wolverhampton, England', *J. Glaciol.*, **9,** 125–33.

27 OSTREY, R. C. and DEANE, R. E. 1963. 'Microfabric analyses of till', *Bull. Geol. Soc. Am.*, **74,** 165–8.

28 PARIZEK, R. R. 1969. 'Glacial ice-contact rings and ridges', *Geol. Soc. America Special Paper* **123,** 49–102.

29 ROSE, J. 1974. 'Small-scale spatial variability of some sedimentary properties of lodgement till and slumped till', *Proc. Geol. Assoc. Lond.*, **85,** 239–58.

30 SAURAMO, M. 1923. 'Studies on the Quaternary varve sediments in southern Finland', *Comm. Geol. de Finlande Bull.*, No. 60.

31 SISSONS, J. B. 1960–1. 'Some aspects of glacial drainage channels in Britain', *Scot. Geogr. Mag.*, **76,** 131–46; **77,** 15–36.

32 STUPAVSKY, M., GRAVENOR, C. P. and SYMONS, D. T. A. 1974. 'Paleomagnetism and magnetic fabric of the Leaside and Sunnybrook tills near Toronto, Ontario', *Geol. Soc. America Bull.*, **85,** 1233–6.

33 SUGDEN, D. E. 1969. 'The age and form of corries in the Cairngorms', *Scot. Geogr. Mag.*, **85,** 34–46.

34 WEST, R. G. and DONNER, J. J. 1956. 'The glaciations of East Anglia and the East Midlands; a differentiation based on stone-orientation measurements of the tills', *Q. J. Geol. Soc. Lond.*, **112,** 69–91.

35 WESTGATE, J. A. 1968. 'Linear sole markings in Pleistocene till', *Geol. Mag.*, **105,** 501–5.

CHAPTER 4

NON-GLACIAL SEDIMENTS AND STRATIGRAPHY

PLEISTOCENE sediments come under the geologist's heading of 'soft rocks'. They are usually unconsolidated and can readily be broken up by hand. Two categories of sediment can be distinguished: inorganic, composed of mineral sediments, and biogenic, composed of the products of plant and animal life.

The biogenic fraction can be divided into an organic component of humus and decayed parts of plants and animals, and an inorganic of diatom frustules, mollusc shells and sponge spicules. The term organic is often loosely used in place of biogenic; but it is used here in the stricter sense to denote the organic component of biogenic sediments. In the biogenic category it is convenient to include those sediments which are a mixture of inorganic and biogenic matter.

Biogenic sediments are treated here at much greater length than the inorganic. The latter and their environments of deposition, are fully described in the standard geological textbooks (e.g. Reineck and Singh, 1973), and they receive only very brief treatment here. Biogenic sediments have not received such a detailed treatment, even though their characterisation and interpretation are highly significant for the study of environmental history.

Inorganic sediments

The British Standard definitions for the size grades of the inorganic fractions of sediments are given in table 4.1. There are many other definitions of the different grades. As non-glacial sediments, the coarser grades are found in sediments deposited by water of high energy environments of rivers, or as beaches in coastal areas. They form a more important component of the glacial sediments than of the nonglacial sediments.

The degree of rounding of stones in river and glaciofluvial environments has been much studied.[1] It depends on a number of factors including hardness of the stones and the extent to which they have been subject to rolling, and may be a useful character to separate sets of river or glaciofluvial gravels of different origin and history. Electron-microscopic examination of quartz sand grains has indicated that certain surface features of the grains are a result of shaping in glacial, glaciofluvial, dune, beach and estuarine environments.[8]

The finer grades, sand, silt and clay, are most commonly associated with fluviatile or estuarine deposition under low energy conditions. Floodplain alluvium is mainly composed of such fine grades in varying proportions. The variation results from the differences in load and velocity of the water depositing the alluvium. The term 'brickearth' covers a wide variety of unconsolidated fine-grained sediments, but many brickearths are ancient alluvial deposits of Pleistocene rivers.

Under estuarine or fully marine conditions, which can often be determined palaeontologically, the finer grades predominate in the low energy environments of the tidal flats and deeper water regions.

Chemical precipitates, such as calcareous tufas, vivianite (ferrous phosphate) and limonite (bog-iron) are sometimes found as Pleistocene deposits. The first are often important because of their richness in fossil molluscs.

Table 4.1 Size grades of inorganic sediments

		Size of particle (*mm*)
Cobbles		200–60
Gravel	coarse	60–20
	medium	20–6
	fine	6–2
Sand	coarse	2·0–0·6
	medium	0·6–0·2
	fine	0·2–0·06
Silt	coarse	0·06–0·02
	medium	0·02–0·006
	fine	0·006–0·002
Clay		smaller than 0·002

Biogenic sediments

Biogenic sediments are commonly deposited during the non-glacial phases and they are frequently richly fossiliferous. Their study is therefore very important for investigations of environmental history during these phases. There are a number of classifications of biogenic sediments; most useful are a genetic classification, explaining the origin of different types of biogenic deposit, and a descriptive classification for use in the field.

A genetic classification. Biogenic deposits form where plant (and, more rarely, animal) detritus is produced, and preserved from decay by the prevention of oxidation, usually because of the presence of water. For example, biogenic, chiefly organic, deposits will form in lake basins, as a result of the deposition of plant and animal skeletons (inorganic biogenic component) and of fine organic matter derived from the breakdown of aquatic plants, both large and microscopic; or peat forming plants will build up organic deposits in wet and rainy upland areas where the drainage is poor.

A biogenic deposit forms from the detritus of particular plant communities, and these plant communities are characteristic of particular environments. A genetic classification therefore takes into account the chief environmental factors which control the distribution of the plant communities forming the biogenic deposit (fig. 4.1).

Several criteria are used in a genetic classification:

1. Position of the deposit in relation to the free water surface or groundwater.

Limnic deposits (inorganic and biogenic) formed below low water level.

Telmatic deposits (usually chiefly organic) formed between low and high water levels.

Semi-terrestrial deposits (usually chiefly organic) formed at about high water level.

Terrestrial deposits (organic only) formed above high water level. Terrestrial organic deposits can form as a result of impeded drainage, where ground water level is high. Such deposits have been termed soligenous. Or they may be formed as a result of high rainfall allowing plant communities to grow up beyond the influence of the ground water level. These are known as ombrogenous deposits.

2. Manner of deposition. Biogenic deposits are autochthonous (or sedentary) where the plant detritus is deposited where the plant is growing. Autochthonous biogenic, chiefly organic, deposits are called peats. The plant material is usually coarse and well-preserved, and identifiable to the peat-forming plants.

When the plant detritus is transported and then deposited, as in a lake, an allochthonous (or sedimentary) deposit is formed. Such deposits are often termed muds (in contrast to the geologist's definition of mud—an unconsolidated rock of clay grade). The plant material is usually finely divided and much of it is unidentifiable to particular species.

3. Supply of nutrients to the plant communities producing the deposits. With a rich supply of mineral nutrients, eutrophic conditions are said to be present. They favour the growth of fen, and autochthonous fen peats are formed as a result. In open water under these conditions there is a rich aquatic flora and much plankton, and gyttja* (nekron mud) is formed as an allochthonous deposit.

With a poor supply of nutrients, oligotrophic conditions are present. Valley bogs are formed where water level is high in the low-

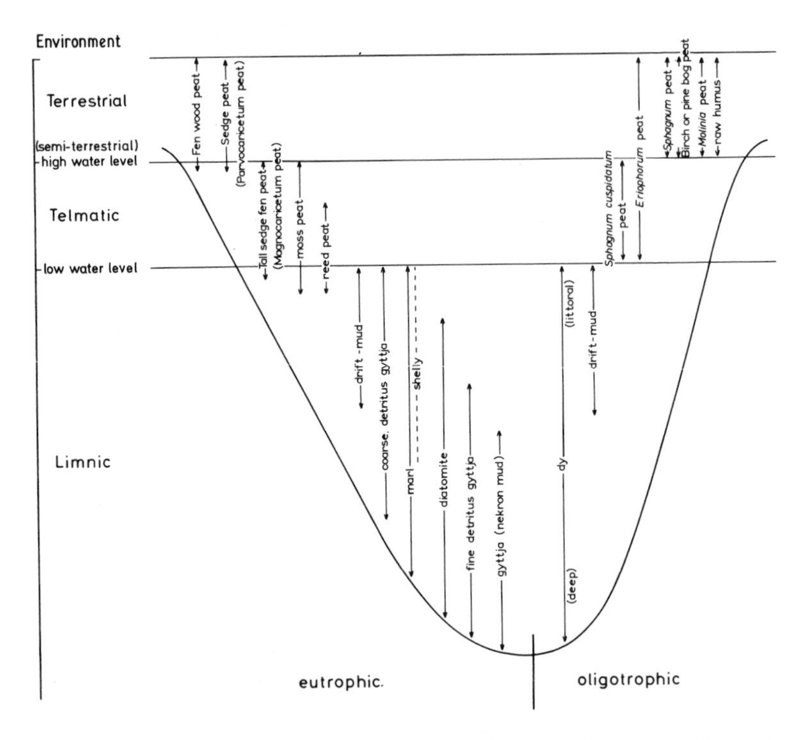

Fig. 4.1 Relation between environment and types of sediment formed (after Lundquist).

* Originally the term gyttja was coined by H. Von Post[9] to describe a coprogenous deposit formed by animals on and within a lake bottom, consisting of plant fragments, diatom frustules, quartz grains, pollen and spores. Its present-day use has been changed and takes note primarily of the organic fraction.

† Originally the term dy was coined by H. von Post[9] to describe a gyttja (von Post's sense) mixed with unsaturated humic colloids. It is now usually used to describe a deposit of humus colloids.

lands, and raised and blanket bogs where the rainfall is sufficiently high. Acid bog peats are formed under these conditions. In open waters, there are poor crops of plankton and most organic deposition is of amorphous humus, called dy† (gel-mud).

4. Topographical and drainage conditions. Three categories of mire, each associated with particular peat types, were described by L. von Post.[10]

Topogenous mires. Topography is the determining factor and produces water-holding basins. If such a mire is eutrophic, it is termed a fen. If oligotrophic, a valley bog.

Ombrogenous mires. The peat-forming plants (mostly *Sphagnum* mosses) are sustained by high rainfall, and are not dependent on ground water supply. The two main types of ombrogenous peats are those formed on raised bogs, where the *Sphagnum* peat grows in a massive dome above the ground water level in flat areas (plate 8), and blanket bog, which forms where the rainfall is high enough to allow the growth of the peat-forming plants even on slopes. As a result the peat completely blankets the landforms. Ombrogenous peats are of oligotrophic type, as their water supply is derived from the rainfall.

Soligenous mires. Peat formation occurs as a result of the concentration of surface drainage. Topography is not so important, but there are all transitions to the topogenous type. The peats are usually of an oligotrophic type.

The principal biogenic sediments

Fig. 4.1 gives the principal types of biogenic sediment and their relation to water level and to eutrophic or oligotrophic conditions. In addition to the sediments described below chemical sediments such as calcareous tufas, vivianite and limonite sometimes occur. Naturally there are transitions between the principal types of sediment now described. The terms are applicable as a result of field inspection of sediments, but if necessary the identifications can be confirmed by laboratory analysis. The following arrangement is after Osvald.

A. *Sedimentary* (allochthonous deposits). These are all limnic.
Gyttja (nekron mud). Usually greenish, amorphous jelly, with a C/N (carbon/nitrogen) ratio lower than 10.[4] May be largely algal (*algal gyttja*), or mixed with calcium carbonate (*calcareous gyttja*, marl), or with clay (*clay gyttja*); the last can be moulded in the hand, unlike gyttja.

Detritus gyttja (detritus mud). This sediment contains small fragments of plants, and varies from fine to coarse depending on the depth of deposition. In the fine type (plate 9a), a deep water deposit, the plant remains are usually unidentifiable, but in the coarse type (plate 9b), formed in shallower water, the plant remains may be identifiable and usually include abundant fruits and seeds.

Diatomite (kieselgur). Rich in diatoms, usually mixed with gyttja or fine detritus gyttja.

Drift mud (drift peat). A shallow water sediment containing much drifted coarse plant material, including leaves and twigs. A fine grained component is lacking or small.

Shell marl. A gyttja or detritus gyttja rich in shells.

Dy. A dark brown sediment of amorphous humus colloids. The C/N ratio is greater than 10.[4]

B. *Sedentary* (autochthonous deposits). These are usually known as peats, and may be limnic, telmatic or terrestrial. The peats are named after the principal peat-forming plants found in the sediment.

1. *Limnic.* These peats are formed by plants growing in water and include such types as *Phragmites* peat (reed swamp peat), *Scirpus lacustris* peat and *Typha latifolia* peat.

2. *Telmatic.* Formed by plants growing between high and low water levels. The most important eutrophic types are:

Cladium peat. The reddish rhizomes of *Cladium* characterise this peat.

Magnocaricetum peat. Formed from the tall fen sedges.

Moss peat. Containing fen mosses such as *Amblystegium.*

Oligotrophic types include:

Sphagnum cuspidatum peat. This moss is common in pools on *Sphagnum* bogs.

Eriophorum vaginatum peat. Characterised by the shiny fibrous tussocks of *E. vaginatum.*

3. *Terrestrial.* Form above high water level, or at high water mark (semi-terrestrial). Eutrophic types are:

Parvocaricetum peat. Formed by the small sedges of terrestrial and semi-terrestrial habitats (plate 9c).

Fen-wood peat. Characterised by much wood, usually alder or birch.

Oligotrophic types are:

Sphagnum peat. Formed by hummock-forming species of *Sphagnum* (plate 9d).

Sphagnum-Calluna peat. Contains the black finely ridged twigs of *Calluna*.
Shrub bog peat. Contains the twigs of *Myrica, Vaccinium* and other shrubs found on bogs.
Birch bog peat. Contains birch twigs and leaves with *Sphagnum*.
Eriophorum vaginatum peat (see above).
Scirpus caespitosus peat. With the fibrous tussocks of *S. caespitosus*.
Molinia peat. With tussocks of the grass *Molinia coerulea*.
Raw humus. Sometimes develops to considerable depths in wet sandy heaths.

A descriptive classification for field use

In the field it is necessary to describe organic sediments in as much detail as possible, yet without precise laboratory analysis. Troels-Smith (1955) has provided the most thorough system of field description, on which the following is based.

Three characters of the sediment must be observed: physical properties, components, and degree of humification.

A. *Physical properties*

1. *Colour*, whether uniform or mottled, whether changes colour on drying (change of colour on drying differentiates gyttja from dy; the former, green or brown when fresh, pales on drying, while the latter remains dark as it dries). Colour can be objectively described with the aid of the Munsell soil colour code. An impression of the colour of a sediment can be preserved by smearing some sediment on the page of a notebook.
2. *Structure*, whether granular, felted or fibrous.
3. *Nature of contact between different sediment types*. Important in interpretation of succession of sediments. May be sharp, irregular, or with a slow or rapid transition.
4. *Stratification*, whether homogeneous or stratified. If stratified, whether marked, or with alternating strata of different sediment types.
5. *Content of macrofossils*, plant or animal.
6. The following may be recorded on a 0 to 4 scale: *calcareousness*, by degree of fizz with 10 per cent HCl; *elasticity*, by squeezing; this deforms inorganic sediments, but organic sediments, especially peats, can recover their form after squeezing and thus have greater elasticity.

B. *Components*.

Apart from the inorganic fractions already described three categories of organic sediment may be distinguished. The latin terms in brackets are those suggested by Troels-Smith, and when abbreviated form his system of notation.

1. *Peat* (*Turfa*, *T*). Coarse macroscopic structure with large fragments of plants. Three main components are mosses (*T. bryophytica*, *Tb*), herbaceous plants (*T. herbacea*, *Th*) and woody fragments (*T. lignosa*, *Tl*).

2. *Detrius* (*Detritus*, *D*). Small fragments of plants. Again three components can be distinguished: small fragments of wood (*D. lignosus*, *Dl*), small fragments of herbaceous plants (*D. herbosus*, *Dh*), and smaller fragments of size 2 mm to 0·1 mm of herbaceous and/or woody plants (*D. granosus*, *Dg*).

3. *Mud* (*Limus*, *L*). Fragments smaller than 0·1 mm, i.e. with microscopic structure only. The fragments may be organic, from plants or animals (*L. detrituosus*, *Ld*), or inorganic but biogenic, e.g. of diatom frustules (*L. siliceus organogenes*, *Lso*), or calcium carbonate or marl (*L. calcareus*, *Lc*).

C. *Humification*.

Humified components in organic sediments are soluble in dilute potassium hydroxide. The amount of humification may be measured in the field by adding a little sediment to a few ml of dilute KOH, and allowing some of the resulting liquid to be taken up on filter paper. The browner the liquid, the more highly humified is the sediment. A 0–4 scale of humification may be used. A similar scale may be used with von Post's squeezing method. Wet unhumified peat squeezed gives clear water. If the peat is humified amorphous humus can be squeezed out and colours the water, and the degree of humification may be assessed by the amount of amorphous humus which can be squeezed through the fingers and by the darkness of the water.

In the field it may be difficult to distinguish some terrestrial and telmatic humified sediments from limnic sediments. This difficulty is dealt with in Troels-Smith's classification of the components by having two categories, humous substances (*Substantia humosa*) and humous mud (*Limus humosus*). The first indicates the presence of humified substances regardless of origin, the second is a dy.

A summary of the elements distinguished in Troels-Smith's classification of sediments is given in Table 4.2. The symbols are used in a shorthand method of notation. The proportions of the component parts of a sediment are given on a five-point scale: 4 indicates maximum presence, 0 implies absence. Some examples of the notation follow:

Unhumified moss peat	$Tb^0 4$
Humified moss peat	$Tb^2 4$
Marl, calcareous clay	*Lc*2, *As*1, *Ag*1
Sandy coarse detritus mud	*Dg*3, *Ag*1
Fine detritus gyttja (unhumified)	$Ld^0 4$

The superscripts indicate a five-point (0–4) scale of humification.

Troels-Smith's system provides a concise and objective method for the description of sediments in the field (and in the laboratory), and has the advantage of drawing the observer's attention to the most important properties to be examined.

Class	*Symbol*	*Element*	*Description*
Turfa	Tb^{0-4}	*T. bryophytica*	Mosses, +/− humous substance.
	Tl^{0-4}	*T. lignosa*	Stumps, roots, intertwined rootlets, of woody plants +/− trunks, stems, branches, etc., connected with these. +/− humous substance.
	Th^{0-4}	*T. herbacea*	Roots, intertwined rootlets, rhizomes, of herbaceous plants +/− stems, leaves, etc., connected with these. +/− humous substance.
	Dl	*D. lignosus*	Fragments of woody plants > 2 mm.
Detritus	Dh	*D. herbosus*	Fragments of herbaceous plants > 2 mm.
	Dg	*D. granosus*	Fragments of woody and herbaceous plants, and, sometimes, of animal fossils < 2mm > *c.* 0·1 mm.
Limus	Ld^{0-4}	*L. detrituosus*	Plants and animals or fragments of these; particles < *c.* 0·1 mm. +/− humous substance.
	Lso	*L. siliceus organogenes*	Diatoms, needles of sponges, siliceous skeletons, etc., of organic origin, or parts of these. Particles < *c.* 0·1 mm.
	Lc	*L. calcareus*	Marl, not hardened like calcareous tufa. Particles < *c.* 0·1 mm.
	Lf	*L. ferrugineus*	Iron oxide. Particles < *c.* 0·1 mm.
Argilla	As	Clay	Mineral particles < 0·002 mm.
	Ag	Silt	Mineral particles 0·002 to 0·06 mm.
Grana	Ga	Fine sand	Mineral particles 0·06 to 0·6 mm.
	Gs	Coarse sand	Mineral particles 0·6 to 2 mm.
	Gg	Gravel	Mineral particles > 2 mm.

Superscripts on symbols Tb, Tl, Th and Ld indicate the degree of humicity.

Table 4.2. Elements distinguished in Troels-Smith's classification of sediments

Sediment symbols

Symbols customarily used for the commoner sediments are shown in fig. 4.2. In general, crossing lines indicate limnic sediments, vertical lines telmatic deposits and horizontal lines terrestrial deposits. In *Sphagnum* peats the closer the lines the higher the degree of humification indicated, and in gyttjas the closer the squares, the finer the sediment.

Troels-Smith (1955) has given a comprehensive set of symbols, and sets of symbols have also been given by Jessen[5] and by Faegri & Iversen (1975).

Limnic sediments

Limnic sediments, such as those of Windermere, provide an informative record of environmental history in a lake's catchment area.[13] Forest clearance may be indicated by increase in inorganic components brought in by soil erosion. The stratigraphy of chemical elements such as phosphorus, sodium, potassium, iodine and carbon can indicate changes of organic productivity, soil erosion and pollution, in the same way that analyses of fossil assemblages reveal environmental history. Figure 4.3 shows such a series of analyses from recent sediments of a lake.

Redeposited sediments

Reworking and subsequent incorporation of older sediments into younger ones may occur in lake deposits as a result of changes of water level (fig. 4.4.), by slumping of sediments on steep lake floors, or in a mire by slumping and erosion. Where the sediments are fairly compacted and clayey, the redeposited material may form a breccia (plate 7a). Careful observation of the sediments should reveal reworking, but in more organic or more sandy deposits, it is often very difficult or impossible to recognise. Other features may suggest erosion and the possibility of redeposition, e.g. an unconformity between two strata found in a line or borings, or a very sharp contact between two different sediments. The danger of redeposition is that the redeposited material contains fossils of two different ages, one part of the assemblage from the redeposited material, the other contemporary with the redeposition. In clay-gyttja breccias, lumps of the redeposited material may be analysed separately, and the analyses may indicate their provenance. At the same time redeposition in lake deposits is useful in giving evidence of the history of water levels.

Laminated sediments

Laminations in biogenic limnic sediments are known from tem-

Fig. 4.2 Sediment symbols commonly used.

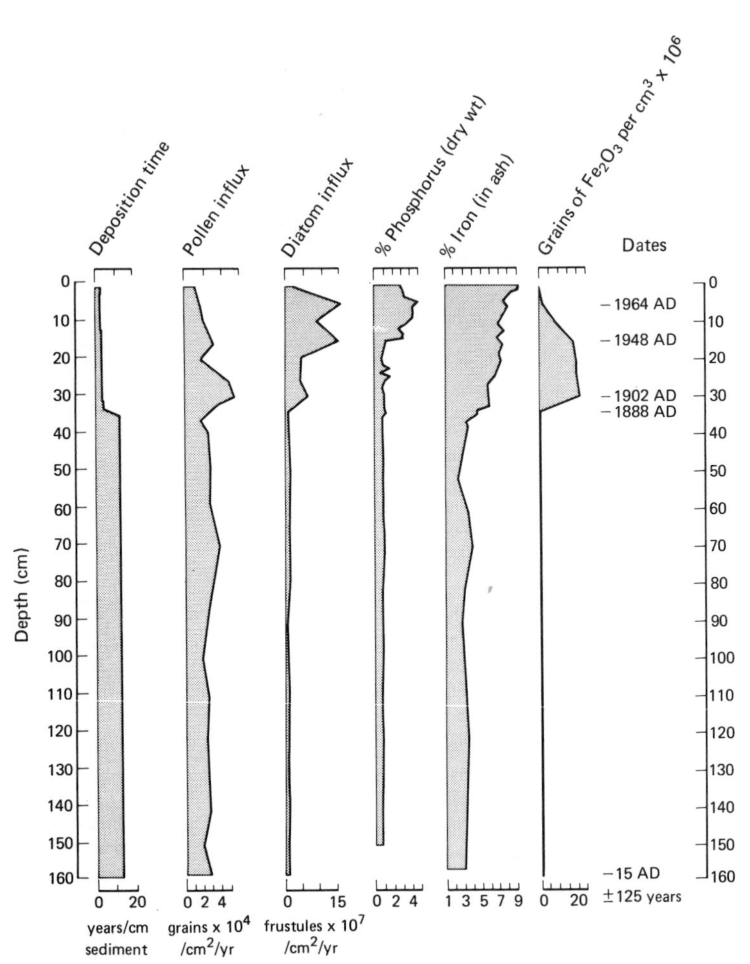

Fig. 4.3 Analyses from the top 50 cm of lake sediment of Shagawa Lake, Minnesota (after Bradbury and Waddington, 1973). The lake is next to haematite mines which started extensive production in 1889. The peak in the haematite grain curve corresponds to the peak in mining production in 1902. Mining activity has declined in the last 20 years, but the total iron remains high, probably through redeposition of fine-grained sediment derived from mining waste. The phosphorus rise at 14 cm may be associated with the time (1948) when detergents containing phosphates became widely used; the recent phosphorus decline may be a result of the recent installation of improved local sewage works. The increase of diatoms above 30 cm is related to the increased nutrient input into the lake through soil erosion, run-off and sewage (eutrophication). An increase in clastics at 20–30 cm is associated with mining activity, soil erosion and run-off, and is reflected in the increased sedimentation rate in the lake.

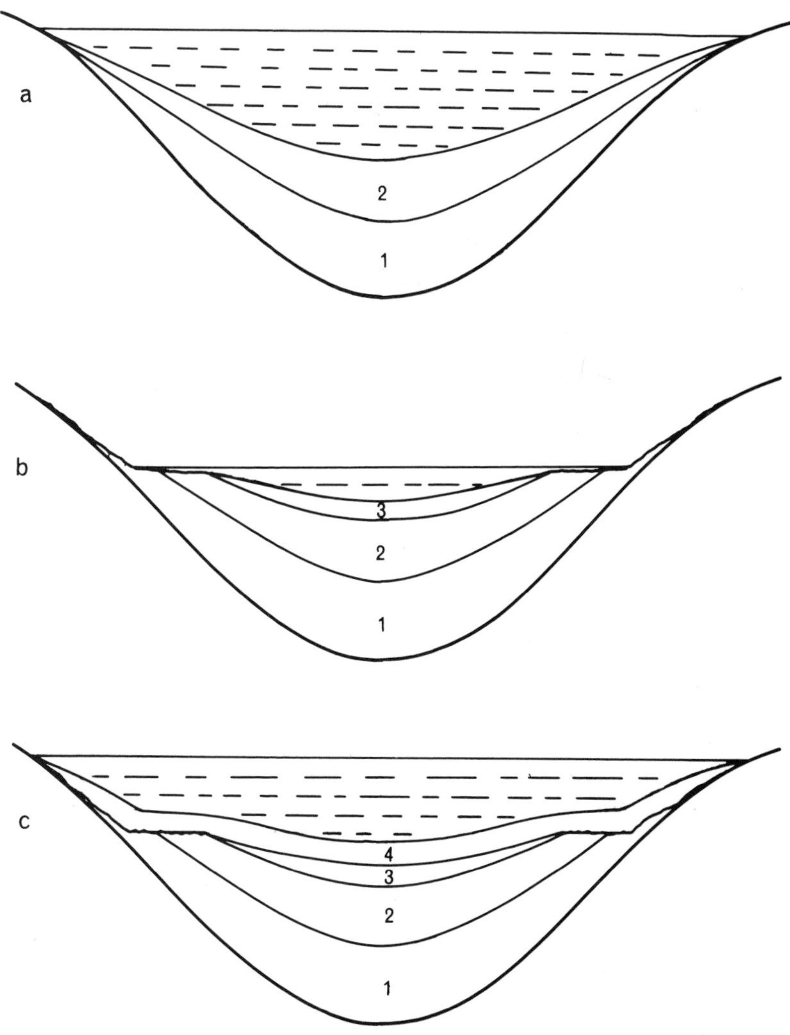

Fig. 4.4 Origin of redeposition in lake sediments.

a. Older (1) and younger (2) sediments in a lake basin.
b. The water level falls and erosion of 1 and 2 takes place at the margins. Redeposition of 1 and 2 takes place and forms 3, together with any contemporary lake sediments.
c. The water level rises to its previous level. The sediment 4 seals off the largely redeposited sediment 3. Sharp contacts occur between sediments 2 and 4, and 1 and 4. Below the contacts the surfaces of 1 and 2 may be reworked and hardened by drying out in the interval when the lake level was low.

perate stages of the Pleistocene (plates 4d, 7b). Light (summer) and dark (winter) layers alternate, the former being rich in calcium carbonate and the latter in organic matter. The precipitation of calcium carbonate in the summer is a result of removal of carbon dioxide and bicarbonate ions from the water by photosynthesis of aquatic plants. The annual origin of a pair of layers has also been demonstrated by studies of variation in the pollen and diatom contents between the light and dark layers.[11, 12]

Such laminated sediments are found in deep lakes where, beneath the thermocline in an oxygen-poor environment, sediments are not disturbed because of the absence of a bottom fauna.

Other types of lamination may result from variations in diatom production or variations in the introduction of silt or clay into water bodies from streams draining into them.[2] In such examples studies of the environment of deposition and the detail of the laminations are necessary to draw conclusions about the periodicity of the laminations.

Succession of non-glacial deposits

In any basin of non-glacial deposition, many types of sediment are usually found. The variety results from changes in the vegetation contributing the organic sediments and on the degree of subaerial denudation in the area of drainage contributing sediment to the basin. Both of these depend on the climatic conditions and on the water regime. A study of the succession of sediments will therefore reveal much of the environmental history of the times when the deposits were formed. The following are examples of successions and their interpretation. Illustrations of these successions are given in fig. 4.5.

A. *Solsö, Jutland.*[7] These deposits were formed in a lake basin in the ground moraine (till) A of the last but one glaciation in Denmark. The oldest sediment in the lake is a late-glacial grey clay (B) with the remains of arctic plants. The clay is probably derived from the washing of the local till. Then follows a brown sandy gyttja (C) formed during the temperate part of the last (Eemian) interglacial, when the surrounds were rich in vegetation and forested. A layer of grey-brown clay with some gyttja (D) and a marginal sand (E) follows this. These sediments again contain an arctic flora. They are predominantly inorganic and were washed into the lake from the surrounding area, which was barely covered with vegetation, and thus subject to erosion in the cold climate. A grey sandy stony bed (F) follows and is best interpreted as a solifluction deposit at a time of severe climate. There is little sorting, which suggests the water level was low at this time. The grey clay G, the grey-brown sand H and the grey clay I are again limnic deposits formed under severe climatic conditions. The beds

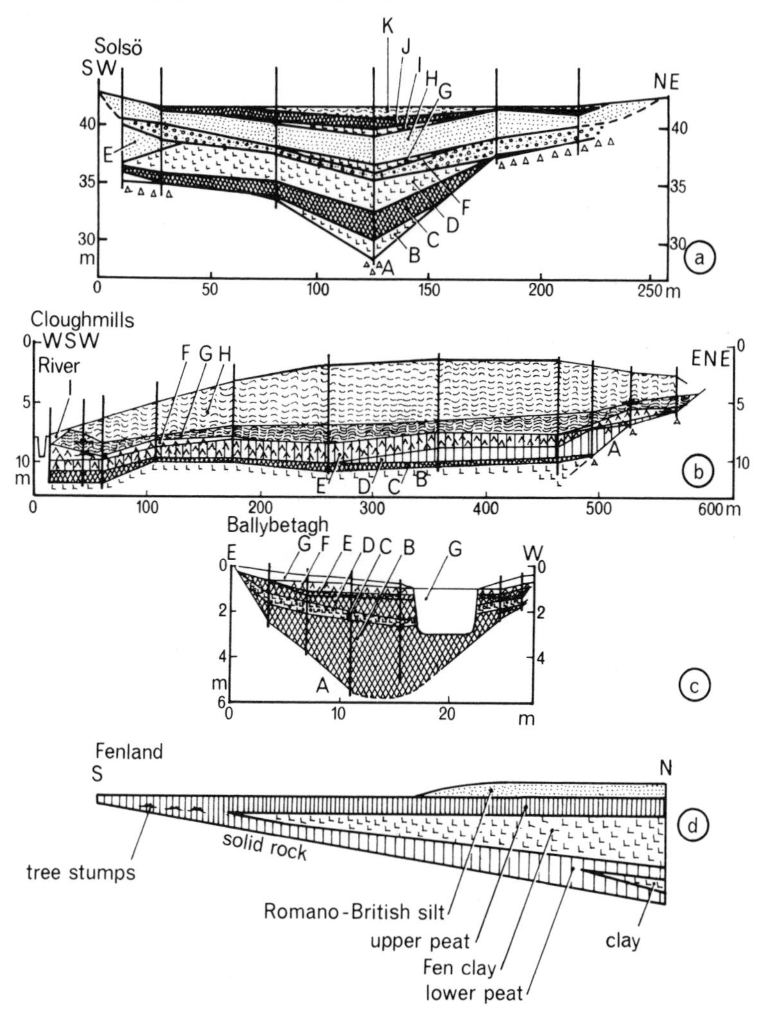

Fig. 4.5 Four examples of sequences in non-glacial deposits. For descriptions see text.

a, Solsö, Jutland; b, Cloughmills, Co. Antrim, Ireland; c, Ballybetagh, near Dublin, Ireland; d, Fenland of East Anglia.

D to I cover the period of the last (Weichselian) glaciation. In the final layers J and K the sediments revert to an organic type, and they were formed during the temperate Flandrian, when the region was forested. J is a brown gyttja, and K a brown *Sphagnum* peat deposited as a bog grew over the surface of the lake.

B. *Cloughmills, Co. Antrim, N. Ireland.*[5] This is a section through a raised bog which developed over a lake. The grey limnic clay B is a late-glacial deposit and rests on a till A. An olive-brown algal gyttja (C) was then formed in the lake during the beginning of the temperate Flandrian. This is followed by a grey-brown *Phragmites* peat (D) and a *Phragmites*-wood peat (E), as the hydrosere developed with the lake becoming shallower and conditions changing from limnic to telmatic. This is succeeded by dark brown birch-wood peat (F) in the marginal parts of the basin of deposition, indicating the growth of birch wood over the previous sediments, as water level fell, and conditions became oligotrophic. Pine also grew at this time, as shown by the pine stumps recorded in the section in the wood peat F. Under ombrogenous conditions *Sphagnum* grew over the entire surface and a raised bog started to form. The dark brown *Sphagnum* peat G was deposited, and was humified in an interval of cessation of growth when pine again grew on the margins. The light brown fresh *Sphagnum* peat H was then formed after renewal of the bog growth. The sandy clay I is a flood deposit of the river.

C. *Ballybetagh, near Dublin.*[6] Here a series of late-glacial and post-glacial deposits is seen lying in a hollow in stony clay (A), probably reworked till. In this lake was deposited a light brown gyttja (B) during the interstadial conditions of zone II (Allerød) of the Late Midlandian towards the end of the last (Midlandian) glaciation. During the colder zone III solifluction occurred and deposited in the lake a grey stony sand (C). During the subsequent temperate forested conditions of the Flandrian a light brown gyttja (D) was first deposited in the lake, then a thin layer of dark brown drift mud (E) when the water was shallow. Over this grew an alder-birch wood which formed the dark brown wood peat F. G is disturbed and recent peat covering the surface and filling peat diggings. The black mark in B on the right marks the position of a find of a skeleton of the Giant Irish deer.

D. *The Fenland of East Anglia.*[3] This schematic north-south section through the Fenland is about 40 kilometres long. The deposits vary in thickness from a few metres in the south to ten or more in the north. On the solid of the Fenland basin lies a series of temperate Flandrian deposits, both freshwater and brackish to marine. At the base lie freshwater peats, mostly telmatic (the lower peat). Freshwater deposition is interrupted by brackish to marine deposition as the sea-level rose in relation to the land. The wedges of blue-grey fen clay were formed as a result of the marine transgression. They are thickest on the seaward side. Regression of the sea-level resulted in an upper freshwater peat being formed. Finally, the Romano-British silt was deposited seawards as the upper peat was flooded.

REFERENCES

BRITISH STANDARDS INSTITUTION. 1963. *Methods of testing soils for civil engineering purposes.* B.S. 1377: 1961.

DEEVEY, E. S. 1953. 'Palaeolimnology and climate'. In *Climatic change*, ed. H. Shapley. Harvard University Press.

FAEGRI, K. and IVERSEN, J. 1975. *Textbook of pollen analysis.* 3rd edn. Copenhagen: Munksgaard.

LUNDQUIST, G. 1927. Bodenablagerungen und Entwicklungstypen der Seen. *Die Binnengewasser*, **2.**

REINECK, H-E. and SINGH, I. B. 1973. *Depositional sedimentary environments.* Berlin: Springer-Verlag.

TROELS-SMITH, J. 1955. Characterization of unconsolidated sediments. *Danm. Geol. Unders.*, IV Raekke, 3, No. 10.

1 BARSCH, H. and BRUNNER, H. 1963. 'Vergleichende untersuchungen zur morphometrischen analyse fluvialer gerölle', Report VIth International Congress of Quaternary, **3,** 21–38.

2 CALVERT, S. E. 1964. 'Factors affecting distribution of laminated diatomaceous sediments in Gulf of California'. In *Marine geology of the Gulf of California—a Symposium.* Memoir No. 3, Amer. Assoc. Petrol. Geologists, 311–30.

3 GODWIN, H. 1940. 'Studies of the Post-glacial History of British Vegetation. III. Fenland pollen diagrams. IV. Post-glacial changes of relative land- and sea-level in the English Fenland', *Phil. Trans. R. Soc. Lond., B,* **230,** 239–303.

4 HANSEN, K. 1959. 'Sediments from Danish lakes', *J. Sed. Petrol.*, **29,** 38–46.

5 JESSEN, K. 1949. 'Studies in late Quaternary deposits and flora-history of Ireland', *Proc. R. Irish Acad.*, **52,** B, 85–290.

6 JESSEN, K. and FARRINGTON, A. 1938. 'The bogs at Ballybetagh, near Dublin, with remarks on late-glacial conditions in Ireland', *Proc. R. Irish Acad.*, **44,** B, 205–60.

7 JESSEN, K. and MILTHERS, V. 1928. 'Stratigraphical and palaeontological studies of interglacial fresh-water deposits in Jutland and

north-west Germany', *Danm. Geol. Unders.*, II Raekke, No. 48.

8 KRINSLEY, D. H. and FUNNELL, B. M. 1965. 'Environmental history of quartz sand grains from the Lower and Middle Pleistocene of Norfolk, England, *Q. J. Geol. Soc. London*, **121,** 435–61.

9 POST, H. VON. 1862. 'Studier öfver nutidens koprogena jordbildningar, gyttja, dy, torf och mylla', *Kungl. Svenska Vetenskab. Akad. Handlingar*, **4,** 1–59.

10 POST, L. VON. 1926. 'Einige aufgaben der regionale moorforschung', *Sver. Geol. Unders.*, Ser. C, No. 337.

11 TIPPETT, R. 1964. 'An investigation into the nature of the layering of deep-water sediments in two eastern Ontario lakes', *Canadian J. Bot.*, **42,** 1693–1709.

12 TURNER, C. 1970. 'The Middle Pleistocene deposits at Marks Tey, Essex', *Phil. Trans. R. Soc. Lond., B*, **257,** 373–440.

13 PENNINGTON, W. 1973. 'The recent sediments of Windermere', *Freshwat. Biol.*, **3,** 363–82.

14 BRADBURY, J. P. and WADDINGTON, J. C. B. 1973. 'The impact of European settlement on Shagawa Lake, Northeastern Minnesota, U.S.A.', pp. 289–307. In *Quaternary Plant Ecology*, ed. H. J. B. Birks and R. G. West. Oxford: Blackwell.

CHAPTER 5

THE PERIGLACIAL ZONE

UNDER present-day conditions of climate in high latitudes and in mountain regions in lower latitudes frost action produces structures in the soils and solifluction (flow-earth) deposits. Likewise during the cold stages of the Pleistocene the same processes occurred in the zone marginal to the advancing and retreating ice sheets. In addition, during the cold stages, wind was responsible for the deposition of sands and silts in this marginal zone. All these processes have been grouped together under the term periglacial, signifying that they are characteristic of the climatic conditions of a zone around the ice sheets. The term is a useful one to name this group of processes and the climatic conditions associated with them, but it is necessary to remember that the processes may occur at the present time in arctic, sub-arctic or montane non-glaciated regions, and in the case of the aeolian sediments, also in climates warmer than these regions.

The periglacial features of the Pleistocene were formed in latitudes lower than those where similar features are now being formed, so that they are fossil relic features in the temperate regions. Each periglacial process occurs under particular conditions of soil, climate and drainage. Their study is therefore important for the interpretation of the environment in the regions around the ice sheets. In fact they provide one of the best means to climatic and chronological sequences in the cold stages of the Pleistocene, particularly because freeze–thaw effects have left their mark in the superficial deposits over a very wide area of the country.

The principal periglacial effects can be divided into two. Those dependent on freezing and thawing in the soil induced by climatic conditions, including solifluction, structure in soil, and landforms resulting from the melting of frozen ground (thermokarst), and those dependent on wind or fluviatile action in sorting sediments (e.g. outwash deposits) produced at the margins of the ice sheet where little vegetation may be present to bind the soil.

The study of periglacial features can be conveniently divided into the study of the sediments themselves, of ground ice, of the structures which result from frost action, and of the landforms produced by periglacial processes.

DEPOSITS OF THE PERIGLACIAL ZONE

Periglacial sediments can be divided into unsorted and sorted types. The sorting occurs through wind action to produce sediments of predominantly sand or silt grade, respectively cover-sands (flug-sand) and loess, or through fluviatile action in the periglacial region. Sorting as a result of freeze–thaw action has also been observed in the laboratory,[5] but it is not known how far the processes involved are active in nature.

Unsorted sediments

These are solifluction sediments (head, coombe rock, flow-earth) produced by the downslope movement of sediments saturated with water. Solifluction is not confined to periglacial regions. It occurs in landslips and under wet conditions in temperate and tropical regions. Solifluction associated with frozen ground has been called gelifluction, a useful term giving an unambiguous association with the periglacial area. However, even in periglacial areas, solifluction may occur independently of frozen ground, so the more general term is retained here. Solifluction is very common in periglacial regions, first because of the unconsolidated nature of most of the superficial sediments, and secondly because during the spring the frozen ground melts down from the surface, and the upper layers, being badly drained because of the impermeable frozen ground beneath, move down slope, even very shallow slopes of 2° to 5°.

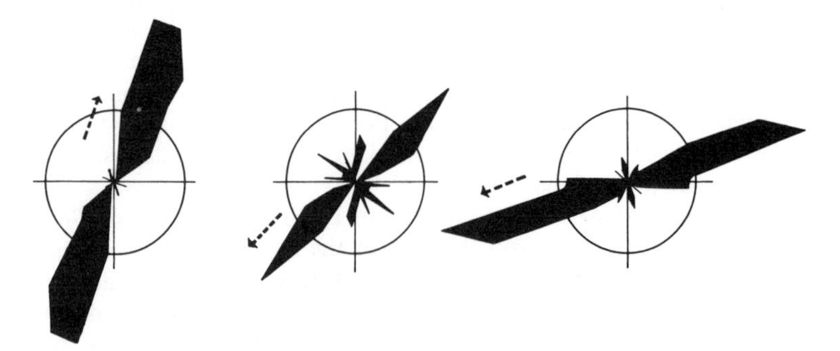

Fig. 5.1 Rose diagrams showing the orientation of long axes of stones in three solifluction deposits in Sweden. The arrows indicate direction of flow (Lundquist, 1948).

The flowing motion produces a preferred orientation of the long axes of stones in the solifluction deposit parallel to the direction of flow.[14] with the dip of the long axes downslope. Figure 5.1 shows the orientation of long axes of stones in some solifluction deposits.

Solifluction deposits may be difficult to distinguish from tills. Their local distribution as sheets on slopes, as lobes, fans and stone streams, the downslope orientation of the stones and the local provenance of the stones may prove a deposit to be formed by solifluction. But where the landforms have changed considerably since deposition and where the deposit may have been derived from a glacial deposit and thus contain its erratics, it may be difficult to distinguish a till from a solifluction deposit.

A number of types of solifluction deposit have been distinguished. Massive solifluction leads to considerable thicknesses of deposit, often contorted and then known as turbulent solifluction. Solifluction on a smaller scale may appear to be laminar, with thin threads of sediment drawn out downslope. The intensity of solifluction depends on the slope, the nature of the source material and the frequency of freezing and thawing. Diurnal changes across the freezing point, such as occur in tropical mountains, produces solifluction on a smaller scale than that caused by the seasonal changes (often accompanied by shorter period changes) occurring in northern latitudes.

Sorted sediments

Aeolian sediments. That wind played an important part in the periglacial environment is shown by the wide distribution of wind-faceted stones (ventifacts, dreikanter) and wind-polished stones in the periglacial areas. The aeolian periglacial sediments are derived chiefly by deflation from areas of outwash, from surfaces of glacial deposits left freshly exposed by ice retreat, from the fine sediments deposited by proglacial rivers in the periglacial areas, and from frost-shattered rocks rich in sand and silt. The deflation may remove the finer fractions of sediments, leaving pavements of the coarser fractions, often with ventifacts.[32]

Loess and cover-sands are two distinct types of aeolian sediment characterised by their grain size; loess is predominantly silt while cover-sands vary from fine to coarse sand. Mechanical analyses of typical examples are shown in fig. 5.2.

Loess is usually not stratified except where it has accumulated in wet depressions or ponds. The grains are mostly quartz, and the mineral content is naturally closely related to that of the sediments from which the loess originated by deflation. The grain size, though predominantly silt, may vary according to the distance of deposition

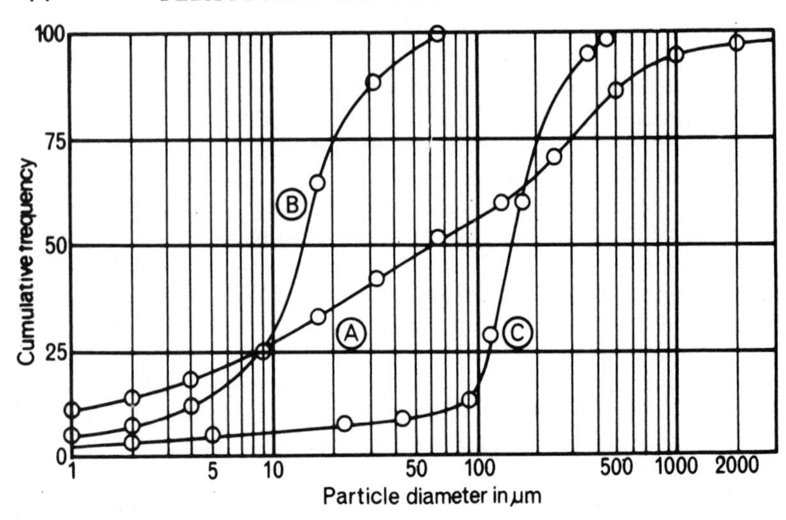

Fig. 5.2 Cumulative frequency curves of different sediments. A, till; B, loess; C, cover-sand.

from the deflation source. The more distant deposits have a finer grade. Loess, being aeolian, forms sheets of varying thickness which blanket the landforms. The thickness varies according to the closeness of the source and the direction of the prevailing wind. It may be so thin as to be hardly recognisable except by mechanical analyses,[22] or near the source it may form sheets up to 20 m thick or more. Such thick sections of loess show vertical cleavage and form steep cliffs. These variations in the composition and thickness of loess make it possible to reconstruct prevailing wind directions at the time of loess formation. In the north European plain such evidence suggests prevailing east winds.

Unweathered loess is generally calcareous and effervesces with acid. Calcium carbonate may be segregated into nodules. A halt in deposition with or without climatic change results in soil formation with humus enrichment and loess sections showing such weathering horizons form the basis for Quaternary stratigraphy in many parts of central Europe (fig. 5.22). Unweathered loess may contain faunal remains including mammals and land and marsh shells. Plant remains are much scarcer as the conditions of deposition do not favour preservation.

Loess formation is not related to a very specific climatic environment; but its deposition in general, as shown by areas where it is now forming, requires a dry and windy climate, at least in the sum-

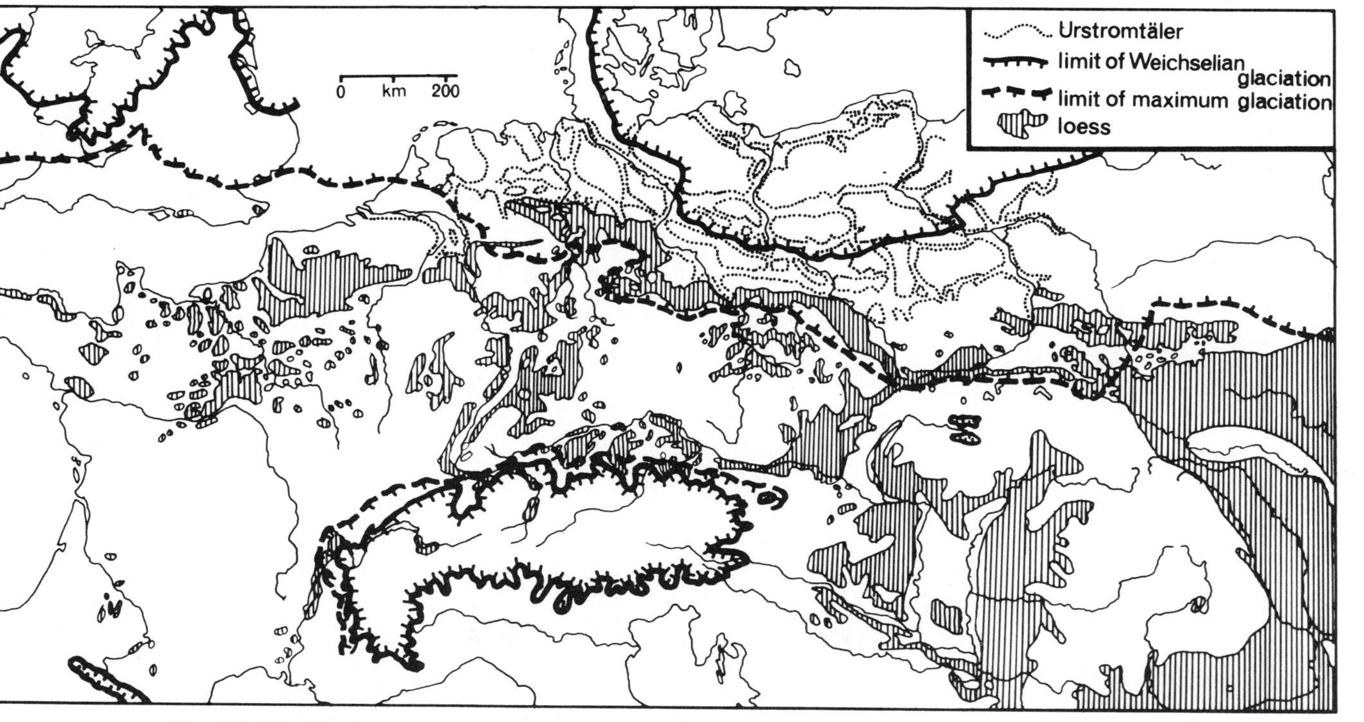

Fig. 5.3 Distributing of urstromtäler and loess in Europe (after Grahmann, 1932, and Woldstedt).

mer when melt-waters release much outwash and the ground is unfrozen. The fossil loess fauna indicates open conditions of vegetation such as now may be found in cold steppe regions. Similarly the distribution of Pleistocene loess in Europe (fig. 5.3), shows that it mainly accumulated in regions with a continental climate. The greatest thicknesses are associated with outwash sources in northern and central Eastern Europe. In the west the distribution becomes patchy. Presumably the oceanic climatic did not favour loess formation.

In Britain the brickearth common as a superficial deposit in the south and east is in many places loess.[25] and a loess fraction has been isolated from soils in places where loess is not apparent.[22]

In the same way that loess blankets the landscape so does the coarser grade cover-sand. Investigations of this periglacial aeolian originated in the Netherlands, where it is a common superficial deposit up to a few metres thick.[16, 34] In Europe it tends to occur between the loess belt and the limits of the Weichselian (Last) Glaciation. The mechanical analyses (fig. 5.2) indicate an aeolian origin of the sand, by deflation from unconsolidated glacial, proglacial and fluviatile deposits. The sand grains show the characteristic frosted surface of wind-blown grains, contrasted with the polished surface of water-worn grains,[4] and they may also have a characteristic surface texture under the electron microscope.[13]

Cover-sands have a predominant grade of 105 to 210 μm, though finer and coarser examples are known. They usually have a horizontal stratification, but may show sloping stratification related to accumulation as dunes or ridges. Alternating loamy (ill-sorted) and sandy (better-sorted) layers occur in cover-sands and these have been interpreted as being niveo-aeolian in origin, the loamy layers being transported by snow, the others by wind. Such alternating layers presently forming have been described from northern Europe.[29] Strings of pebbles in cover-sands have been interpreted as being deposited by wind during snow storms, following observations on the movement of fine gravel under such conditions. Snow melt-waters may also deposit coarser pebbly horizons with cross-bedding, associated with cover-sands. Such deposits have been termed niveo-fluviatile.

Fluviatile sediments. Sorted periglacial sediments are also produced by fluviatile action; for example, sands and gravels deposited by summer snow melt-waters. Such fluviatile (or niveo-fluviatile) deposits may form spreads at the foot of erosional features or in the wide valleys which carry the melt-water floods of the spring. Such floods have been well described in Siberia by Haviland (1926). In the course of deposition rafts of frozen material may be incorporated, especially frozen peat beds cut from eroding river banks (fig. 13.5a). The rafts of organic material known as the Arctic Plant Bed in the terrace sands and gravels of the lower Lea Valley northeast of London originated in this way.

PERMAFROST, SEASONAL FROZEN GROUND AND THE ACTIVE LAYER

Permafrost is the perennially frozen layer of the earth's surface.[2] It is found where ground temperatures remain lower than 0°C over a period of years. Permafrost covers about 20 per cent of the land

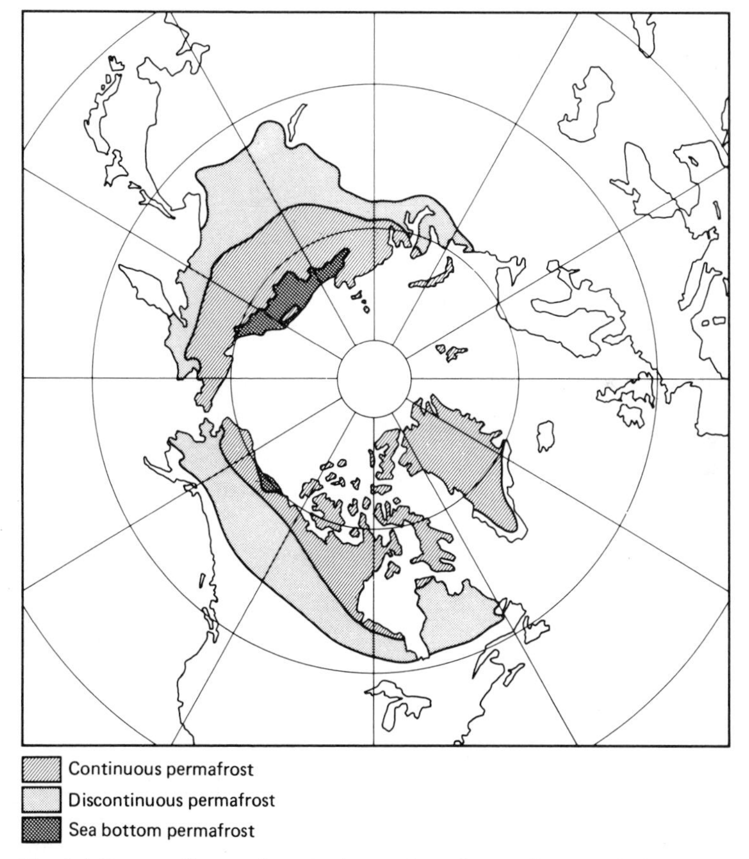

Fig. 5.4 Extent of permafrost in the northern hemisphere, excluding alpine areas (Mackay, 1972).

area of the northern hemisphere; it also occurs under parts of the Arctic Ocean. A distribution map of the continuous and discontinuous zones of permafrost is shown in fig. 5.4. It is thought that permafrost in the continuous zone is more or less in equilibrium with the climatic regime, but in the distontinuous zone is out of balance with the climate and may be in part relic from past times of colder climate, since there is evidence of retreat and degradation of permafrost during the last hundred years in the northern hemisphere. The depth of permafrost reaches over 300 m in parts of Alaska and Siberia and where it is so thick reaches temperatures of −5°C to −12°C at depth. Where permafrost is discontinuous, the thickness may be down to a few metres, with temperatures from above −1°C to −5°C.

Outside the permafrost area, seasonal frozen ground may occur in climates with severe winters, and is characteristic of regions with continental climates. The layer subject to annual freeze and thaw is called the active layer. Figure 5.5 shows the relation between

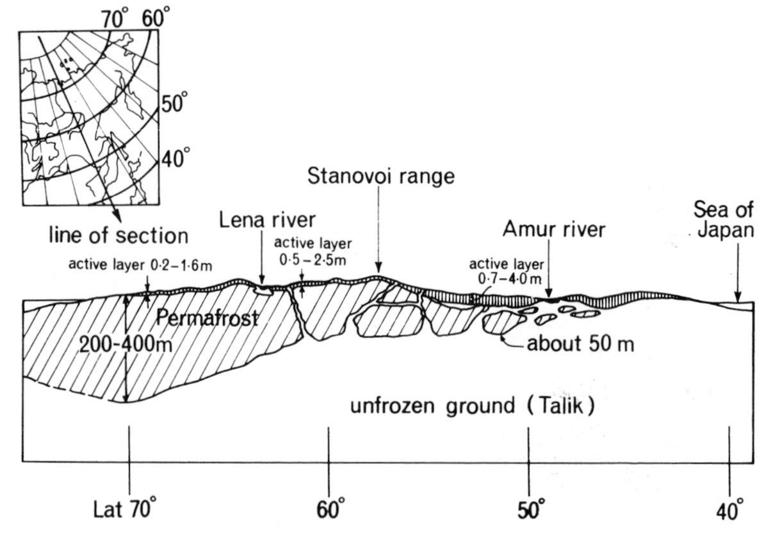

Fig. 5.5 Sketch section through Siberia showing relative thicknesses of permafrost and active layer (Muller, 1947).

permafrost and active layer in Siberia. Many of the periglacial structures observed are associated with cryostatic pressures (pressures developed on freezing) in the saturated active layer as down-freezing takes place. The depth of the active layer depends on the extent of the temperature changes, the nature of the superficial soil, drainage conditions and on the type of vegetation cover. At the base of the active layer, the temperature is equivalent to the mean annual temperature if the permafrost is in equilibrium with climate.

Table 5.1 shows the depth of the active layer at various latitudes in Siberia under different ground conditions. For example a terrace of saturated sandy clay, covered by 0·5 m peat with sparse larch shows an active layer of 0·5 to 0·8 m. Under similar conditions, but with grasses and broadleaf trees, there is an increased depth of active layer, 1·5 to 2·5 m, and a higher drier terrace with lichens, moss and pine shows a further increase to 2·5 to 3·5 m.

Table 5.1 Depths in metres of the active layer in Siberia (from Muller, 1947)

	Ground conditions		
	Sandy	*Clayey*	*Peaty*
Coast of arctic ocean (continuous zone of permafrost)	1·2–1·6	0·7–1·0	0·2–0·4
At the latitude of Yakutsk (62° N) (discontinuous zone of permafrost)	2·0–2·5	1·5–2·0	±0·5
In area south of 55° N (southern part of discontinuous zone of permafrost)	3·0–4·0	1·8–2·5	0·7–1·0

GROUND ICE

Ground ice has a very wide distribution in the permafrost area of the northern hemisphere, so that 'underground glaciation' is an important result of the world's present climate, even though the Pleistocene ice sheets and glaciers of the northern hemisphere have largely disappeared for the moment.

The effects of the growth and decay of ground ice are seen in various fossil periglacial structures, and it is therefore necessary for the interpretation of these structures to know the possible origins of ground ice. Two types are important. The first occurs as a result of freezing of standing or running water; the second as a result of the freezing of water already held in sediments (segregation ice). The former type includes ice formed from spring waters (extruded ice), perhaps later buried by sediments, and ice formed by the annual freezing of melt-waters draining into contraction cracks in frozen ground. This latter has been called perennial vein ice and forms the foliated ice-wedge networks formed in the fissure polygons. The second type, segregation ice, occurs as lenses or veins of ice formed by the segregation of ice when ground freezes, perhaps after the expulsion of water from saturated sand as freezing proceeds, or is the result of the freezing of water or liquefied sediment intruded under pressure along boundaries of impermeable deposits (injection ice). Today the most widespread ground ice is that of ice-wedge networks and segregation ice.

Segregation ice can develop in many different patterns, such as foliar, reticulate, and irregular. Some of these types and the sediments in which they form are seen in fig. 5.6. On melting, signs of structure, e.g. lamination, will be left in the thawed sediment, and may be seen in some Pleistocene sediments.

The formation and melting of ground ice produces distortion and faulting in sediments, and characteristic land forms (thermokarst). These effects are considered later in relation to fossil periglacial structures and land forms.

PERIGLACIAL STRUCTURES

Frost action working in the ground at high latitudes and altitudes produces patterned ground and isolated features. Some of these are seen as fossil structures, and as they are related to particular environmental conditions they can be valuable for the reconstruction of past environmental conditions. Present-day occurrences and processes will be described first and then fossil occurrences.

Patterned ground

Patterned ground is not confined to surfaces affected by frost; it is, for example, seen in the desiccation patterns of arid lands. But it is highly characteristic of arctic, sub-arctic and montane regions. Washburn[36, 37] has produced a most useful descriptive classification of patterned ground. It is based on the shape of the pattern and the degree of sorting in the pattern. In general, the shape is related to the slope. On flat ground, circles, nets and polygons develop, on moderate slopes steps, and on steeper slopes stripes (plate 10a). The transition from polygon to stripe may be direct without steps. The magnitude of the patterns varies from 0·5 m or even smaller, to 3 to 10 m with a much greater size, up to tens of metres, with ice-wedge polygons. The size is related to the frequency of the freeze–thaw cycle, the availability of water and the severity of the climate. The smallest forms are found in montane regions, and the largest in continental arctic- or sub-arctic regions.

Table 5.2 indicates the variety of patterns; clear distinctions be-

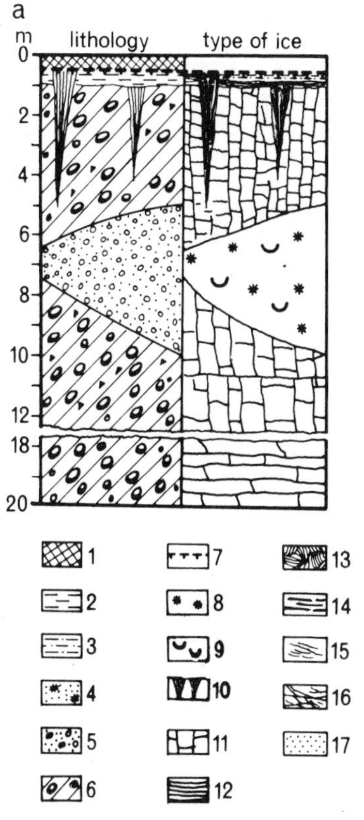

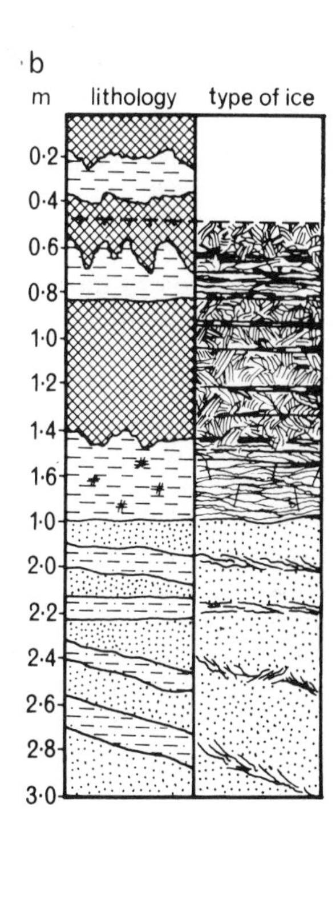

Fig. 5.6 Sections of perennially frozen upper Pleistocene deposits in the Anadyr region, east Siberia (after Vtjurin 1964), showing types of ground ice in relation to sediment type.

a. Section through frozen till, Kanchalan estuary.
b. Section through frozen floodplain deposits of the river Kanchalan.

Sediments

1. peat
2. loam
3. sandy loam
4. peaty sand
5. gravel
6. till
7. limit of seasonal thaw

Ground ice types

8. sublimation ice (crystallisation from water vapour)
9. groups of ice-cemented stones
10. foliated ice of wedge

11–16. segregation ice of different textures
- 11. coarse reticulate
- 12. partly foliated
- 13. irregular confused
- 14. horizontally foliated
- 15. incompletely foliated
- 16. reticulate-foliate

17. massive frozen sediment with pore ice

tween the types are not always possible, as is to be expected with phenomena produced by a number of processes.

Isolated structures

In addition to those periglacial features organised into patterns, there are those which occur in an irregular or isolated fashion, such as frost mounds.[15, 17] Their formation results from local peculiarities of drainage and sediment. With such local conditions, e.g. badly drained sand and silt and peat, local ice lenses may develop in the permafrost or above it by segregation of ice, perhaps during the down freezing of the active layer, and on melting these produce horizontal veinlike structures or depressions in the ground. Freezing-down of the active layer in the autumn or the development of permafrost over water-bearing sediments may produce injections of saturated slower freezing clays or silts under pressure into frozen overlying coarser sediments (fig. 5.16).

A special form of frost mound resulting from the bulging-up of frozen ground under hydrostatic pressure from below is the hydrolaccolith. Here the pressure of water trapped between the freezing-down of the active layer or the enveloping permafrost above and impermeable rocks or permafrost below causes the ground to swell and form mounds which in size may vary from a few metres across for annual mounds (mostly resulting from freezing-down of the

---

Fig. 5.7 Mounds resulting from frost action.

a. Pingo of Greenland type (after Müller 1959).
b. Pingo of Mackenzie type (after Müller 1959). 1–4 show stages of trapping of unfrozen saturated sediment by enveloping permafrost. 1. Lake and related unfrozen ground beneath. 2. Permafrost spreads as a result of lake shallowing and freezing of infill. 3. Hydrostatic pressure of water in unfrozen saturated sediments at depth causes bulging of ground. Ice begins to form as water is expelled upwards from the saturated ground beginning to freeze. 4. Envelopment by permafrost complete.
c. Hummock formed by local presence of segregation ice. Localisation may result from presence of sediment favouring segregation of ice (silt, clay), from thick peat cover and hydrology favouring total heat loss over the year (pals), or from degradation of widespread permafrost and ice-wedge systems leaving frozen ground nuclei in protected areas (thermokarst hummocks).
d. Formation of a mound as a result of seasonal freezing of a saturated terrace sand over impermeable rock. Arrows show direction of encroachment by permafrost and resulting upward hydrostatic pressure.

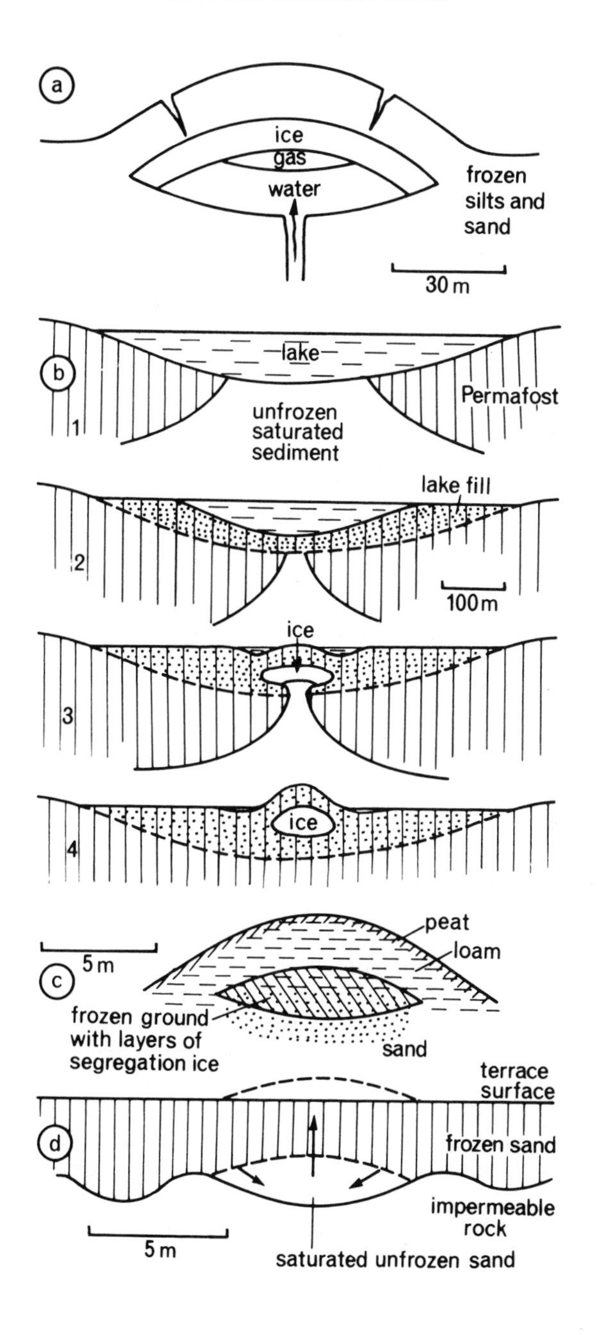
a
ice
gas
water
frozen silts and sand
30 m
b
lake
Permafost
1
unfrozen saturated sediment
lake fill
2
100m
ice
3
ice
4
peat
loam
5 m
c
frozen ground with layers of segregation ice
sand
terrace surface
d
frozen sand
impermeable rock
5 m
saturated unfrozen sand

Table 5.2 Types of patterned ground

Pattern	*Sorted (s) and non-sorted (ns) types*	*Remarks*
1. Circles	s stone circles, debris islands ns peat and tussock rings, spot medallions	isolated or grouped, circular or near circular features. The non-sorted types manifest in vegetation patterns. 1–10 m diameter
2. Nets or hummock fields	s patterns intermediate between 1s and 3s ns earth hummocks, frost mounds, palsar; patterns intermediate between 1ns and 3ns	more closely spaced than 1, sizes similar
3. Polygons	s stone polygons ns frost-crack and ice-wedge (tundra) polygons (plate 10b); tussock-birch heath polygons, desiccation polygons	patterns 1–10 m diameter, but non-sorted may be up to 100 m or more across. Non-sorted polygons manifest in vegetation patterns. Low centre (active growth) polygons, and raised centre (wedge degradation) polygons
4. Steps	s stone garlands ns peat or turf terraces	on moderate slopes, and may form elongated polygonal patterns
5. Stripes	s stone stripes ns vegetation stripes (plate 10a)	occur on slopes 2° or steeper, 1–10 m wide; non-sorted types manifest in vegetation patterns

active layer) to hundreds of metres for the perennial types. Typical structures are shown in fig. 5.7. The large perennial hydrolaccoliths are known as pingos, [18, 21] derived from the Eskimo for a large hillock.

The large pingos are of two types. The Greenland ('open system') type results from hydrostatic pressure of water moving upwards from under or in the permafrost and there freezing causing a conical swelling of the ground (fig. 5.7a); springs may be found in the crater at the top of the swelling. The Mackenzie ('closed system') type results from the formation of massive ice from water trapped between enveloping permafrost and permafrost already present at depth (fig. 5.7b). This type of pingo develops characteristically in lake basins.

Hydrolaccolith hillocks develop as outlined above and eventually a number of processes leads to their degradation. Loss of the covering sediment by solifluction, for example, leads to the melting of the ice mass by the penetration of summer warmth. If a spring is present in

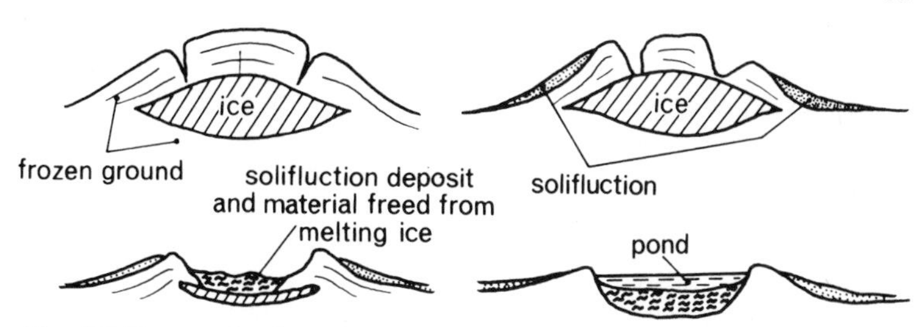

Fig. 5.8 Decay of a frost mound. Slumping and solifluction degrade the raised mound as the ice melts. If the ice core is substantial and the water table high, a pond may be formed in the centre.

the crater, the outflowing water may breach the wall and degradation result. Figure 5.8 portrays the degradation of a frost mound.[17]

Causes of structure formation

Very many hypotheses have been put forward to explain the origin of the structures already described (Washburn, 1973). Several processes can be recognised:

1. Sorting. The possibility of sorting by freeze–thaw has already been mentioned. A commonly observed sorting process is the upward

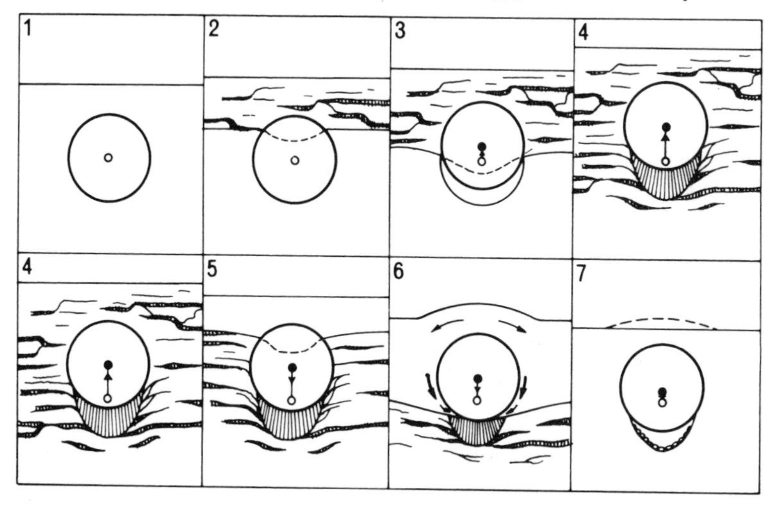

Fig. 5.9 Stone movement during freeze-thaw (Beskow 1930). 1–4 on freezing; 4–7 on thawing. The thin lines and vertically lined layers represent ground ice. The arrows between the black and outline small circles show the resulting movement of the centre of the stone, the outline circle giving the original position.

movement of stones by continued freezing and thawing (fig. 5.9). On freezing a fine-grained water-saturated sediment containing stones expands; embedded stones are thrust upwards. On thawing from the surface down the stone does not settle back to its original position because it will be held by its base in frozen sediment while thawed sediment collapses around it. The result is the tendency of stones to become oriented with their long axes vertical and finally thrust up out of the mass of sediment. In this way sorting the stones from such a sediment can occur.

2. Movements resulting from the (cryostatic) pressures exerted by the volume expansion of some 10 per cent which takes place on the freezing of water. If the sediment mass does not change volume much on freezing, as with sand, water will be expelled. Ice lenses may result.
3. Differential heaving produced by the effect of alternate freezing and thawing on sediments of different grain size. The finer sediments will expand more and thus low mounds or boils of fine sediment will rise to the surface. The amount of segregation ice formed will determine degree of heaving. Maximum segregation tends to occur in silts (fig. 5.6).
4. Pressure effects such as those produced by down-freezing of the active layer in the autumn. The fine sediments will contain a greater amount of water and will stay unfrozen longer than the coarse sediments and relative movements between the fine or coaser sediments are liable to occur. Considerable hydrostatic forces will be exerted on the frozen sediment. Unfrozen sediment may be trapped between the down-freezing layer and the permafrost or other impermeable sediments. Liquid fine-grained sediments may then be injected upwards under heavy pressure as the sediments expand on freezing or a hydrolaccolith may be formed. Freezing of sediments may be delayed by the lowering of freezing point under pressure.
5. Contraction of ground results in fissuring. Contraction is a result of a number of processes:
   a. desiccation, caused by evaporation, drainage and the transfer of water to places of ice formation.
   b. thermal contraction when freezing of ground takes place. It may be seasonal, producing cracks and fissures in the active layer in winter; or perennial, as permanently frozen ground contracts with the formation of deep fissures.

   Polygonal patterns of fissuring develop in permafrost areas, the fissures often becoming filled with water in the spring thaw, water which later freezes. Fissures filled with ice (ice-wedges) are formed in this way (fig. 5.15a).

6. The effect of peat and vegetational cover on the rapidity of freezing and thawing is important. A vegetation pattern may develop as soil structures develop and one will enhance the other.

   As an example of this type of effect we may take the formation of peat hummocks (palsar) which occur near the southern border of the permafrost region, e.g. northern Scandinavia. Frozen saturated peat has a higher thermal conductivity than saturated peat, which in turn has a higher thermal conductivity than dry peat. In the summer a surface layer of dry peat prevents the warming up of the soil, but in winter when the surface peat becomes wet and frozen, cold penetrates deeply. As a result, segregation ice forms in the peat and hummocks grow. The effects of frost heaving on vegetation also controls the expression of features. Very fine sediments which readily react to freeze–thaw by heaving will not be colonised by higher plants (e.g. grasses) easily, since their roots will be disturbed. Coarser sediments show less heaving and will more easily be colonised. With no vegetation the fine sediment becomes even more susceptible to temperature variation.
7. Solifluction. This has already been considered.

Usually it is not one of these effects that explains patterned ground, but a combination. For example, in stone circles, 1 will be important (and the sorting will be aided by eluviation of fines and gravity), in stone stripes, 1 and 7, and in non-sorted polygons 5 may be important but 1 weak. In palsar formation 6 is important, and in hydrolaccoliths 2 and 4.

The magnitude of these effects is dependent on a number of factors: the aspect and slope of the surface, the sediment composition and stratigraphy, the magnitude and frequency of the freeze–thaw temperature fluctuations, mean temperature, and the availability of water. This last factor relates to whether the system is closed, when any effects taking place are limited by the available water (and its expansion on freezing), or open, when water can be drawn in from outside, as in open system hydrolaccoliths, so giving greater effects.

All in all, it can be seen that there is tremendous opportunity for variation in process and effect, which of course leads to difficulties in trying to produce an artificial classification of periglacial phenomena.

Relation of structures to permafrost and climate

The structures mentioned above are not all characteristic of areas where permafrost is present. Many occur under climates with sea-

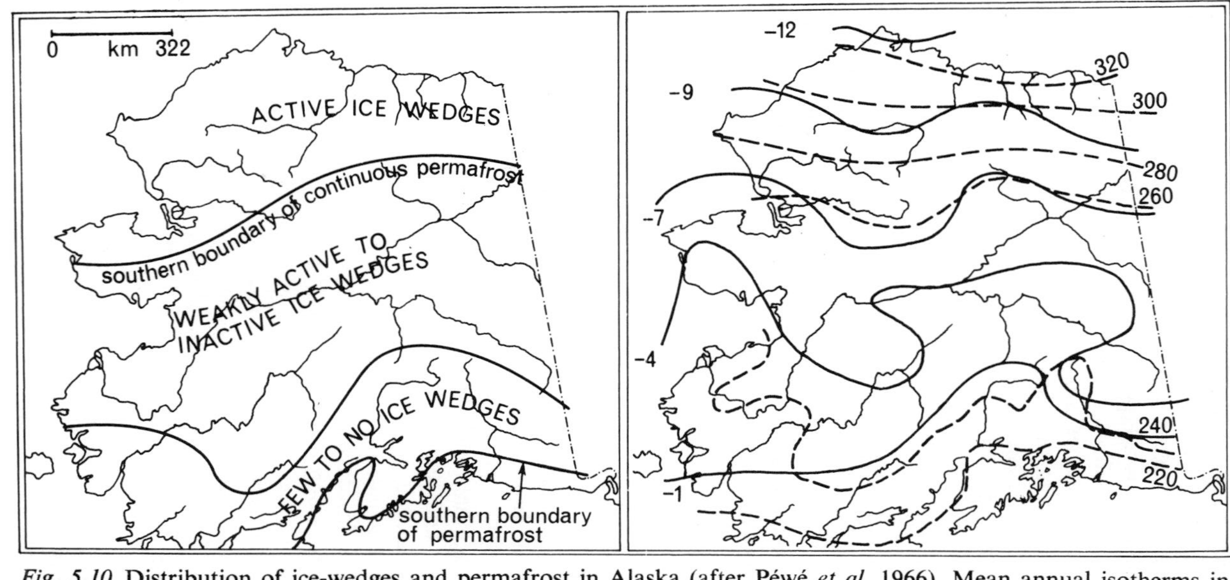

Fig. 5.10 Distribution of ice-wedges and permafrost in Alaska (after Péwé *et al.* 1966). Mean annual isotherms in Centigrade are shown, also number of days in the year with freezing temperature.

sonal or diurnal freeze–thaw changes. The following in particular appear to be characteristic of permafrost areas: large sorted polygons, ice-wedge polygons, certain non-sorted circles and polygons (e.g. spotted tundra) and certain large sorted or non-sorted stripes. Ice-wedge polygons in particular are a sure indication of permafrost. The distribution of ice-wedges in relation to temperature and permafrost in Alaska is shown in fig. 5.10, and in northern Eurasia in relation to permafrost in fig. 5.11.

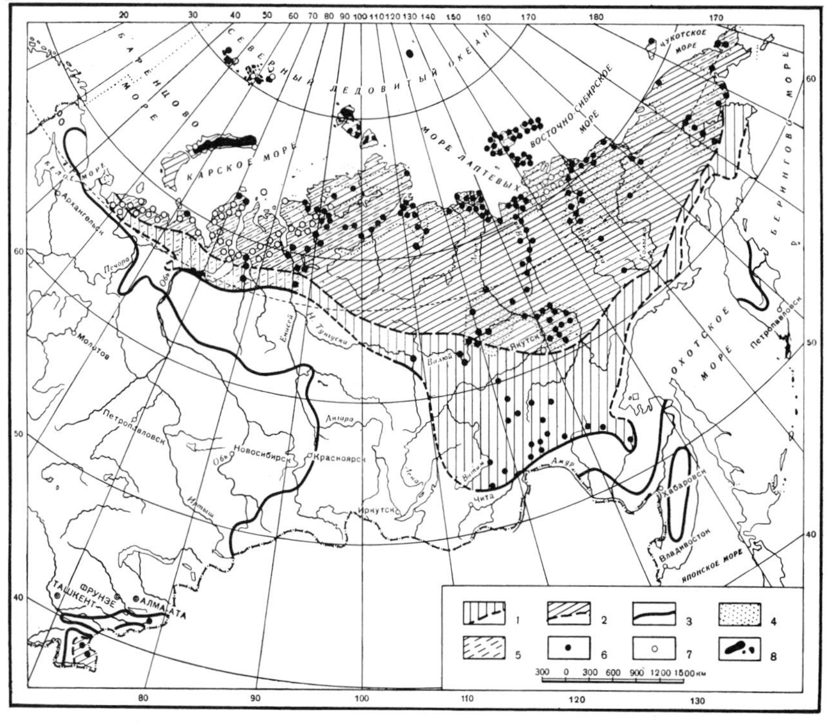

Fig. 5.11 Schematic map of the distribution of underground perennial vein ice (mainly ice-wedges) in the U.S.S.R. (Shoezov, 1959).

1. Area and southern limit of fossil and contemporary perennial vein ice.
2. Area and southern limit of fossil vein ice.
3. Southern limit of permafrost.
4. Regions with vein ice of considerable thickness.
5. Regions with vein ice of small thickness.
6. Sites where vein ice known from ground observation.
7. Sites where vein ice known from air observation.
8. Glaciers.

The distribution in relation to climate of the structures is not known in great detail.[33] The difficulty of climatic interpretation of periglacial features is that vegetation, soil type and drainage play such a large part in determining freeze–thaw effects, as has been already mentioned. Drainage is a particularly significant factor in determining freeze–thaw effects. Thus where seasonal freezing occurs, but not permafrost, the effect will be much more in evidence in badly-drained than well-drained ground. If permafrost is prevalent then freeze–thaw effects will be more widespread, occurring, but with different result, on both badly- and well-drained ground.

In Eurasia it appears that large sorted structures are characteristic of high arctic and oceanic sub-arctic regions, e.g. Spitsbergen and the Scandinavian mountains respectively, while palsar and ice-wedge polygons are better developed in the more continental sub-arctic climates, where deep freezing takes place before much snow accumulates in winter. Frost mounds are frequent in the boreal continental climate with deep summer thawing and thin permafrost, as seen in fig. 5.12, showing the distribution of mounds associated with injection ice in northern Eurasia. Mounds associated with seasonal freezing are also shown on this map.

Washburn (1973) has given an outline of the ranges of periglacial processes and features in terms of lowland and highland areas of the polar and subpolar regions and of middle and low latitudes.

Fossil periglacial structures

Many of the structures already described occur in the fossil state, that is, they occur in regions where climatic conditions no longer cause their formation. The following notes describe the most frequently found fossil structures and discuss their interpretation:

Upright stones, a result of process 1 above. Indicative of freeze/thaw but not necessarily permafrost.

Involutions and injections. In many sections of Pleistocene deposits contorted structures are seen in the superficial sediments. These are known as involutions. The contortions most often involve the interdigitation of lobes or flames of sediment of different grain size. The involutions may have a horizontal and vertical regularity making them appear closely related to structural surface patterns of the kind which develop in the active layer. Or they may be more irregular and form flames which are probably the result of injection processes (fig. 5.16c). Various types of involution are shown in fig. 5.13. The contortions are chiefly the result of process 2, 3 and 4 above; where

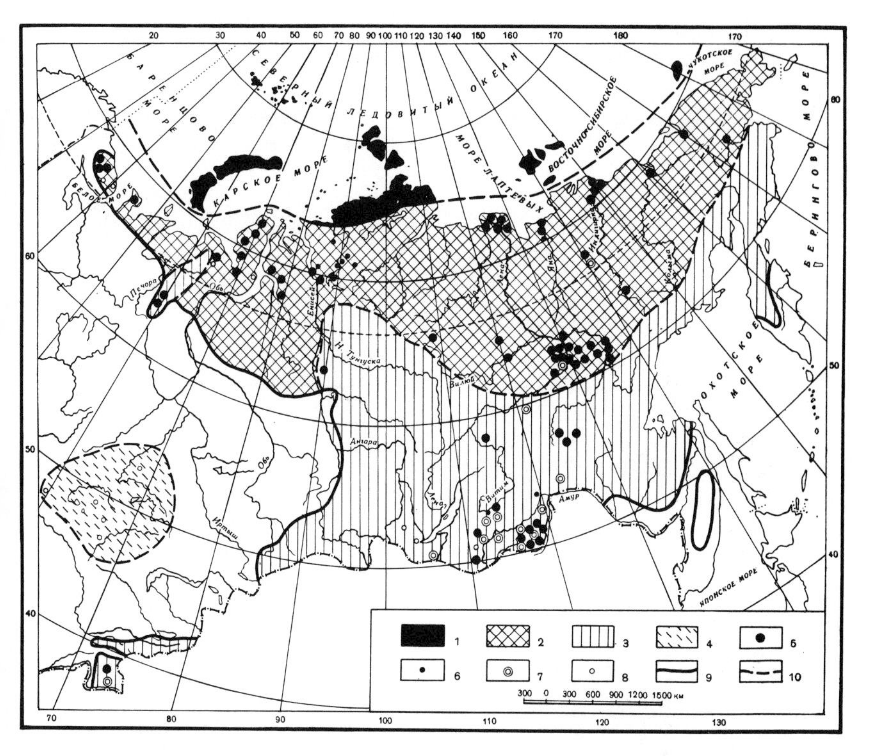

Fig. 5.12 Schematic map of the distribution of mounds associated with injection ice in the U.S.S.R. (Shoezov, 1959).

1. Arctic area, mounds not found.
2. Area with perennial (rarely seasonal) ice mounds, mainly by freezing of ground beneath lakes.
3. Area with perennial and seasonal ice mounds, mainly where springs occur.
4. Area with isolated seasonal mounds at spring outlets, outside the permafrost limit.
5. Groups of perennial ice mounds.
6. Isolated perennial ice mounds.
7. Groups of seasonal mounds.
8. Isolated seasonal mounds.
9. Southern limit of permafrost.
10. Boundaries of areas.

they occur on a slope they may be drawn out down-slope by solifluction.

Involutions and injections are not necessarily indicative of permafrost conditions. They can be formed where the active layer overlies

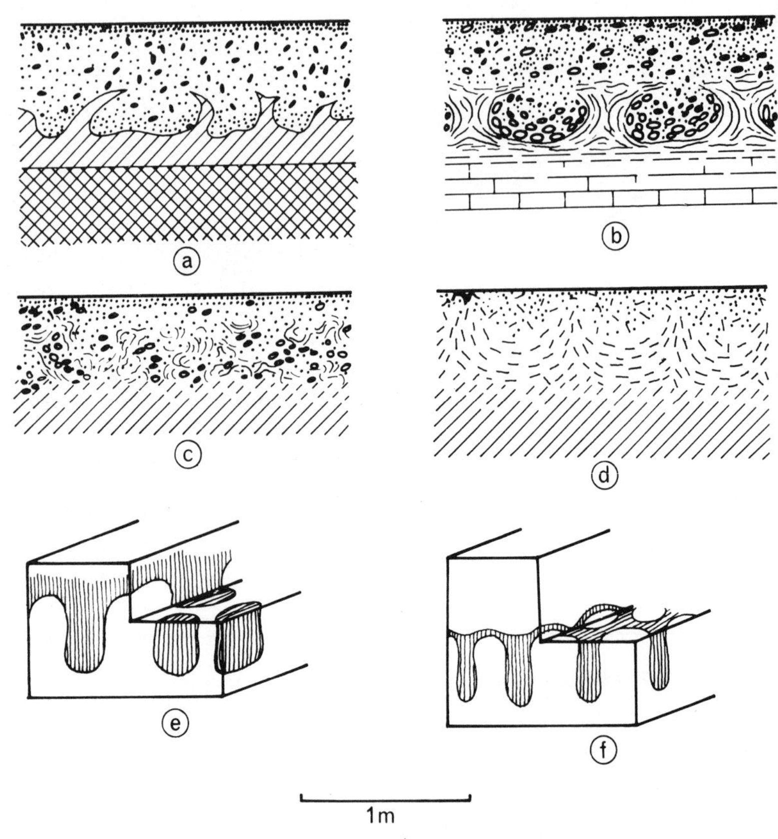

Fig. 5.13 Various types of involution (a–d Sekyra 1961).

a. Injection of tongues of finer sediments (silt, clay) into overlying sand and gravel.
b. Frost pockets, frost kettles, or pillar involutions, with ascending tongues of finer sediments.
c. Irregular or amorphous involutions.
d. Festoons, with tongues of frost-shattered disintegrating rock rising into finer sediments.
e. Pocket involutions (finer sediments shaded).
f. Drip or plug involutions (finer sediments shaded).

an impermeable sediment. Regular involutions caused by freeze–thaw should not be confused with load casts in water-laid deposits.[6, 7, 11] The geological context of the contortion should provide a distinction between the two.

Frost-cracks and ice-wedges. Wedge-shaped fissures or linear near-vertical cracks or fissures are commonly seen in sections of Pleistocene sediments. The detailed interpretation of these is difficult because of the variety of ways in which such cracks and fissures can develop. When ground freezes cracks develop as desiccation and contraction proceeds. The depth to which these cracks penetrate, the width to which they grow and whether they become filled with ice (ice-wedges, fissure ice) or sediment as they form, depends on the composition of the sediment, the moisture content of the sediment, the degree of saturation of the sediment with ice and, most important, the temperature regime. The spacing of cracks appears to depend on the magnitude of the maximum annual gradient of temperature in the profile, and the width on the magnitude of autumnal temperature decrease compared with mean annual temperature, as well as on the age of the fissure.

Various forms of contemporary cracks and fissures have been described from permafrost areas[28] (fig. 5.14):

1. Seasonal cracks in the active layer, which may become filled with thin ice, which melts in the summer so that the cracks become filled with sediment.
2. Deeper fissures penetrating the permafrost, in which ice forms and in which ice-wedges (fissure ice) may grow. In the active layer above, cracks of type 1 may be present.
3. Casts of fissures of type 2, with sediment replacing the fissure ice after thawing.
4. Fissures which fill with sediment, often wind-blown, as they form, the so-called sand wedges.

Types 2, 3 and 4 are indicative of permafrost. Their size is partly a function of age, as ice will be formed annually in the seasonally developed crack, and their depth indicates the minimum permafrost depth, as they can only grow in frozen ground. Type 1 can occur on a small scale in seasonally frozen ground,[38] but on a larger scale is very probably indicative of permafrost, though this does not appear to have been proved.

Probably all these types occur fossil in Pleistocene sections. The problem is recognising the distinctions between them, which would be valuable because each develops under particular environmental conditions.

Fig. 5.14 Various types of fissure (after Romanovskij, 1973).

1.

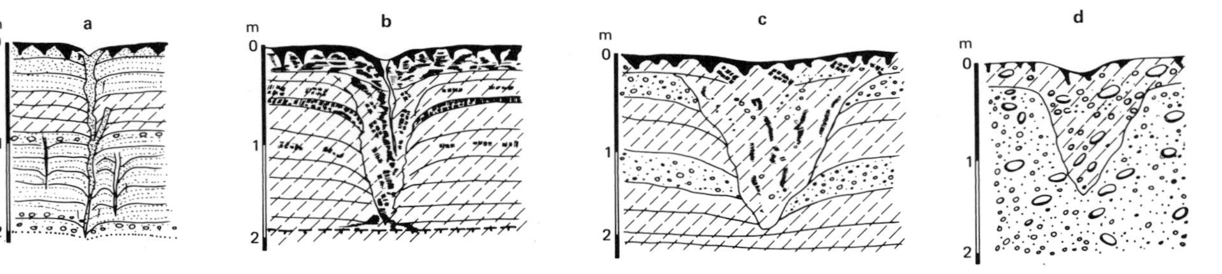

1. Development of fissures.
   a. In active layer, not reaching permafrost table.
   b. In active layer, when fissures begin to penetrate permafrost.
   c. In frozen ground at low temperature.

2.

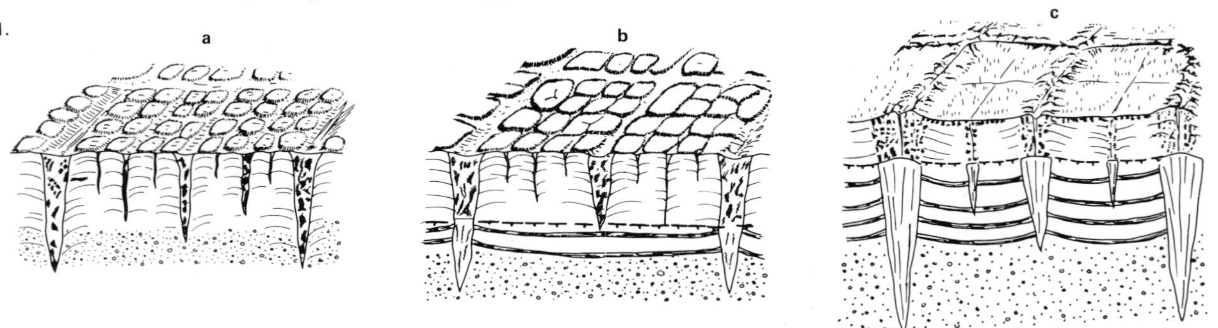

2. Fissures in the active layer. a. Initial stage. b, c. Later stages. d. In gravel.

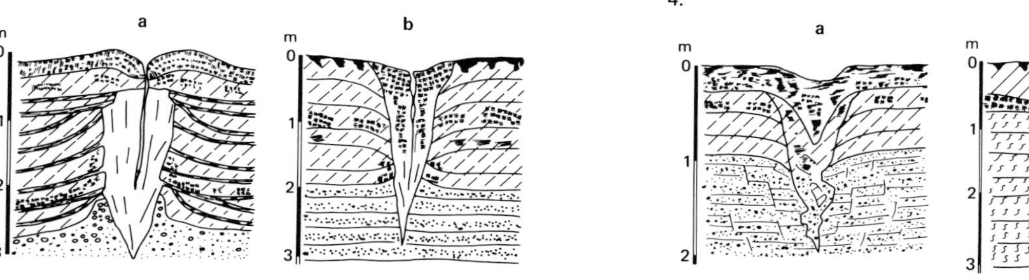

3. Fissures with ice (ice-wedges).
   a. Syngenetic.
   b. Epigenetic.

5. Fissures filled with sediment while forming.
   a. In sand.
   b. In till.

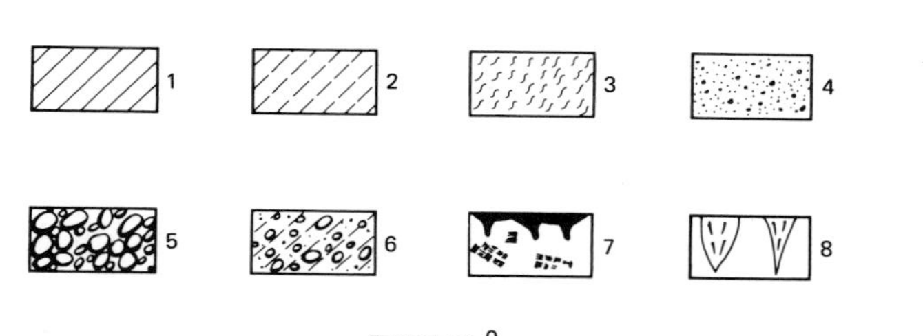

4. Fissures filled after thawing of ice.
   a. With faulting.
   b. 'Two-stage', upper part in active layer, lower part in permafrost.

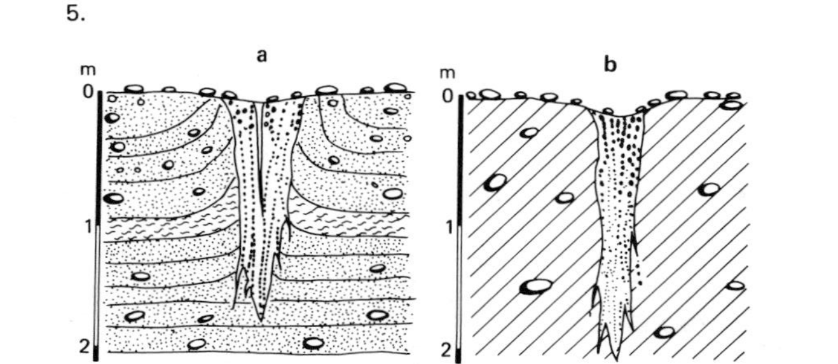

Key to symbols: 1, silt; 2, clayey sand; 3, loess; 4, sand; 5, gravel; 6, till; 7, peat, humus; 8, ice; 9, lower limit of active layer, permafrost table.

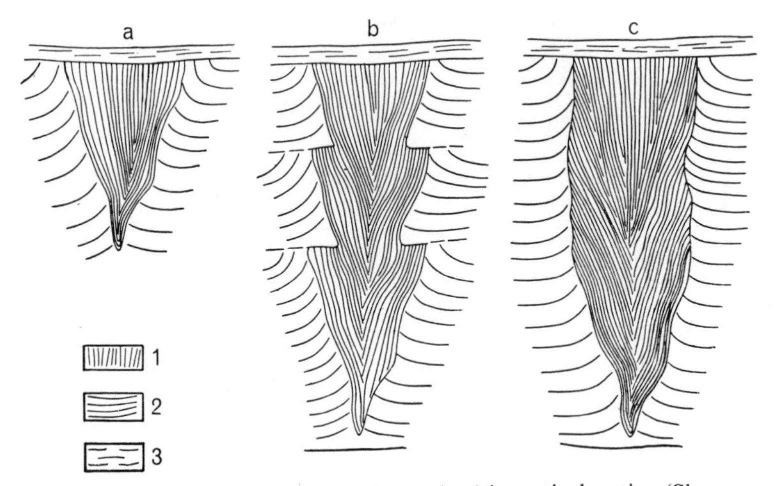

Fig. 5.15 Three types of fissure ice (ice wedges) in vertical section (Shoezov, 1959). a, epigenetic wedge; b, epigenetic wedge showing renewed growth; c, syngenetic wedge.
1. annual layering of the ice in the wedges.
2. frozen ground, stratified.
3. active layer.

Type 4 sand wedges would appear easiest to distinguish, since they contain no or little ice and on thawing retain their near vertical structure in the filling and the upturned stratification near the fissure. The filling can only take place when there is little snow cover and wind-blown sediment available. They have been observed in the arid regions of Antarctica, but may well have formed in Eurasia under cold very continental conditions.

The 'two-stage' structure of type 3 (fig. 5.14.4b) indicates thawing across the active layer/permafrost boundary, with the upper part, corresponding to the active layer level, showing wider collapse. Downturning of the stratification near the fissure boundary is common to types 1 and 3, but fissures of type 1 may show flowage of sediment as the fissure is in the active layer.

The structures above and below the permafrost table in types 2 and 3 will differ because of the growing ice body in the fissure below the permafrost table. Remnants of upturned stratification due to ice pressure in type 2 may persist on thawing, but will not be so obvious as in type 4.

When the ice of fissures melts, it will be replaced by the collapse of adjacent sediments and by younger sediments above it. The cast of the fissure may then be seen in vertical section (plate 14a, fig. 5.22) or in horizontal view (plate 10b). The sediment in or near the apparent cast may be faulted, through collapse of blocks of sediments, or it may be contorted, either as a result of collapse or partly because of earlier distortions resulting from the growth of the fissure ice.

Fissures may be epigenetic, penetrating the ground to form a wedge, or they may be syngenetic, growing vertically up as sediment accrues. Three types of wedge known from present-day observation are shown in fig. 5.15. Whether the epigenetic and syngenetic type can be distinguished in the fossil state it is difficult to say, but the syngenetic types have more parallel sides and blunter bases, and these properties may be preserved in the fossil state. The importance of the difference is that syngenetic wedges imply a long period of permafrost while sediment builds up, while epigenetic wedges give evidence for a period of permafrost at the particular horizon marked by their upper limit.

Fossil fissures are of particular importance in 'climatic' stratigraphy. They are characteristic of perennially frozen ground, e.g. it is thought that a mean annual air temperature of −6 to −8°C is required for ice wedges to grow.[23, 24] They are thus indicators of the past existence of permafrost at times which can often be very clearly identified within a stratigraphical sequence. The sediment penetrated by the wedge will be of importance for climatic interpretation. For example in the continuous permafrost zone in Alaska active wedges occur in sand, gravel and silt, but in the discontinuous zone they are confined to silt.

Polygons and stripes. These have been recognised in vertical section and also in surface view. Sections showing these structures are not very common; examples are shown in fig. 5.16. Surface expression of polygons and stripes is best seen in aerial photographs, where the distribution of vegetation, dependent on soil differences resulting from the structures, clearly reveals the nature of the patterns. Such patterns are widely distributed in the Breckland of East Anglia (plate 10a).[40, 41] Small-scale polygons and stripes do not necessarily indicate permafrost, but the large ones do. In some cases, e.g. tussock–birch heath polygons, it appears that the depth of the active layer is marked by the depth of disturbance of the soil, so that the depth of the active layer at the time of formation of the similar fossil structure may be estimated.

Frost mounds. [15, 17] The surface form taken by decayed frost mounds is mentioned later, but occasionally such structures are seen in section but not at the surface.[42] The presence of frost mounds does not neces-

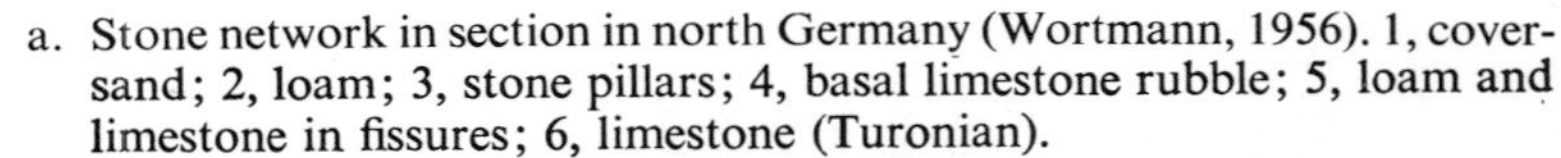

Fig. 5.16 Three sections showing periglacial structures.

a. Stone network in section in north Germany (Wortmann, 1956). 1, cover-sand; 2, loam; 3, stone pillars; 4, basal limestone rubble; 5, loam and limestone in fissures; 6, limestone (Turonian).

b. Non-sorted stripes in section in East Anglia (Watt, Perrin and West, 1966). An aerial view of such stripes as these is shown in Plate 10a. Chalky drift overlies *in situ* Chalk. The stippled areas are sand and sandy soil. The more intense stippling in the middle indicates the presence of calcium carbonate. The parallel lines at the base of the stippled layer and the single lines within it indicate zones with clay enrichment. Flints are shown black, and erratics shown open. Injections of very flinty chalk drift are seen rising towards the surface.

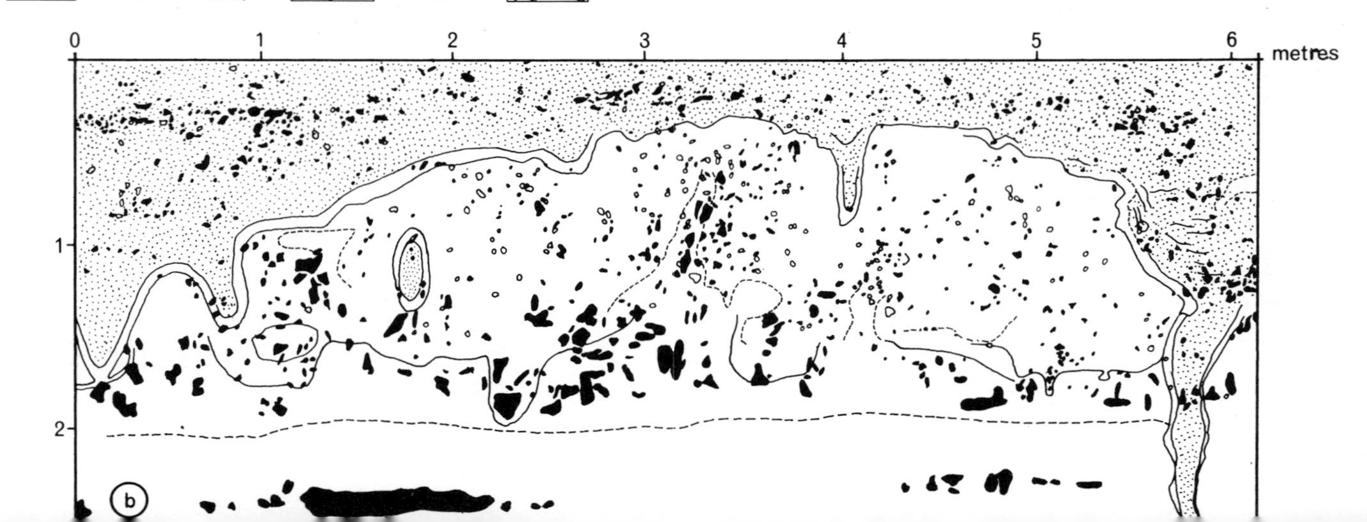

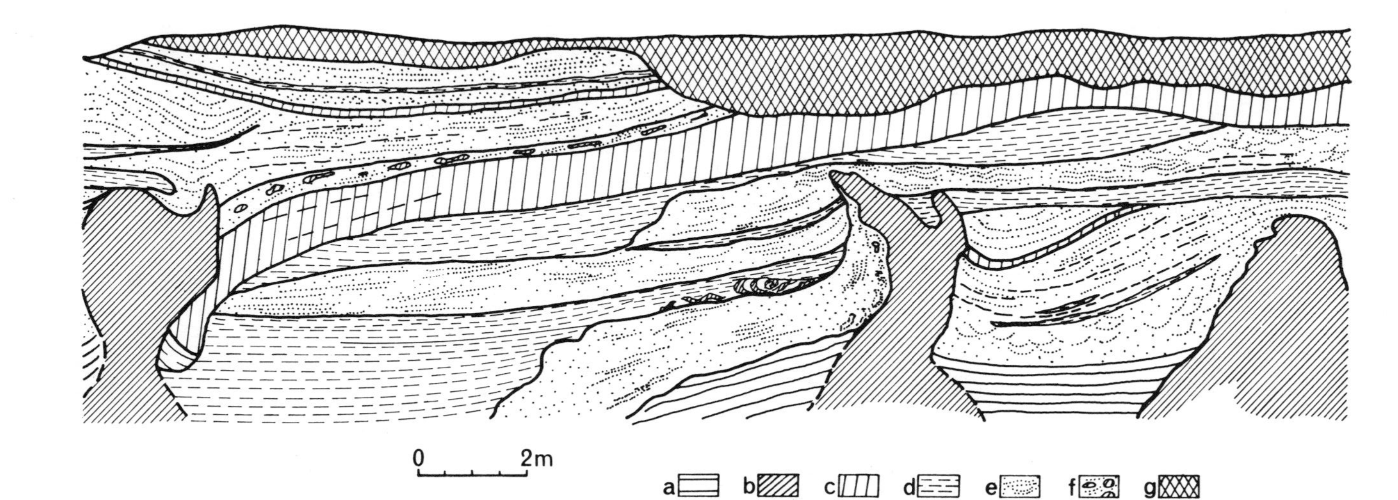

c. Section showing injections of clay into overlying silts and sands near Warsaw (Makowska 1961). The relations of the injected clay and basin deposits suggest injection and deposition of the latter alternated. a, varved clays; b, injected clay; c, clay; d, silt; e, sand; f, sand with clay-balls; g, talus.

sarily indicate permafrost. Small mounds (a few metres across) are usually annual and may develop by freezing down towards an impermeable unfrozen layer. The larger ones (pingos) are probably perennial and may then be associated with permafrost. Decayed frost mounds have been recorded from East Anglia, Wales and Ireland.

Faulting in superficial sands and gravels has been associated with tensions induced by the cooling of frozen ground in winter.[31]

LANDFORMS

Summer melt-water, solifluction and wind produce characteristic landforms, both constructional and erosional, in the periglacial region. The prevalence of solifluction will lead to a smoothed landscape. Such a landscape, smoothed, say, during the last glaciation in a periglacial region, contrasts greatly with the fresh landforms of the last glaciation deposits. This contrast has led to the concept of older and newer drift,[9] the boundary between them marking the limit of the last glaciation. Hills become flat-topped, their superficial sediments becoming affected by freeze–thaw processes. Laterally solifluction deposits mantle the slope, becoming thicker on the lower slopes.

Constructional features arise by the action of wind, melt-water and solifluction. Sand-dunes were built up in the great 'Urstromtäler' of northern Europe (fig. 5.3), wide valleys draining the ice fronts, and they were also formed by the blanketing cover-sands. The directions of winds prevalent during dune formation has been inferred from the orientation of the dunes (fig. 5.17) and the bedding of the dune sands.[16] Although longitudinal and transverse dunes of Pleistocene age are known, most dunes in the periglacial region are U-shaped with their convex side facing downwind, and also their steeper side and steep dip of foreset beds also facing downwind. Dunes imply strong prevalent winds and an abundant source of sand, but they are not necessarily associated with dry conditions and a lack of vegetation, though these of course greatly favour dune formation.

Constructional features are also formed by the lobes of fluviatile sediments deposited by the summer melt-waters. Similarly, solifluction lobes may be formed, sometimes very coarse and forming stone streams.

Erosional features are formed by frost action, solifluction and snow melt-waters. The term nivation includes these processes and describes the erosive effects associated with patchy snow cover. In the periglacial region, the open vegetation, the sudden bursts of melt-water liberated in the spring, and the impermeable frozen ground, combine

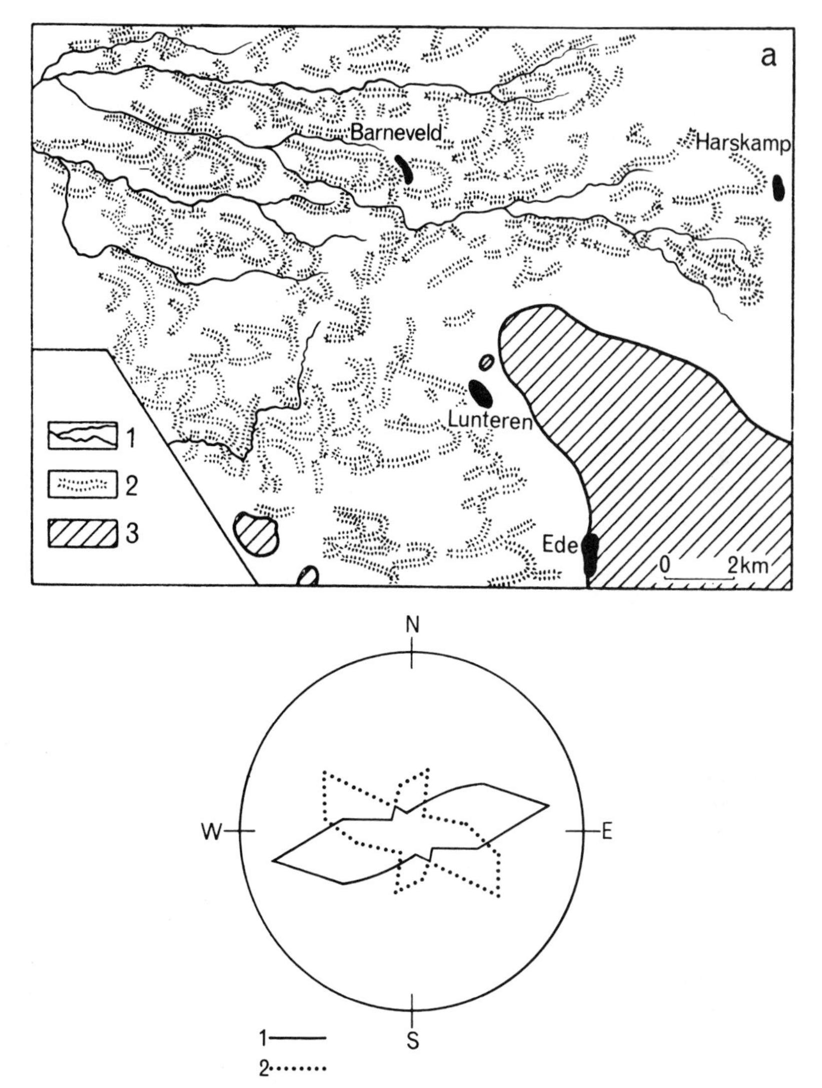

Fig. 5.17 Late Weichselian cover-sands in the central Netherlands (Maarleveld, 1960).

a. Ridges of younger cover-sand I (older Dryas, Zone I). 1, rivers; 2, dune ridges; 3, push-moraines of Saale age.
b. Orientation of dunes. 1, ridges of younger cover-sand II (younger Dryas, Zone III); 2, ridges of younger cover-sand I (older Dryas, Zone I).

to make surface drainage an important erosive agent. Channels may be formed in fluviatile deposits, and valleys in areas of greater relief. These valleys are often now dry and cannot carry water because they are cut in permeable sands and gravels. During periglacial times the frozen ground held up percolation and surface drainage followed. The frozen ground need not have been perennial, though the size of many dry valleys suggests that it often was.

The upper ends of the valleys often form wide shallow hollows which probably originated by nivation and were enlarged by melt-waters. Further down the valley the sides are steeper and the strength of the melt-waters seems to have prevented the degrading of slopes by solifluction. The valleys again become shallow and wider as they flatten out. Small flat hollows were also formed by nivation at the angle of slope at the top of the valley sides. Such hollows presumably represent an early stage of valley formation.

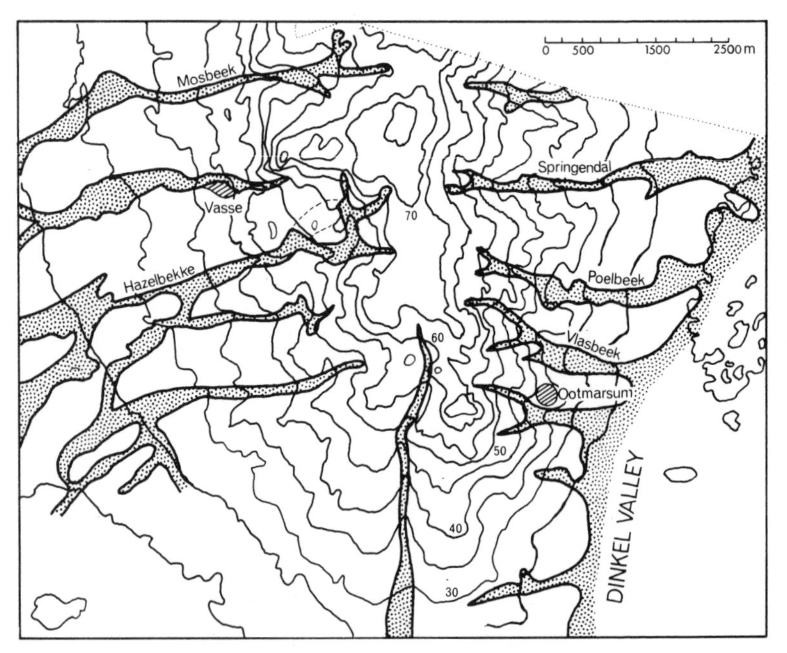

Fig. 5.18 Map of ridge dissected by periglacial valleys near Dotmarsum, south-east Netherlands (van der Hammen 1951). The ridge is an ice-pushed feature of the Saale glacial stage, and the valley fills are Late Weichselian and Flandrian. The valleys were cut during the mid-Weichselian glacial stage. Contours at 5 m vertical interval.

The arrangement of periglacial valleys may often take on a radial form, valleys occurring at regular intervals at right angles to the face of the slope (fig. 5.18).

In certain valleys, especially those trending north–south, the angle of slope of each side may differ, giving an asymmetrical cross-section. There are a number of causes of this assymmetry.[27] In the upper parts of the valleys where no fill is present and erosion is active, the south- and west-facing slopes may be shallower than those facing north and east, as they will be subject more to insolation and thus be more liable to solifluction as the soil warms in summer. On the other hand, in the lower parts of the valley insolation may produce more melt-water at the base of the south- and west-facing slopes, and the stream so formed may undercut the slope and make it steeper than those facing north and east.

Terraces cut in bedrock, with a veneer of nivation products, and occurring at varying altitudes and irregular intervals, have been described as altiplanation terraces and ascribed to a periglacial origin. The formation of tors, protruding masses of bedrock isolated by differential weathering, may also be in part the result of periglacial mass wasting, as may be the case on Dartmoor, where they are associated with stone streams and altiplanation terraces.[39]

Besides dry valleys, other valleys of 'older drift' areas, carrying small misfit streams and substantial fills of gravel and sand, may have been widened by periglacial conditions; such valleys are not usually found in newer drift areas. The spreads of gravel and sand were probably largely deposited by swiftly moving snow melt-waters during the spring thaw. An example of such a wide valley carrying sand and gravel and with a small present-day stream is the river Lea valley, tributary to the Thames, northeast of London.

Irregular landforms produced by the melting of ground-ice has been termed thermokarst.[3] The melting of such ice procudes thaw depressions or lakes, isolated or in groups. A special case are the walled depressions, perhaps infilled by sediments, formed by decayed pingos.

Valley-bulging and cambering

Superficial disturbances of rocks in the valleys of the East Midlands and southern England have been ascribed to the establishment and melting of permafrost.[10] The disturbances take the form of cambering of strata into the valleys, the development of sags and gulls, and bulging at valley bottoms (fig. 5.19). Such disturbances are associated with favourable hydrological conditions for the saturation of rocks and for their early freezing as permafrost develops, e.g. where

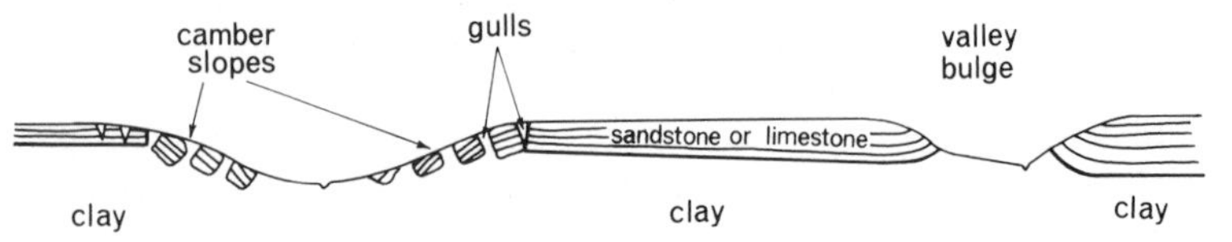

Fig. 5.19 Transverse section through valleys showing cambering and bulging of strata.

aquifers (e.g. Northampton Sand) rest on impermeable clays (e.g. Lias Clay). As a result of freezing, ice formation in saturated clays results in expansion and bulging of the rocks near the valley floor. On melting, the bulged material is subject to removal by streams, and as a result of the disappearance of the ground ice collapse occurs, thus leading to sagging and cambering of the overlying more competent strata.

DISTRIBUTION AND STRATIGRAPHY OF PERIGLACIAL DEPOSITS

Figure 5.20 shows schematically the distribution of various periglacial effects across a north–south section in central Europe. It is evident that changes in climatic regime will produce differing periglacial effects, so leading to a sequence of periglacial deposits. Thus sections of superficial deposits often show disturbances by frost

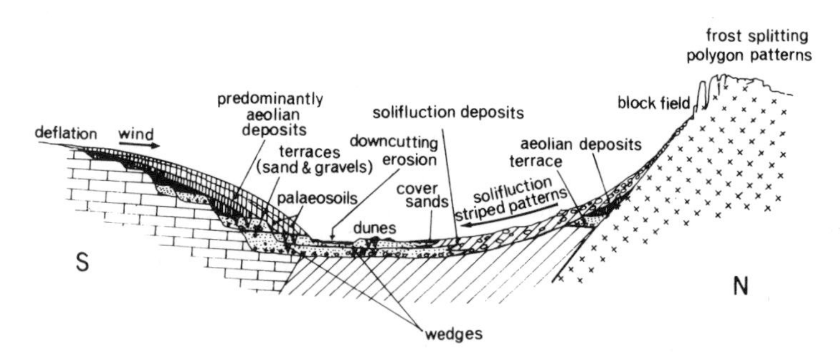

Fig. 5.20 Scheme of periglacial conditions, structures, and deposits in Europe (after Sekyra, 1961).

action, i.e. involutions and casts of ice-wedges, intercalated in undisturbed sediments which may be aeolian, fluviatile or lacustrine in origin, perhaps containing plant and animal remains.[42] Such sequences of periglacial deposits are very important for the study of climatic and environmental changes, for they give a sequence of periods of permafrost, involution formation, solifluction, aeolian sedimentation and interstadial conditions.

Figure 5.22 shows two sections giving evidence of changes in periglacial regime.

The mapping of periglacial features is a means of reconstructing Pleistocene climates. The difficulty is the synchronisation of periglacial deposits and structures across a region. An approach to such a reconstruction for the Weichselian periglacial climate in northern Europe is shown in fig. 5.21.

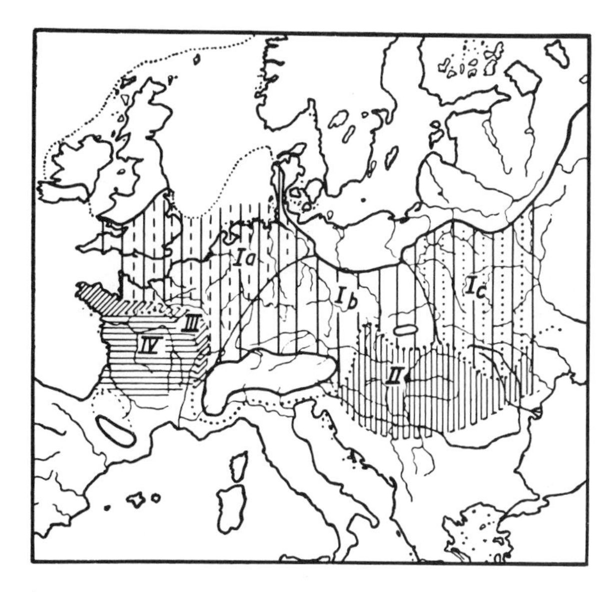

Fig 5.21 Climatic provinces of the Weichselian glacial stage in northern Europe (Poser 1948). The differentiation into provinces is based on the areal distribution of ice-wedge casts, involutions, frost heaving, asymmetric valleys, and forest and steppe-tundra pollen spectra, and on the summer thaw depth indicated by involutions.

I. Permafrost-tundra climate. Permafrost, little summer warmth.
   a. glacial-maritime province.
   b. province between the Scandinavian and Alpine ice. The coldest area, with the least summer thaw depth.
   c. glacial-continental province
II. Permafrost-forest climate. Permafrost, warmer summers.
III. Maritime-tundra climate. Transition between Ia and IV. Winters not cold enough for permafrost, little summer warmth.
IV. Maritime forest climate.

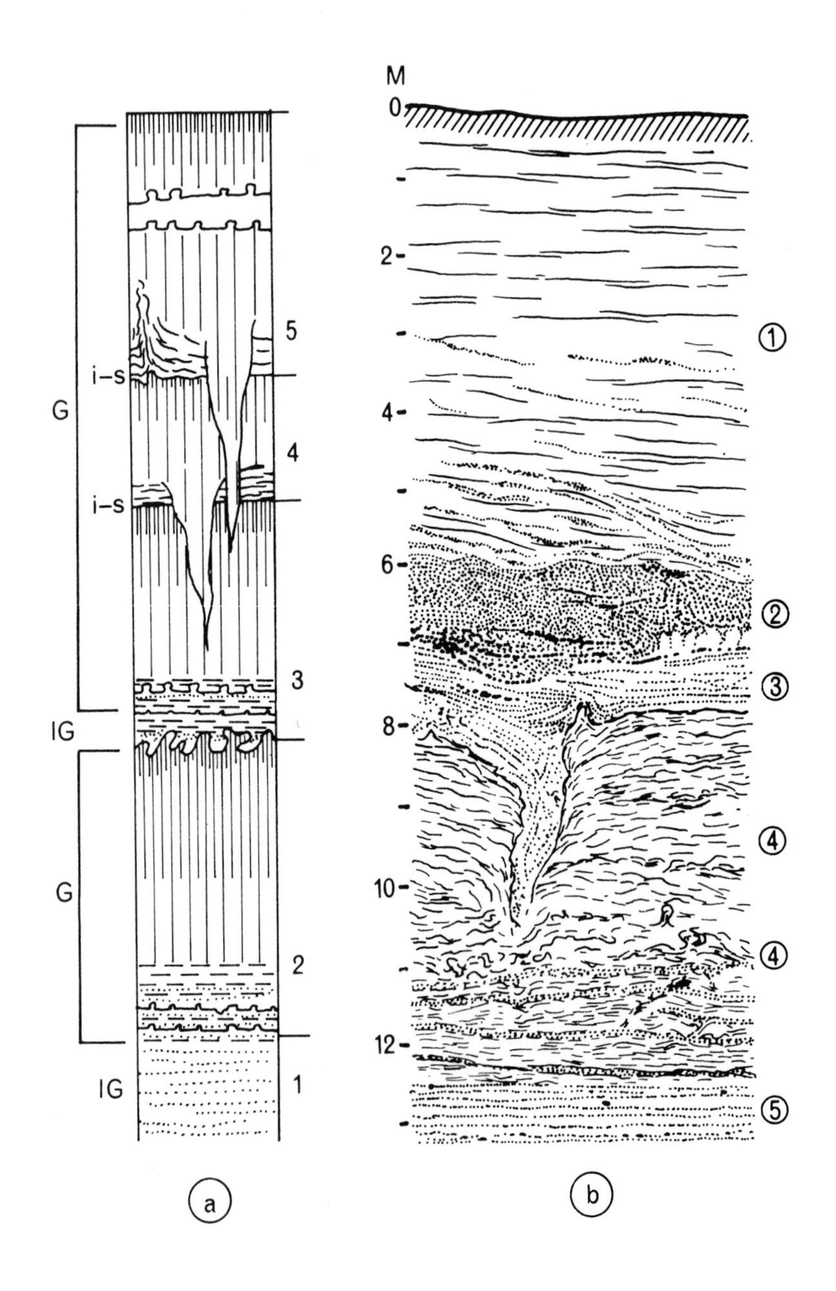

M
0
2
4
6
8
10
12
G
i–s
i–s
IG
G
IG
5
4
3
2
1
①
②
③
④
④
⑤
a
b

Fig. 5.22 Sections through periglacial deposits with structures.

a. Loess-stratigraphy at Hrubieszow in eastern Poland (Mojski, 1961). IG, interglacial; G, glacial; i–s, interstadial. Two permafrost periods and several times of involution formation are shown in the upper glacial stage.
   1. Fluviatile sands (Holstein interglacial).
   2. Loess, alluvial and with involutions at base. Chernozem soil developed at top (Eemian interglacial soil), disturbed by involutions.
   3. Loess, alluvial at the base, and disturbed by involutions and solifluction at the base. Brown earth soil at top (Weichselian interstadial soil).
   4. Loess, disturbed by solifluction at the base, and with ice-wedge casts near the base. Thin brown earth soil at the top (Weichselian interstadial soil?).
   5. Loess, disturbed by solifluction and involutions at the base and in the upper part. Ice-wedge casts in the lower part. Present soil, degraded chernozem, at the top.

b. Section at Topola in southern Poland (Klajnert, 1961).
   1. Loess (Weichselian glacial).
   2. Humus-rich soil (Eemian interglacial).
   3. Limestone gravel, sand and silt (Eemian interglacial).
   4. Lacustrine silts and silty sands, with involutions and pierced by ice-wedge cast (Saale glacial).
   5. Fluviatile gravels (Holstein interglacial).

REFERENCES

Biuletyn Peryglacjalny, Lodz.

EMBLETON, C. and KING, C. A. M. 1968. *Glacial and periglacial geomorphology*. London: Arnold.

FRENCH, H. M. 1976. *The periglacial environment*. London: Longman

HAMELIN, L-E. and COOK, F. A. 1967. *Illustrated glossary of periglacial phenomena*. Quebec: Les Presses de l'Université Laval.

HAVILAND, M. D. 1926. *Forest, steppe and tundra*. Cambridge University Press.

MACKAY, J. R. 1972. 'The world of underground ice', *Annals Assoc. American Geogr.*, **62,** 1–22.

MULLER, S. W. 1947. *Permafrost or permanently frozen ground and related engineering problems.* Ann Arbor: Edwards Inc.

Permafrost: the North American contribution to the 2nd International Conference on Permafrost, Yakutsk, 1973. Washington: U.S. National Academy of Sciences.

PÉWÉ, T. L. 1969. *The Periglacial Environment.* Montreal: McGill-Queen's University Press.

SHOEZOV, P. F. 1959. *Osnovy Geokriologij.* Part I. Akademia Nauk. S.S.S.R.

SHUMSKI, P. A. 1964. *Principles of structural glaciology.* New York: Dover Publications.

WASHBURN, A. L. 1973. *Periglacial processes and environments.* London: Arnold.

1 BESKOW, G. 1930. 'Erdfliessen und strukturböden', *Geol. Fören. Stockh. Förh.*, **52,** 622–38.
2 BLACK, R. F. 1954. 'Permafrost—A review', *Bull. Geol. Soc. America*, **65,** 839–56.
3 BROWN, R. J. E. 1974. 'Ground-ice as an initiator of landforms in permafrost regions'. In *Résearch in polar and alpine geomorphology*, ed. B. D. Fahey and R. D. Thompson, 25–42. Norwich: Geoabstracts Ltd.
4 CAILLEUX, A. 1969. 'Quaternary periglacial wind-worn sand grains in U.S.S.R.'. In *The Periglacial Environment*, ed. T. L. Péwé, 285–301. Montreal: McGill-Queen's University Press.
5 CORTE, A. E. 1966. 'Particle sorting by repeated freezing and thawing', *Biul. Peryglacjalny*, **15,** 175–240.
6 DZULYNSKI, S. and WALTON, E. K. 1963. 'Experimental production of sole markings', *Trans. Edinb. geol. Soc.*, **19,** 279–305.
7 EMERY, K. O. 1950. 'Contorted Pleistocene strata at Newport Beach, California', *J. Sedim. Petrology*, **20,** 111–15.
8 GRAHMANN, R. 1932. 'Der Lösz in Europa', *Mitt. Gesell. f. Erdkunde zu Leipzig* (1930–31), 5–24.
9 GRIPP, K. 1924. 'Über die äusserste Grenze der letzten Vereisung in nordwest Deutschland', *Mitt. geogr. Ges. Hamburg*, **36,** 161–245.
10 HAINS, B. A. and HORTON, A. 1969. *British Regional Geology. Central England.* London: H.M.S.O.
11 JARDINE, W. G. 1965. 'Note on a temporary exposure in central Glasgow of Quaternary sediments with slump and load structures', *Scottish J. geol.*, **1,** 221–4.
12 KLAIJNERT, Z. 1961. 'Topola'. In Guide Book of excursion Baltic to Tatras, Part 2, vol. 2, VI Congress International Association of Quaternary Research, Warsaw, 95–6.
13 KRINSLEY, D. H. and FUNNELL, B. M. 1965. 'Environmental history of quartz sand grains from the Lower and Middle Pleistocene of

Norfolk, England', *Quart. J. Geol. Soc. Lond.*, **121,** 435–61.

14 LUNDQUIST, G. 1948. 'Blockens orientering i olika jordarter', *Sver. geol. Unders.*, Årsbok **42** (C), no. 497.

15 LUNDQVIST, J. 1969. 'Earth and ice mounds: a terminological discussion'. In *The Periglacial Environment*, ed. T. L. Péwé, 203–15. Montreal: McGill-Queen's University Press.

16 MAARLEVELD, G. C. 1960. 'Wind directions and coversands in the Netherlands', *Biul. Peryglacjalny*, **8,** 49–58.

17 MAARLEVELD, G. C. 1965. 'Frost mounds', *Med. Geol. Stichting*, n.s. **17,** 1–16.

18 MACKAY, J. R. 1963. *The Mackenzie Delta, N.W.T.* Memoir No. 8. Geographical Branch, Mines and Technical Surveys, Canada.

19 MAKOWSKA, A. 1961. 'Diapir forms in varved clays at Baniocha near Warsaw'. In *Prace o Plejstocenie Polski Srodkowej.* Ed. S. Z. Rozyckiego. P.A.N. Komitet Geologiczny, 159–76.

20 MOJSKI, J. E. 1961. 'Periglacial deposits and structures in the stratigraphy of the Quaternary in Poland', *Instytut Geologiczny* (*Warszawa*), *Prace*, **34,** 675–96.

21 MÜLLER, F. 1959. 'Beobachtungen über pingos', *Medd. om Gronland*, **153,** No. 3.

22 PERRIN, R. M. S. 1956. 'Nature of "Chalk Heath" soils', *Nature*, **178,** 31.

23 PÉWÉ, T. L. 1966. 'Palaeoclimatic significance of fossil ice wedges', *Biuletyn Peryglacjalny*, **15,** 65–73.

24 PÉWÉ, T. L. 1973. 'Ice wedge casts and past permafrost distribution in North America', *Geoforum*, **15,** 15–26.

25 PITCHER, W. S., SHEARMAN, D. J. and PUGH, D. C. 1954. 'The loess of Pegwell Bay, Kent and its associated soils', *Geol. Mag.*, **91,** 308–14.

26 POSER, H. 1948. 'Boden- und Klimaverhältnisse in Mittel- und Westeuropa während der Würmeiszeit', *Erdkunde*, **2,** 53–68.

27 POSER, H. and MÜLLER, T. 1951. 'Studien an der asymmetrischen Tälern des Niederbayrischen Hügellandes', *Nachrichten der Akad. Wiss. in Göttingen, Math.-Phys. Klasse*, 1951, 1–32.

28 ROMANOVSKIJ, N. N. 1973. 'Regularities in formation of frost-fissures and development of frost-fissure polygons', *Biul. Peryglacjalny*, **23,** 237–77.

29 SAMUELSSON, C. 1926. 'Studien über die Wirkungen des windes in den kalten und Gemässigten Erdteilen', *Bull. Geol. Institute Upsala*, **20,** 57–231.

30 SEKYRA, J. 1961. 'Periglacial phenomena', *Instytut Geologiszny*, (*Warszawa*), *Prace*, **34,** 99–107.

31 SHOTTON, F. W. 1965. 'Normal faulting in British Pleistocene deposits', *Q. J. Geol. Soc. Lond.*, **121,** 419–35.

32 THOMPSON, D. B. and WORSLEY, P. 1967. 'Periods of ventifact formation in the Permo-Triassic and Quaternary of the north-east Cheshire basin', *Mercian Geologist*, **2,** 279–98.

33 TROLL, C. 1944. 'Strukturböden, solifluktion und frostklimate der Erde', *Geol. Rundschau*, **34,** 545–694 (In English translation: U.S. Army Snow, Ice and Permafrost Research Establishment, translation No. 43, 1958).

34 VAN DER HAMMEN, T. 1951. 'Late-glacial flora and periglacial phenomena in the Netherlands', *Leidse Geol. Meded.*, **17,** 71–183.

35 VTJURIN, B. I. 1964. *Kriogennoe stroenie chetvertichnyx otlozhnenij.* Akademia Nauk. S.S.S.R.

36 WASHBURN, A. L. 1956. 'Classification of patterned ground and review of suggested origins', *Bull. Geol. Soc. America*, **67,** 823–66.

37 WASHBURN, A. L. 1970. 'An approach to a genetic classification of patterned ground', *Acta Geogr. Lodziensia*, **24,** 437–46.

38 WASHBURN, A. L., SMITH, D. D. and GODDARD, R. H. 1963. 'Frost cracking in a middle-latitude climate', *Biul. Peryglacjalny*, **12,** 175–89.

39 WATERS, R. S. 1971. 'The significance of Quaternary events for the landform of south-west England'. In *Exeter essays in geography*, ed. K. J. Gregory and W. L. D. Ravenhill, 23–31. University of Exeter.

40 WATT, A. S., PERRIN, R. M. S. and WEST, R. G. 1966. 'Patterned ground in Breckland: structure and composition', *J. Ecol.*, **54,** 239–58.

41 WILLIAMS, R. B. G. 1964. 'Fossil patterned ground in Eastern England', *Biul. Peryglacjalny*, **14,** 337–49.

42 WEST, R. G., *et al.* 1974. 'Late Pleistocene deposits at Wretton, Norfolk. II. Devensian deposits', *Phil. Trans. R. Soc. London, B*, **267,** 337–420.

43 WORTMANN, H. 1956. 'Ein erstes sicheres Vorkommen von periglazialem Steinnetzboden im Norddeutschen Flachland', *Eiszeitalter u. Gegenwart*, **7,** 119–26.

CHAPTER 6

STRATIGRAPHICAL INVESTIGATIONS

The history of the Pleistocene is based on the stratigraphy of the deposits. From the point of view of gross stratigraphy, the interpretation of a succession of deposits is not normally difficult. For example, two tills separated by a thick sand indicate two ice advances, each laying down a till, separated by a phase of ice retreat during which the sand was deposited; and the occurrence of a peat deposit with remains of temperate trees between two tills indicates a period of ice retreat with a temperate climate between two ice advances.

The finer details of stratigraphy are just as important as the gross stratigraphy, for they give the clue to important details of environmental history. Climatic changes, from, say, arctic to temperate, are known to have taken place during the deposition of a few tens of centimetres of sediment or less. All changes of sediment, both in structure and composition, have significance. This is especially evident in the study of biogenic deposits and periglacial deposits. With biogenic deposits it is necessary to link the changing sediments with the contained fossil flora and fauna, because only in this way can the most complete environmental analyses be obtained. Taking a single bulk sample from, say, an open peat section, may give an overall view of the fossils over the whole period during which the peat was laid down, but will say nothing about environmental changes during this time. Careful stratigraphical study followed by close sampling is necessary to show this. Similarly with the study of periglacial deposits. Alternating conditions of, say, frost-heaving, polygon formation and aeolian deposition of sand will only be revealed by careful stratigraphical work.

Stratigraphical investigations of Pleistocene deposits will include the study of open or excavated sections and of underground stratigraphy by boring. The methods and techniques associated with each of these will be treated in turn.

SECTIONS

The study of open sections is far more satisfactory than that of borehole samples, in particular because Pleistocene sediments are apt to vary considerably over short distances and borehole records may give an unreliable picture of the whole sequence; for example, the non-sequences may not be apparent. The first step in studying a section is to clean the face thoroughly to produce a vertical section, or if verticality is not possible, as is often the case in a talus slope, to make a series of vertical sections separated by horizontal steps. The horizontal steps may also be useful in giving a three-dimensional view of any structures present in the sediments and this may be important for the interpretation of the structures, e.g. those of periglacial origin. The recording of a section is most complete if done by both drawing and photography. Carefully drawn records (e.g. fig. 5.16) are just as valuable as photographs, and drawing a section in detail invariably brings to notice detail which would escape casual examination and photography. The most suitable method is to mark out a grid of metre squares on the face to be studied, and the details of stratigraphy can then be drawn to a suitable scale. This will vary according to the section being studied, but a scale of 4 cm to a metre is often satisfactory.

Sampling from open sections is far safer than from boreholes. The risk of contamination is low if suitable precautions are taken to clean the face and the sampling instrument. The distance apart of samples will vary according to the section. A 5-cm interval is usually adequate, but it is best to take additional samples near contacts of different sediment types, as the change of sediment type may result from a significant environmental change. If the sediments are compacted and varying rapidly vertically, it is best to use a closer interval than 5 cm. In the laboratory, samples wider apart than this can be analysed first and, if necessary, the full closer sequence can then be done. The vertical position chosen to sample should show the most complete series of sediments possible, but it may be necessary to take subsidiary series in other places to make the sampling complete.

For pollen analysis samples a small spatula or penknife will be found most useful, with the sample of about 5 cm^3 being placed in a glass or plastic tube about 5 cm × 2 cm, which is then labelled with the site and depth. For analysis of macroscopic plant and animal remains it is best to take a series of bulk samples, in the same place as the pollen samples, so that the whole column of sediment is represented in the samples. The relation of the macroanalyses to the microanalyses will then be beyond doubt.

It may be more convenient to take a monolith of the sediments back to the laboratory for study and sampling there. A convenient size of monolith is 15 cm wide and 15 cm deep and in segments about 40 to 50 cm long. Such may be obtained in the softer sediments (e.g. mud and peat) by hammering a metal box of these dimensions into the face with a sledge hammer after lining the box with polythene sheeting. The sediment column can then be stored until ready for study. The same method may be used with clays, e.g. varved clays, but here, if it is just the varving which is to be studied, a much smaller monolith may be all that is necessary.

A method of preserving a thin film of sediments has been developed for use with exposures of clays, sands and fine gravels. This is the lacquer method, described in Appendix 2 (p. 426). There are also methods of impregnating unconsolidated sediments so that their microscopic structure can be studied in thin section.[1]

BORING

Methods

No one sampling device can sample with the same degree of success in all sediments. Here we are concerned with sampling in relatively unconsolidated sediments varying from till to gravel to sand to biogenic deposits of one sort or another. There are samplers which work in each of these sediments, and the choice of a particular sampling tool will depend on the type of sediment to be cored, the depth to which it is necessary to go, the manpower required to work the tool, the ease of transport, and the cost. The most difficult problem encountered when boring in Pleistocene sediments is that of penetrating loose wet sand. There is usually no alternative to casing the borehole in this event, and a rig is necessary to do this at depth. Thus hand-operated methods are usually stopped by such sediment.

It is essential to choose the sampling methods best fitted to the job in hand. Each site has its own characteristics in terms of sediment, depth of sediment, ease of approach and so on. The investigator should be prepared to adapt the methods to his own particular problem, bearing in mind that the chief features of a sampler should be simplicity, ease of operation, robustness and ease of repair. It is always best to carry a tool kit so that repairs can be made in the field.

It is not possible to give here a complete description of the many boring tools which have been used for investigation of Pleistocene stratigraphy. A good comprehensive account has been given by Wright, Livingstone and Cushing[14] to which reference should be made for greater detail; and Lammers[6] has described hand-drilling tools suitable for Pleistocene investigations. The most useful and commonly used sampling devices are described below. The principal properties of each sampling method are summarised in table 6.1.

Three classes of boring methods concern us. The first, for use in lakes or in the sea, are the free-fall samplers attached to a cable. The second are hand-operated devices working with rods, used on water or land, and the third are the heavier boring devices, usually with rigs, for use on water or land. The second group are most commonly used and more attention will be given to this group than the others.

Free-fall samplers

These are simple short lengths of tubing of 2 to 5 cm diameter. They are attached to a cable and penetrate after free-fall into sediment to a distance determined by the height of the fall, the weight of the sampler, the diameter of the tube and the thickness of its wall and the amount of friction between the sediment and the tube wall. Core retaining springs filled at the base of the tube help to keep the core safe as the sampler is surfaced. This type of sampler is useful for sediments in deep water, but the core obtained is short and of course the depth of sampling is limited to the amount of penetration at one free-fall.

Hand-operated samplers

These are attached to a string of rigid rods rather than to wire and can be used on land or on water if it is shallow enough. The diameter and composition of the rods is determined by the cost, the strength they are required to be and the weight they can be for a given method of transport. Light weight alloy rods are most convenient when the sampler has to be man-carried, but where weight is not so important and cost is, standard water pipe can be used. The rods may be 1, $1\frac{1}{2}$ or 2 m in length and they can be joined either by a male and female coupling held by a set-screw as in the standard Hiller sampler, or by a coupling enclosed by a sliding tube held in place by spring-loaded pins.[7] Simple mechanical aids have been used to assist penetration or withdrawal of hand-operated samplers, such as chain-hoists, car jacks working on mole-grips attached to the rods and other jacking devices such as the one described later working with a screw auger.

Hand-operated samplers take in cores from the side as with a Hiller, or at the base as in piston or percussion corers, or take samples by screw or other type of auger. The piston corers are of two types, those where the piston is retractable and those where the piston remains stationary as the corer is driven.

Table 6.1 Comparison of some hand-operated samplers

Type	*Sample*		*Deformation of sediment*	*Compression of sediment*	*Suitability for sediments*	
	Length	*Width*			*Good for*	*Bad for*
Hiller	50 cm	3 cm	much	none	peat, compact mud	loose organic sediment, inorganic sediment
Russian	50 cm	5 cm	none	none	peat, mud	loose organic sediment, inorganic sediment
Dachnowski	30 cm	5 cm	little	little	non-fibrous peat, mud	fibrous peat, loose organic sediment
Livingstone	50 cm	4 cm	little	some	non-fibrous peat, mud	fibrous peat, loose organic sediment, inorganic sediment
Punch	50 cm	6 cm	little	some	compact organic sediment, clay, silt, fine sand	fibrous peat, loose organic sediment
Screw auger	25 cm	4 cm	much	much	compact organic sediment, clay and silt	soft organic and inorganic sediment
Other augers	10–30 cm	5–10 cm	much	little	compact organic sediment, clay, silt, sand, gravel	soft organic and inorganic sediment

Hiller sampler (fig. 6.1a). This sampler is the classical peat borer. It consists of a sampling chamber 50 cm long and about 3 cm diameter with a slit about 2 cm wide down one side. This is covered by a close-fitting outer tube with a similar sized slit bordered by a flange 2 or 3 cm wide set at a tangent. An inner removable liner of light alloy can be used inside the chamber, so that the whole core obtained can be removed and taken back to the laboratory.

The sampler is pushed down with a slight clockwise pressure in the closed position, i.e. with the outer tube covering the slit of the inner tube. On reaching the sampling depth the sampler is rotated anti-clockwise, the two slits coincide, and on further rotation (2 or 3 times) the flange forces sediment into the inner tube. Clockwise

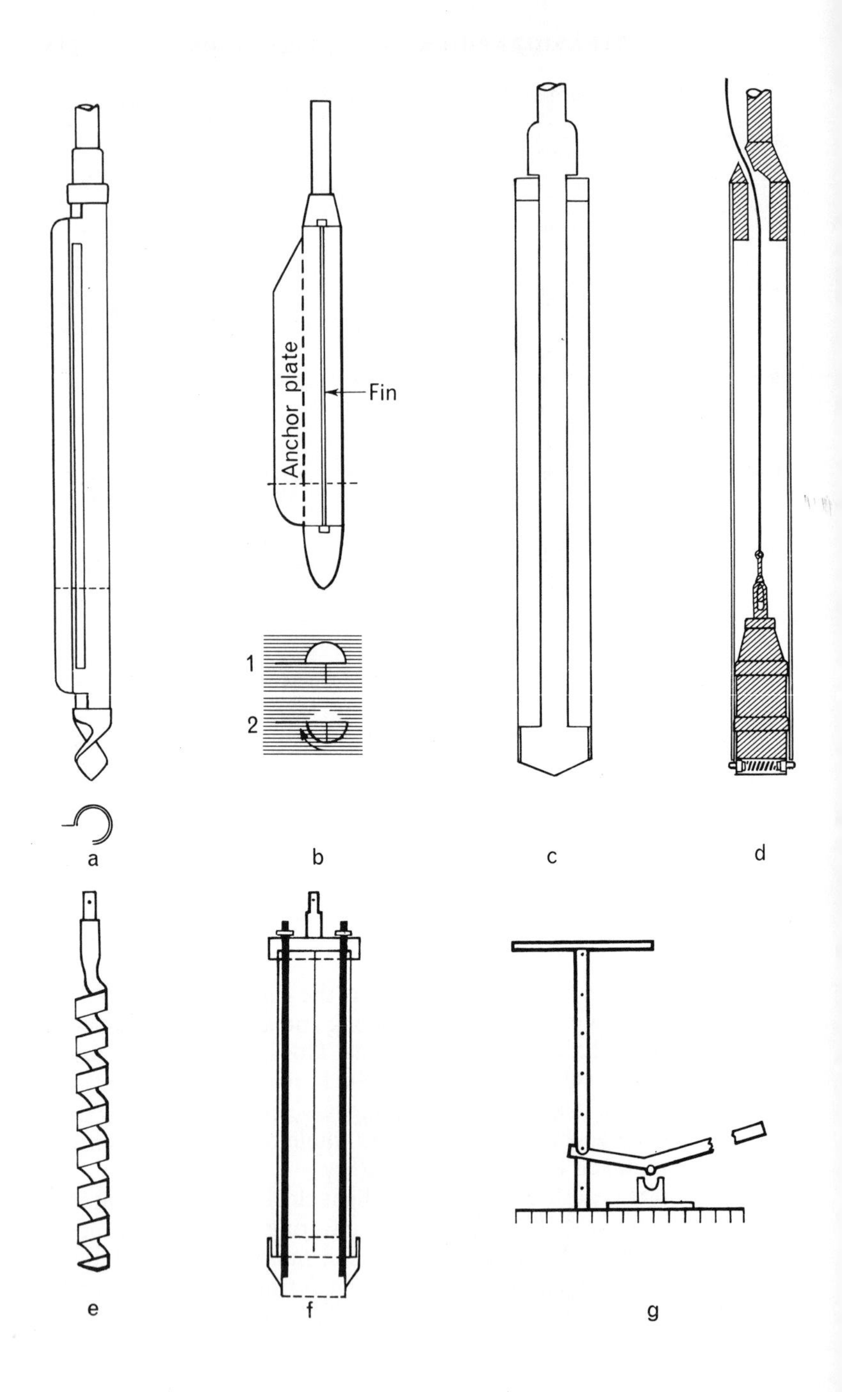
Anchor plate
Fin
1
2
a
b
c
d
e
f
g

rotation closes off the inner tube again and the sampler is withdrawn with a slight clockwise pressure to keep it shut. The samples are taken with a clean spatula from the centre of the chamber after opening the sampler by anticlockwise rotation, and after scrupulous cleaning off of the exposed sediment surface. The advantages of a Hiller are these: it is easy to use, there is no compression of the sediments, and it takes reasonable samples in fibrous, mossy, or wood peat and in fairly stiff lake muds. The disadvantages are: it distorts sediments as the flange forces them into the tube—thus microstructures are obscured, it is jammed by sand, it will not hold loose sediments such as watery muds, or penetrate very compact sediments, the samples are liable to contamination by sediment and water getting into the tube as the sampler descends or ascends, and it is difficult to thoroughly clean the inside except where water is freely available.

Russian peat sampler (Fig. 6.1b).[4] This sampler consists of a half-cylinder (shuttle) 50 cm long and 5 cm diameter, fixed to the sampler head, and which can be rotated at depth through 180° to retain a half-cylinder of peat against a central anchor plate. The shuttle rotates round this anchor-plate, which projects into the sediment on one side and thus remains stationary while the shuttle rotates. The half cylinder of peat obtained by rotating the shuttle is bisected lengthways by a fin-plate at right angles to the anchor plate. The advantage of this sampler over the Hiller is that sediment structures are preserved, the sampler is more easily cleaned, and the sample obtained can be easily removed and stored in polythene sheeting.

Dachnowski sampler (Fig. 6.1c). This is a piston corer with a retractable piston. It has a tube about 30 cm long and 5 cm diameter, with a piston attached to the end of the rod bearing the tube. The

---

Fig. 6.1 Samplers

a. Hiller, with section in open position.
b. Russian, with sections in open (1) and closed (2) position.
c. Dachnowski.
d. Livingstone.
e. Punch.
f. Screw auger.
g. Jack for screw auger.

sampler is lowered with the piston in position at the end of the tube. When the sampling depth is reached, the piston is raised to the top of the tube by withdrawing the rods till it is held in this position by a spring catch on the rod engaging a collar at the upper end of the tube, or by giving the rods a half-twist and engaging splines cut in the rod, and working in grooves at the top of the tube, against the top of the tube. The sampler is then pushed down the length of the tube and a core is taken. After raising, the core is extruded by pushing back the piston, and can either be sampled on the spot or taken back to the laboratory. The advantages of a Dachnowski sampler are that it is a simple design, light and easy to use, penetrates well in fairly stiff fine-grained sediments and takes undisturbed cores. Its disadvantages are that it will not retain loose sediments though core-retaining springs if fitted will help, it may be jammed by sand, it is liable to contamination from sediment and water leaking in through the top, it will take only a short sample, and compression of sediments is liable to occur.

Livingstone sampler (Fig. 6.1d).[8] In this type of piston sampler, the piston is held stationary while the coring tube is driven. Thus hydrostatic pressure above the core is eliminated and entry of unwanted sediment into the upper part of the tube before sampling is prevented. A core up to a metre or a metre and a half can be taken at one drive, the limit being set by the friction of the sediment sample with the corer wall, increasing as penetration increases.

There are many types of this sampler.[14] Basically the sampler consists of a removable tube about a metre long and 4 cm wide fixed by screws to a head. The piston consists of two rubber stoppers on a threaded bolt, and the fit to the tube, which is critical, can be adjusted by tightening or loosening nuts bearing on the stoppers. The piston is held in the base of the tube by spring-loaded pins, and is attached to a wire running to the surface. The sampler is pushed to the required depth, the wire is then fixed firmly by mole-grips, and the sampler is driven down for the length of the tube past the stationary piston. The sampler is then raised and the core either extruded or removed in its tube back to the laboratory after corking the ends. A new tube is easily fitted and the operation repeated. A modification[12] makes the basal rod pass through the head and bear against the piston in its basal position, thus preventing unwanted movement of the piston while the sampler is being lowered. On reaching the required depth the rods are raised and the base of the basal rod returns to its normal position near the head where it is held by a spring catch working on a collar. The piston is held in position by a wire and the sampler is then driven down to take the sample. This sampler is more robust than the original and may be driven through compact sediments with a hammer. The hammering may result in more deformation of the core than usual.

A 10 cm diameter corer of Livingstone type,[2 9] has been used to raise larger cores for macrofossil analysis and radiocarbon samples. In this modification the piston is held in position not by a wire but by a string of rods working inside the main rods. Mechanical aids, such as the use of chain hoist and a lightweight rig, are usually necessary to drive and raise the sampler. Cores 1·5 m long have been obtained using this sampler.

The advantages of a Livingstone sampler are these: it takes long cores, the cores can be taken in rapid succession once the field technique is mastered, the sediments are not or little deformed except perhaps at the base of each core where friction may be great at the end of a sampling drive, it is ideal for lake sediments which are not too compact or contain too much coarse inorganic sediment, and it can be easily used from a boat through casing (10 cm piping). Its disadvantages are that it will not sample coarse fibrous material which tends to become jammed in the mouth of the tube, the sediments are liable to be compressed, and if the sediments are very loose they may not be recovered. The last is unlikely with a long core as the friction is enough to hold the core. Loss can be prevented by pumping air down an air line to the base of the sampler and keeping pressure up as the sampler is withdrawn,[9] or by the use of a core-retaining device fitted to the corer shoe. The most convenient of such devices for loose sediments is a flap valve cut from a cylinder of such a size that it fits closely in the side of the shoe, and shaped so that it closes the cylinder when it is forced down by the weight of sediment.

Punch sampler (Fig. 6.1e). This sampler was devised to take cores 6 cm in diameter and 50 cm long in compacted sediments containing much inorganic sediment. The tube is split into two lengthways and the two halves are held together by insertion at each end into the head and the shoe. Bolts run from the shoe to the head just inside the core tube so that the whole assembly can be bolted tight. The sampler is driven by sledgehammer working on a striking platform at the top. It can be raised by the jacking system shown in fig. 6.1g. The core obtained can be removed by slackening the bolts and removing one side of the tube. Good recovery of sediments has been obtained from interglacial deposits with this sampler, but compression and some deformation occur during sampling. It will not raise cores of uncompacted wet sediment.

Screw sampler (Fig. 6.1f). A single-spiral shipwright's auger has been successfully used to sample compacted clays and clay-muds. The diameter should be from 3 to 5 cm and the length about 40 cm. Such an auger can be screwed down through clays or clay-muds about 25 cm at a time. This depth of sediment will usually fill the auger. The samples taken from the filled auger should be from the centre of the spaces between the spiral, and the outer sediments and those touching the spiral should be removed before taking them. Naturally much deformation occurs with this sampler, but it does enable samples to be taken in deposits which could not be worked by hand with other types of sampler. Several metres can be penetrated by one man unaided, though the work is slow, but if wet loose sediments are met, the method fails, as samples will be washed out by raising the auger.

Other samplers, Any type of post-hole or bucket or power-driven auger will take disturbed samples if the sediments are compact enough, but the process of augering breaks up the sediments and samples for analysis should only be taken from unbroken lumps of material coming up in the auger. Thus it will not usually be possible with these types of auger to take samples at close intervals.

Large capacity hand-operated peat samplers, taking samples adequate for radiocarbon dating, have been designed by Digerfeldt[3] and Smith, Pilcher and Singh.[10]

Heavier sampling devices

These samplers are costly to run, more difficult to transport, and require more than one man to operate. They include commercial drilling rigs and other rigs especially designed for specific problems. Of the commercial rigs, the standard 10 cm percussion system with core samplers as used by soil mechanics investigators, is very useful in compacted sediments, especially in the older Pleistocene deposits where gravel or till has to be penetrated before sampling of organic sediment can take place. A special type of sampler using a metal foil liner extruded from the base of the sampler as it is driven down has been used to take continuous cores in varved clays.[5] The foil effectively reduces the friction between the sediment and sampler wall, so it is possible to take cores of up to 20 m long.

Rotary coring devices are less useful except in compact sediments and usually the thickness of a deposit does not merit the setting up of such expensive rigs. If they are used, however, great care should be taken that any drilling mud used is free of microfossils.

Two other types of sampler are available for use on water. One is the Kullenberg sampler, famous for its use in raising cores up to

20 m long in deep water. The sampler is of a stationary piston type, weighted, and it is lowered till a few metres from the sediment surface when it is released and the piston held stationary. In deep water mechanical aids are necessary to control and effect the lowering and raising of the sampler. The other type of sampler for use in water is the Mackereth sampler.[11] This is easier to use than the previous one, takes a single core up to about 8 m, and can be worked by two men in a boat. The sampler is of the stationary piston type. It is lowered by rope to the surface of the sediment, and then driven down by compressed air supplied by an air line. The reaction is provided by a drum anchored to the bottom hydrostatically. The sampler is then raised by pumping air into the anchor drum, which rises carrying the sampler with it.

Field operations

In the investigation of extensive fossiliferous deposits, e.g. those of a lake, a raised bog or an interglacial deposit, it is first necessary to carry out a programme of boring to determine the extent and stratigraphy of the deposits. In order to get a profile across the deposit lines of boreholes should be made and the stratigraphy of the whole reconstructed, as shown in fig. 4.5. Once the stratigraphy is clear the best site for detailed sampling can be chosen where the deposits are most completely preserved and thickest.

It is worth pointing out that though the time and energy spent in obtaining a profile is substantial, if much more time is to be spent on sedimentary and palaeontological analyses from a single borehole, it is clearly best to analyse a sequence from a site carefully chosen after the profiles and stratigraphical sequence of the lake infill have been obtained and considered. The final site chosen will depend on the nature of the problem under investigation; for example, whether it is a study of the history of local taxa or of regional vegetation. For the latter the best site is usually near the centre. Near the edge fluctuations of water level may cause periods of erosion and non-deposition, and then the record of sedimentation will be incomplete. In boring in sediments of lakes still existing, a recording echo-sounder may help to locate the most suitable place for boring, as there will be reflections from the sediment/water interface and from denser layers within the sediment.

On a land surface it is necessary to level the tops of the boreholes to obtain a satisfactory profile. A simple highway level is quite accurate enough for this task. In boring in a lake the water line is obviously a good datum point. The difficulty here is the erection of a stationary boring platform. The borers can be operated from a boat or, more conveniently, from a platform erected between two boats or two inflatable rafts. These must be securely anchored by at least three anchors. If the boring can be done while the lake is frozen, it is much easier to work through the ice than from boats.

In boring to obtain samples for analysis it is of prime importance to secure completely pure uncontaminated samples. Some samplers are more liable to lead to contamination than others, as discussed in the descriptions of the various sampling devices. A few general rules apply to all the devices. Ideally, the boring should be done on a day and a time of year when pollen production and dispersal are low or absent and when the weather is dry. Notes on the local vegetation should be made at the time of boring so if any contamination by recent pollen arises, it may be related to local conditions. The sampling device should be thoroughly cleaned by washing between each sample being taken. If possible samples should be taken alternately at successive depths in two closely adjacent boreholes, to lessen the risk of carrying material down the borehole and to prevent disturbance of the sediments where one sample ends and the next begins. When boring from land surfaces, a sod should first be removed from the top soil, and the first samples taken from this. Then the boring is carried out, sampling alternately from two holes. The sampler should be pushed down (except of course where augering is concerned) rather than screwing it in, so keeping the hole as clean as possible, and in soft deposits where sinking of the sampler may occur under its own weight, care should be taken to prevent any error in determining the sampling depth.

The actual taking of samples from different types of sampler has already been mentioned, and the same general rules about the distance between samples apply as to sampling from open sections. Samples (about 5 cm^3) for microanalysis are taken with a clean spatula or penknife, put in plastic or glass tubes, and their depth clearly marked on the tube.

Before the sampling is made it is best to describe the sediment in the sampler in as much detail as is possible in the field (see ch. 4). The samples are then taken, either in the field or in the laboratory if the core can be transported home; the sediment is then removed from the sampler and if macrosamples are required, placed in a polythene bag. Further useful notes can often be made from the sediments after examination of hand specimens after removal from the sampler, e.g. the nature of any macrofossils present, or the degree of humification.

SAMPLES AND STORAGE

Back in the laboratory, it is necessary to seal the microfossil samples with paraffin wax. Both these and the macrofossil samples should be kept damp and at a low temperature. Damp samples are far more easy to work and the low temperature reduces any chemical changes tending to occur.

REFERENCES

FAEGRI, K. and IVERSEN, J. 1975. *Textbook of pollen analysis.* 3rd edn. Copenhagen: Munksgaard.

1 CATT, J. A. and ROBINSON, P. C. 1961. 'The preparation of thin sections of clays', *Geol. Mag.*, **98,** 511–14.
2 CUSHING, E. J. and WRIGHT, H. E. 1965. 'Hand-operated piston corers for lake sediments', *Ecology*, **46,** 380–4.
3 DIGERFELDT, G. 1965. 'A new type of large-capacity sampler', *Geol. För. Stockh. Förh.*, **87,** 425–30.
4 JOWSEY, P. C. 1966. 'An improved peat sampler', *New Phytol.*, **65,** 245–8.
5 KJELLMAN, W., KALLSTENIUS, T. and WAGNER, O. 1950. 'Soil sampler with metal foils: device for taking undisturbed samples of very great length', *Royal Swedish Geotechnical Institute*, Proc., No. 1. 75 pp.
6 LAMMERS, J. 1965. 'Hand-drilling tools for geological investigation', *Geologie en Mijnbouw*, **44,** 94–5.
7 LICHTWARDT, R. W. 1952. 'A new light-weight shaft for peat samplers', *Palaeobotanist*, **1,** 317–18.
8 LIVINGSTONE, D. A. 1955. 'A light-weight piston sampler for lake deposits', *Ecology*, **36,** 137–9.
9 SEROTA, S. and JENNINGS, R. A. 1957. 'Undisturbed sampling techniques for sands and very soft clays', *Proc. IV Int. Conference on soil mechanics and foundation engineering*, **1,** 245–8.
10 SMITH, A. G., PILCHER, J. R. and SINGH, G. 1968. 'A large capacity hand-operated peat sampler', *New Phytol.*, **67,** 119–24.
11 SMITH, A. J. 1959. 'Description of the Mackereth portable core sampler', *J. Sed. Petrol.*, **29,** 246–50.
12 VALLENTYNE, J. R. 1955. 'A modification of the Livingstone piston sampler for lake deposits', *Ecology*, **36,** 139–41.
13 WALKER, D. 1964. 'A modified Vallentyne mud sampler', *Ecology*, **45,** 642–4.
14 WRIGHT, H. E., LIVINGSTONE, D. A. and CUSHING, E. J. 1965. 'Coring devices for lake sediments'. In *Handbook of palaeontological techniques*, eds. B. Kummel and D. M. Raup, San Francisco: Freeman.

CHAPTER 7

BIOLOGICAL INVESTIGATIONS

THE study of stratigraphy and of sediments leads to the reconstruction of the history of the geological environment, while the study of plant and animal remains leads to the reconstruction of the history of vegetation, flora and fauna. As life is intimately related to and dependent upon environment, so the knowledge of plant and animal remains leads also to reconstructions of climate and other factors of the environment, such as soils or the salinity of water.

Both geological and biological studies are necessary towards the building of the history of the Pleistocene. Each provides information which is complementary, and in particular circumstances one may provide much more information than the other. For example, in the study of the glacial stages evidence from the glacial deposits will give information about the distribution and movement of ice and the processes of its melting, but little climatic detail will emerge. On the other hand, the study of plant and animal remains from sediments formed in a periglacial area during a glacial stage will tell much about conditions of life and climate during the glacial stage.

The Pleistocene is characterised by climatic change and detail of climatic change can be deduced from biological studies. For this reason such studies become very important in the investigation of the Pleistocene. Reconstructions of past climatic change from fossils depend on how much is known of the ecology and distribution of the species found as fossils. In the Pleistocene many of the fossils are closely allied to or appear identical to present species, so that valuable reconstructions of past plant and animal life and climate can be made.

Many groups of animals and plants are found fossil with their harder parts preserved. Two broad categories of remains can be made, macroscopic fossils or macrofossils, large enough to collect and study by naked eye or under low magnifications, and microscopic fossils or microfossils, which need special methods for extraction and a high-power microscope for observation.

Taking the plants first, the most common identifiable macrofossils are seeds, fruits, leaves, and wood fragments. The plant microfossils comprise the pollen grains of higher plants, the spores of lower plants, less commonly leaf hairs and epidermis, and the skeletons of diatoms and desmids. Among animals, macrofossils include the skeletal remains of vertebrates, the shells of molluscs, and the exoskeletons of insects and crustaceans, especially cladocerans and ostracods. The microfossils include rhizopods and foraminifers.

Each of these groups of organisms has advantages and limitations when used as evidence for environmental conditions, and it will first be useful to discuss each group briefly on its own. A later section deals with the interpretation of assemblages.

Plant macrofossils[2, 4, 22] (plate 11). These are mostly derived from plants growing locally near the site of deposition. In autochthonous deposits, the plants will be growing at the place they are found fossil; in allochthonous deposits they may be more distantly derived by, say, running water or solifluction. Plant remains from different plant communities will collect in allochthonous deposits and it may be difficult to separate them into the communities from which they originated, but the broad categories of community and thus environment, e.g. aquatic, marsh, woodland, are usually evident. Plant macrofossils are thus good indicators of local environment and vegetation.

Plant microfossils[10, 12] (plate 12). Pollen grains and spores in sediments mostly come from the air by fall-out (pollen-rain) and they are thus derived from the regional and local vegetation. Regional and local vegetational changes produced by climatic, edaphic or biotic changes cause changes in the pollen-rain and thus in the microfossils incorporated in sediments. Changes in the processes of dispersal and sedimentation will also cause changes in the assemblages in the sediment. Other plant microfossils, e.g. skeletons of diatoms[11, 30] and desmids and dinoflagellate cysts,[39] common in marine sediments, originate in the water body depositing sediment, and as the populations of these in lakes is dependent on nutrients in the water as well as on climatic changes, they give information on conditions within the water-body.

Vertebrates.[9, 35, 46] Remains of vertebrates are not so common as are other animal remains, except perhaps in cave deposits, where they may be abundant. Autochthonous fossils include fish, amphibian and rodent remains of lake deposits, whereas allochthonous vertebrate faunas may be found in sands and gravels, and in cave deposits where animal bones may be accumulated by carnivores. Valuable evidence of both climate and environment may be established from studies of vertebrates, as their present-day ecology and distribution is usually rather well-known, e.g. the climatic indications of cold conditions given by the presence of lemmings and reindeer, and the open vegetational conditions indicated by horse. On the other hand, many important Pleistocene vertebrate species are extinct, e.g. species of elephant and rhinoceros, and these are more difficult to use in palaeoecological reconstructions. Another difficulty is caused by the robustness of vertebrate remains, which may lead to their redeposition in sediments much younger than the time of original entombment.

Molluscs[25, 27, 32] (plate 16). Molluscs, both marine and non-marine molluscs, have been successfully used to bring out details of climate and local ecology. The freshwater molluscs give indications of water conditions and salinity. The marine molluscs have been less well studied and their ecological interpretation is more difficult, mainly because of lack of detailed knowledge of the present-day ecology, but also because many species are extinct, especially in the early Pleistocene.

Insects[8, 31] (plate 16) *and crustaceans.*[14] Fossil insects also give good indications of climate and environment. Many have a narrow feeding range of plants and are associated with particular plant communities so their presence can be used to demonstrate the existence of such communities. Fossil crustaceans, especially cladocerans and ostracods, are often abundant in lake sediments, and besides climatic indications which the species may give, they also are useful in providing information on water conditions and nutrients, and the productivity of lakes in the past.

Microscopic animals. Foraminifera[13] are important in providing local and regional climatic information and their tests are the chief source of material for palaeotemperature determinations by the oxygen isotope method. They may also indicate local conditions of salinity and water-depth. Rhizopods[14] are found in bog and lake deposits, and are useful indicators in bogs of changing hydrological conditions, as the abundance of certain species appears to be related to the wetness of the bog surface.

Preservation of fossils

Most Pleistocene fossils are the harder parts of the animals or plants, having the same composition in the fossil state as when they were originally formed. There are exceptions, such as the mineralised bones of the lower Pleistocene (e.g. the black bones of the Netherlands Scheldt, which contain a large amount of cryptocrystalline pyrites). Usually little compression has occurred, so that the fossils are of the same dimensions as the harder parts of the living organism. An exception here is wood which is often found to be compressed and is difficult to identify in older Pleistocene deposits.

Impressions of fossilised plants may be found in sandstones, tufas and cave stalagmites. Here the original organism has disappeared, leaving its form impressed as a cast on the surface of the sediment in which it was buried.

Preservation is best where the fossils have remained under anaerobic and unleached conditions. Oxidation will quickly lead to destruction of fossils, particularly plant fossils, while leaching will dissolve calcareous fossils. Where some oxidation and some leaching has gone on, there may be differential destruction of fossils of different species and this may confuse the interpretation of fossil lists. Fossils are best preserved in unweathered fine-grained marine or freshwater sediments, or in terrestrial deposits which remain waterlogged. Where the grain size of the sediment is larger, percolation and oxidation may quickly destroy fossils.

Primary and derived fossils

It is obviously necessary to know whether a fossil is in its primary position or has been derived from an older sediment and redeposited. Redeposition will result from reworking of an older deposit, say under conditions of fluctuating water-levels. It is particularly obvious in the case of large mammalian fossils such as teeth and bones, which may first be deposited in gravels, and then reworked as the gravels became subject to processes of erosion. But similar events can occur with microscopic fossils,[18, 24] such as pollen grains, as in the case of fossils introduced into till by ice action, then carried into lake deposits by streams draining the till surface during a subsequent warmer period. Both with micro- and macro-fossils it may be difficult to separate primary from derived fossils. The best clues are to be found in considering the sedimentary environment, that is, whether it is likely that the processes of sedimentation will be liable to cause derivation. Thus in autochthonous and pure biogenic deposits, little or no derivation is to be expected, but in inorganic sediments it may occur if the parent material from which the sediment is derived is fossiliferous. The autofluorescence of pollen and spore exines has been used to distinguish primary from derived fossils, since the colour and intensity of fluorescence varies with age. The method works best where there is a considerable difference in age, e.g. Tertiary pollen in an interglacial deposit.[29]

Chemical methods, such as fluorine determinations of bones, may also reveal derivation. In a particular fauna preserved under the same conditions the fluorine content will be similar in different bones. An older bone which has been subject to a different history of fossilisation will have a different fluorine content (see ch. 9).

Choice of site

It is important to choose the best site available for the particular investigation in hand. For studies of regional vegetational change the centre of a large lake or bog may be most suitable, but for more detailed studies of local conditions of plant and animal life, marginal areas or small lakes or bogs may be best. Thus small areas of accumulation of organic sediment within forest will give the best record for the study of detail of forest history. The taphonomy of the fossils in each of these types of sedimentary basin will vary and must be taken into account in interpretation of palaeoecology.

Obtaining the data

In the collection of sediments for the extraction of fossils, it is all-important to know the exact provenance of the sample and the thickness of sediment it covers. An analysis of a bulk sample whose exact provenance at a site is unknown is not of much use in the reconstruction of past climates and environments. The sample should be large enough (say a kilogram) to provide material for the analysis of various groups of organisms, as it is also important that analyses of these different groups come from the same sediments. In this way a combined picture of the environment at the time the sediment was deposited can be built up.

While the recovery of large vertebrate fossils requires no special techniques, the extraction of smaller fossils depends on the use of a variety of techniques and processes, of which the most important are described in Appendix 1 (p. 414).

Identification[23]

Identification of fossils must be by comparison with collections of reference material, e.g. slides of pollen of living species, shells of modern species of molluscs. It may be necessary to treat reference material (e.g. seeds, by boiling in sodium hydroxide for a short time) to bring it to a similar state of appearance as the fossils. It is obviously important to have reference collections in which the species are known to be correctly determined. Identification of a fossil may be limited because of the present unsatisfactory state of the taxonomy of the nearest living species

Morphological similarities between fossil and type, such as shape, size, and surface patterns are the most certain guide to identity (plates 11, 12). Size-frequency analyses are also used when there is general morphological similarity, but difference in size between closely related species. For example, two species of the spruce, *Picea abies* and *P. omorica-type* have been separated on the basis of size differences (fig. 7.1).[1] However, the interpretation of size differences

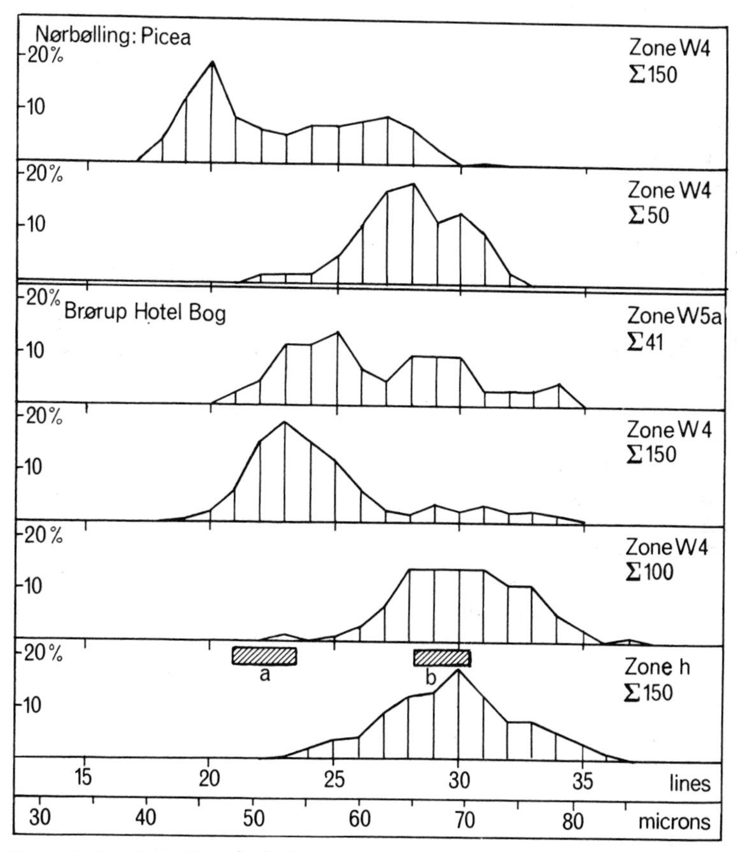

Fig.7.1 Size frequency distribution of the height of the air sacs of fossil *Picea* pollen grains from late Eemian and early Weichselian deposits in Denmark. Mean size of same parameter of recent pollen grains of *P. omorica* (a), and *P. abies* (b) (after Andersen, 1961). The curves indicate that the two species of *Picea* are present fossil.

has to be done with care, for different methods of preservation in sediments, different histories since incorporation in sediments, and different methods of fossil preparation (e.g. in making-up pollen slides) may all lead to size variation in the one species.

Qualitative and quantitative analyses

A qualitative analysis of fossils from a particular sediment will give a list of taxa represented, while a quantitative analysis[38] will give frequencies of the taxa. The frequency of a taxon may be expressed in a relative way, as a percentage of the total number of fossils, or in an absolute way, as a frequency based on the total number of fossils in a given volume of sediment or on the total number of fossils in sediment (pollen concentration), surface area cm^2, accumulated in one year (pollen influx). The advantage of absolute frequencies is that the frequency of the taxon is independent of the frequency of other taxa; with relative percentages, the frequencies of all taxa used in the basis for calculation are interdependent. Absolute frequencies based on unit time can of course only be made in sediments which can be adequately time-correlated by radiocarbon or other dating method or by the presence of annual laminations.

The difference between qualitative and quantitative analysis is thus analogous to the difference between a simple plant list from a plant community and a detailed plant-sociological description of the same community. So while the presence of a particular species, e.g. that of a plant or animal of arctic distribution, will give information on local environmental conditions, once the relative or absolute frequencies are known, we are in a much better position to make reconstructions of past plant and animal life.

Before undertaking frequency analyses, it is worthwhile considering their significance. The main question is to what extent an assemblage of fossils from a given horizon represents a random sample of local or regional life. With pollen and foraminifers the fall-out from the air and from the sea respectively make them well suited for frequency analyses, as the counts of different species are related to the regional living communities which gave rise to the fossils. With larger fossils, however, local effects may predominate, and quantitative analyses will give information on the types of community found locally. For example, in the case of freshwater and land molluscs, the frequencies of the different groups of molluscs requiring running water, marsh or dry land conditions, will give information on the local habitats present near the scene of deposition (fig. 7.2, table 7.1),. Beetle remains and macroscopic plant remains have a similar type of significance. With even larger fossils, such as

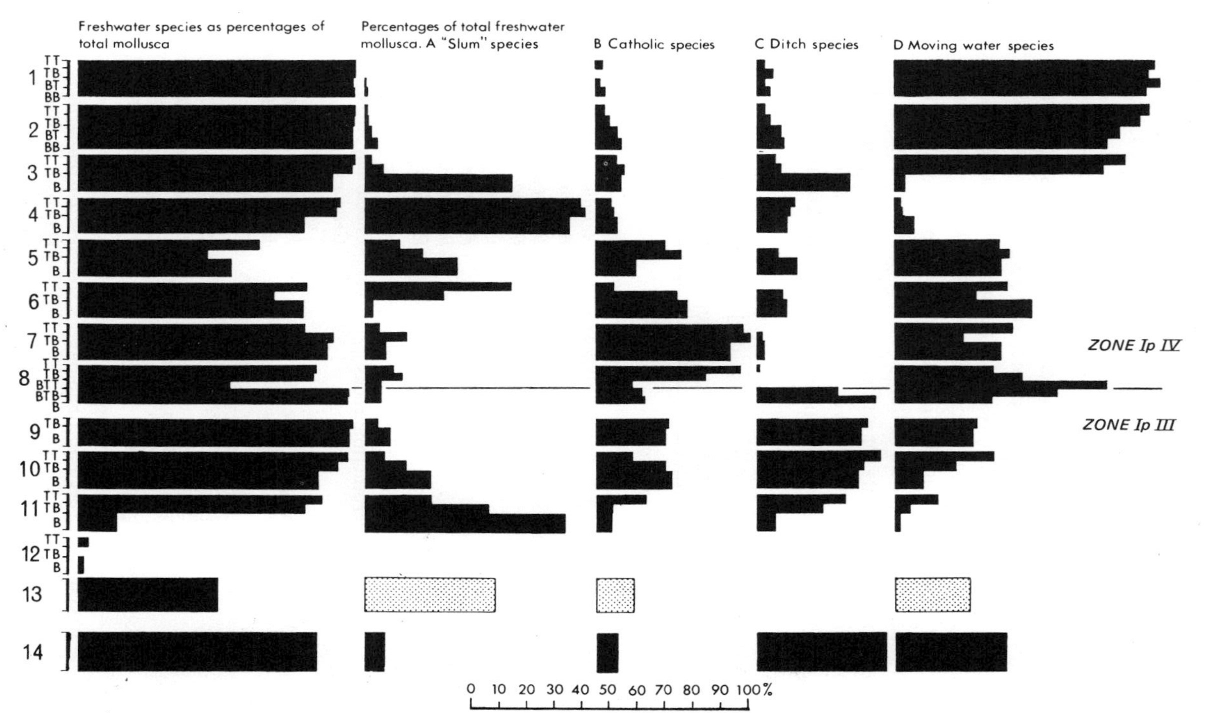

Fig. 7.2 Frequencies of ecological groups of freshwater molluscs in the late Ipswichian silty marls forming a river terrace at Histon Road, Cambridge. Dots are used where numbers are scarcely adequate for analysis. The relative changes in the groups are a key to the changing aquatic environment. The depth of sediment analysed is 6 m (Sparks and West 1959).

those of vertebrates, the frequency of fossils represented must be carefully considered, as it cannot be said that in any particular environment of deposition, e.g. a bog or a cave, the vertebrate skeletons are representative of regional or even local faunas. In the case of bogs, only those animals large enough to sink in and be trapped may be represented and in the case of caves the list of larger mammals may represent a hyaena's choice of food.

As well as considering the analyses from an ecological point of view, we have also to consider their use as indicators of climate. Qualitative studies will give an idea of climate, and may be preferred in some instances, but quantitative lists are here again important, for it is the frequencies of plants or animals which require particular climatic conditions which will give the most information about climate (fig. 7.3, table 7.2), particularly since there are numerical

Table 7.1 Ecological groups of molluscs used as a basis for environment analysis in central Europe (after Lozek, 1965)

Eclogical groups		*Index molluscs*
A Woodland Species	1 Closed forest 2 Woodland-Scrub (→ Timbered Steppe) 3 Moist woodland	*Acicula, Aegopis, Ruthenica, Bradybaena fruticum, Cepaea hortensis, Perforatella bidentata, Pseudalinda turgida*
B Grassland Species	4 Steppe—xerothermic rocks 5 Open country (in general)	*Chondrula, Abida, Gelicopsis, Cecilioides, Pupilla muscorum, Vallonia pulchella*
C Indifferent Species	woodland → grassland / woodland ← grassland: 6 dry 7 medium (+ indifferent) 8 damp	*Cochlicopa lubricella, Cochlicopa lubrica, Euconulus fulvus, Vertigo substriata*
D Aquatic and marsh species	9 Marshes—very moist habitats 10 Water (in general)	*Vertigo antivertigo, Oxyloma, Carychium minimum, Lymnaea, Viviparus, Unio, Pisidium*

methods available for comparing fossil assemblages or taxa with modern assemblages or taxa living under known climatic requirements.

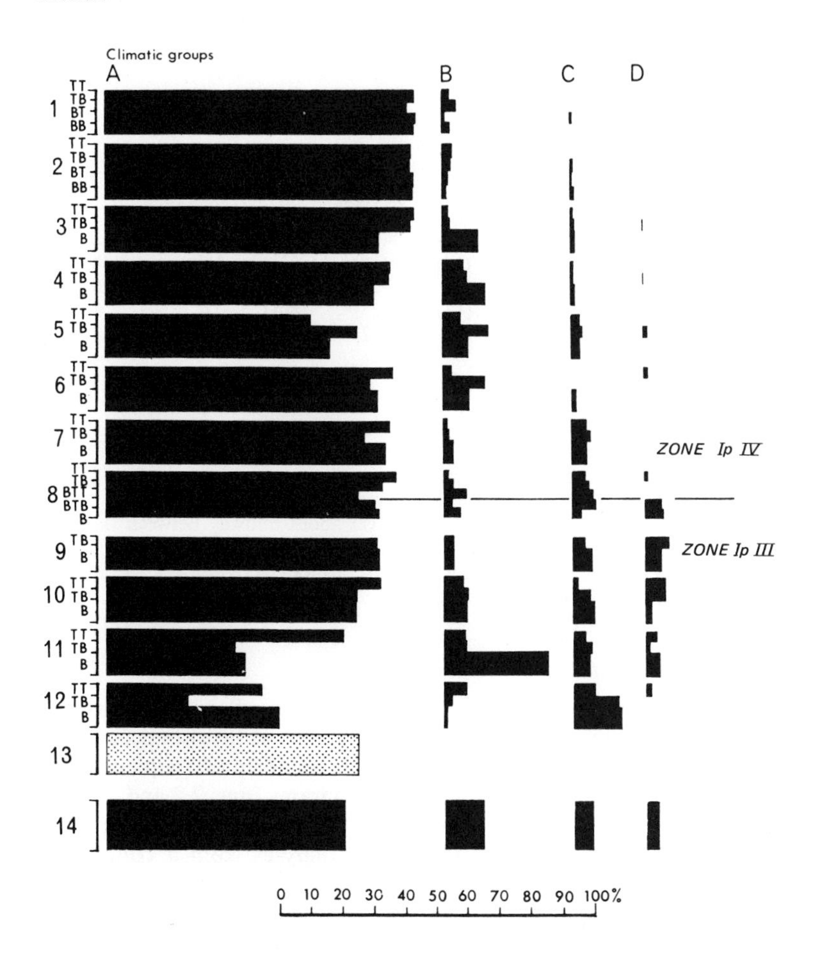

Fig. 7.3 Frequencies of distribution (climatic) groups of molluscs in the Histon Road interglacial deposit (see fig. 7.2) (Sparks and West, 1959).

A. species reaching to or almost to the Arctic Circle in Sweden.
B. species reaching approximately 63° N in Sweden.
C. species reaching approximately 60°–61° N in Sweden.
D. species found only in the very south of Scandinavia or confined to the European mainland.

Table 7.2 Analysis of plant-geographical categories in the list of macroscopic plant remains from interglacial deposits at Bobbitshole, Ipswich (West 1957)

The figures give the number of species (or genera) in each category in each zone; figures in parentheses are the percentages of the zone totals in each category.

	Category						
Zone	1	2	3	4	5	6	*Zone total*
f	—	13(25)	8(15)	8(15)	19(37)	4(8)	52
e	—	12(32)	7(18)	6(16)	12(32)	1(3)	38
d	1(5)	5(25)	3(15)	5(25)	5(25)	1(5)	20
c	—	4(50)	1(13)	2(25)	1(13)	—	8
b–c transition	1(14)	2(28)	—	3(42)	1(14)	—	7

Definition of categories:

1. Arctic-alpine plants.
2. Plants distributed throughout Scandinavia.
3. Plants with northern limits in Scandinavia near the Arctic Circle.
4. Plants with northern limits in Scandinavia about midway between those in categories 3 and 5.
5. Plants with northern limits in the south of Scandinavia.
6. Plants with northern limits in northwest Europe south of Scandinavia.

Presentation of data

The data can be presented as qualitative lists of species, as lists of species with subjective estimates of frequency, as lists of species with absolute numbers of fossils found, or in the form of frequency diagrams. Frequency diagrams may be histograms (figs. 7.2 and 7.3), or the frequencies may be used to form curves as in the familiar pollen diagram (figs. 7.4, 7.5). In a series of analyses from a column of sediment, there is much to be said for the frequencies at each level (e.g. the pollen spectrum at that level) being kept separate, the thickness of the line representing the thickness of sediment sampled at each level. Where large numbers of species are concerned it may be more convenient to group the species into ecological or climatic groups, and then graph the frequencies of these groups (fig. 7.3).

There are a variety of ways in which pollen data can be presented in graphical form, depending on the sum of pollen used as a base for the calculation of pollen frequencies.[45] In an absolute pollen diagram (fig. 7.6) the data are presented in terms of pollen grains of each taxon

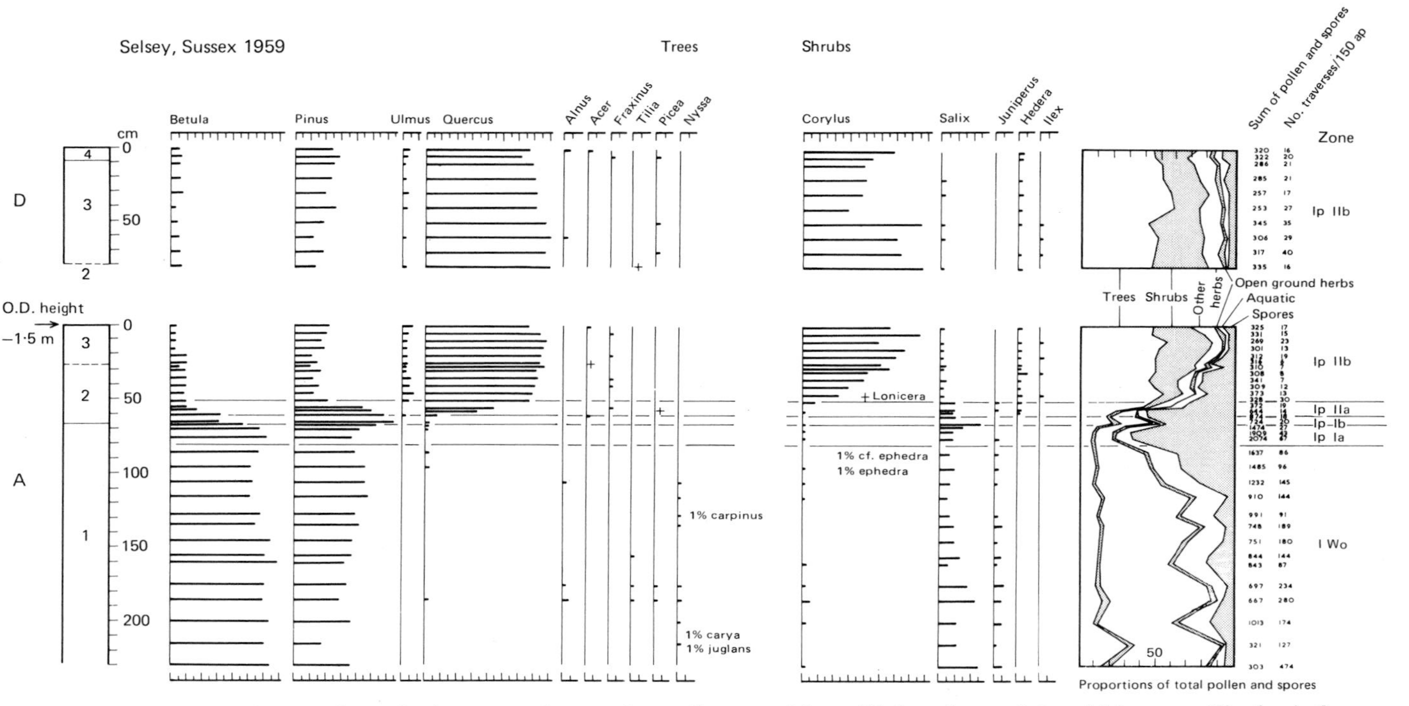

Fig. 7.4 Pollen diagram from freshwater and estuarine sediments of Late Wolstonian and Ipswichian age. The basis for calculating the percentages is the total tree pollen. Proportions of the various categories of pollen and spores are shown on the right. Sediments: 1, freshwater silt and clay; 2, detritus mud with wood; 3, estuarine grey silty clay; 4, estuarine brown-grey silty clay.

deposited per cm^2 per year. In percentage diagrams the base for percentage calculation will vary. If the pollen diagram is used primarily as a means of investigating regional vegetational change, the base of the calculation should be related to the regional vegetation. The pollen sum in a forested area may be the tree pollen only, pollen of non-trees being expressed as a percentage of total tree pollen (fig. 7.4). The significant changes in forest composition over a period of time are thus clearly expressed. A second type of diagram (fig. 7.5), used where there is much non-tree pollen, uses as a pollen sum the total pollen of terrestrial plants. The difficulty here is recognising which pollen types are of terrestrial and regional origin and which are derived from local communities in the bog or around the lake from which the sediments are taken. Whatever the pollen sum used, it is important to state clearly in the diagram how it is constructed, and to give all the factual data without any element of interpretation,

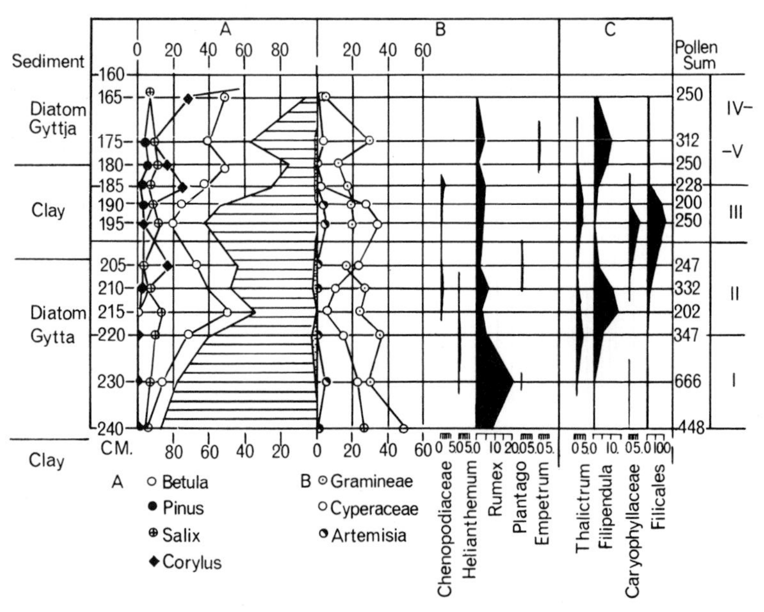

Fig. 7.5 Pollen diagram through Devensian late-glacial and Early Flandrian lake sediments at Kentmere, Westmorland, showing the late-glacial interstadial. The basis for calculating the percentages is the sum of pollen types in category **A** (tree pollen) and **B** (anemophilous terrestrial herbs). Category **C** contains other herbs, probably local and marsh species, and spores (Walker, 1955).

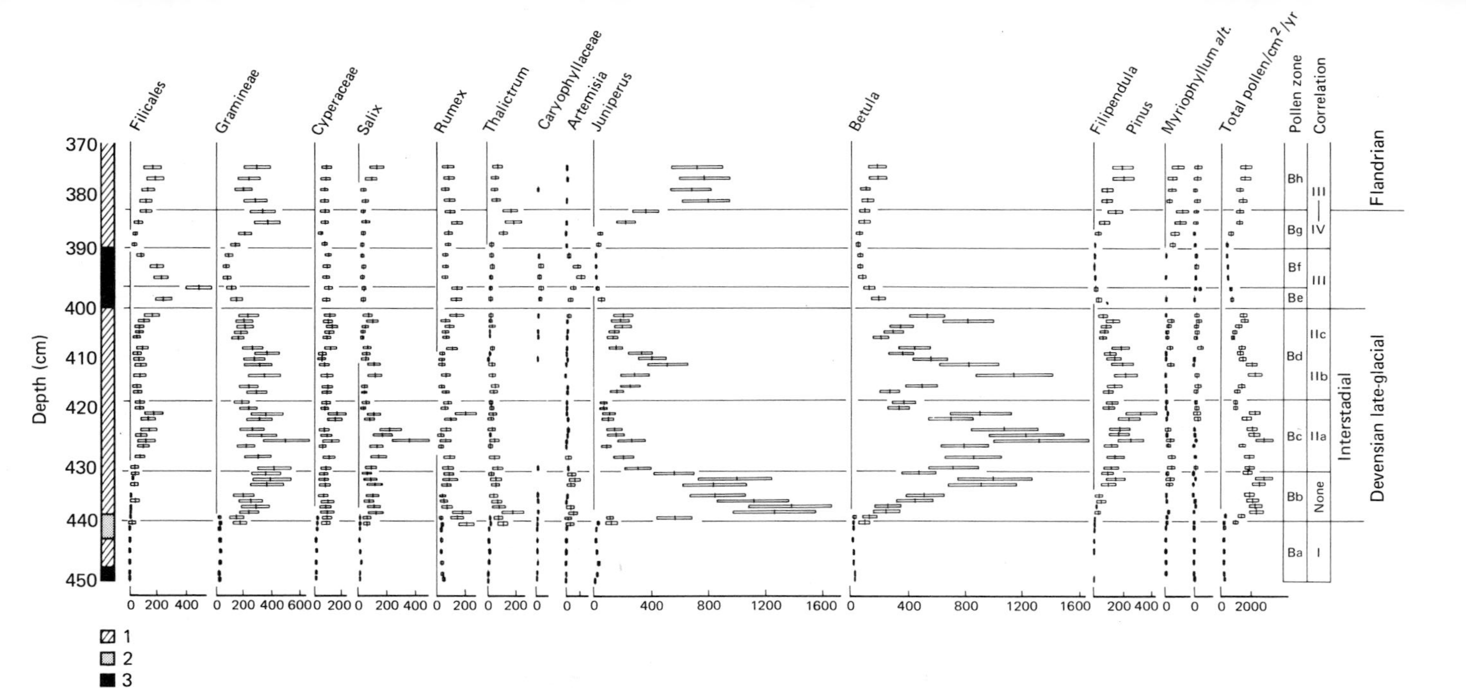

Fig. 7.6 Absolute pollen diagram from the basal (Devensian late-glacial) sediments of Blelham Bog, near Windermere (Pennington and Bonny, 1970). The counts are expressed as numbers of grains deposited per cm^2 per year. The plot for each value shows the 90 per cent confidence interval for the counting error. 1000 fossil grains counted for each sample. Sediment symbols: 1, organic mud; 2, organic silt; 3, clay.

so that the reader can understand what the curves represent, and will be able to draw his own conclusions from the data provided.

Likewise, in quantitative lists of other organisms, the nature of the individual counted (fragments of shell, apex of shell, elytra of beetle, head of beetle) should be mentioned, again so that the reader can draw his own conclusions about the significance of the counts.

Interpretation of data[32, 41]

There are three aspects of interpretation to consider.

1. The taphonomy of the fossil assemblages, that is to say, the study of the processes which govern the production of fossil assemblages

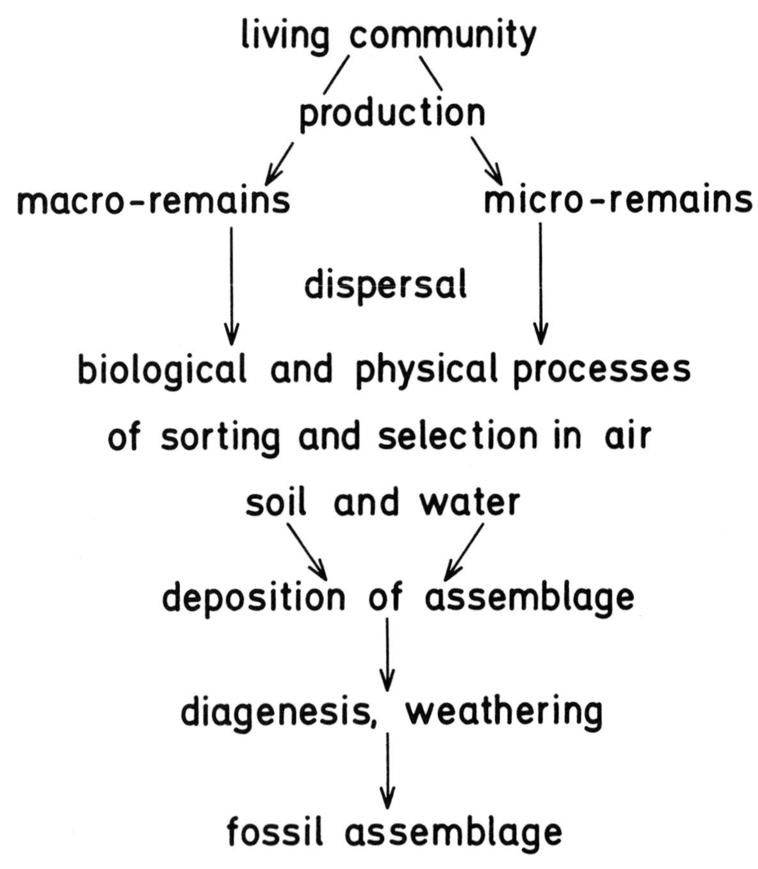

Fig. 7.7 Processes involved in the production of fossil assemblages from living communities.

from living assemblages. Figure 7.7 summarises the processes involved. Variation in kind and degree of the processes will inevitably alter the composition of the assemblage as it moves towards and is incorporated in the sediment.
2. A knowledge of the taxonomy of the fossil assemblage and its relation to the taxonomy of living species.
3. A knowledge of the ecology of living species as a basis for ecological reconstruction from fossil assemblages.

In all these aspects the study of present conditions and processes is the key to successful interpretation of the past. Processes important in taphonomy can be studied in the field. The empirical relation between vegetation and pollen rain, as well as the physical processes involved in dispersal and deposition can be studied in the field, using strict quantitative methods.[7, 44] Thus pollen rain on unit area over unit time has been measured and related to neighbouring vegetation, and plant macro assemblages in lake surface sediments have been analysed in relation to the surrounding vegetation.[5]

In making deductions from lists of fossils it has to be assumed that in the past flora and fauna have been related to climate and topography as they are at the present day. There is ample support for this in the fact that reconstructions based on plant and on animal life are not at odds with one another. However, there is the possibility that some species have changed their ecological requirements while the morphology of their parts found fossil have not changed. For example, the gastropod *Potamopyrgus jenkinsi* appears to have spread from brackish to freshwater in the last 100 years. Taking an assemblage as a whole, however, if the species present point to similar conditions of environment and climate, there is no reason to suppose that ecological requirements of the species have significantly changed since the time of their fossilisation. It is thus safer to use assemblage rather than individual species as indicators of past conditions.

The limitations of reconstruction are placed by lack of knowledge about taphonomy, taxonomy and ecology of the fossil taxa and assemblages concerned. Even so, there is a prodigious amount of information available on the history of flora and fauna, and many detailed reconstructions have been made of the history of plant and animal communities.

Evidence of local and regional conditions of life

Some categories of fossils give evidence of local environmental conditions, others of regional conditions, as already mentioned. Microfossils, such as pollen, spores and foraminifers provide information of regional and local conditions, as they are derived from a wide source area. Larger organisms, such as molluscs and plant macrofossils, provide information on more local conditions, as their remains are usually not distributed far from their source. Two examples can be quoted to show the value of micro- and macro-analyses in the interpretation of phenomena as being of local or regional significance.

In fig. 7.8, it is seen that Cyperaceae macroscopic remains increase in quantity towards the top of the succession of lake sediments analysed. This change was accompanied by rising percentages of Cyperaceae pollen in the pollen diagram from the same sequence, most of which is therefore probably derived from local aquatic and marsh species of Cyperaceae, and does not reflect a regional change of vegetation.

The other example, also palaeobotanical, is taken from the

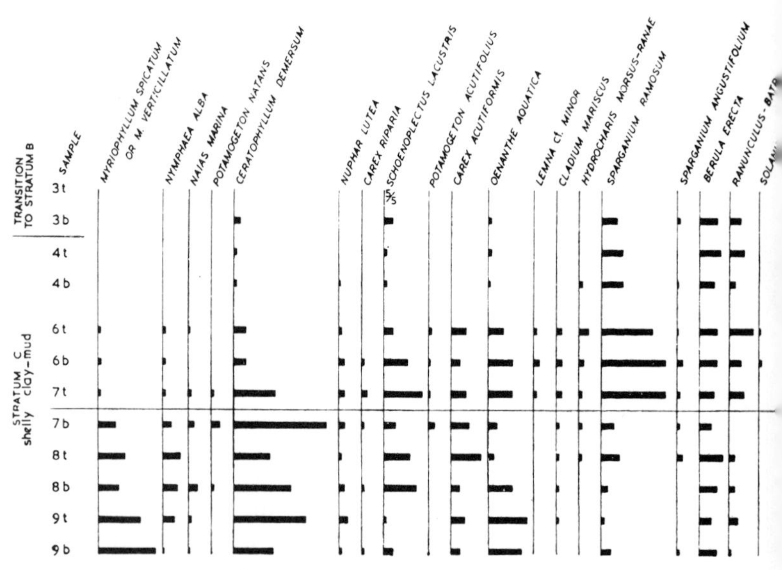

Fig. 7.8 Diagram of the changing frequency of various macroscopic remains in the Ipswichian interglacial deposits at Bobbitshole near Ipswich. Aquatics to the left, reedswamp species in the centre and marsh plants to the right. The sequence of changes of the plant remains indicates change from open water to reedswamp to marsh, corresponding to the change in sediment from clay-mud to mottled alluvial clay (West, 1957).

vegetational history of the middle part, zone II, of the Ipswichian (last) interglacial in southeast England. In a number of pollen diagrams through zone II, *Alnus* pollen is very scarce, usually 1 to 5 per cent of the total tree pollen. The deposits concerned were formed in lakes or slowly-flowing rivers or streams. This scarcity of *Alnus* might suggest that the tree was absent or rare in the regional vegetation and in the local swamp communities bordering the sites of deposition. But at some levels on a zone II riverine deposit at Wretton, Norfolk, *A. glutinosa* fruits have been found abundantly and at the same levels *Alnus* pollen is 60 per cent of the total tree pollen. It seems that communities with *Alnus* were not widespread in this interglacial and that only locally do they contribute to the pollen rain. In the Flandrian (post-glacial) on the other hand, *Alnus* pollen becomes frequent throughout southern England at the beginning of the Atlantic, zone VII, though in a few diagrams it appears earlier. It

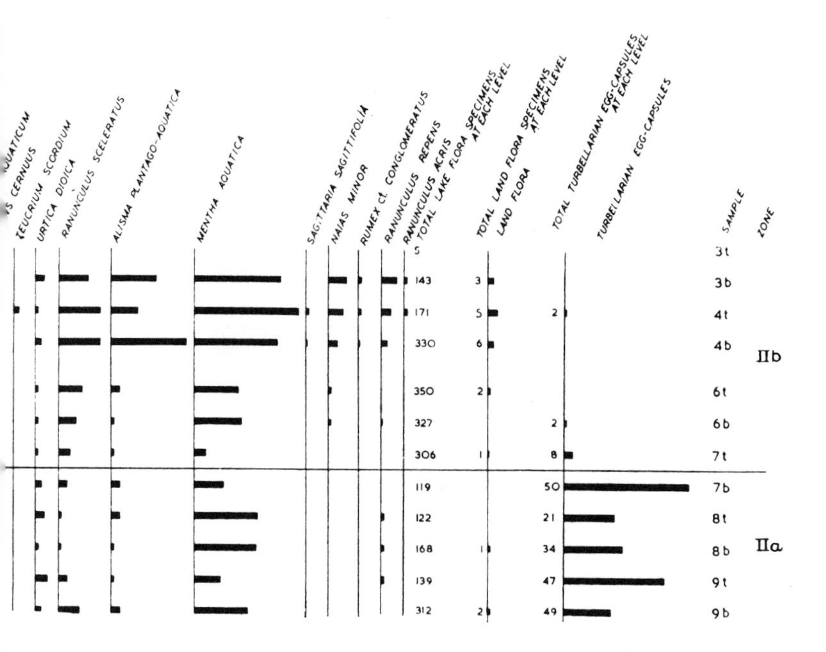

appears that locally distributed communities with *Alnus* became widespread as a result of climatic changes occurring near the zone VI/VII boundary.

The macrofossils alone, both plant and animal, are useful in building up a picture of a changing local environment, as shown in fig. 7.8, which portrays the change from open water to reed swamp to marsh vegetation, and fig. 7.9 which shows a change from freshwater to brackish water conditions as indicated by molluscs.

The microfossils are usually most useful for regional changes, as indicated in a typical pollen diagram. But in some cases, with very

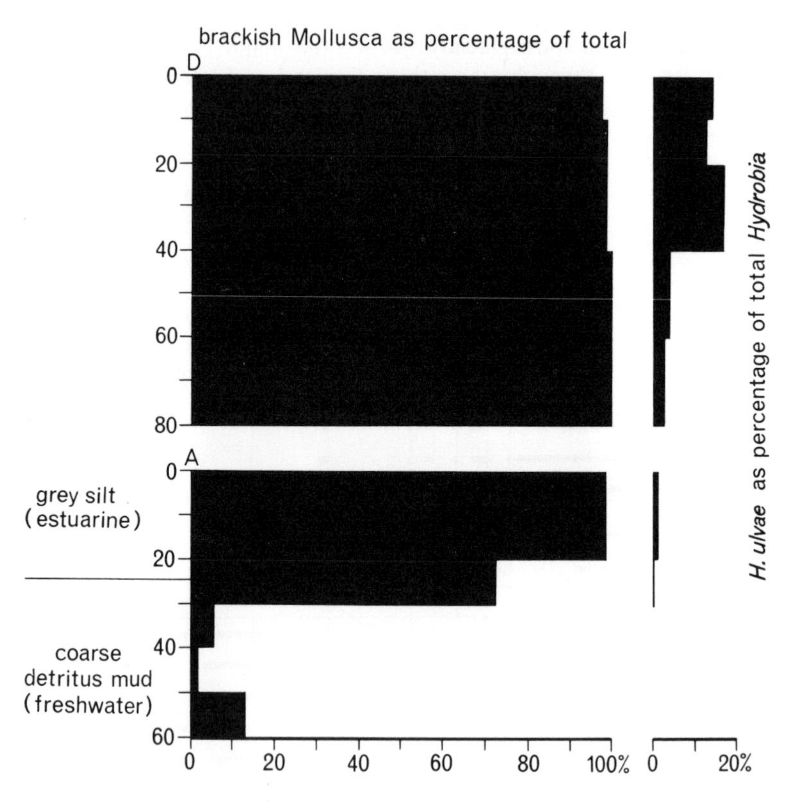

Fig. 7.9 Percentage frequencies of *Hydrobia ventrosa* and *H. ulvae* at the change from freshwater to estuarine deposition during the Ipswichian interglacial at Selsey, Sussex. *H. ventrosa* prefers less saline water than *H. ulvae*. The pollen diagram from this deposit is shown in fig. 7.4 (after West and Sparks, 1960).

detailed pollen diagrams, it is possible to indicate the plant communities involved in the changes, as for example shown by Iversen[19, 21] in his studies of the phases of deforestation associated with Neolithic cultures in Denmark.

In interpretation we should also be aware of the possibility that past communities are not necessarily represented today. For example, it has been considered by some that under the stress of climatic change vegetational zones and even plant communities have migrated as a whole, e.g. the southward movement of vegetational zones in Europe during the Weichselian (last) glaciation. But floral lists from deposits of this age show mixtures of floristic elements, such as thermophilous water plants, tundra plants, steppe plants, alpine plants, dune plants and weeds.[3] It seems that these grew in the same area at the same time. This is a clear indication that past vegetation types are not necessarily in existence today. A similar demonstration follows from a study of Russian periglacial floras,[16] which indicates new combinations of vegetational units, not found in the present vegetational zones (fig. 7.10).

Similar changes have been demonstrated in faunas, such as a mixture of arctic (e.g. lemming) and steppe (e.g. jerboa) species of rodents in cold Pleistocene faunas.[46]

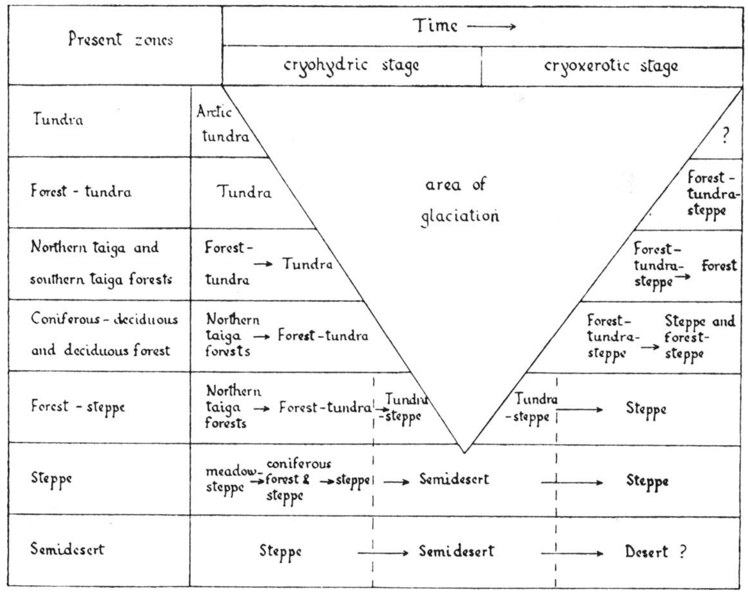

Fig. 7.10 Scheme of vegetational change in the Russian Plain at the time of maximum glaciation (after Grichuk and Grichuk, 1960).

Interpretation of pollen diagrams

Pollen diagrams form such an important part of Pleistocene investigation that it is necessary to consider their interpretation separately. The results of pollen analysis are presented in curves or histograms. With relative (percentage) analyses, these show the percentage frequency of the pollen and spore taxa identified at successive levels (figs. 7.4 and 7.5). An absolute pollen diagram (fig. 7.6) will show the numbers of grains or spores deposited on unit surface area of sediment in unit time (pollen influx), or the numbers in unit volume of sediment (pollen concentration). Each analysis (pollen spectrum) is a count of a sample of the pollen sedimentation at the site during the time in which the thickness of sediment sampled was deposited, except that there may be selective decay of certain pollen taxa after deposition. For example, *Quercus* pollen is susceptible to, and *Pinus* is resistant to, weathering.[17]

The difficulty in interpretation of pollen diagrams results from not being able to determine the relation between pollen sedimentation at a site and the vegetation producing it, because the kind and degree

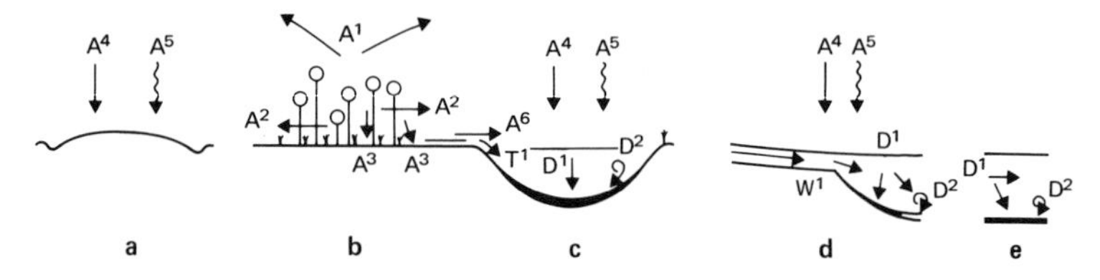

Fig. 7.11 Paths of pollen dispersal and deposition in various environments.
a. Pollen deposition on a growing raised bog surface.
b. Pollen dispersal from vegetation.
c. Pollen deposition in a lake.
d. Pollen deposition in an estuary.
e. Pollen deposition in the sea distant from land; low atmospheric contribution.

A^1 atmospheric dispersal.
A^2 dispersal in the trunk space.
A^3 dispersal to the ground.
A^4 atmospheric fall-out.
A^5 atmospheric rain-out.
A^6 filtering effect in lateral dispersal.
T^1 terrestrial dispersal of component A^3 (and components A^4 and A^5).
W^1 transport by water in rivers.
D^1 deposition in water, accompanied by sorting by current action and differential flotation.
D^2 recirculation of pollen already once deposited by reworking of sediments.

of pollen production is subject to so many influencing factors and because the dispersal processes are also very variable. Figure 7.11 shows how pollen dispersal paths vary according to the environment of deposition, and each process of dispersal may act selectively on the dispersing assemblage.[26] Thus, for example, in marine sediments the assemblage often shows an over-representation of *Pinus*, the pollen of which floats more easily because of its wings.

Each plant community will contribute pollen to the pollen rain, the amount depending on the pollen production of each species in the community, the frequency of each species in the community, the size of the area covered by each community, and also on the structure of each community, important when considering the processes of the movement of pollen from the community into the atmosphere.[36]

The pollen production varies from species to species, so that low pollen producers, such as insect-pollinated species like *Acer*, may be little represented, even though frequent, while great pollen producers like the wind-pollinated *Pinus* may be very much over-represented in pollen analyses. Thus in interpreting pollen spectra it is important to remember that absence or low frequencies of pollen do not necessarily mean absence or rarity of a tree, and the converse. Also where tree cover is low and non-tree pollen frequent, tree pollen types having an ability to travel far, e.g. the winged *Pinus* grain, may be well represented because of long-distance transport.

Many detailed studies have been made recently of the relation between plant communities and the pollen rain they produce,[7, 43] especially in forest and tundra regions of the northern hemisphere. They provide information on the representation of particular species in pollen rain, and allow some degree of generalisation useful for the interpretation of pollen diagrams. Because the relation between pollen sedimentation and vegetation at a particular site is probably unique, care has to be exercised in applying generalisations. It may help, during a study of vegetational history of a particular area, to take recent samples from moss-polsters or surface muds of lakes, and consider their pollen content in relation to the make-up of the vegetation surrounding each sampling site. Some insight is then gained into the origin of the pollen rain in the area.

In any pollen diagram it is important to include the stratigraphy of the sediments, because the sediment will tell the reader the type of environment which received the pollen rain, and thus aid the interpretation. For example, the presence of raised bog peats will suggest the local origin of pollen types liable to be derived from the bog communities.

The stratigraphy will also indicate whether changes in the curves are of local or regional origin. Thus a hydrosere may be indicated in the sediment by a change from lake sediments to fen peats, then to raised bog peats. Maxima of *Alnus* or *Betula* pollen may be associated with the formation of wood peat at the fen stage rather than indicate any climatic change. If the course of the pollen curves is independent of sediment change they are probably of regional origin, and may then be governed by regional plant succession, resulting from climatic change, or merely from succession following competition after arrival and spreading of the species in the area without any specific climatic change.[21] It is important to bear these possibilities in mind when interpreting pollen curves. Synchroneity of changes over a wide area may indicate a climatic origin for the change, while trends such as one particular tree appearing later the further north the site may indicate that the time of arrival and spread of the species is important. Edaphic change, such as regional podsolisation of soils,[20] would be indicated by the increased frequency of heath plant pollen types.

Pollen diagrams may be divided into a sequence of pollen biozones.[42] Each biozone should be a recognisably distinct assemblage characterised by its constitution of taxa and named after the characteristic

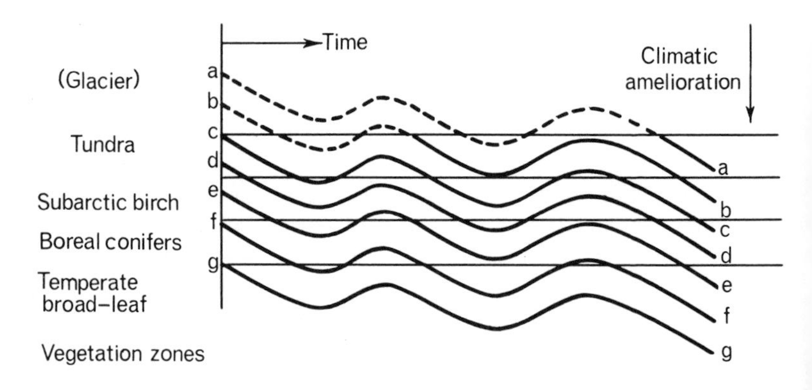

Fig. 7.12 Parallelism of climatic and vegetational change in northwest Europe. The curves show the course of vegetational change, as deduced from pollen analysis of sediments at the latitudes or altitudes corresponding to the named vegetational zones. In curve b, from a sub-arctic area, the oscillations are difficult to detect. In curve c, from the boreal coniferous zone, the oscillations are recorded by interchange between tundra and subarctic birch, and finally to coniferous forest. In curve f, from the temperate zone, the same oscillations are recorded as interchange between boreal coniferous forest and temperate forest (after Faegri and Iversen).

or dominant pollen taxa present. It may then be termed a pollen assemblage biozone (other types of biozone are mentioned in Ch. 11). A pollen assemblage biozone may have a local or regional significance, its constitution resulting principally either from local or from regional vegetation. When considering the zonation of a pollen diagram it is essential to zone on the evidence of the diagram alone, then at a later stage of the interpretation to correlate this zonation to regional or standard sequences. Otherwise biozones of a preconceived nature may be forced on to the diagram and one of the prime reasons for investigating vegetational history—the elucidation of regional and local diversity and development—will be lost. Recent developments in numerical methods have made it possible to apply

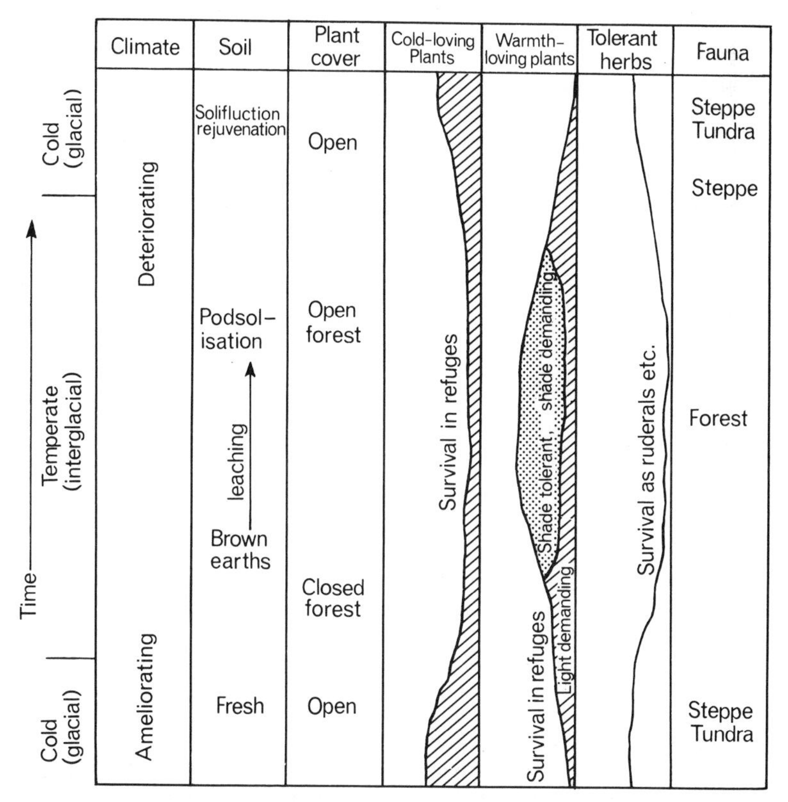

Fig. 7.13 A scheme of environmental and vegetational changes in temperate north-west Europe during an oscillation from cold (glacial) to temperate (interglacial) and back to cold (glacial).

zonation procedures using standard criteria for the segregation of pollen spectra into assemblage units, allowing common methods to be used by different interpreters.[6, 38] Methods have also been devised for comparison of pollen diagrams.[15]

On the question of correlation between pollen diagrams over wide areas, it is essential to remember that similar climatic changes will have different effects on vegetation according to latitude and longitude. Possible effects of synchronous changes are illustrated in fig. 7.12. The figure shows that an oscillation is not manifested at all the latitudes shown. Similarly, if we considered an oceanic–continental (west–east) parallelism the same oscillation would not appear at all longitudes.

The climatic cycle

Biological investigations of the Pleistocene have resulted in the clear association of climatic change with changes in flora, fauna and soils. Some of the changes associated with the cycle of climatic amelioration and deterioration in northwest Europe are shown in fig. 7.13. These are the gross changes which are evident from studies of the flora and fauna. Minor fluctuations of climate and the detailed behaviour of particular species and of communities are much more difficult to define, though equally important from the biological point of view, and much further study is required for their elucidation.

REFERENCES

1 ANDERSEN, S. T. 1961. 'Vegetation and its environment in Denmark in the Early Weichselian Glacial', *Danm. Geol. Unders.*, II Raekke, No. 75.

2 BEIJERINCK, W. 1947. *Zadenatlas der Nederlandsche Flora*. Wageningen: Veenman.

3 BELL, F. G. 1970. 'Late Pleistocene floras from Earith, Huntingdonshire', *Phil. Trans. R. Soc. London*, B, **258**, 347–78.

4 BERTSCH, K. 1941. *Früchte und Samen. Handbücher der praktischen vorgeschichtsforschung*, **1**. Stuttgart: Enke.

5 BIRKS, H. H. 1973. 'Modern macrofossil assemblages in lake sediments in Minnesota', 173–89. In *Quaternary Plant Ecology*, ed. H. J. B. Birks and R. G. West. Oxford: Blackwell.

6 BIRKS, H. J. B. 1974. 'Numerical zonations of Flandrian pollen data', *New Phytol.*, **73**, 351–8.

7 BIRKS, H. J. B. and WEST, R. G. (editors). 1973. *Quaternary plant ecology*. Oxford: Blackwell.
8 COOPE, G. R. 1970. 'Interpretation of Quaternary insect fossils', *Ann. Rev. Entomol.*, **15,** 97–120.
9 CORNWALL, I. W. 1956. *Bones for the archaeologist*. London: Phoenix House.
10 ERDTMAN, G. 1943. *An introduction to pollen analysis*. Waltham, Mass.: Chronica Botanica.
11 EVANS, G. H. 1970. 'Pollen and diatom analyses of Late-Quaternary deposits in the Blelham Basin, North Lancashire', *New Phytol.*, **69,** 821–74.
12 FAEGRI, K. and IVERSEN, J. 1975. *Textbook of pollen analysis*. 3rd edn. Copenhagen: Munksgaard.
13 FEYLING-HANSSEN, R. W. 1964. 'Foraminifera in late Quaternary deposits from the Oslofjord area', *Norges Geol. Unders.*, No. 225.
14 FREY, D. G. 1964. 'Remains of animals in Quaternary lake and bog sediments and their interpretation', *Arch. Hydrobiol. Beih.*, **2,** 1–114.
15 GORDON, A. D. and BIRKS, H. J. B. 1974. 'Numerical methods in Quaternary palaeoecology'. II. Comparison of pollen diagrams, *New Phytol.*, **73,** 221–49.
16 GRICHUK, M. P. and GRICHUK, V..P. 1960. 'O prilednikovoi rastitelnosti na territorii S.S.S.R'. In *Periglacial phenomena of the U.S.S.R.* eds. Markov, K. K. and Popov, A. I. Moscow.
17 HAVINGA, A. J. 1971. 'An experimental investigation into the decay of pollen and spores in various soil types', 446–79. In *Sporopollenin*, ed. J. Brooks *et al.* London: Academic Press.
18 IVERSEN, J. 1936. 'Sekundäres pollen als Fehlerquelle', *Danm. Geol. Unders.*, IV Raekke, **2,** No. 15.
19 IVERSEN, J. 1941. 'Land occupation in Denmark's Stone Age', *Danm. Geol. Unders.*, II Raekke, No. 66.
20 IVERSEN, J. 1948. 'The bearing of glacial and interglacial epochs on the formation and extinction of plant taxa', *Uppsala Universitets Årsskrift*, **6,** 210–15.
21 IVERSEN, J. 1960. 'Problems of the Early Post-glacial forest development in Denmark', *Damn. Geol. Unders.*, IV Raekke, **4,** No. 3.
22 KATZ, N. J., KATZ, S. V. and KIPIANI, M. G. 1965. *Atlas and keys of fruits and seeds occurring in the Quaternary deposits of the U.S.S.R.* Moscow: Publishing House Nauk (in Russian).
23 KERRICH, G. J., MEIKLE, R. D. and TEBBLE, N. 1967. *Bibliography of key works for the identification of the British Fauna and Flora.* 3rd edn. London: Systematics Association.
24 LEROI-GOURHAN, A. 1965. 'Les analyses polliniques sur les sediments des grottes', *Bull. Assoc. française pour l'Etude du Quat.*, 1965, 145–52.
25 LOZEK, V. 1965. 'Problems of analysis of the Quaternary non-marine Molluscan fauna in Europe', *Geol. Soc. Amer. Speckal Paper* **84,** 201–18.
26 MULLER, J. 1959. 'Palynology of recent Orinoco delta and shelf

sediments', *Micropalaeontology*, **5,** 1–32.

27 NORTON, P. E. P. 1967. 'Marine molluscan assemblages in the Early Pleistocene of Sidestrand, Bramerton and the Royal Society borehole at Ludham, Norfolk', *Phil. Trans. R. Soc. London*, B, **253,** 161–200.

28 PENNINGTON, W. and BONNY, A. P. 1970. 'Absolute pollen diagram from the British Late-Glacial', *Nature* (*Lond.*), **226,** 871–3.

29 PHILLIPS, L. 1972. 'An application of fluorescence microscopy to the problem of derived pollen in British Pleistocene deposits', *New Phytol.*, **71,** 755–62.

30 ROUND, F. E. 1957. 'The Late-glacial and Post-glacial diatom succession in the Kentmere Valley deposit', **56,** 98–126.

31 SHOTTON, F. W. 1965. 'Movements of Insect Populations in the British Pleistocene', *Geol. Soc. Amer. Special Paper* **84,** 17–33.

32 SPARKS, B. W. 1961. 'The ecological interpretation of Quaternary non-marine Mollusca', *Proc. Linn. Soc. London*, **172,** (1959–60), 71–80.

33 SPARKS, B. W. and WEST, R. G. 1959. 'The palaeoecology of the interglacial deposits at Histon Road, Cambridge', *Eiszeitalter und Gegenwart*, **10,** 123–43.

34 SPARKS, B. W. and WEST, R. G. 1960. 'Coastal interglacial deposits of the English Channel', *Phil. Trans. R. Soc. London*, B, **243,** 95–133.

35 STUART, A. J. 1974. 'Pleistocene history of the British vertebrate fauna', *Biol. Rev.*, **49,** 225–66.

36 TAUBER, H. 1965. 'Differential pollen dispersion and the interpretation of pollen diagrams', *Danm. Geol. Unders.*, II Raekke, No. 89.

37 WALKER, D. 1955. 'Studies in the Post-glacial history of British vegetation. XIV. Skelsmergh Tarn and Kentmere, Westmorland', *New Phytol.*, **54,** 222–54.

38 WALKER, D. 1972. 'Quantification in historical plant ecology', *Proc. Ecol. Soc. Australia*, **6,** 91–104.

39 WALL, D. and DALE, B. 1968. 'Early Pleistocene dinoflagellates from the Royal Society borehole at Ludham, Norfolk', *New Phytol.*, **67,** 315–25.

40 WEST, R. G. 1957. 'Interglacial deposits at Bobbitshole, Ipswich', *Phil. Trans. R. Soc. London*, B, **241,** 1–31.

41 WEST, R. G. 1964. 'Inter-relations of ecology and Quaternary palaeobotany', *J. Ecol.*, **52** (suppl.), 47–57.

42 WEST, R. G. 1970. 'Pollen zones in the Pleistocene of Great Britain and their correlation', *New Phytol.*, **69,** 1179–83.

43 WEST, R. G. 1971. *Studying the past by pollen analysis*. Oxford Biology Readers, No. 10. Oxford University Press.

44 WEST, R. G. 1973. 'Introduction', 1–3, in *Quaternary Plant Ecology*, ed. H. J. B. Birks and R. G. West. Oxford: Blackwell.

45 WRIGHT, H. E. and PATTEN, H. L. 1943. 'The pollen sum', *Pollen et spores*, **5,** 445–50.

46 WRIGHT, W. B. 1937. *The Quaternary Ice Age*. 2nd edn. London: Macmillan.

CHAPTER 8

LAND/SEA-LEVEL CHANGES

In coastal areas around the world there is ample evidence of changes of ocean-level relative to land. Old shore-lines, such as those marked by raised beaches, are found at considerable heights above present sea-level, indicating former sea-levels higher than the present in relation to the land, while drowned valleys indicate flooding by a rise of sea-level relative to the land. It is necessary to use the term relative in this context, as there is no certainty that the effects are caused solely by rise or fall in sea-level or by rise or fall of land. Both may be involved in any changes to produce the net result seen today. Similar ancient shore-lines, resulting from the same type of causes, are also found around the inland waters.

The interest of the study of past sea-levels is twofold. It gives us an insight into the mechanisms which are now important in determining the sea-level relative to that of the land, mechanisms which must have operated over the whole of geological time. Secondly, land/sea-level changes are important in studying the movements of faunal and floristic elements in the past.

Changes in land/sea-level result from two principal causes. Local upwarping or downwarping of rocks and world-wide (eustatic) changes of sea-level. We shall discuss the role of each of these two in causing changes of relative land/sea-level.

Local effects[1]

These effects are the result of movements of the earth's crust brought about by the deformation or displacement of rocks. The movements associated with the concepts of plate tectonics, with sea-floor spreading and the drift of continental plates, must be especially important, since the resulting displacements are obvious in the coastal regions of the earth. In addition to this long-term tectonic movement, there are also isostatic movements—those thought to result from adjustment in the balance between different parts of the crust, so that if for example a load of sediment is added in one area of ocean depression occurs concurrently in that area, while a corresponding rise will occur where the load of rock is being reduced by erosion. Adjustment of levels by flowage of material deep in the earth follows until isostatic equilibrium is re-established (fig. 8.1).

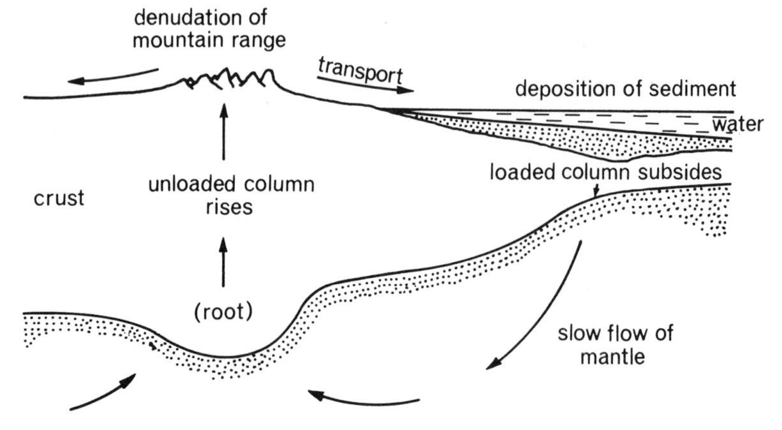

Fig. 8.1 Diagram illustrating isostatic readjustment in response to denudation and deposition (after Holmes).

A special case of isostasy, and one important in the present context, is where the load is of ice, and depression of the crust by deformation follows the build-up of an ice sheet (fig. 2.7b). On removal of the load during deglaciation, there is a slow recovery of the crust and an updoming of the land surface results. That the removal of ice load is indeed the cause of the isostatic adjustment which leads to updoming is suggested by the close relation between the limits of ice sheets and the limits of updoming in Scotland, Scandinavia and Canada, by the signs of uplift near the margins of presently glaciated areas (Greenland, Spitsbergen, Antarctica), by the concentric arrangement of isobases (lines joining points of equal uplift) in relation to the centre of dispersion of the ice, and by gravity anomalies indicating deficiencies of mass in formerly glaciated areas showing present uplift, e.g. Scandinavia and Canada. The amount of depression by ice load will depend on the density and thickness of the ice, the density of the material beneath and its rate of deformation. If we estimate as an approximation that the density of the flowing material is three times that of ice, the amount of depression will be a third the maximum thickness of the ice. Niskanen[24] has made some calculations regarding total crustal depression and thickness of the Scandinavian ice. He estimated 730 m of depression and

2650 m of ice. There appears to be a minimum ice load causing depression, as there is little evidence that the ice bodies of small glaciers produce an effect on the earth's crust. Thus there is evidence of such effects with ice caps greater than 500 km in diameter and with ice more than 1 km thick.[3]

An interesting question in relation to isostatic depression by ice load is whether there is a compensatory uplift in areas marginal to the ice sheet, or alternatively, a compensatory downwarping in marginal areas accompanying the uplift of a deglaciated area. Such has not yet been conclusively demonstrated, but there are indications as seen in the present downwarping of the southern margin of the Baltic shown in fig. 8.2. In any case, the transfer of water from the oceans to the ice load should result in an upward movement in the oceanic mantle.

The evidence for isostatic recovery after deglaciation comes from observations of heights of raised beaches. From these, isobases (lines joining points of equal uplift) can be drawn, showing the form and degree of the uplift (fig. 8.10), with the zero isobase marking the outer limit of uplift. There has been much discussion over whether uplift is regular and the updoming simple, whether discontinuities occur in the rise, whether block uplift occurs, and whether hinge lines occur dividing areas of unequal rate of uplift (fig. 8.6c).[30] This last point is discussed later in relation to the Baltic area.

A further type of isostatic response follows the rising sea-level which accompanies deglaciation. The ocean floor of coastal regions may be depressed by the greater water load, thus giving a greater degree of submergence of the coast than that which is solely the result of sea-level rise. Thus there is evidence that the water load of the post-glacial rise in sea-level has isostatically deformed the Atlantic coast of the United States in proportion to the average weight of water at the present shore-line.

Eustatic movements of sea-level

Eustatic movements of sea-level can result from changes in the form of ocean basins and consequently the thickness of the oceans, from accumulation of sediments displacing sea-water, from displacement of sea-water by isostatic recovery or other sorts of uplift, and from changes in the temperature of sea-water causing volume changes. The last is of little significance; it has been estimated that a rise of 1°C would raise sea-level by 60 cm. The magnitude of the other factors in effecting movements of sea-level is unknown, but certain of them will be of great importance in the long term. For example, it has been estimated that ocean basins may be widening by sea floor

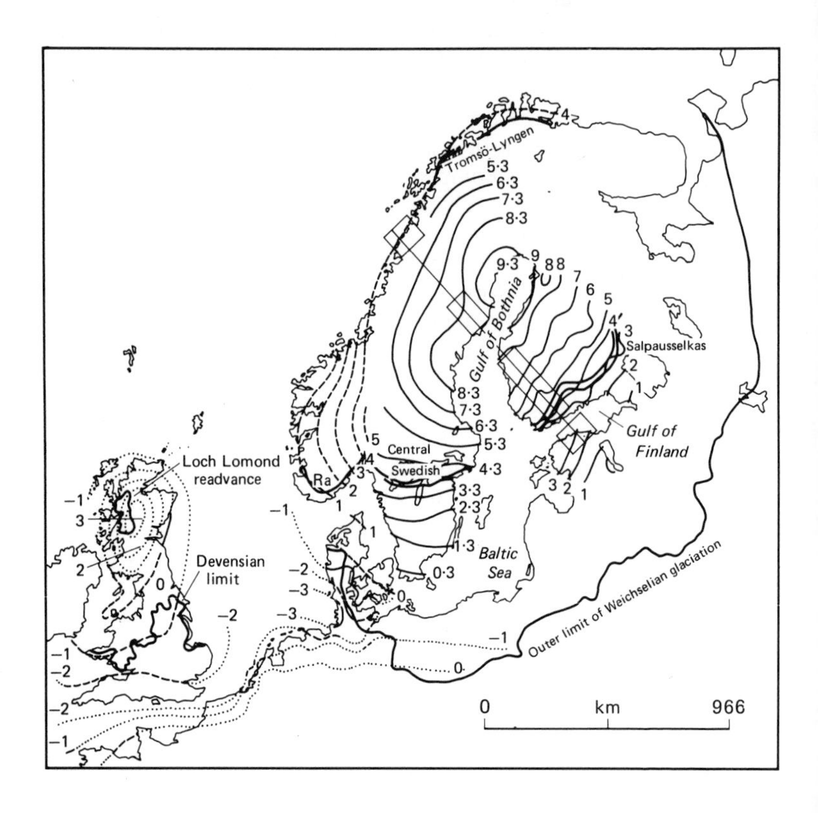

Fig. 8.2 Contemporary uplift and downwarping in north-west Europe. The isobases represent rate of change (+ or −) in mm/year, determined from tide-gauge records. Dashed isobases are less certain, and the dotted isobases are based on interpolation (after Kukkamäki, 1968 and Valentin, 1954). Limits of the last glaciation and positions of the Fennoscandian moraines are shown, and the baseline for the shore-line diagram in fig. 8.9.

spreading up to 16 cm per year; using a figure of 10 cm per year it has been calculated that post-glacial shore-lines would be some 8 m lower than interglacial shore-lines of 100,000 years ago. [2]

But the most immediate cause of eustatic sea-level fluctuation in the Pleistocene has been the glacial control of sea-level. The level of the oceans is related to the balance between the volume of water in the oceans, the amount of moisture in the atmosphere and the volume of water on the land surface. During a glaciation, atmospheric temperatures fall, and water accumulates on the land surface as a result of increased snowfall, reduced ablation in the summer, and freezing of water on and in the ground. Thus sea-level will fall because of the increase in water held on the land surface, principally in the form of ice. Conversely, when temperatures rise, the ice will start to melt, water will return to the oceans and sea-level will rise.

Various calculations have been made to estimate the amount of fall of sea-level during the glaciations, based on estimates of the volume of ice formed. They range from 80 to 150 m. This figure is not very different from figures obtained from geological evidence of lowered sea-levels.

It has also been estimated that the sea-level would rise by 40 to 60 m if all the world's ice were melted. But shore-lines of more than twice this height are known and other effects than glacial control must be involved, as discussed above. Small changes in continental area and ocean depth may produce large changes in the flooding of continental margins,[11] and we have no way yet of estimating the importance of this kind of factor. It is thought that there is an overall trend of eustatic fall in the Pleistocene, a conclusion reached from observations that shore-lines seem progressively younger as they get lower. This could be caused by sea floor spreading, marginal uplift of continents and possibly by build-up of Antarctic ice. But much remains to be learnt of the non-glacial causes of eustatic changes of sea-level before the contribution of these factors to the overall changes can be understood.

Elucidation of the course of the eustatic changes in sea-level can only result from shore-line studies and from the dating of submerged freshwater or shallow marine sediments and of emerged marine sediments or organisms. Ideally such studies should be made in areas of crustal stability, at least in the short term, for long-term stability on the continental margins seems unlikely. But most detailed studies of the eustatic rise since the last glaciation have in fact been made in areas sensitive to the effects of isostasy or tectonic movement. The course of this last (Flandrian) eustatic rise has been charted many times (fig. 8.7a, b), and the general course of the rise is quite clear. But particular points about the curve are a matter for discussion and further work. Has the Flandrian sea-level ever risen above its present level? When was the present sea-level reached? Has the rate of rise been steady or are there many fluctuations to be superimposed on the generalised curve? Obviously the answers to these questions will vary according to the coastal region studied, because of the many other factors which affect the relation of a land mass to the eustatic sea-level.

The relation of the eustatic curve to the disappearance of the major Pleistocene ice sheets some 7000 radiocarbon years ago indicates a rise of some 10 m since this deglaciation was completed.[2] Part of this may be accounted for by the melting of smaller ice sheets and of Greenland and Antarctic ice, as well as by isostatic depression in coastal areas.

The Isokinetic Theory of W. B. Wright and the causes of shore-line displacement

W. B. Wright[37] in 1914, in considering the shore-lines of northern Europe, showed that a combination of isostatic recovery with a eustatic rise of sea-level gave an explanation of the shore-lines formed after the retreat of the ice of the last glaciation. He suggested that shore-lines were formed as a result of sea-level standstills when rate of eustatic rise was the same as the rate of isostatic uplift. This theory is a useful basis for interpretation of shore-lines in recently glaciated areas, but it should be remembered that even in these areas other possible factors, such as tectonic movements and downwarping of marginal areas, may be effective.

Outside these areas, both eustatic changes and tectonic movements may combine or act singly to produce shore-lines. The diverse causes of shore-line displacement and changes of relative land/sea-level are best clarified by examining later in this chapter the evidence from three areas, in each of which a different combination of factors resulted in shore-line displacement. These are the Baltic, where eustatic and isostatic factors combine, the Netherlands, where eustatic rise is accompanied by tectonic sinking, and the Mediterranean, where longer-term eustatic effects may be seen.

Contemporary and recent changes in relative land/sea-level

These changes can be demonstrated by studies of tide-gauge records, as has been done in Scandinavia, the British Isles and the North Sea area (fig. 8.2), and North America (fig. 8.3). The records need careful analysis since the secular variation of sea-level is accompanied by variations caused by atmospheric and tidal effects.[26] The records show that updoming is still continuing in areas subject to heavy glaciation during the Pleistocene. The maximum rate of uplift observed is in the northern part of the Gulf of Bothnia, nearly 1 cm a year, and in each area the place of maximum uplift corresponds well with the centres of ice dispersal.

It has been estimated that the uplift in Scandinavia will not be complete for many thousands of years and this implies that there is a very considerable time lag between the end of isostatic recovery and the final disappearance of the Scandinavian ice cap about 6000 years ago.

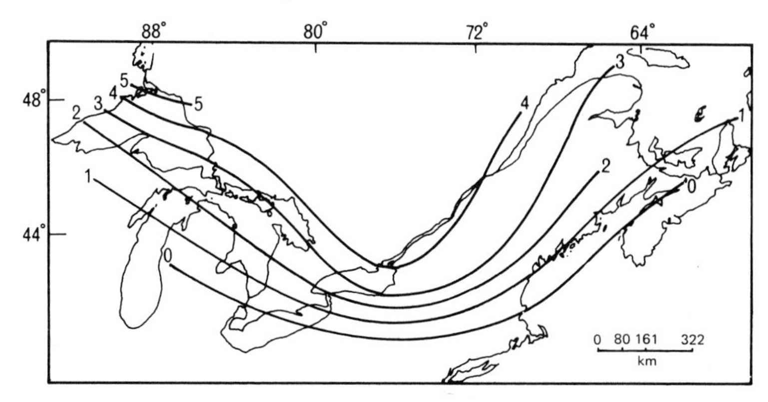

Fig. 8.3 Contemporary uplift in eastern north America. The isobases represent rate of uplift in mm/year, determined from tide-gauge records (after Gutenberg, 1941).

Outside the zero isobase for the present uplift, as seen in fig. 8.2, downwarping has been observed in the southern Scandinavian region and in the British Isles and North Sea region. It is not possible to say how much this is associated with independent tectonic changes and how much with any isostatic compensation following updoming further north.

In Britain, the most reliable values of secular variation during this century have been obtained from tide-gauge records as follows (mm/year):[26] Aberdeen +0·8, Dunbar +0·1, Felixstowe +1·6, Southend +3·4, Sheerness +3·3, Newlyn +2·2, with values at Brest of +2·1 and in the Netherlands of +2·0. These results indicate that Scotland is rising relative to southern England by at least 1·5 mm/year. If the eustatic rise is, say, +1 mm/year, the figures suggest slight uplift in Scotland and slight subsidence in southern England and the shores of the Channel, with the highest rate of subsidence around the Thames estuary.

Evidence for changes of relative land/sea-level in the past [12]

Evidence for higher sea-levels than at present (plate 13). Both erosional and constructional forms give evidence of higher sea-levels. Features associated with shore-lines are shown in fig. 8.4. Sea-cliffs, wave-cut platforms or marine terraces, wave-cut notches at the base of sea-cliffs, sea-cave systems and lines showing past activity of rock-boring molluscs have all been used to indicate past positions of sea-level. Off-shore deposits, off-shore bars, spits, beach gravels, all can give evidence of former marine transgression over the land. So does the alternation between estuarine and/or marine deposits and freshwater deposits. The contact between freshwater and brackish-water or marine deposits is called a transgression (freshwater to brackish) or regression (brackish to freshwater) contact, and these points of contact are important in determining how former sea-levels changed, as they register salinity changes of the water at particular heights, which can often be dated and related to vegetational history.

Aggradation terraces of the lower reaches of rivers are also important for determining former sea-levels, since the terrace profile tends to flatten out when near sea-level (fig. 8.5).

Evidence for lower sea-levels than at present. Submerged shore-lines, submerged deltas and the extension of valley systems well into the sea, give evidence of former lower sea-levels. As an example of the last, the Elbe valley is recognisable in charts to a distance of 500 km from the shore, and to a depth of −80 m suggesting a former relative sea-level of at least that depth. The lowered sea-level would also be likely to change the regimen of the rivers concerned, which would start downcutting in their lower reaches, the downcutting working back as a nick-point in the long profile of the river (fig. 8.5). Many such nick-points have been identified in the Thames profile.[39]

Sediments also give clear evidence of former low sea-level in the occurrence below present mean sea-level of brackish or freshwater sediments intercalated in marine or estuarine series of deposits (fig. 8.4c).

Fig. 8.4a Elements of the shore zones on a well-developed coastline (after Johnson 1919).
Fig. 8.4b Section at Porthleven, Cornwall, showing re-excavation of 'fossil' wave-cut notch containing cemented beach material and exhumation of 'fossil' cliff by the sea. The present notch re-occupies the excavated 'fossil' notch, and the old wave-cut platform forms the present one (plate 13a).
Fig. 8.4c Schematic section through western part of the Netherlands, showing sediments formed during the Flandrian time of rising sea-level behind beach barriers (van Straaten 1961).

1. Sands of young coastal dunes (after twelfth century).
2. Tidal channel, flat and salt marsh deposits (Sub-Atlantic).
3. Spit barrier deposits (Sub-Boreal), beach sands, grading downwards into finer sands with mud layers, upwards into dune sands.
4. Relatively coarse grained sands of North Sea floor.
5. River deposits, partly contemporaneous with peat.
6. Clay of salt and brackish marshes and swamps (Atlantic and early Sub-Boreal).

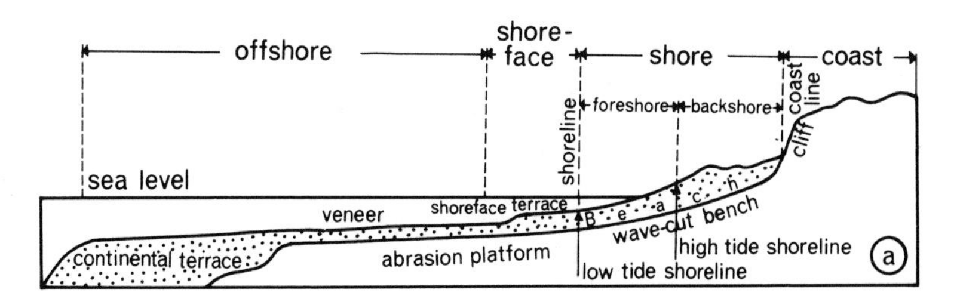

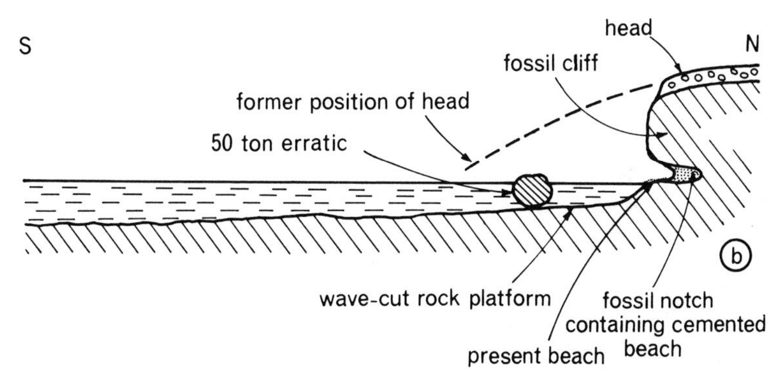

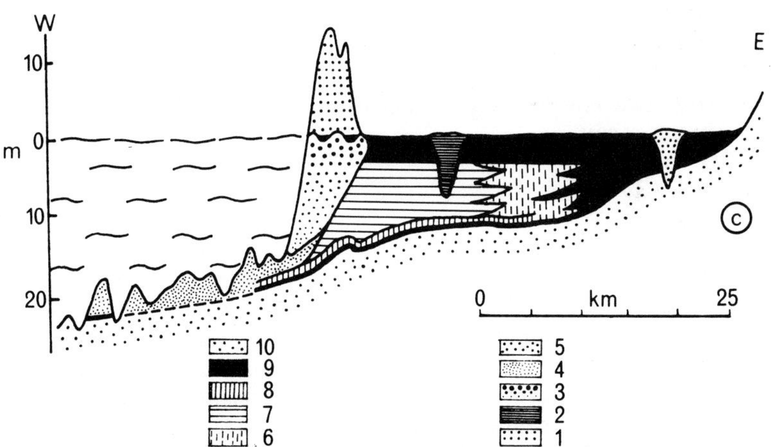

7. Sediments of tidal flat environment, channels, flats and marshes (Atlantic and early Sub-Boreal).
8. Lagoon clays (Atlantic).
9. Peat (late Boreal, Atlantic, Sub-Boreal and Sub-Atlantic). Much of this peat has been dug away by man. The diagram gives roughly the original situation.
10. Eolian and fluvial sands (Pleistocene and Pre-Boreal).

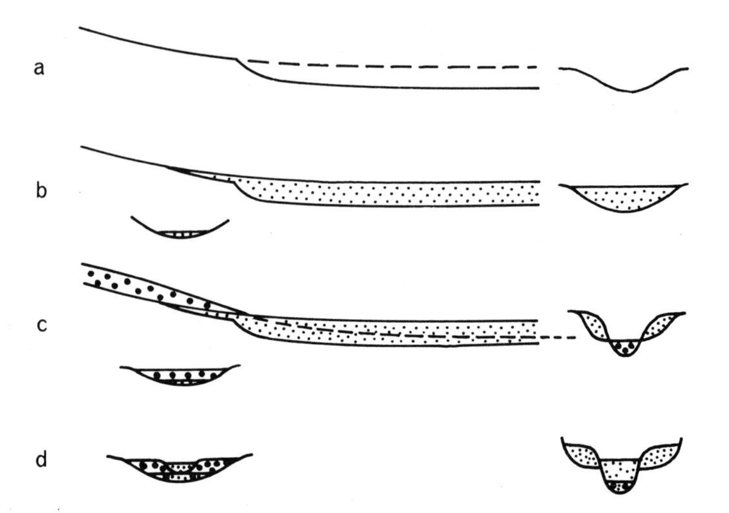

Fig. 8.5 Long profiles of rivers with down-cutting and aggradation phases associated with changing sea-levels. Fine dots, alluvium of temperate stages; large dots, sands and gravels of cold stages.

a. Preglacial valley, river rejuvenated by low sea-level of glacial times. Nick-point marks the head of rejuvenation.
b. Aggradation as a result of a rising interglacial sea-level.
c. A further glacial low sea-level results in down-cutting in the lower parts of the valley and aggradation of outwash and weathering debris in the upper part.
d. Further aggradation during a second interglacial stage of higher sea-level.

The interpretation of evidence concerning former sea-levels

It was considered at one time that a world-wide chronology of the Pleistocene could be based on eustatic changes of sea-level, as indicated by former shore-lines. Such a chronology may be possible in the most general terms, but there are two major groups of difficulties affecting its attainment in detail. One is the difficulty of determining the eustatic component of any changes which have resulted in shore-line displacement, and adjusting for local isostatic and tectonic effects. The older the displaced shore-line, the more difficult it is to isolate the effective factors. It must be known whether the area under study is a relatively stable area, or one subject to isostatic recovery or to the tectonic activity. If a shore-line is level over a long distance it is probably in an area of relative stability. But there are few areas where such stability can be well demonstrated.

The other group of difficulties concerns the actual height relation of a geological feature to the sea-level at the time it was formed.[13] It is not easy to relate the position of the notch, the height of the wave-cut platform, the break of slope at the back of a platform, to a particular sea-level, as the position will vary with tidal range, the nature of the rock, whether it is well bedded horizontally or not, and the exposure of the coast. The relation of raised beach heights to sea-level is also difficult, as witness the height of the present Chesil beach up to 12 m above sea-level. There is also the possibility of sheets of beach gravel being formed as the regression of the sea occurs. These beach gravels will then not represent high still-stands of sea-level. Neither has any absolute datum been used for determination of former sea-levels over wide areas. These uncertainties can be reduced by accurate measurements and by reports on the reliability and meaning of the particular features measured, but substantial sources of variability remain.

With transgression and regression contacts, the evidence for salinity change and its relation to height is more objective, but here again there are such sources of error as compaction of sediments and the possibility of changes in salinity caused by coastal changes rather than by changes in sea-level.

Though these difficulties remain, much has been done to elucidate relative land/sea-level changes, as will be seen in the subsequent examples. It is in the sphere of a eustatic chronology that the difficulties are most apparent. The problem is to relate the high interglacial still-stands of sea-level with particular interglacials, and in doing this the cycle of change of sea-level related to climatic change must be remembered. There will be transgression, still-stand and regression in each cycle, and it is difficult and important to relate particular features to particular parts of the cycle.

Dating of displaced shore-lines

Height. Height has been used extensively in correlating shore-lines, especially in areas thought to be so stable that eustatic changes of sea-level are closely related to the displacement of the shore-lines.

Stratigraphy. The relation of shore-line deposits to solifluction sheets (head), tills and interglacial deposits has been one of the most effective methods of dating shore-lines.[36]

Contained fossils. These include pollen grains, molluscs, vertebrates, and even artefacts, as for example, in the dating of the raised beach

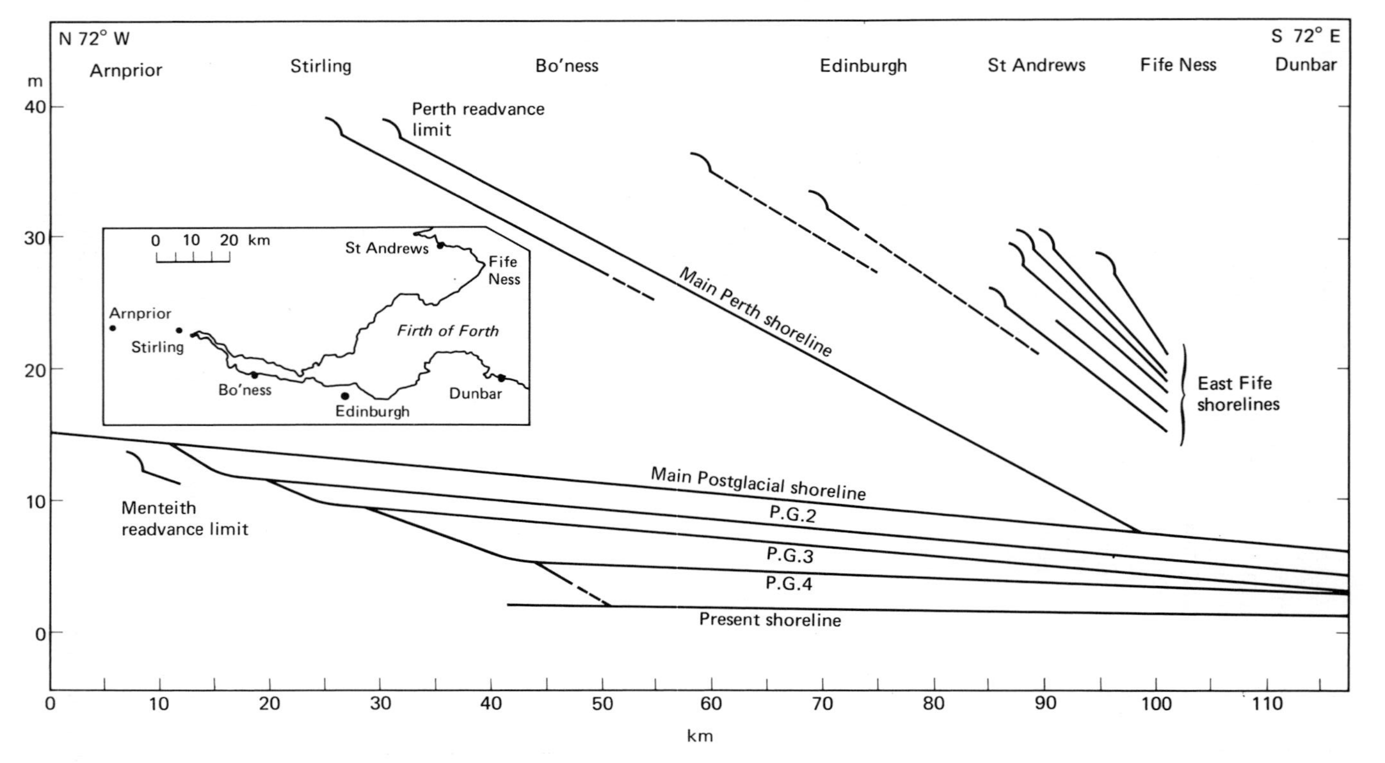

Fig. 8.6a Shore-line sequence in south-east Scotland (after Sissons *et al.* 1966).

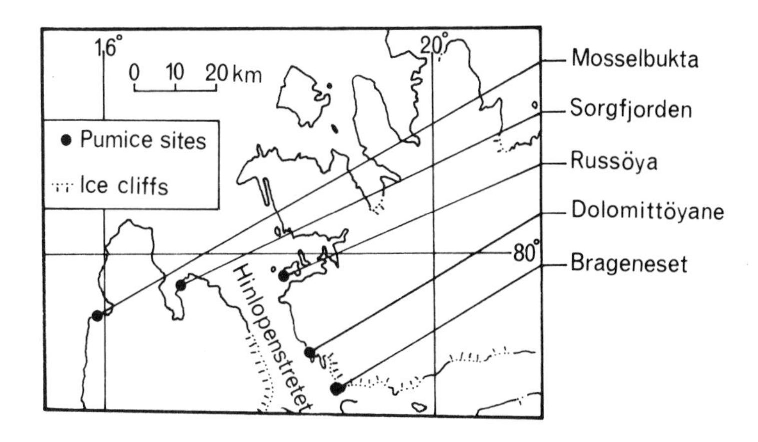

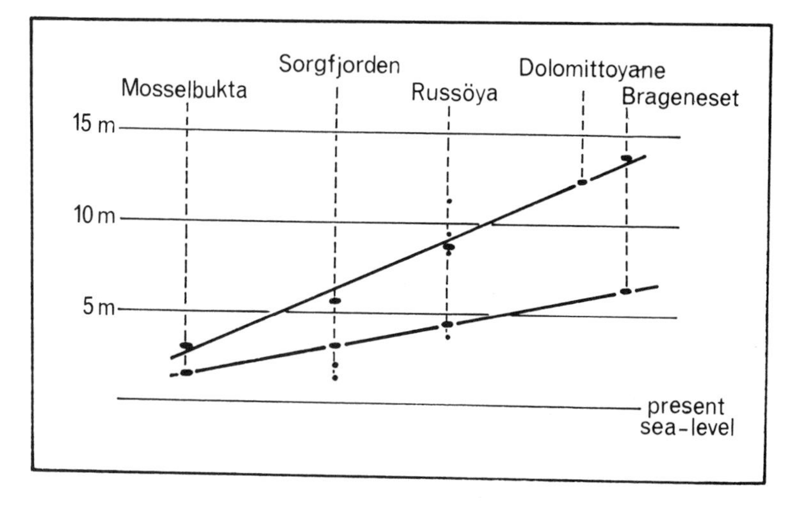

Fig. 8.6b Shore-line relation diagram of pumice levels in the northern part of Hinlopenstretet, Nordaustlandet, Spitsbergen. The position of the sites is shown on the map (Donner and West, 1956).

in Sussex containing Acheulian implements to the interglacial (Hoxnian) in which these artefacts are characteristic. Changes of pollen content at transgression and regression contacts between terrestrial and marine or estuarine sediments have proved useful in dating sea-level changes in the Flandrian[14] and in earlier interglacials.[35] Likewise, land uplift in Scandinavia has been documented by pollen studies of marine and freshwater sediments associated with particular shore-line stages.

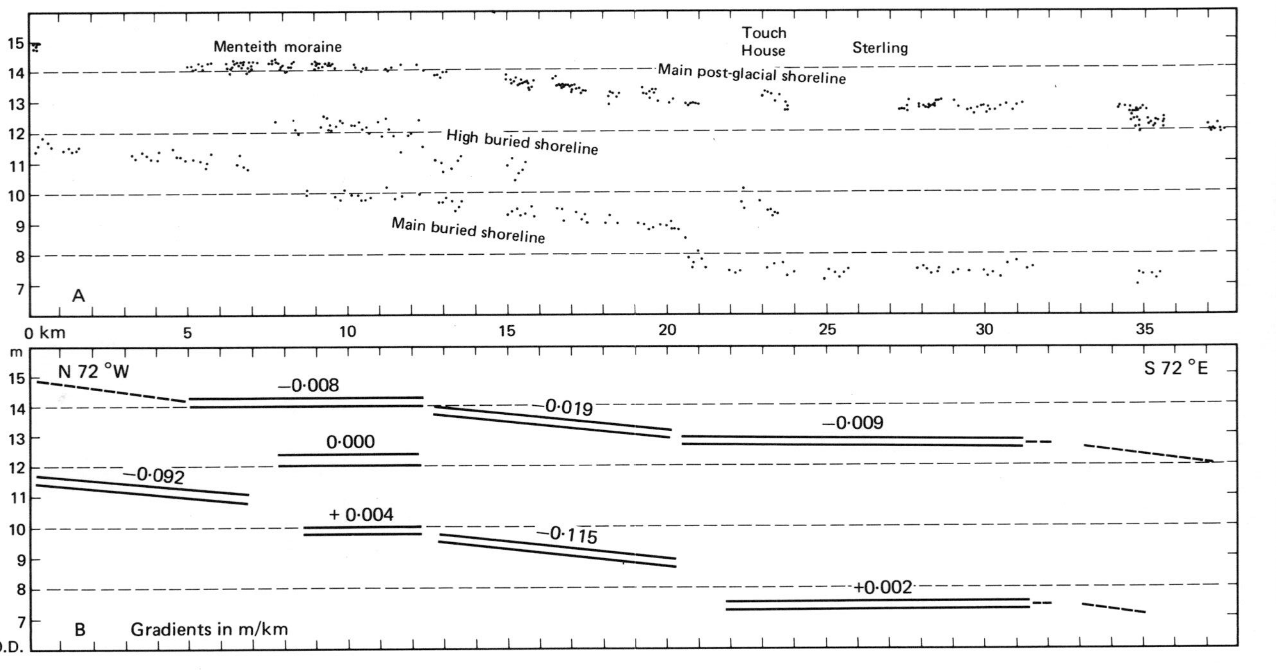

Fig. 8.6c A, Altitudes of three shorelines in the western part of the Forth Valley. B, Gradients of the shorelines; the double lines are one standard deviation on either side of the calculated line (after Sissons, 1972).

Absolute dating.[2a, 15, 25, 32] Radiometric dating, especially radiocarbon dating, of biogenic materials found in close relation to transgression and regression contacts of organic sediments has been successfully used to determine past sea-levels (fig. 8.7). The geological context of any sample to be dated should be quite clear. Ideally there should be no evidence of unconformity between the two sides of the regression or transgression contact, and that the sediment dated is close in age to the time of the transgression or regression. Shells have also been used to determine ages of raised beaches, especially in the Arctic.

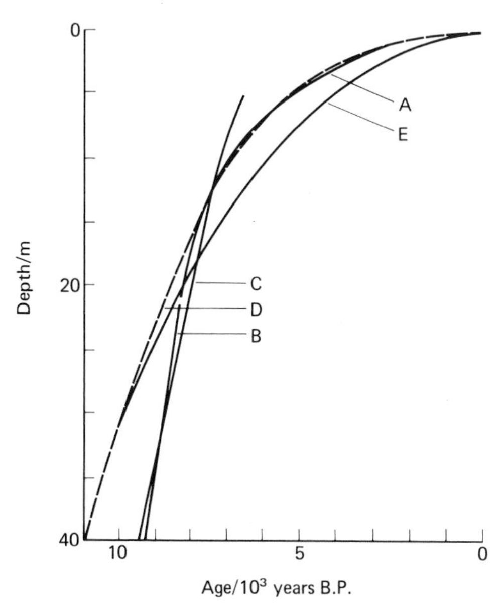

Fig. 8.7a Curves for the Flandrian eustatic rise of sea level. A, in the Netherlands (Jelgersma, 1961); B, in the southern North Sea (Jelgersma, 1961); C, off the coast of south-east Devon (Clarke, 1970); D, curve of Shepard (1963); E, curve of Segota (1973) based on 147 worldwide radiocarbon-dated levels.

The graphical expression of changes in relative land/sea-level

Isobase maps (fig. 8.10). In these maps, lines joining points of equal uplift at particular times show the process of deformation of the earth's crust. The points are based on shore-line displacement studies. In the construction of such a map, the dating of the shore-lines is obviously of prime importance.

Shore-line diagrams (fig. 8.6).[5] These diagrams show the height of distinct shore-lines (y-axis) at different distances (x-axis) from the margin or centre of the area of uplift (distance diagram) (fig. 8.6a). Or they may be drawn so that a well-developed and well-dated shore-line is used as a reference level, drawn in as a straight tilted line, and other shore-lines then inserted in relation to the reference level (relation diagram). Figures 8.6a and c are diagrams of the former type from the British Isles.[31] Figure 8.6b is an example of the latter type,[9] of two raised beaches in Spitsbergen, identified as belonging to particular times by the presence of pumice which had drifted on to them when they were formed. The reference level used is the straight tilt of the lower of the two pumice horizons. Figure 8.9 shows a more complicated shore-line diagram from Fennoscandia.

Examples of studies of past sea-level changes

The Netherlands. Relative land/sea-level changes in the Netherlands during the Flandrian have been studied by Jelgersma[18, 19] in particular detail by means of radiocarbon dating of horizons significant for tracing the relative rise of sea-level at this time. Three factors are important in the Netherlands in determining the rate of Flandrian sea-level rise. Eustatic change of sea-level, tectonic downwarping during the subsidence of the North Sea basin, and the compaction of the soft Flandrian sediments, which cover a large part of the west of the country. It is difficult to separate the first two factors and to decide what part of the rise in sea-level is caused by tectonic subsidence known to occur in the Netherlands in the pre-Flandrian, and what part by eustatic rise. Only the net effect of these two can be measured, and this has been done by making radiocarbon datings of fen peat sediments lying immediately on top of the older sands whose surface dips seawards. The formation of this fen peat results from a rise of water-level in the sand following rise in sea-level, though it is possible that precipitation changes will also affect the position of the water level. Peat samples from local depressions in the basal sands or ombrogenous peats whose formation is not related to ground water are avoided. In this way the results of compaction of the soft Flandrian sediments (which may amount to as much as 80 per cent of the original thickness) are avoided, as the basal sand is hardly affected by compaction.

Figure 8.7.b shows the time/depth relations as determined by datings of the lowermost peats, and it covers the period 9300 years B.P. to 2500 years B.P. The first curve is based on the assumption that peat growth may have started at different heights above high tide

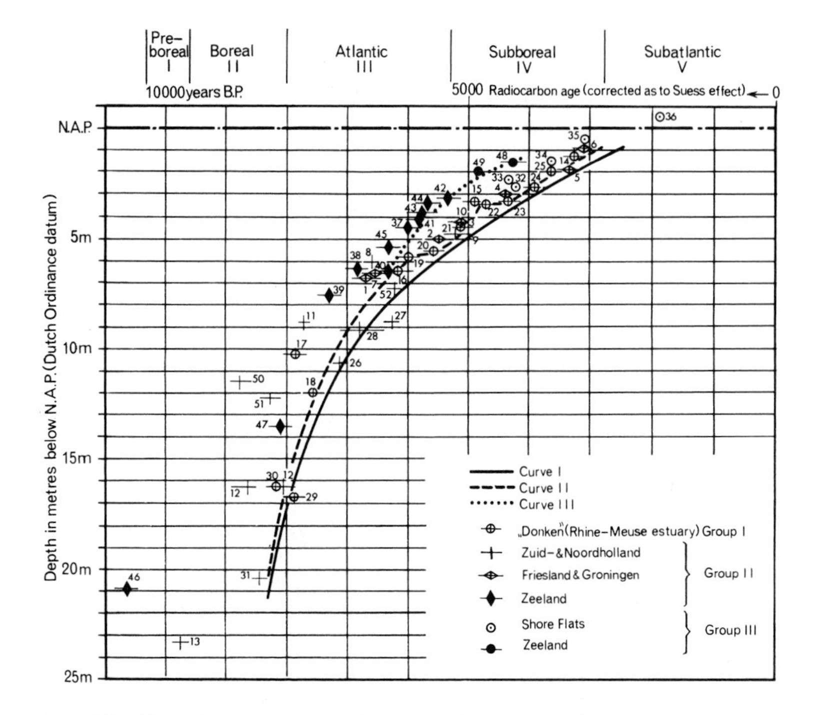

Fig. 8.7b Time-depth graph with curves showing relative changes in sea-level in the Netherlands (Jelgersma 1961). Each point represents a peat at depth, radiocarbon dated. The peats dated are divided into three groups. Group I, peat-beds in the Rhine-Meuse estuary; Group II, peats on Pleistocene aeolian sands in Zeeland, Holland, Friesland and Groningen; Group III, peats on sandy marine facies. The horizontal bar through the points gives the size of the standard deviation in age. For explanation of curves, see text.

level and is drawn through the lowermost points of any given age. The second curve connects datings from the Rhine–Meuse estuary as it is believed the peats concerned were deposited close to sea-level. The third curve shows datings from the province of Zeeland, and is higher than the others, perhaps due to greater tidal range in this area or because tectonic subsidence has been less.

The geological section in fig. 8.8 shows the accumulation of sediments during the Flandrian in the Netherlands. The accumulation of sediments closely followed the sea-level rise, so that alternations of tidal flat and freshwater peats form the succession. It should again be stressed that the positive change in sea-level has a component of

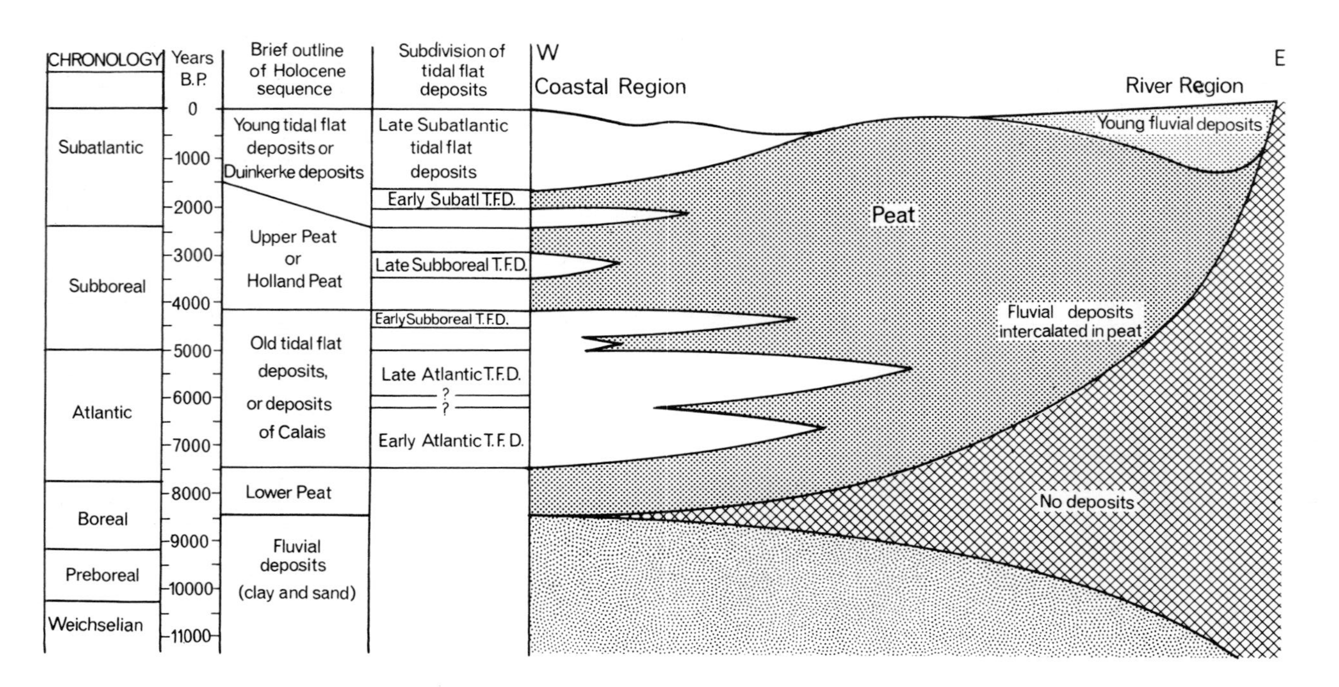

Fig. 8.8 **Schematic section through Late Pleistocene (Weichselian and Flandrian) deposits of the western part of the Netherlands. Horizontal distances not to scale; vertical distances based on a time-scale (Jelgersma, 1961).**

subsidence as well as eustatic rise. If subsidence is small, it appears from the time-depth curve that the eustatic rise continued till recently and that there have been no substantial eustatic fluctuations during the time represented, nor any rise much above present sea-level. However, as the Netherlands is an unstable area, the curves shown in fig. 8.7b should not be regarded as representative of a wider area, and certainly not a definitive curve for the Flandrian eustatic rise.

Relative land/sea-level changes in the Baltic area. This is the classic area for the study of displaced shore-lines of the Late Weichselian and Flandrian. These past shore-lines of the Baltic are widely distributed but usually discontinuous, so that it may be difficult to trace the same shore-line for any distance. The shore-lines are present as beaches fronting slopes of till or of rock washed free from drift, and as shore bars and terraces in front of glacial sand and gravel deposits such as eskers and deltas. As the tidal ranges in the Baltic are small the old sea-levels can be determined by the position of the beaches and wave-cut features to accuracies of less than one metre.

The shore-line displacement reflects the changes in the height and area of the Baltic as deglaciation proceeded, and as lake levels were affected by the isostatic uplift following deglaciation and by the Flandrian eustatic rise. The isostatic component is the most important and it has been estimated that the total uplift at the centre of uplift has been in the order of 500 m.

The data required to reconstruct the history of the Baltic are mapping of the shore-lines, the measurement of their heights, and their dating into the Flandrian pollen-zone system. The relation of the ice retreat to the Baltic stages is shown by determining whether a shore-line is terminated by the line marking the ice margin at any particular time.

The dating of the shore-lines and correlation of disjunct stretches of shore-line are the main difficulties in the reconstruction of isobase maps showing the warping of shore-lines formed at different times in the past, examples of which are shown in fig. 8.10. Such maps demonstrate a dome-like upwarping with a centre in the northern part of the Gulf of Bothnia, similar in kind to the upwarping trend demonstrated by the isobases of contemporary uplift (fig. 8.2). The isobases may be somewhat closer near the centre of upwarping, implying increased uplift nearer the centre and so crustal bending. In a shore-line diagram of the shore-lines of the principal Baltic stages (fig. 8.9) the vertical profiles of each shore-line show that differences in the angles of slope occur, indicating that the rate of uplift has differed from time to time. Sauramo[27] and Donner[8] have

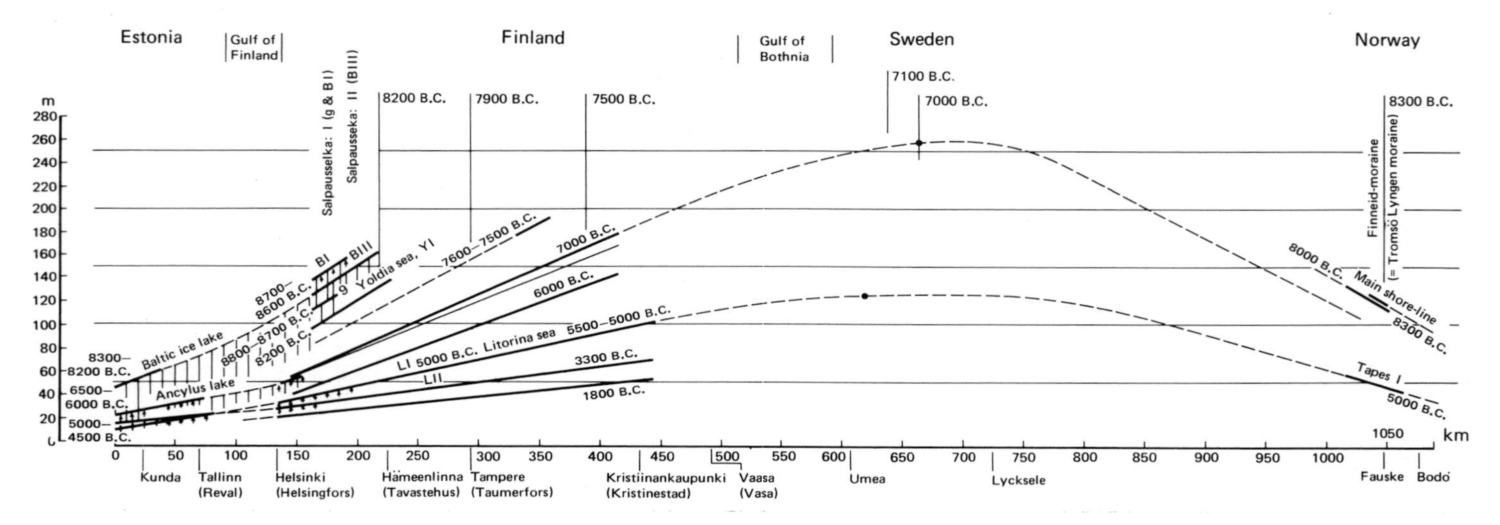

Fig. 8.9 Shore-line diagram across Fennoscandia, along the baseline shown in fig. 8.2. Positions of the retreating ice margin are shown at the top of the diagram. The continuous lines show determined shore-lines and broken lines the connection between them. The positions of the water levels of the Baltic Ice Lake and the Ancylus Lake above sea-level are shown by vertical lines, and transgressions by arrows underneath the levels reached by them (Donner, 1969).

described sudden changes in the slope of some shore-lines which they ascribed in movement at a hinge or hinge-zone. The demonstration of such a hinge-line requires accurate dating of the shore-lines on either side of the hinge-line.

We may now follow through the history of the Baltic, as described by Sauramo.[28] The stages are here simplified. Most of them have several episodes of shore-line displacement within them. The sequence of shore-lines within the older stages are based on the form of the shore-lines and on the varved clay chronology, whereas those of the younger stages are dated by shell beds, peat stratigraphy, pollen analysis and archaeology. The isobase maps for the different stages are shown in fig. 8.10.

a. *The late-glacial Yoldia Sea* (*or Karelian Ice Sea*). This is a marine stage, during which the Baltic was connected with the White Sea in the east and the North Sea to the west. (The *Yoldia* sea is named after the marine mollusc *Yoldia arctica*.)

b. *The Baltic Ice Lake*. Uplift in the areas where there had formerly been ocean connections resulted in the isolation of the Baltic Ice Lake, which drained westwards through central Sweden at Billingen, and northwards marginal to the eastern limit of the ice sheet at this time. The ice lake was contemporary with the ice-marginal moraines of the Salpausselkä in Finland and the middle Swedish moraines.

c. *The Yoldia Sea*. Retreat of the ice led to the rapid fall of the lake level as new outlets were exposed in central Sweden. Eustatic rise in sea-level brought marine conditions to the Baltic again.

d. *The Ancylus Lake* (named after the freshwater mollusc *Ancylus fluviatilis*). The period of submergence was brought to an end by the rapid rise of the land, and again the Baltic became a freshwater lake, with its first outlet through the Svea river in south Sweden. Progressive tilting of the land transferred the outlet south to the Danish Sound. During this time the Ancylus Lake waters transgressed far more in the south of the Baltic than in the north, where land uplift was greater than in the south.

e. *The Littorina Sea* (named after the winkle *Littorina*). The sea then transgressed the Danish Sound bringing marine conditions to the Baltic again. This is the fore-runner of the present Baltic, which is less saline than the Littorina Sea, perhaps because its greater shallowness, related to uplift since *Littorina* times, prevents penetration of undercurrents of saline water far into the Baltic.

The relation of these Baltic stages to the chronology of the Flandrian is shown in table 8.1.

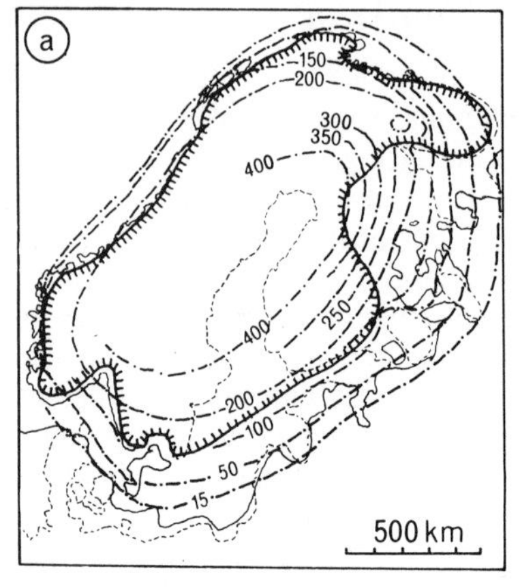

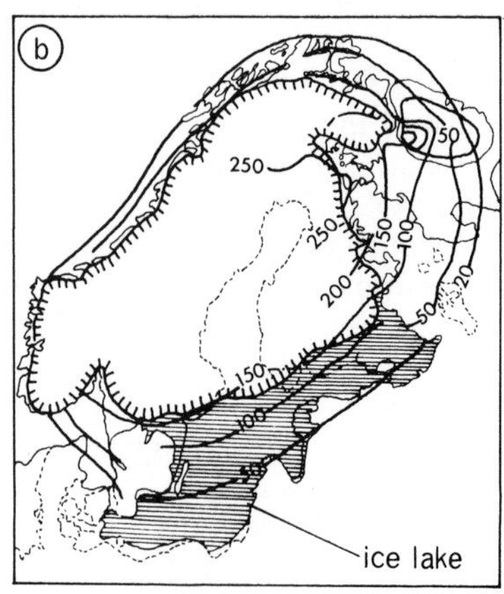

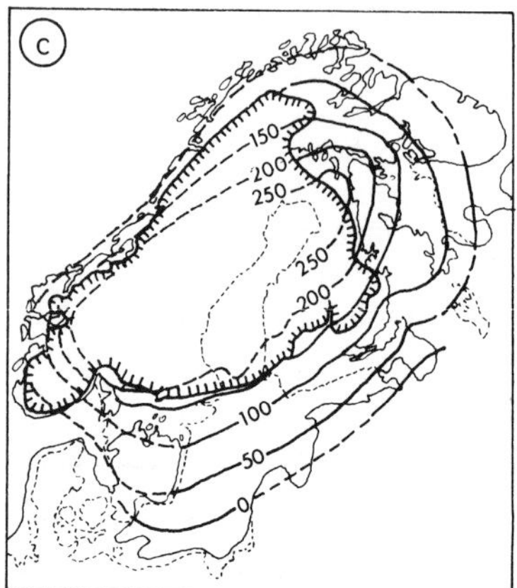

Fig. 8.10 **Iso**base maps (m) of stages in the history of the Baltic. The maps also show the extent of the ice and dry land at each stage (Sauramo).

a. Late-glacial Yoldia **Sea.**
b. Baltic Ice Lake.
c. Yoldia Sea (first stage).
d. Ancylus Lake.
e. Littorina Sea.

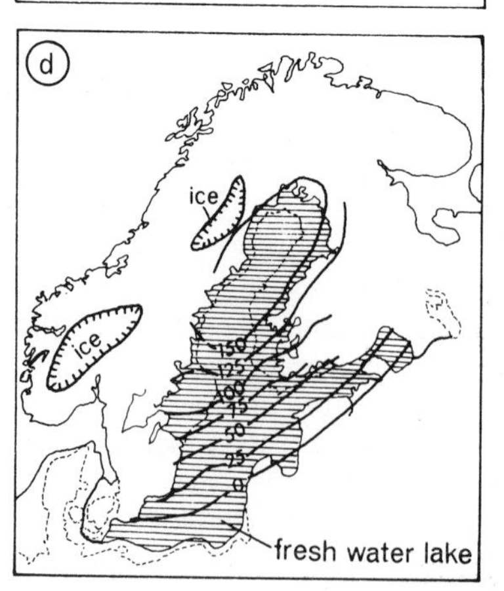

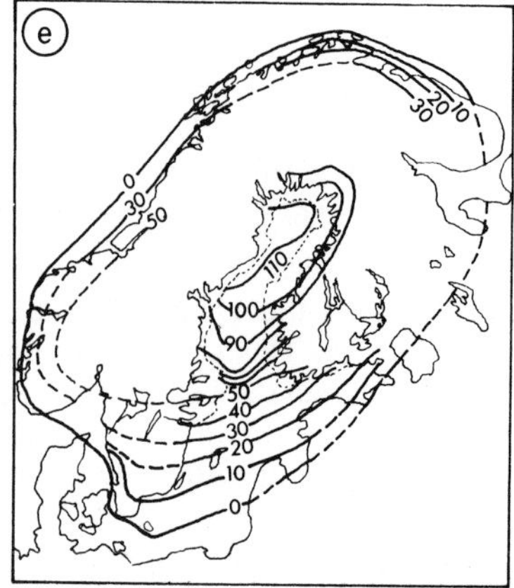

Table 8.1 Correlation of Baltic stages with ice retreat stages and pollen zones in Finland (Donner, 1965b)

Years A.D./B.C.	*Stages of ice retreat*	*Pollen stratigraphy (southern Finland)*		*Stages in the development of the Baltic Sea*
1000				Mya Sea
			IX Spruce—pine	
0				Limnaea Sea
1000			VIII Birch–pine–spruce–elm, lime	*LVI* *LV* *LIV*
2000				Littorina Sea
				LIII
3000	postglacial	Flandrian		*LII*
4000			VII	
			Birch–pine–elm–oak–lime	
5000				
				Mastogloia Sea
			VI	
6000				
				A Ancylus Lake
			V Pine	
				E Echineis Sea
7000				
	Finiglacial		IV Birch	*YV* ↓ *YI* Yoldia Sea
8000				
	Salpausselkä III Salpausselkä II Salpausselkä I		III Younger Dryas (tundra)	*BVI* ↓ *BI* Baltic Ice Lake
	Gothiglacial	Late Weichselian	II Alleröd	Late-glacial Yoldia Sea
10,000				

The Mediterranean shore-lines. The Mediterranean is a classic area for the study of Pleistocene marine faunas and of raised shore-lines. Both lines of evidence have been used to work out sequences of marine stages and shore-lines, and as the same names have some-

Table 8.2 Shore-line and faunal stages in the Mediterranean and Black Sea

Faunal stages		Mediterranean Altimetric stages (*Depéret, Zeuner*)	Libya shore-lines (*fig. 8.11*)	Black Sea (*Fedorov*) shore-lines	Black Sea (*Fedorov*) Faunal stages
Versilian	F			2 m	New Black Sea
	LG	Epi-Monastirian 3 m		4–5 m	New Black Sea
		Late Monastirian 6–8 m	6 m	LG	
				12–14 m	Karangat
		Main Monastirian 18–20 m	15–25 m		
Tyrrhenian					
		Tyrrhenian 28–30 m			
			35–40 m	35–37 m	Uzunlar
			44–55 m	48–50 m	Uzunlar
				42–44 m	Old Euxinian
Milazzian		Milazzian 55–60 m		60–65 m	Old Euxinian
			70–90 m		
Sicilian		Sicilian 90–100 m		95–105 m	Chaudian
Emilian					
			140–200 m		
Calabrian					

F, Flandrian.LG, Last Glaciation

times been used to denote faunal stages and altimetric stages, a certain confusion has resulted. Moreover, because of the prevalence of tectonic activity around the Mediterranean, the correlation of shore-lines and their relation to eustatic changes of sea-level has proved difficult to support.

The shore-lines vary in height from a few metres to 325 metres above sea-level. The highest are believed to be late Tertiary or early Pleistocene. Generally the older the shore-line, the higher it is, but evidence for a younger shore-line being formed above an older shore-line has been described in the Lebanon and elsewhere. Many of the shore-lines are deformed, and even if only the undeformed shore-lines are taken, there seems little evidence for a grouping into the classic altimetric stages put forward by Depéret[4] and Zeuner[38, 39] (table 8.2), except for those at 60 m and 100 m, with shore-lines at around these heights found at both ends of the Mediterranean and on the Caucasus coast.[17] As a result, the Depéret–Zeuner nomenclature seems to have fallen largely out of use.

Table 8.2 also shows faunal stages identified from the Mediterranean. The Calabrian and Sicilian faunas are characterised by immigrants from the north Atlantic, the Emilian by a more temperate fauna. The Milazzian stage, originally identified as an altimetric unit, contains a modern Mediterranean fauna, while the Tyrrhenian contains warmer-indicating faunas with *Strombus bubonius*, now found no further north than the coast of Senegal. The final stage, the Versilian, corresponds to the transgression period following the last glaciation. The relation between these faunal stages and the altimetric stages is a matter for discussion. The Tyrrhenian fauna is found from below 3 m to higher shore-lines between 20 and 30 m, and there are isolated occurrences of higher shore-lines with faunas from the older stages. The difficulty is that the older faunas are either not very distinctive or sites are rather scarce.

On the Caucasus coast of the Black Sea a series of faunal stages has been worked out in some detail and associated with particular shore-lines.[10] The sequence and its relation to that of the Mediterranean is shown in table 8.2. Some of the levels are common to both areas, but the Black Sea shows two low postglacial shore-lines. The type of evidence for shore-line history most acceptable is from shore-line diagrams. Few of these have been produced in the Mediterranean. Figure 8.11 shows such a diagram for Cyrenaica. It shows some areas of warping, but with the lower shore-lines there is little warping.

For successful correlation, and subsequent eustatic interpretation, there must be much better chronological data. Some has started to be forthcoming in the form of Thorium–Uranium dates of mollusc

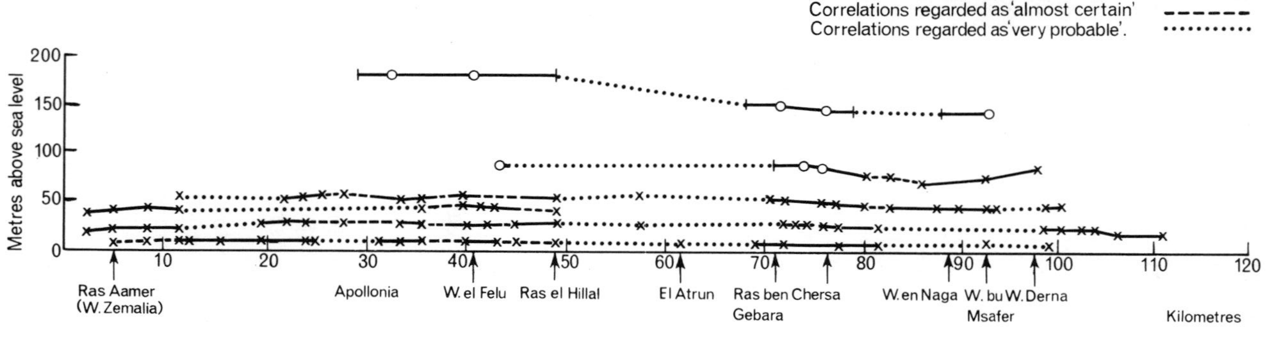

Fig. 8.11 Raised shore-lines between Ras Aamer and Derna, Libya (McBurney and Hey, 1955).

shells, indicating ages for the *Strombus* fauna 80,000–82,000, 115,000–120,000 and 160,000–170,000.[32]

There is much doubt about the age of the levels in terms of the glacial/interglacial sequence. The Milazzian has been said to be Günz/Mindel interglacial, the Tyrrhenian Mindel/Riss interglacial, and the Monastirian I and II Riss/Würm. As might be expected the most widespread shore-line is also one of the most recent. It is the 5–8 m shore-line (including the so-called Monastirian II). Shore-lines at this level have been observed in southern France, North and South Africa,[21] southern England, and eastern North America (the Suffolk Scarp). In southern England there are associated interglacial deposits of Ipswichian (Eemian)age.[36] But the dating of the other shore-lines is far from evident.

REFERENCES

GUILCHER, A. 1969. 'Pleistocene and Holocene sea level changes', *Earth-Sci. Rev.*, **5**, 69–97.

STEPHENS, N. and SYNGE, F. M. 1966. 'Pleistocene Shorelines'. In *Essays in Geomorphology*, ed. G. H. Dury. London: Heinemann.

1 BLOOM, A. L. 1967. 'Pleistocene shorelines: a new test of isostasy', *Geol. Soc. America Bull.*, **78**, 1477–94.

2 BLOOM, A. L. 1971. 'Glacial-eustatic and isostatic controls of sea-level since the Last Glaciation', 355–79, in *The Late Cenozoic Glacial Ages*, ed. K. K. Turekian. New Haven: Yale University Press.

2a BROECKER, W. S. and BENDER, N. L. 1972. 'Age determinations on marine strandlines'. In *Calibration of Hominoid evolution*, ed. W. W. Bishop and J. A. Miller, 19–36. Wenner-Gren Foundation: Scottish Academic Press.

2b CLARKE, R, H. 1970. Quaternary sediments off south-east Devon. *Q. 5e. geol. Soc. Lond.*, **125**, 277–318.

3 DALY, R. A. 1938. *Architecture of the Earth*. New York: D. Appleton-Century.

4 DEPÉRET, C. 1918. 'Essai de coordination chronologique générale des temps Quaternaires', *C.R. Acad. Sc. Paris*, **167**, 418–22.

5 DONNER, J. J. 1965a. 'Shore-line diagrams in Finnish Quaternary research', *Baltica*, **2**, 11–20.

6 DONNER, J. J. 1965b. 'The Quaternary of Finland'. In *The Quaternary*, vol. **1**, ed. K. Rankama. New York: Interscience.

7 DONNER, J. J. 1969. 'A profile across Fennoscandia of Late Weichselian and Flandrian shore-lines', *Soc. Scient. Fennica, Comm. Physico-Math.*, **36**, 1–23.

8 DONNER, J. J. 1970. 'Deformed Late Weichselian and Flandrian shore-lines in south-eastern Fennoscandia', *Soc. Scient. Fennica, Comm. Physico-Math.*, **40**, 191–8.

9 DONNER, J. J. and WEST, R. G. 1956 'The Quaternary geology of Brageneset, Nordaustlandet, Spitsbergen', *Norsk Polarinstitutt Skrifter*, No. 109.

10 FEDOROV, P. V. 1969. 'The marine terraces of the Black Sea coast of the Caucasus and the problem of the most recent vertical movements', *Dokl. Acad. Nauk USSR*, **144,** 661–3 (in Russian).

11 FLEMMING, N. C. and ROBERTS, D. G. 1973. 'Tectono-eustatic changes in sea level and sea floor spreading', *Nature (Lond.)*, **243,** 19–22.

12 GILL, E. D. 1967. 'Criteria for the description of Quaternary shorelines', *Quaternaria*, **9,** 237–43.

13 GILL, E. D. 1972. 'The relation of present shore platforms to past sea levels', *Boreas*, **1,** 1–25.

14 GODWIN, H. 1945. 'Coastal peat beds of the North Sea region, as indices of land- and sea-level changes', *New Phytol.* **44,** 29–69.

15 GODWIN, H. 1960. 'Radiocarbon dating and Quaternary history in Britain', *Proc. Roy. Soc. London*, **B 153,** 287–320.

16 GUTENBERG, B. 1941. 'Changes in sea level, postglacial uplift, and mobility of the earth's interior', *Bull. Geol. Soc. America*, **52,** 721–72.

17 HEY, R. W. 1971. 'Quaternary shorelines of the Mediterranean and Black Seas', *Quaternaria*, **15,** 273–84.

18 JELGERSMA, S. 1961. 'Holocene sea level changes in the Netherlands', *Med. Geol. Stichting*, Serie C-VI, No. 7.

19 JELGERSMA, S. 1967. 'Sea-level changes during the last 10,000 years', *Royal Meteroological Society Proceedings of the International Symposium on World Climate from 8000 to 0 B.C.*, 54–71.

20 JOHNSON, D. W. 1919. 'Shore processes and shoreline development'. New York: Wiley.

21 KRIGE, A. V. 1927. 'An examination of the Tertiary and Quaternary changes of sea-level in South Africa, with special stress on the evidence in favour of a recent world-wide sinking of ocean level', *Ann. Univ. Stellenbosch*, **5,** 1–81.

22 KUKKAMÄKI, T. J. 1968. 'Report on the work of the Fennoscandian Subcommission', *Third Symposium of the C.R.C.M.*, Leningrad, 1968. 4 pp.

23 MCBURNEY, C. B. M. and HEY, R. W. 1955. *Prehistory and Pleistocene geology in Cyrenaican Libya.* Cambridge University Press.

24 NISKANEN, E. 1939. 'On the upheaval of land in Fennoscandia', *Ann. Acad. Sc. Fennicae*, Ser. A, **53,** No. 10.

25 OLSSON, I. and BLAKE, W. 1961. 'Problems of radiocarbon dating of raised beaches based on experience in Spitsbergen', *Norsk Geografisk Tidsskrift*, **18,** 1–18.

26 ROSSITER, J. R. 1972. 'Sea-level observations and their secular variation', *Phil. Trans. R. Soc. Lond.*, A, **272,** 131–9.

27 SAURAMO, M. 1955. 'Land uplift with hinge-lines in Fennoscandia', *Ann. Acad. Sc. Fennicae*, Ser. A III, No. 44.

28 SAURAMO, M. 1958. 'Die geschichte der Ostsee', *Ann. Acad. Sc. Fennicae*, Ser. A III, No. 51.

29 SEGOTA, T. 1973. 'Radiocarbon measurements and the Holocene and Late Würm sealevel rise', *Eiszeitalter u. Gegenwart*, **23/24,** 107-15.

29a SHEPARD, F. P. 1963. Thirty-five thousand years of sea level. In *Essays in marine geology in honor of K. O. Emery*, pp. 1–10. Los Angeles: University of Southern California Press.

30 SISSONS, J. B. 1972. 'Dislocation and non-uniform uplift of raised shorelines in the western part of the Forth Valley', *Trans. Inst. British Geographers*, **55,** 145–59.

31 SISSONS, J. B., SMITH, D. E. and CULLINGFORD, R. A. 1966. 'Late-glacial and Post-glacial shorelines in south-east Scotland', *Trans. Inst. British Geographers*, **39,** 9–18.

32 STEARNS, C. E. 1967. 'Th^{230}–U^{234} dates of late Pleistocene marine fossils from the Mediterranean and Moroccan littorals', *Progr. Oceanogr.*, **4,** 293–305.

33 VALENTIN, H. 1954. 'Gegenwärtige Niveauänderungen im Nordseeraum', *Petermann's Geographische Mitteilungen*, **98,** Jahrgang (1954), 103–8.

34 VAN STRAATEN, L. M. J. U. 1961. 'Sedimentation in tidal flat areas', *J. Alberta Soc. Petroleum Geologists*, **9,** 203–26.

35 WEST, R. G. 1972. 'Relative land/sea-level changes in southeastern England during the Pleistocene', *Phil. Trans. R. Soc. Lond.*, A, **272,** 87–98.

36 WEST, R. G. and SPARKS, B. W. 1960. 'Coastal interglacial deposits of the English Channel', *Phil. Trans. R. Soc. London*, B, **243,** 95–133.

37 WRIGHT, W. B. 1937. *The Quaternary Ice Age*, 2nd edn. London: Macmillan.

38 ZEUNER, F. E. 1955. 'The three Monastirian sea-levels', *Aetes du IV Congrès International du Quaternaire (1953)*, **4,** 1–7.

39 ZEUNER, F. E. 1959. *The Pleistocene period*, 2nd edn. London: Hutchinson.

CHAPTER 9

CHRONOLOGY AND DATING

CHRONOLOGY concerns the dating of events in geological time and aims at establishing a scale of time to which geological events can be related. The dating of these events varies in accuracy from a few tens of years to a few thousands or more, according to the methods used, so that the building-up of a chronology series is no easy matter. In the determination of the passage of time, measurements of geochemical changes associated with the passage of time are of prime importance. They have largely replaced older methods involving estimates of the passage of time based on the rates of geological processes, such as deposition, weathering and erosion.

The dating of events to years has been called absolute or chronometric dating, in contrast to so-called relative dating, which places one event in time relative to another, but with no reference to absolute time.

Chronology provides a time-scale of events of known position in a stratigraphical succession, so that chronology and stratigraphy combine to give a history of geological events. The chronology of the Pleistocene is far from being known in any detail, except in the most recent parts, but nevertheless more is known than that of earlier geological times. We shall be concerned with relative dating and some methods available for widespread correlation of events, estimates of the passage of time based on geological processes, and measurements based on natural rhythmic processes and on geochemical change. Finally an outline chronology for the Pleistocene is constructed.

DETERMINATIONS OF RELATIVE AGE

These determinations are mostly based on stratigraphical relations. Oakley (1969) has devised a system for making the basis of such determinations clear. He has distinguished four categories of relative age determination, especially in connection with the dating of fossil man: R1, when the object or event is related to the deposit containing

Plate 1. Oblique aerial photograph (20,000 ft.) of glaciers in north-east Baffin Island (69° 25′ N, 68° 52′ W).

There are small ice caps in the middle distance, left and right, with a piedmont glacier in the centre middle distance draining the ice cap to the left, and some valley glaciers. End moraines girdle the termini of the glaciers and are pierced by melt-water streams. Medial moraines are seen on the glacier in the right foreground. The melt-water drainage is north, then east in the middle distance. There is a small corrie glacier in the centre near distance. Note the different degrees of end-moraine development at the different glacier termini. *Copyright, Mines and Technical Surveys, Canadian Government. Photo T 219 R (101), July 1948.*

Plate 2. Oblique air photo of the margin of the Vestfonna ice sheet in Nordaustlandet, Spitsbergen, near the North Franklin Glacier. The end-moraine is well developed here, and is associated with shear structures in the ice, shown as slight lines parallel to and close to the edge. A braided stream system is seen in the foreground, and the surface drainage channels of the ice sheet can also be seen. The topography of the inland ice can be clearly seen in the middle and far distance. *Copyright, Norsk Polarinstitutt, Oslo. Photo Le 77a, 691.*

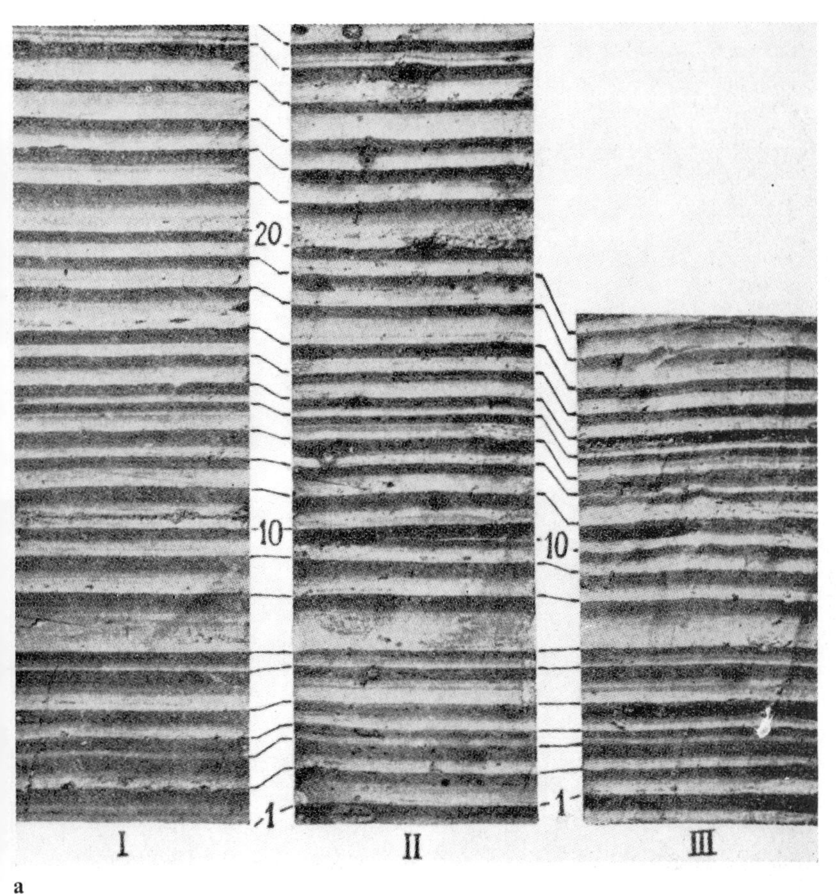

a

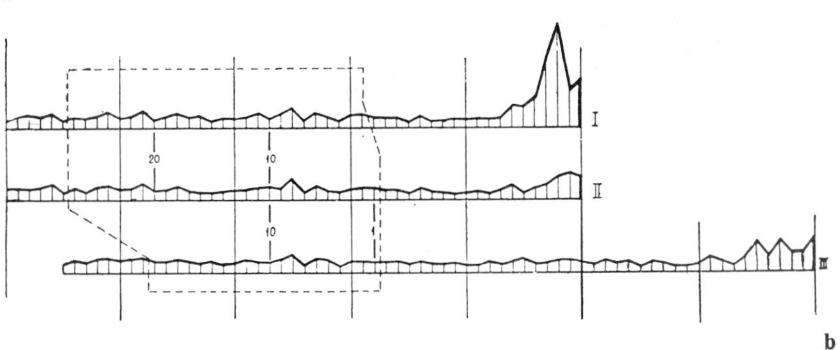

b

Plate 3. Finnish varves and their connection (from Sauramo 1929).
a. Three samples of the varved Late-Glacial clay in the valley of the Kalajoki River in Ostrobothnia, correlated with each other. Sample I from a locality lying 6 kilometres, and sample II 2 kilometres northwest, and sample III 8 kilometres southeast of the church of Haapajärvi.
b. Diagram of the connected clay samples named above showing the variations of the thickness of the varves. The bottom of the sections to the right. The dotted line bounds that part of the samples shown above in **a.**

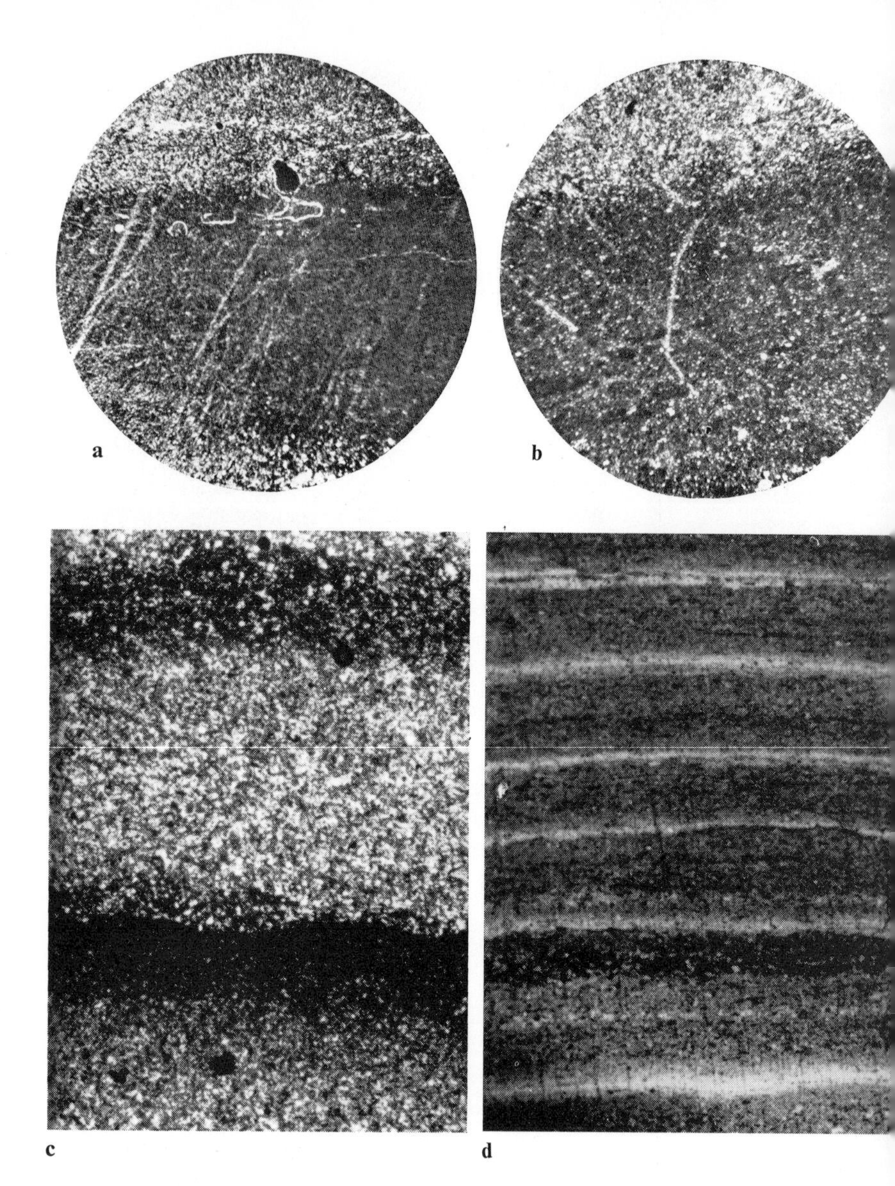

Plate 4.
a. Thin section of a diatactic varve from Leppäkoski, Finland, ×12 (from Sauramo 1923).
b. Thin section of a symmict varve from Mattila, Finland, ×12 (from Sauramo 1923).
c. Thin section of early glacial (Gippingian) varved clay at Marks Tey (Essex). ×31.
The lamination is caused by strongly graded bedding of inorganic sediment. The lower part of each lamina contains abundant angular quartz grains, and the upper part clay minerals. *Photo by C. Turner.*
d. Thin section of laminations in interglacial lacustrine clay-mud of Hoxnian age (zone II). ×8. The light layers are very rich in diatoms (principally *Stephanodiscus astraea*) packed in a calcareous matric. This layer grades upwards into a darker more organic sediment. The pairing may be annual, the light layer resulting from a spring flush of the diatom *Stephanodiscus*. *Photo by C. Turner.*

Plate 5.
a. Fresh morainic landforms and kettle holes, part of the Southern Irish end-moraine (Weichselian maximum), near Blackwater, Co. Wexford, Eire.
b. Drumlins submerged by the Flandrian rise in sea level. Clew Bay, Co. Mayo, Eire.
Both photographs by J. K. St. Joseph. Cambridge University Collection.

a

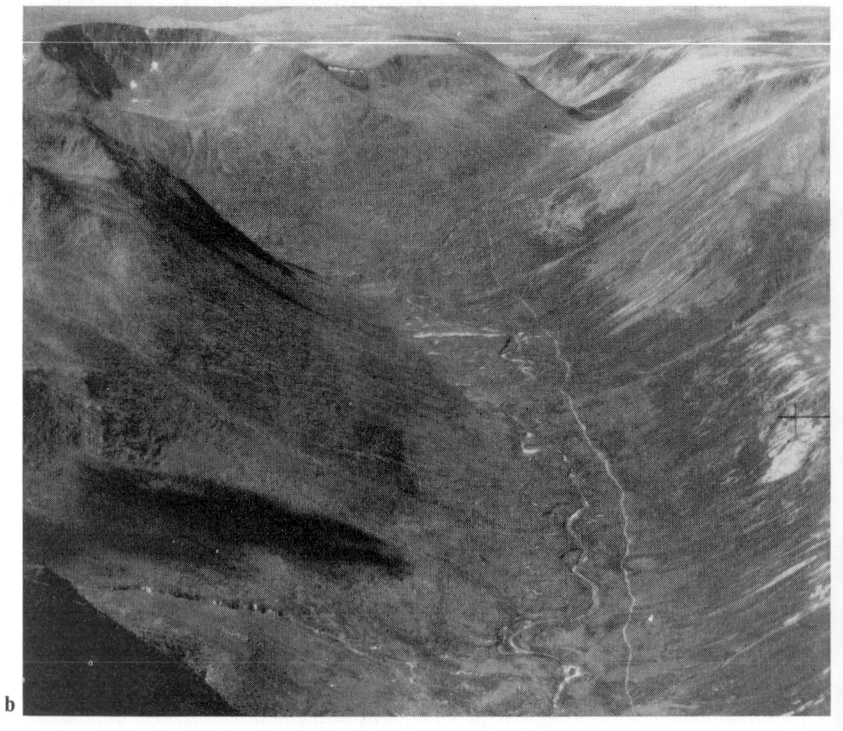
b

Plate 6.
a. Esker and kames, near Clonmacnoise, Co. Offaly, Eire.
b. U-shaped valley and corries. Glen Dee, Aberdeen, looking north to the Pools of Dee. *Both photographs by J. K. St. Joseph. Cambridge University Collection.*

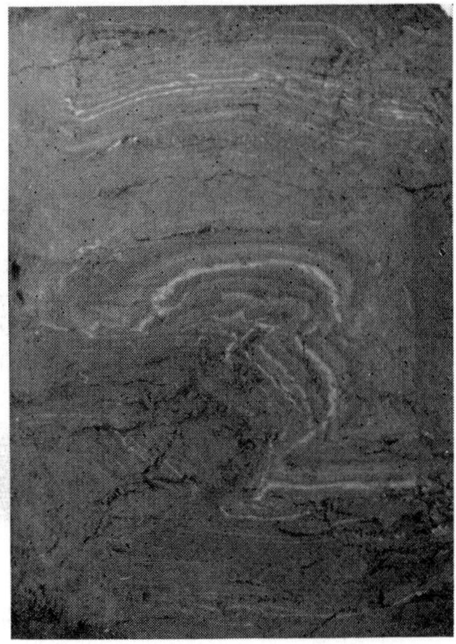

a

b

Plate 7.

a. 4″ diameter core through reworked and brecciated lake deposits at Hoxne, Suffolk, $\times\frac{1}{2}$. The mottled sediments are largely clay-mud fragments reworked from marginal deposits and removed to the centre of the basin by solifluction; they also contain pebbles of chalk from the surrounding till on which the lake deposits lie. Pale bedded lacustrine silts were deposited at periods of higher water level.

b. Slump structures in laminated interglacial lacustrine clay-mud of Hoxnian age at Marks Tey, Essex. $\times\frac{1}{2}$. *Photo by C. Turner.*

Plate 8.

a Section in a peatdigging at Westhay (Godwin's Piece), Somerset Moors (photo by H. Godwin). Changes in peat composition indicate two horizons of flooding (layers 2 and 5) within a sequence of raised bog peats. Depth of section 1 metre. Section as follows:

1. Highly humified *Sphagnum-Calluna* peat.
2. Aquatic *Sphagnum* peat, with some *Eriophorum vaginatum* (precursor peat). This represents a flooding horizon in the peat sequence and its formation precedes the renewed growth of fresher *Sphagnum* peat.
3. Fresh *Sphagnum* peat. Renewed growth of the bog surface ('regeneration-complex' peat).
4. Humified *Sphagnum-Calluna* peat, with *Eriophorum vaginatum.*
5. Laminated peat with aquatic *Sphagna* and *Scheuchzeria* rhizomes, the latter prominent in the peat face. A second flooding horizon.
6. Mouldered humified *Sphagnum* peat.

The peat blocks cut from the face (from layer 3) show the banding characteristic of 'regeneration-complex' peat, resulting from alternate layers of darker more humified *Sphagnum-Calluna* peat with much fresher *Sphagnum* peat. These variations in peat type are associated with the pool and hummock complex of regenerating bog surfaces.

b. Interglacial raised beach gravel at Selsey, Sussex. This beach gravel is of the same age (Ipswichian) as that under the head at Brighton (Plate 13b). It is overlain by brickearth of aeolian origin, probably formed during the Weichselian.

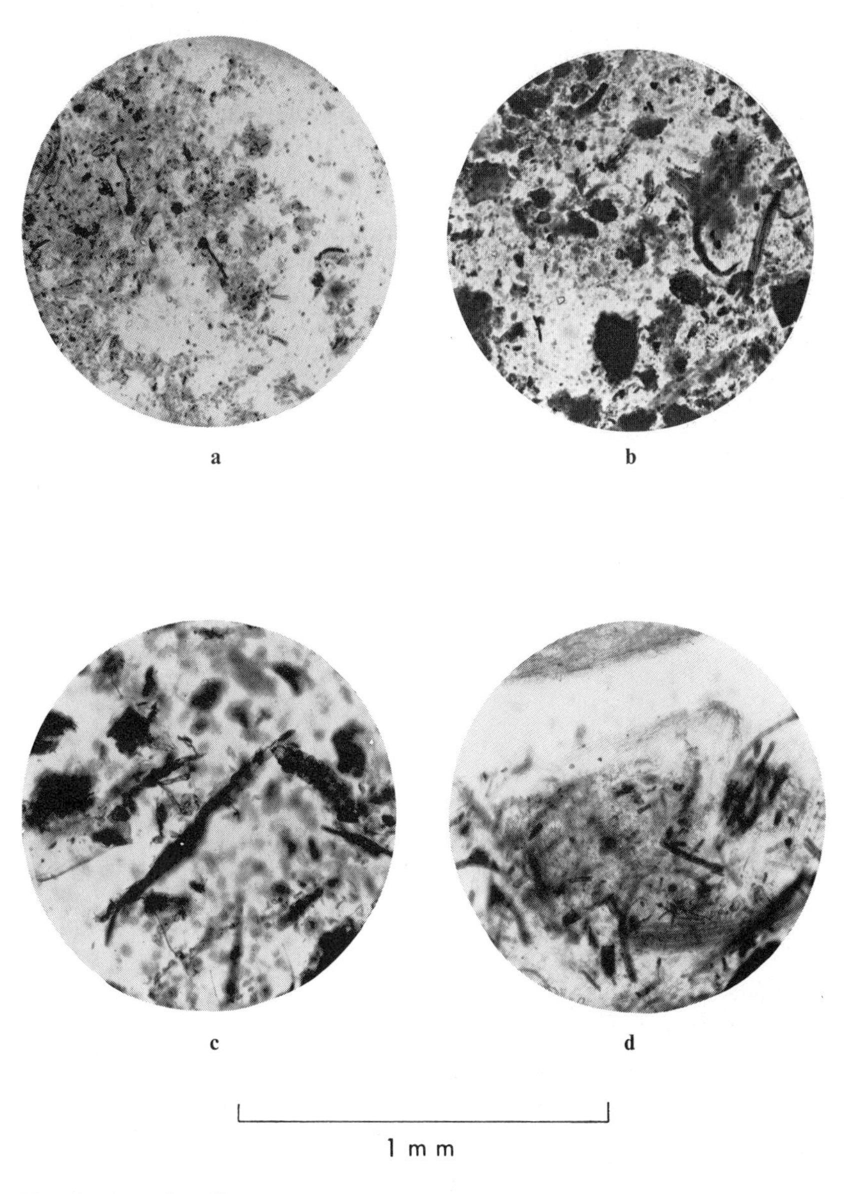

Plate 9. Organic sediments.
a. Fine detritus gyttja.
b. Coarse detritus gyttja.
c. Parvocaricetum peat. The radicells (cells with thickened or protruding walls) characteristic of Cyperaceae roots are seen near the base of the photo.
d. *Sphagnum* peat. The peat is unhumified and shows the characteristic structure of the leaves of *Sphagnum*.

a

b

Plate 10.

a. Non-sorted networks and stripes manifested as vegetation patterns. Brettenham Heath, Thetford, Norfolk. The networks occur on level ground, the stripes on slopes. Age Weichselian.

b. Network of tundra polygon type on surface of a Weichselian terrace, near Colne, Hunts.

Both photographs by J. K. St. Joseph. Cambridge University Collection.

Plate 11. Macroscopic plant remains.
a. *Saxifraga oppositifolia*, leafy shoot.
b. *Onobrychis vicifolia*, pod.
c. *Armeria maritima*, calyx.
d. *Naias marina*, fruit.
e. *Salix polaris*, leaf.
f. *Helianthemum canum*, capsule.
g. *Scabiosa columbaria*, fruit.
h. *Naias flexilis*, fruit.
i. *Naias minor*, fruit.
a, b, c, e, f, g, from Weichselian deposits at Earith, Hunts. (*Photos by F. G. Bell*). *All to same scale.*
d, h. i, from Ipswichian deposits at Bobbitshole, Ipswich.

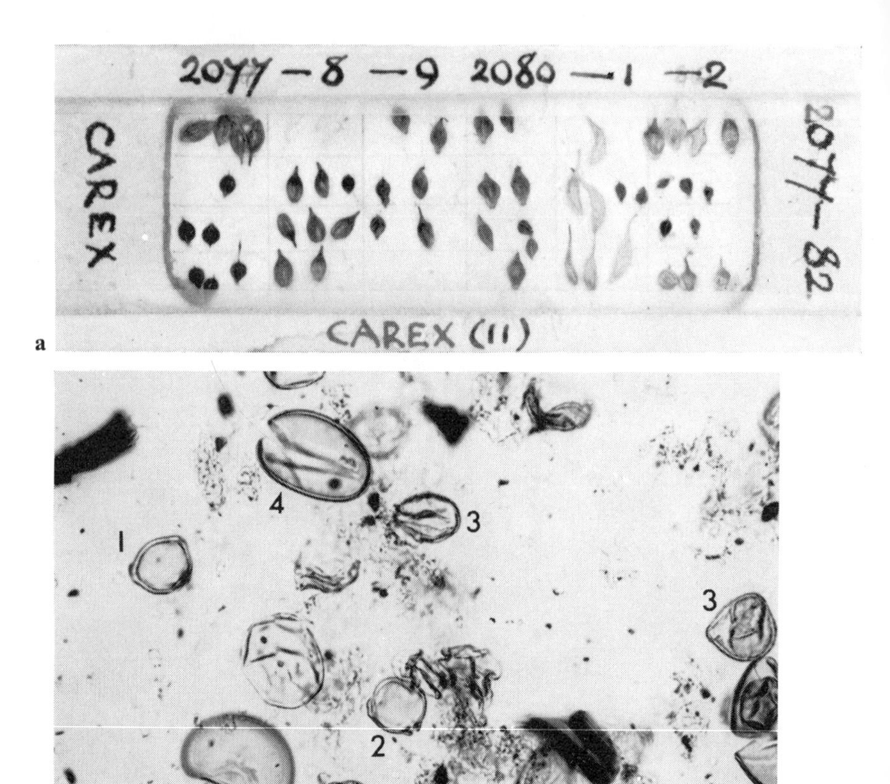

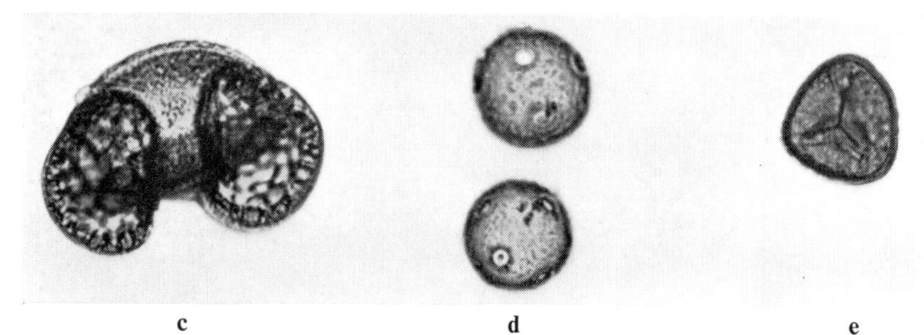

Plate 12. Plant remains.

a. Slide of reference material, fruits and nuts of *Carex* spp. Each column contains examples of a single species, identified by a number (London Catalogue number).

b. A pollen preparation from a late Flandrian peat in Scotland. ×260. 1, 3-pored grain of *Corylus*; 2, 3-pored (septate pores) grain of *Betula*; 3, 3-furrowed grains of *Quercus*; 4, *Filicales* spores.

c. Pollen grain of *Pinus sylvestris*. ×480. The central body and two wings is characteristic of many species of conifers.

d. Pollen grains of *Plantago lanceolata*. ×480. The grains have a verrucate surface and bordered pores scattered on the surface (periporate). This pollen grain is frequently associated with land-clearance phases of the late Flandrian.

e. Spore of *Sphagnum recurvum*, with the trilete scar often found in lower plants. ×400.

a

b

Plate 13.

a. Ancient rock-platform at Porthleven, Cornwall. The present sea-level closely approximates to that of the time when the platform was cut, and the sea is now re-excavating the notch at the base of the cliff, partly filled with ancient beach material. A large erratic of gneiss is in the right middle distance, possibly dropped by drifting ice. The age of the platform is probably pre-Hoxnian.

b. Black Rock, Brighton. A sea cliff of chalk formed during the Ipswichian interglacial is in the centre of the photo. Solifluction during the Weichselian has led to the formation of the bedded solifluction deposits (head) now resting on the face of the cliff and on the interglacial beach, present at the base of the section on the left.

a

b

Plate 14.

a. Terrace sands and gravels of Weichselian age at Earith, Hunts. The cast of a large ice-wedge is seen in the centre. The surface expression of similar casts on the same terrace in a nearby locality is seen in Plate 10b.

b. Coarse gravels and sand of the crevasse filling at Blakeney, Norfolk. The gravels are largely unsorted and abrupt changes of lithology occur. A period of frost-heaving has resulted in the upright arrangement of stones near the top on the left hand side of the photo.

Plate 15.

a. Cliffs at West Runton, Norfolk. Peat of the Cromerian temperate stage outcrops at the base of the cliff, itself composed of glacial deposits of the earliest glaciation of East Anglia.

b. Cliff section at Beeston, Norfolk. Glacial deposits, mainly till, contorted by ice-push and containing lenses of ice-contact stratified drift, mainly sand.

Plate 16. Fossil non-marine molluscs and insects.
a. *Columella columella* (Benz). ×15. Gippingian, from tjaele gravel at Thriplow, Cambs. Photo by B. W. Sparks (*Geol. Mag.*, **94** (1957), Plate VII).
b. *Belgrandia marginata* (Michaud). ×24. Ipswichian, from organic deposits at Bobbitshole, Ipswich. Photo by B. W. Sparks (*Phil. Trans. R. Soc.*, B, **241** (1957), Plate 3).
c. *Potamida littoralis* (Cuvier). ×1. Ipswichian, from terrace gravels at Barnwell, Cambs. Photo by B. W. Sparks.
d. *Stenus brunnipes* Steph. (Staphylinidae). ×110. Aedeagus, Hoxnian, from interglacial deposits at Nechells, Birmingham. Photo by Professor F. W. Shotton and P. J. Osborne (*Phil. Trans. R. Soc.*, B, **248** (1965), Plate 30).
e. *Hippodamia arctica* Schneid. (Coccinellidae). ×16. Left elytron showing characteristic pattern. Weichselian (Upton Warren Interstadial Complex), from terrace deposits of the Tame Valley, Warwicks. A non-British northern species of Fennoscandia. Photo by G. R. Coope and C. H. S. Sands (*Proc. R. Soc.*, B, **165**, Plate 50).
f. *Chiloxanthus stellatus* Curtis. (Saldidae). ×17. Hemielytron, showing characteristically patterned wing. Weichselian (Upton Warren Interstadial Complex), from terrace deposits of the Tame Valley, Warwicks. A non-British species of bug, with a northern circumpolar distribution. Photo by G. R. Coope and C. H. S. Sands (*Proc. R. Soc.* B, **165,** Plate 50).

it; R2, when the containing deposit is referred to the local geological succession; R3 when the latter is correlated to a wider scheme of stratigraphy; and R4, when the relative age determination is based on form or morphology, e.g. of a bone. The category R1 has been divided into a number of classes intended to indicate whether the fossil is contemporaneous with the containing deposit, or is approximately contemporaneous, or is intrusive and therefore younger than the deposit. This system has the merit of indicating briefly the circumstances of a fossil find and the nature of its relative age determination.

The fluorine method[40] is useful for the relative dating of animal skeletal remains found in sand and gravel. Bones and teeth contain a proportion of hydroxy-apatite (calcium phosphate), which traps fluorine in groundwater and forms stable fluorapatite, the quantity of which can be measured by X-ray diffraction methods. The fluorine to phosphate ratio in bones and teeth is in part a measure of the time the fossils have been able to trap fluorine, so that bones in the same deposit which have suffered the same groundwater conditions should show similarities in fluorine content. Thus a mixture of bones in the same deposit but of different age may be revealed by measurement of the fluorine–phosphate ratio. The fluorine content of bone and antler is greater than that of teeth of the same age found in the same deposit, so measurements are best made on the same type of skeletal material.

Other methods, such as nitrogen and uranium analyses[40] have also been used to determine relative ages of bones and teeth, e.g. in the study of the Piltdown fossils which resulted in the exposure of the hoax.

DATING BY FAUNA AND FLORA

Changes in the composition of faunas and floras have been used as a means of relative dating in the Pleistocene as with the earlier geological epochs. But the effects of the rapid climatic changes on fauna and flora tend to mask the evolutionary changes important for the stratigraphy of older deposits, and the shortness of Pleistocene time, as well as the greater detail of chronology required, tend to make the criteria used for the older rocks inapplicable. However, evolutionary changes have been used as a basis for chronology in micropalaeontological studies of deep-sea sediments.[16]

Because of the widespread occurrence of pollen in Pleistocene sediments, pollen analysis became an important method of relative dating. The development of pollen analysis of Flandrian peats and other sediments in northwest Europe led to the subdivision of Flandrian time into a number of pollen-analytical zones, each characterised by a particular assemblage of forest trees (table 12.10). This succession has been related to the climatic succession of Blytt-Sernander in Scandinavia (see ch. 10) and has become a basis for relative dating on the assumption that factors (e.g. climatic change) determining pollen zone succession in neighbouring areas have common effects over wide areas, and therefore pollen zones can be considered synchronous to a certain extent. This method of relative dating has been much used in northwest Europe, in relation to vegetational history, climatic episodes and archaeology. But the basis for subdivision of Flandrian time has now passed to radiocarbon dating, though in many instances pollen analysis is still a useful method of relative dating within restricted areas. With deposits beyond the range of radiocarbon dating, for example interglacial or interstadial deposits, pollen analysis is still of prime importance in revealing the vegetational histories characteristic of each interglacial, and in this field it remains an important method of relative dating.

TEPHROCHRONOLOGY[63]

This is a method of establishing a chronology based on ash falls. A volcanic eruption produces a widespread deposition of ash, and if such a layer becomes preserved in a sedimentary sequence (continental or marine), its wide occurrence will permit correlations. Thus sequences of ash layers in Iceland, Patagonia and elsewhere have been extensively used as marker horizons. Some ash falls have been dated, such as the Laacher See ash of the Eifel, found in Allerød (zone II, Late Weichselian) sediments in southwest Germany, or the Mazama ash from Crater Lake in Oregon, dated to about 6600 years ago, which covered much of northwestern United States and adjacent Canada. A 9300-year-old horizon of disseminated volcanic ash in North Atlantic sediments between 45°N and 65°N has provided a reference level for the study of the warming of ocean surface waters.[48] An important older ash fall is the Pearlette ash of late Kansan age in the United States which spread over large parts of the Great Plains.

PALAEOMAGNETISM CHRONOLOGY[10]

Parameters of the earth's magnetic field, such as polarity, declination and inclination, change in time. Many rocks, including both igneous and sedimentary rocks, become magnetised by the earth's field at the time they are formed or shortly after, and this natural remanent magnetism in the rock may be analysed. There are several possible sources of error in the method, including self-reversal during initial magnetisation, slow magnetisation after initial magnetisation, and magnetisation changes imprinted during chemical alteration, such as haematite formation during weathering. However, with care, the results of such palaeomagnetic analyses can be used to produce a sequence of the palaeomagnetic properties of a rock succession, and this can be dated by radiometric methods; e.g. radiocarbon dating of lacustrine sediments and potassium/argon dating of volcanic rocks. The major changes of the earth's magnetic field are those of polarity. There are epochs of normal and reversed polarity, which may themselves contain short events of reversed or normal polarity respectively. The palaeomagnetic polarity chronology of the Pleistocene as so far analysed and dated by the radiocarbon and potassium–argon methods is shown in fig. 9.1. This figure shows correlations based on measurements from different continents. Several measurements

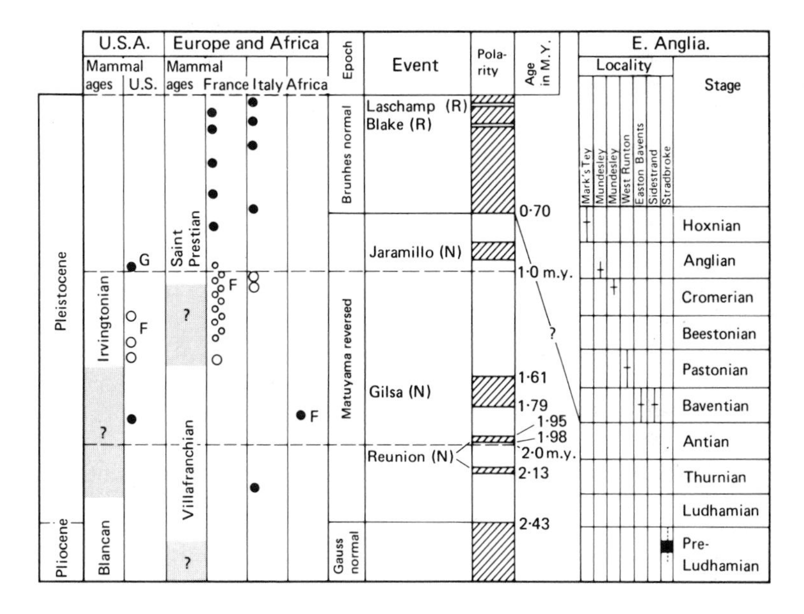

Fig. 9.1 Correlations between the western United States, north-west and southern Europe, and Africa, based on palaeomagnetic and radiometric data.

Closed circles, normal polarity; open circles, reversed polarity; large circles, radiometric dates; small circles, no radiometric dates; G, glacial deposits; F, fossils. Stippled areas indicate range of uncertainty in mammalian stage boundaries. In the diagram on the right, the crosses indicate normal polarity of samples from pits, the black square from a bore hole sample (after Cox, Doell and Dalrymple, 1965; Cox, 1969; Cooke, 1973; Montfrans, 1971; Noel and Tarling, 1975).

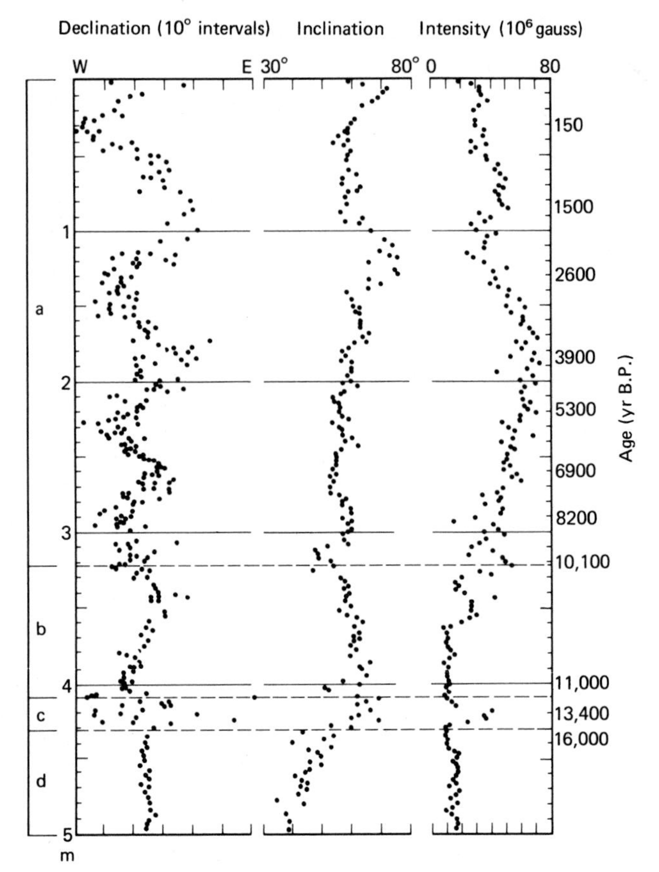

Fig. 9.2 Secular variation of palaeomagnetic properties of sediment from four 5 m cores from Lake Windermere (Thompson, 1973). Declination values are relative rather than absolute. a, Flandrian organic sediments; b, Late Devensian laminated clays; c, amelioration period; d, glacial varved clay.

from East Anglian sediments have been made by Montfrans[37] and a tentative correlation based on these is also shown in this figure.

It can be expected that more detail will be added to this type of chronology as research proceeds. Obviously a chronology of this kind is extremely important for the Pleistocene. It provides a world-wide framework for dating, dependent for its degree of resolution on potassium/argon dating.

On a finer scale, variation in declination, inclination and intensity of remanent magnetism have been analysed in lacustrine sediments of various age.[13, 59a] In Flandrian sediments the curves so obtained can be dated by the radiocarbon method and the resulting chronology of magnetic variation can be used as a basis for correlation, in addition to its intrinsic value for geomagnetic studies of the earth. Figure 9.2 shows results of this kind from sediments of Lake Windermere.[59] Analyses have also been made from lacustrine sediments older than the Flandrian. At Hoxne, Suffolk, type site for the Hoxnian temperate stage, the variations found are similar in scale to those of Windermere.[60] Further work should demonstrate the use of such analyses as these for wide correlation.

A further interest of the palaeomagnetic chronologies lies in the possibility that variation of the earth's magnetic intensity show correlation with fluctuations in atmospheric radiocarbon activity and climatic change.[65] If so, the linking of these distinct expressions of environmental history will be clearly of fundamental importance.

GEOLOGICAL PROCESSES INVOLVING PASSAGE OF TIME

Estimates of time based on these processes had been made in several areas before the advent of more rigorous methods of dating. We may take some examples to illustrate the methods involved.

Penck[44] made estimations of the duration of interglacial stages in the Alps. Comparing the intensity of the weathering of the Deckenschotter of Mindel age around Münich with that of the Riss gravels, he concluded that the interglacial between the two was four times as long as during the Riss–Würm interglacial. On a similar basis, he estimated that the latter was three times as long as the post-Würm interval. From the observation that the two Deckenschotters (Günz and Mindel) bear the same relation to one another as the Riss gravels do to the Würm gravels, the duration of the Günz–Mindel interglacial was thought to be similar to that of the Riss–Würm interglacial. The relative lengths of the interglacials thus computed are shown in table 9.1.

Table 9.1 Penck's estimates of interglacial time in the Alps

	Günz-Mindel	*Mindel-Riss*	*Riss-Würm*	*post-Würm*
Relative duration	3	12	3	1
Duration in years	60,000	240,000	60,000	20,000

Estimations of the length of the post-Würm interval, based in part on the growth of a delta into Lake Lucerne after the retreat of the Bühl-stadium towards the end of the Würm glacial stage, gave a time span of 20,000 years for this interval, thus giving the duration figures in table 9.1 for the interglacials. This was the basis for the length of the so-called Great Interglacial, the Mindel/Riss.

Estimates of the durations necessary for interglacial weathering of tills were made by Kay[32] in Iowa. He assumed that the time available for leaching of the Wisconsin glacial deposits was some 25,000 years and estimated that the average depth of leaching was 750 mm in this time. The depths of leaching he found on older glacial gravels were 3 m on Illinoian deposits, 9 m on Kansan deposits and 6 m on Nebraskan deposits. Assuming that the depth of leaching is proportional to time, he suggested that the lengths of the post-Illinoian (Sangamon) interglacial was 120,000 years, of the post-Kansan (Yarmouth) interglacial 300,000 years, and of the post-Nebraskan (Aftonian) interglacial 200,000 years.

Other estimates have been based on the rates of recession of waterfalls, and on accumulation of peat, travertine and ocean sediments. Prestwich's[46] estimate in 1888 of 15,000 to 25,000 years for the duration of the Glacial Epoch and 8000 to 10,000 years for the length of post-glacial time, was based on the dissection of the landscape in northwest Europe. The latter is very near what we now believe to be the truth, but the former, though near to the present estimates of the duration of the most recent extension of the ice sheets in northern Europe and North America, is very much less than the minimum now thought possible for the length of the glacial period. Prestwich's estimates were based on his own observations and on his belief that estimates of the same periods by others at that time were far too great.

In all these estimates based on deposition, weathering and erosion, it should be remembered that time is merely one factor of many that effect rates of the processes concerned. With regard to weathering, factors of climate, soil texture and drainage will affect rates of leaching, and in the question of sediment deposition, the stream gradients, load and surrounding vegetation will all affect the rate of deposition. For these reasons such estimates of time are very liable to error, and in fact, by comparison with more recent methods, have in most instances been proved unreliable. Nevertheless, they were made at a time when no other methods were available, and they at least correctly gave a rough order of magnitude to Pleistocene time.

ABSOLUTE OR CHRONOMETRIC DATING

The measurement of the passage of time is most reliably obtained by counting a sequence which results from the annual recurrence of the natural rhythmic processes of varve deposition and tree-ring formation, or by the measurement of geochemical changes as a result of radioactivity. We first consider the use of natural rhythmic processes.

Dendrochronology[53]

Seasonal changes in the wood growth of trees produce annual rings. In spring wood cell-elements with large lumens are produced, while in the summer and autumn the cells become smaller. Thus a year's growth starts with larger cells and ends with smaller cells and both types comprise an annual ring (fig. 9.3).

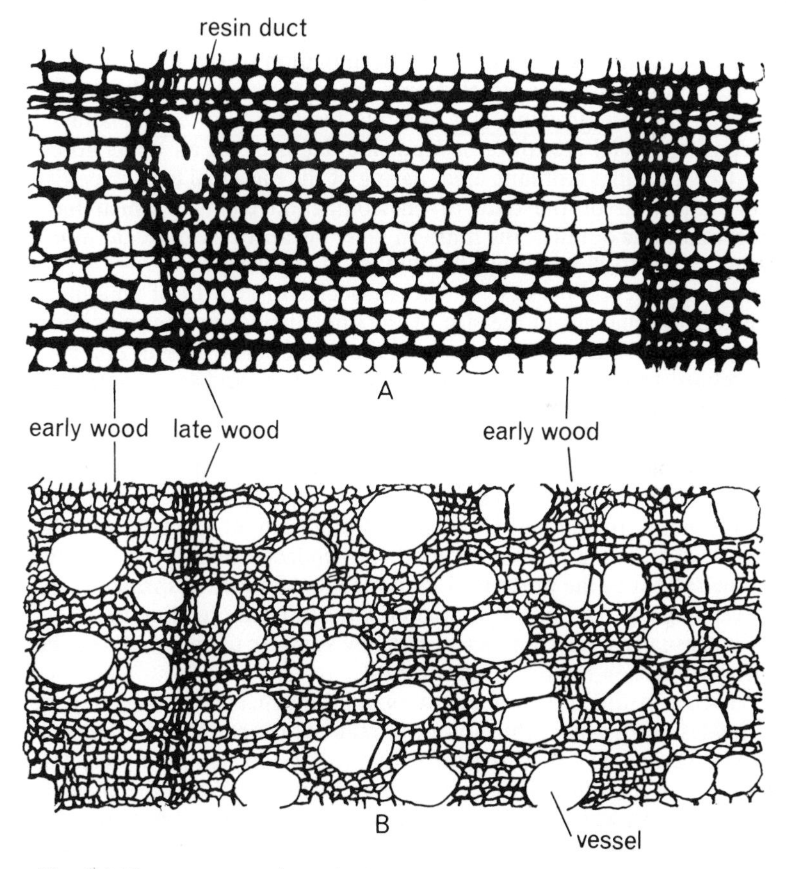

Fig. 9.3 Transverse sections (×35) through A, a conifer, *Pinus strobus*, and B, an angiosperm, *Salix nigra*, showing the cell arrangement in annual rings (after Esau).

Dendrochronology is the study of tree-rings for the purposes of dating. The time range covered has been extended back some seven thousand years in areas where trees of such ages survive. The method has been most widely used as a means of dating prehistoric settlements with buildings constructed partly of wood in the southwest of North America in the last millennium, but it has also been used to date medieval artefacts in Europe. In recent years dendrochronology has provided a scale of 7400 years (*Pinus aristata*, Britlecone Pine, in southwest North America) for the correlation of the radiocarbon time-scale to the secular passage of time, discussed later in the chapter.

Within a particular climatic province dendrochronology is a very accurate method of dating. The determination of the age of trees in forest stands gives means of dating the minimum age of the soil on which they are growing, for example the minimum age of alluvial deposits, glacial moraines or landslide deposits. At an upper forest or tree limit in a montane area, dating of the marginal trees may give an indication of the date of environmental changes which have allowed a rise or fall of the limit. In these instances the age determination results from counting the number of tree-rings.

Environmental factors control the degree of growth of an annual ring. Thus changes in the size of the annual rings are an indication of past environmental changes and their dates. The growth of a ring is related to the food manufactured by the tree, and so is controlled by the climatic factors which limit photosynthesis and thus food reserves. It is therefore necessary to know the relation between ring width of a particular tree species and environmental factors before embarking on an investigation of past climates involving dendrochronology. In semi-arid climates, droughts limit growth, and the width of tree-ring in conifers used for dendrochronology is thought to be related to the precipitation in the late-summer to spring period preceding the start of the annual growth. In northern climates water is not usually a limiting factor, but low summer temperatures in the growing season reduce the ring width.

The main difficulties attendant on building up a chronology of past climates using dendrochronology are the occasional presence of false rings, caused by early season droughts or low temperatures, the absence or discontinuity of a ring in unfavourable years, the changes in ring width which are a function of the age of a tree (growth function) and which have to be allowed for in the interpretation, and finally the fact that there may be a serial correlation between widths of successive rings independent of other environmental effects on ring width.

The trees which have been mostly used in dendrochronology are conifers, especially species of pine and spruce in the north and species of pine (*Pinus aristata*, *Pinus flexilis*) in semi-arid regions in North America. *Sequoia* was extensively used in early studies, but the rings have a high degree of serial correlation. In Europe *Quercus* and *Pinus* have been used for dendrochronological studies.

Methods of obtaining samples for tree-ring analysis have been described by Stokes and Smiley.[53] Either blocks of wood reaching to the centre near the base of a tree or samples taken with a tree auger are used. There are three stages in building up a chronology. First the selection of different sites where the tree-ring sequences or series of particular species appear sensitive to climatic change (sensitive series) and also where they show little variation apart from growth function (complacent series). Then a process of cross-dating is carried out to overcome the difficulties of false and absent rings; a large number of trees of a number of species from different sites are sampled and their tree-ring series worked out with a low-power binocular microscope. In particular, samples are taken from slow-growing trees at stress sites, as these are important for climatic indications, and from quicker growing trees on favourable sites where all rings should be present. Drought years with thin rings are the best indicators for correlation (fig. 9.4). Finally, the tree-ring series from different species and different sites are built up into an overall chronology showing periods of drought or low summer temperature back over as long a time as that covered by the oldest trees. Chronologies may

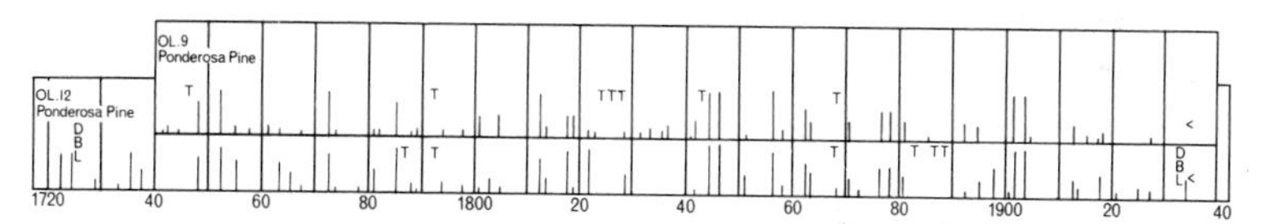

Fig. 9.4 Tree-ring plots from two trees of *Pinus ponderosa* at the forest border in northern Arizona, showing good cross-dating. T, thick ring; DBL, double ring. The longer the vertical lines the narrower is the growth layer, as judged by eye. Rings of normal thickness are not shown, but are counted (Glock, W. S. 1950. 'Tree growth and rainfall', *Smithsonian Misc. Coll.*, **III**, No. 18).

also be erected on samples from prehistoric dwellings or wooden artefacts and such a 'floating' chronology may be tied to an absolute chronology if it overlaps a series based on living trees in the same region.

As regards the distance within which correlations can effectively be made, in western North America, within a particular climatic province, significant correlations can generally be made between chronologies 500 km apart, and some significant correlations have appeared at distances of up to 1750 km. Thus the method can be used to date climatic changes over wide areas, and differences in climatic change in the different areas may reveal changes in past weather patterns. Table 9.2 shows changes during the sixteenth century A.D. in mean June–July temperature in three areas in northern latitudes and changes in precipitation (winter-spring and annual) in the western United States.

Table 9.2 Tree-ring averages in the sixteenth century A.D. expressed as a percentage of the mean for the period 1851–1950 (Fritts, 1965)

A Temperature-sensitive chronologies at the northern tree limit (June–July mean)	*1501–1550*	*1551–1600*	*Mean*	
Alaska (*Picea* spp.)	111	105	108	(≤ + 0·8°C.)
Mackenzie Valley (*Picea* spp.)	93	88	90	(≤ − 1·0°C.)
Northern Fennoscandia (*Pinus sylvestris*)	103	107	105	(≤ + 0·5°C.)

B Precipitation, drought-sensitive chronologies at the lower forest limit in the United States (a, winter-spring total; b, annual total)	*1501–1550*	*1551–1600*	*Mean*
Montana (*Pinus flexilis* v. *reflexa*)	94	86	90
a California (*Pinus aristata*)	97	95	96
b Arizona, Flagstaff area (*Pinus ponderosa*)	83	97	90

In Europe, tree-ring studies have had a much slower development, since the climatic and edaphic factors controlling tree-ring width are complex. However, some floating and some short-period dated chronologies have been obtained from Britain and Europe and current work may result in much more complete dendrochronological series.[19]

Varves and other laminated sediments

Laminated sediments or rhythmites, showing regular and repeated alternations of two sediment types, are commonly found in the Pleistocene. The sediments may be principally inorganic in origin, as in varved clays, or the alternation may be caused by organic-rich and organic-poor layers. The word varve was originally applied by de Geer[23] to describe 'periodically laminated sediments in which the deposition for every single year can be discriminated'. The sediments concerned were the late- and post-glacial varved clays of Sweden. The origin of these clays has already been described (Ch. 3) and the annual repetition of the pairs of sediment types, coarse and fine, is well established. Other rhythmites with an apparent annual origin are those described from Lake of the Clouds, Minnesota,[12] from the Faulenseemoos in Switzerland,[62] and from Marks Tey, Essex.[51] They show alternate light layers, the summer layer, rich in diatoms and calcium carbonate, and dark layers with fewer diatoms and rich in organic matter, the winter layer (plate 4d). The time represented by a column of the Swiss Flandrian sediments, taking the pairs as annual, agrees with estimates of Flandrian time from other sources. The Marks Tey deposits gave a length of *c.* 20,000 years to the Hoxnian temperate stage. Laminated marine sediments with banding caused by alternate diatom-rich and clay-rich sediments have been described from the Gulf of California by Calvert,[7] and these are preserved as well-laminated sediments by the relative absence of burrowing organisms in an oxygen-poor environment. Freshwater diatomites of the Luneburg Heath in north Germany are laminated and estimates of time represented have been made.[2] It is obvious that the annual (or other periodic) origin of the sediment pairs must be established before a chronology can be attempted, and it is often difficult to prove that the pairs represent a particular span of time. Even in the instance of the classic varved clays, minor laminations within a single pair may confuse correlations and so render a chronology doubtful. Such minor laminations may arise from redeposition of sediments after storms.

As varved clays have been extensively used as a basis for a chronology, especially in Scandinavia, it is necessary to consider the methods and difficulties of this type of dating. Each sediment pair (varve) results from a year's deposition, coarse in the spring melt period, to finer in later summer and winter as the clays settle out (plate 4). An exposure of varved clays shows a number of such pairs. Successive pairs will vary in thickness, depending on the amount of material fed into the extra-glacial lake in which the sediments were formed. De Geer, the pioneer of varve chronology, assumed that the climatic variation over a wide area was similar and caused the thicknesses of sediments deposited in any year to vary in a similar way. Thus changes in thickness in successive varves could be used to correlate varve successions from one section to the next (plate 3). Each section shows a set of varves covering a short time, the northern areal limit of each varve being the ice limit at the time of deposition of that varve. Each varve will overlap its predecessor northwards as the ice retreats. By taking a north–south transect at right angles to the direction of ice retreat in Sweden, counting the varves in sections as close as possible, then correlating them by thickness variation de Geer[25] was able to erect a chronology of the retreat of the ice sheet for the last 17,000 years, covering several hundred kilometres of retreat of the ice front of the Weichselian (last) Glaciation. The later part of the chronology agrees with that obtained by radiocarbon dating, but the earlier part has been disputed because of the possibility of extra laminations (not of annual origin) (e.g. 'storm-laminations') being interpreted as varves.

Sauramo[49] did not believe that de Geer's method of correlation based on thickness of individual varves was adequate because of local environmental effects on varve thickness. In producing a varve chronology for Finland he relied mainly on the correlation of varve series, each series showing particular characters of the varves in colour, in coarseness of grain, and in the arrangement of the coarse layers. He thus relied on the physical characters of the varves for his principal correlations. The datings he arrived at for Finland are supported by radiocarbon chronology.

Some attempts have been made to relate varve sequences in the Old and New Worlds, so-called tele-connections[24] as opposed to close connections. It has yet to be proved that such correlations are valid.

Geochemical methods of dating[4, 52]

A number of methods of dating are based on the measurement of amounts of elements which in the process of time are either formed by

Table 9.3 Pleistocene geochemical dating methods (after Broecker 1965)

Isotope	Half-life 10^3 years	Method	Range 10^3 yrs	Materials	Applications: Ocean temp.	Sea level	Glacier extent	Arid lakes	Pollen sequence
^{14}C	5·7	Decay	0–35 35–50	Organic, $CaCO_3$ Organic	+	+	+	+	+
^{231}Pa	32	Decay ^{230}Th normal	5–120 5–120	Red clay or *Globigerina* ooze	+	0	0	0	0
^{230}Th	75	Decay	5-400	Red clay or *Globigerina* ooze	+	0	0	0	0
		Growth	0–200	$CaCO_3$ (biogenic)	0	+	0	+	+
^{234}U	250	Decay	50–1000	Coral	0	+	0	0	0
^{4}He	—	Growth	No limit	Molluscs, coral	0	+	0	+	0
$^{40}Ar(K/A)$	—	Growth	No limit	Volcanic	+	+	+	+	+

or are subject to radioactive decay. The rate of decay being known, the time-interval may be determined between the present and the time when the particular parent element was fixed and its decay began.

Table 9.3 summarises the isotopes concerned in these dating methods and the applications of each method. Radiocarbon and potassium-argon dating are the most important of these methods at present and will be described first. Other methods are in the course of development although some (e.g. the $^{231}Pa/^{230}Th$ method) have been used considerably, and in the future they will assume greater importance.

Radiocarbon dating.[34] This method is based on measuring the radiocarbon (^{14}C) activity of biogenic material such as wood, peat and shells. Radiocarbon is produced in the upper atmosphere by the interaction between nitrogen and cosmic ray neutrons (fig. 9.5). The radiocarbon is oxidised to carbon dioxide and mixes with the atmospheric carbon dioxide. It is then taken up by plants and animals and also becomes dissolved in sea and freshwater. Here it will be taken up by living organisms and exchange reactions will also occur between dissolved carbon dioxide, carbonates and bicarbonates to form inorganic sediments (fig. 9.5). Thus the radiocarbon is incorporated into the biosphere, and reaches equilibrium with the atmospheric, freshwater and ocean carbon reservoir. On the death of the organism the radiocarbon is trapped and will decay with the passage of time. The rate of decay, immutable, is given by the half-life, taken for purposes of age calculation as 5568 years. In that time half the radioactivity will be lost. Therefore by measuring the radioactivity of fossil biogenic material containing carbon, the date at which death took place, and equilibrium not maintained, can be determined.

Counting the radioactivity. The specific activity of modern carbon is extremely low, about 13·8 disintegrations of radiocarbon per minute per gram of carbon. It would be impossible to measure such a small amount in any detecting instrument in the presence of much larger amounts of cosmic ray mesons and soft gamma radiation derived from the surroundings. The effects of this unwanted radiation may be eliminated to a large extent by appropriate shielding of the detector. The soft gamma radiation is severely attenuated by housing the detector in a massive metal shield made of lead or iron. The effect of the cosmic ray mesons may be cancelled electronically by the principle of anticoincidence. The radiocarbon detector is surrounded by a ring of geiger counters such that with the appropriate electronics a pulse produced by a cosmic ray meson in the geiger counters will

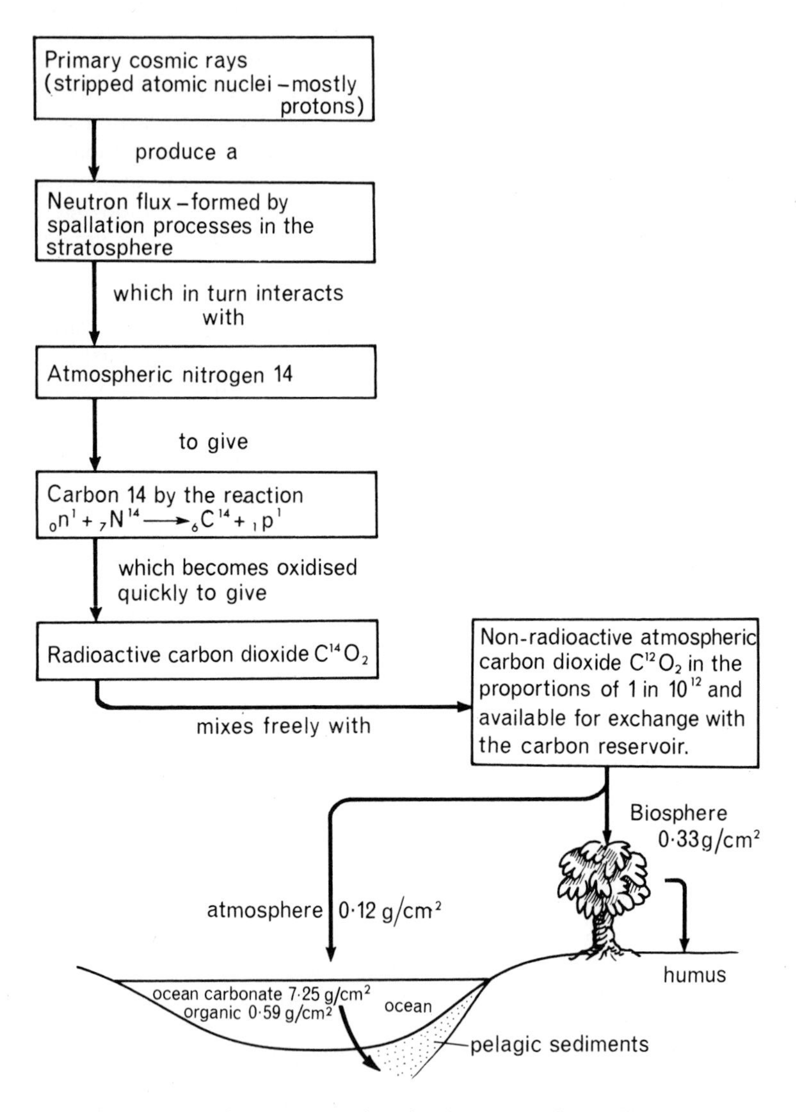

Fig. 9.5 Sketch of ^{14}C production in the atmosphere. The amount is shown of carbon (^{12}C and ^{14}C) in different parts of the exchange reservoir, expressed as grams of carbon in exchange equilibrium with the atmospheric carbon dioxide for each square centimetre of the earth's surface. The main reservoir is in the ocean carbonate.

automatically prevent the record of a pulse produced by the same particle in the radiocarbon detector itself.

Since radiocarbon is a weak β emitter (154 kev) and has low penetrating power, it must be introduced inside the detector itself in order to be detected. This may be achieved in a detector known as a gas proportional counter, where the carbon is introduced as carbon dioxide, methane or acetylene gas. The emission of a radiocarbon β particle causes an ionisation of the gas which is amplified and recorded electronically. The gas is kept at pressure within the counter since it is desirable to get as much carbon into the counter as possible.

The proportional counter is so named because the pulses obtained are proportional to the initial energy expended by the radiation (in contrast to a geiger counter where any discharge which results in the ionisation of the filling gas gives a large pulse). In this way, the particular pulses caused by decay of radiocarbon can be isolated from pulses caused by other radioactive materials, such as radon, which give much higher pulses from the α radiation.

The size of the counter is governed by the size of samples available. Thus for small samples, a small counter at high pressure (2 to 3 atm) and with low background (say 2 to 6 counts per minute) gives the best accuracy. A larger counter, holding more carbon, with a similar background will give higher accuracy. In other words, the 'signal to noise' ratio for a large counter is greater than with a small one.

Another counting technique involves measurement of scintillation in organic liquid made from the sample. Advances in benzene chemistry and liquid counting techniques have resulted in liquid scintillation methods of counting comparable in performance to gas counting techniques, and perhaps better able to deal with larger numbers of samples and to attain a greater dating range.

The age is calculated as follows:

$$\frac{T}{\log_e 2} \times \log_e \frac{(S - b)}{(S_0 - b)}$$

where T = half life,
S count of fossil sample,
b background count, and
S_0 contemporary count.

Assumptions made in radiocarbon assay. Several assumptions have to be made in using the method for dating. In general, these assumptions appear to be valid, subject to comments below, as shown by dating of samples of known age. The assumptions are:

1. That the specific radioactivity of living organic material has not varied in the past; i.e. that the concentration of radiocarbon in the atmosphere has remained constant and that any isotopic fractionation by organisms taking it up has remained the same (in living wood $^{12}C : ^{13}C : ^{14}C = 98{\cdot}9 : 1{\cdot}1 : 1{\cdot}07 \times 10^{-10}$). The constancy of the activity in the atmosphere will depend on the constancy of production of radiocarbon, on the loss of radiocarbon by sedimentation and on the isotopic equilibrium between different parts of the reservoir (fig. 9.5). Variations in the atmospheric activity have been found by radiocarbon dating tree-ring sequences and sediments with annual lamination.[54] Small variations in the order of 1 to 2 per cent in the atmospheric activity have been found by dating tree-rings over the past 1000 years,[64] which lead to dating errors of ± 100 years. Much greater variations has been found in woods, mainly of Bristle-cone Pine (*Pinus aristata*), up to 7400 years old.[42, 55] By radiocarbon dating of tree-rings up to this age, it has been possible to relate sidereal years to radiocarbon years over this long period. Figure 9.6 shows the relation, and it will be seen that there are substantial discrepancies between the two, up to 900 years nearly 7000 years ago. The curve shows short term fluctuations imposed on a longer term trend. Changes in the intensity of the earth's magnetic field may be responsible for the alterations in the cosmic ray flux which have produced the long term trend of variation, while other factors such as sun-spot cycles may be effecting the cosmic ray flux to produce short term cycles.[6, 38] Both the varve chronology and the radiocarbon chronology suggest roughly the same ages for the retreat from the Salpausselka (zone III) moraines of Fennoscandia, so that variation between the sidereal and radiocarbon chronologies

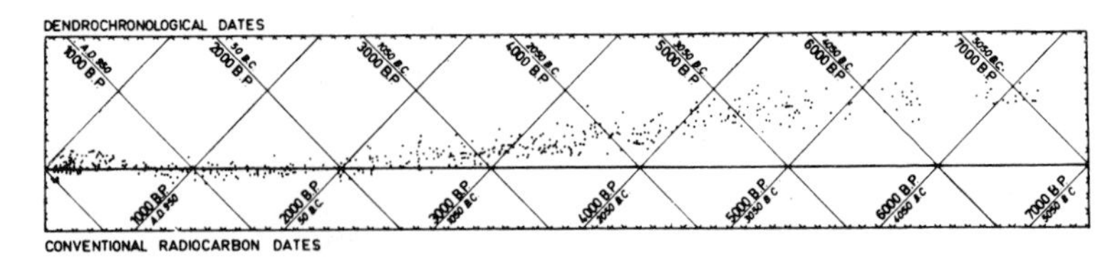

Fig. 9.6 Tree ring (including *Sequoia*, and species of pine) calibration of conventional radiocarbon ages. The radiocarbon dates are from wood in the Northern hemisphere, with radiocarbon ages determined by laboratories in Arizona, Gröningen, Philadelphia, Cambridge and La Jolla (after Olsson, 1970, and Suess, 1970).

may be reduced again at 10,000 years ago.[58] Future investigations should clarify this important point.

The curve on fig. 9.6, based on large numbers of radiocarbon dates of tree-rings, shows that a single radiocarbon age may mean more than one true age, an unfortunate result. Perhaps it is rather premature at present to use the curve to correct radiocarbon to true age in any precise way. In fact, various curves have been produced to demonstrate the relation between sidereal and radiocarbon chronologies, and such calibration curves are being used to convert radiocarbon years to sidereal years.[47, 57] The calibration curves may be useful to indicate further possibilities of correlation, but should not be taken as necessarily applicable on a world-wide scale. If it can be shown that the curve follows a similar course in different continents, with different datable materials from different ecological situations then we may have a secure basis for the conversion. Meanwhile, we should still consider the radiocarbon chronology as the basic independent scale.

2. That the atmospheric source of radiocarbon is geographically constant in its specific activity, so that dates from different continents, say, are comparable.

3. That the half-life has been accurately determined. The half-life has been measured several times and its mean value is taken as 5730 $\pm$ 30 years. Nevertheless, the Libby standard of 5568 years is used to achieve continuity of definitive publication in the journal *Radiocarbon*.[31]

4. That there has been no exchange of radiocarbon since the death of the organism which incorporated it. Here there are possibilities for exchange and contamination, and these are considered next.

Sources of error. Sources of possible error are many, ranging from isotopic fractionation in the plant or shell during life or during measurement, to contamination in sample-taking and storage.

1. Isotopic fractionation during processing occurs; very small amounts of radiocarbon are lost and this may become significant in samples older than a few thousand years. If the proportion in the samples of ^{12}C to ^{13}C is known (measured by mass spectrometry) a correction for this loss can be made.[43] Measurement of the ^{13}C content of samples not only provides a correction for this kind of error, but may also reveal changes in the sedimentary column which result from changes in hardness of water and in organic productivity.[54a]

There is evidence that isotopic fractionation in plants leads to enrichment of ^{12}C during photosynthesis, while in the surface waters of oceans the activity is lower than average because of mixing between the surface radiocarbon atoms and those of greater age in the deeper oceans. Thus in the Atlantic, surface water has an apparent age of about 450 years.

2. Intrusion of younger carbon into samples. Carbon contained in humic acids or in recent roots may give much greater radioactivity and therefore reduce the apparent age of samples (table 9.4). This is particularly important in dating old samples near the limit of radiocarbon dating, where even the introduction of very small amounts of radiocarbon will give an apparent age to a sample too old to have any measurable activity of its own.[14] With humus, the humus fraction may be separated and dated separately to test its age. Where recent roots are found in near-surface samples, it may be necessary to abstract all fossil leaves, twigs, seeds, etc., from a sample and date these. It is sometimes very difficult to separate recent intrusive plant organs from organic sediments. Exchange of carbonate with that in percolating waters may lead to greater radioactivity of the outer rinds of shells; and here it is necessary to wash the shells in dilute hydrochloric acid to remove the outer layers. There may also be contamination by radiocarbon via washing samples during their preparation, as has been shown by dates obtained from Tertiary shells.[43] Introduction of recent radiocarbon may also occur during storage of samples.

Table 9.4 Effects of contamination by modern and dead carbon on radiocarbon ages

A. Contamination by modern carbon

True ^{14}C age	*Approximate age with percent contamination* 1%	5%
600	540	160
1000	910	540
5000	4900	4300
10,000	9800	8700
25,000	23,400	19,000
40,000	32,800	23,200
Infinite	37,000	24,000

B. Contamination by 'dead' carbon

Years older than true age	*Percentage contamination*
400	5%
830	10%
1800	20%
2650	30%
5570	50%

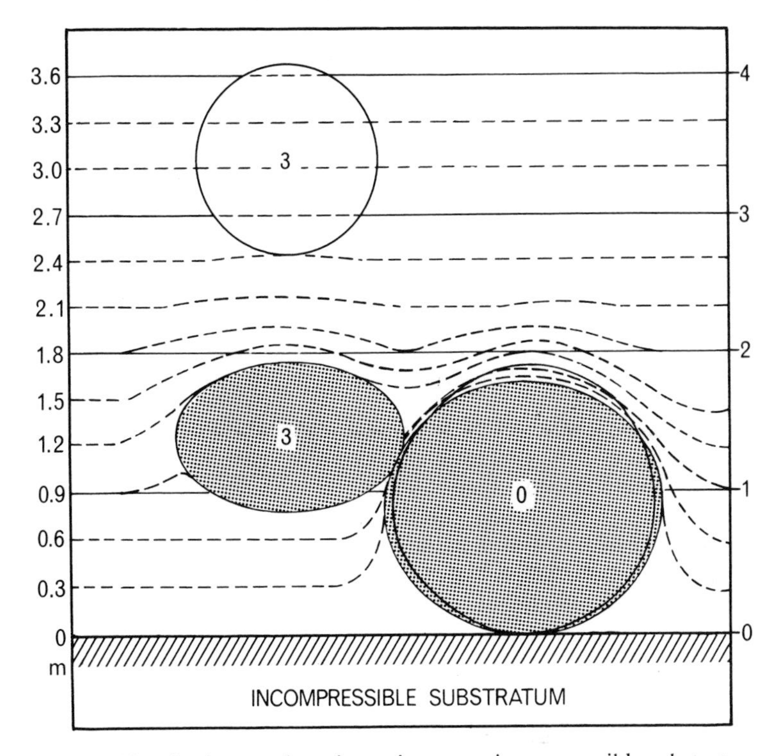

Fig. 9.7 Sketch of a peat deposit, resting on an incompressible substratum, showing compaction effects in relation to two logs of different ages in the peat (after Kaye and Barghoorn 1964). The figures on the right and the solid lines indicate lines of sediment deposited at the same time (depositional isochrons) before compaction. The empty circles show the positions of two logs at time of deposition, the figure in each referring to the time of deposition in relation to the isochrons.

The dashed lines, numbered on the left, are depositional isochrons after compaction of the peat to one-third of its thickness. The stippled logs show their position after compaction.

3. Intrusion of old carbon. Here there is the possibility of older carbon introduced in the form of fragmentary coal in sediments, or introduced into submerged aquatic plants by photosynthesis when the water contains dissolved carbonate from surrounding rocks (hard water error) (table 9.4). Freshwater and land-living molluscs may also incorporate old carbon.

4. Stratigraphical sources of error. The relation between the object

dated and the geological context of the sample must be quite clear, so that the date is meaningful in geological terms. For example, logs in a raised beach may not be contemporaneous with the formation of the beach, and so will not date it. An artefact may not be of the same age as the dated organic sediment containing it. The charcoal in an archaeological site may result from burning wood much older than the time of occupation. And the compaction of peat (varying from one-half to one-seventh of its original thickness) may lead to the relative displacement of peat and its contained wood (fig. 9.7).[33] In the dating of trees it is necessary either to use the outer rings as giving the nearest date to the time the tree died or was felled, or to use young trees or twigs covering a small time span.

This source of error can be greater than any of the above.

Collection and treaıment of samples. In general, the materials used for dating are to be preferred in this order: charcoal, wood, peat, mud,[26] bone (collagen), shell.

The samples for dating should be taken with extreme care, if possible out of range of penetration of recent roots and other types of contamination. The samples should be stored dry in sealed polythene bags. Pretreatment of organic materials[41] is necessary to remove carbonates (hot dilute hydrochloric acid) and humic acids (dilute sodium hydroxide), while with shell samples it is necessary to treat with dilute hydrochloric acid to remove surface layers which may have exchanged carbon with percolating waters. With peat and wood it may be necessary to separate the lignin and cellulose fraction in expectation that a contaminant will be eliminated or will follow either fraction, as will be seen from the separate dating of each fraction. In washing samples to extract organic material it is necessary to use boiled distilled water or dilute hydrochloric acid.[43]

The amount of material required for dating naturally varies with the carbon content of the sample. In general 3 to 5 g of carbon is required, depending on the size of the counter. Thirty g dry weight of sample is easily sufficient where carbon content is high, but where it may be low (e.g. partly inorganic sediments) much more (200 to 300 g may be necessary.

Interpretation of dates. In the calculation of the age of a sample, the half-life for calculation is now taken as 5568 years, and the common standard for recent radioactivity of living material is taken as 95 per cent of the activity of oxalic acid provided by the U.S. Bureau of Standards. The use of this standard avoids the danger of different laboratories using different standards for modern activity, some of which may suffer from increased radioactivity as a result of nuclear weapon tests and others from decreased activity as a result of infusion of dead carbon into the atmosphere by industrial processes using coal and other fuel. The 95 per cent activity of the standard oxalic acid is close to the activity of wood formed towards the end of the last century, prior to the fossil fuel dilution effect.

With marine shells there is a difficulty in obtaining a modern standard activity. Recent marine shells may give an apparent age of a few hundred years,[35] but this varies according to the particular ocean environment in the area, an effect already discussed.

A correction, normally small for wood and peat, has also to be made for isotope fractionation of radiocarbon during assay.

The date itself is stated in terms of measured probabilities because of the random nature of radioactive decay. A date is given as a time interval within which the true date will lie with a certain probability. The error given for each determination is calculated on the counting statistics and refers to one standard deviation, within which there is a 68 per cent probability that the real date lies. The results of the assays are given as radiocarbon age B.P. (before the present), taken as 1950, or as a date A.D./B.C.

The limit of radiocarbon dating is up to about seven half-lives (30,000 to 40,000 years) unless a process of isotopic enrichment[29] has been carried out. Using the latter process, some dates up to 64,000 years have been obtained. At these extremes limit of the radiocarbon method the effects of even minute contamination by active carbon may give a finite date to a sample too old to date, and great care is necessary in the selection and dating of samples.

The main advantage of radiocarbon dating, as opposed to, for example, varve chronology, is that it is a method which is very widely applicable and results in long-distance correlations. It has been effectively and extensively used to date archaeological, vegetational and climatic events revealed in organic sediments, and changes in sea-level as revealed by sections showing transitions between fresh and sea water sediments. Its use is perhaps more limited where shells are concerned[35] because of the possible errors already discussed. The method has also been used to determine the minimum age of soils by the assay of humic materials of B-horizons of podsols.[45] The further development of the method, in terms of refinement and application continues with strength.

Potassium-argon dating.[17, 36] A long-lived and naturally occurring isotope of potassium, ^{40}K, decays to argon, ^{40}A. By measuring the ratio of ^{40}A produced to the remaining ^{40}K, the time of origin of crystals or whole rocks containing potassium, such as those of basalt flows and tuffs, can be dated. The decay product is not a normal constituent of rocks, but assumptions have to be made concerning trapping of argon after the time of crystallisation, no loss of argon in time by diffusion, no changes of potassium content after crystallisation, and no reheating by later intrusions, resulting in the alteration of the K–A ratio. There is also the difficulty that a volcanic rock may contain older rock detritus which would increase the age significantly. These sources of error may lead to much greater percentage errors than with radiocarbon dating.

The method has been developed so that Pleistocene volcanic rocks and their crystal constituents can be dated, and it has achieved the dating of Pleistocene events beyond the range of radiocarbon dating, especially when used in association with argon isotope measurements.

A large number of dates have been obtained from the Lower Pleistocene and these have been used to date palaeomagnetic epochs and events, as well as fossil faunas and floras. The most intensively dated sequence is that from Olduvai Gorge, where over forty potassium/argon dates (from Beds I and II) show a range from 1·5 to 2·1 m.y., with the Olduvai normal palaeomagnetic event bracketed between 1·6 and 1·9 m.y.[3]

Another series of measurements[20] on the Laacher (Eifel) volcano products found in the Rhine terraces has given an indication of the age of these terraces (figure 9.8).

It is clear that potassium/argon dating will provide the basic chronology for the Lower Pleistocene, and perhaps also the Middle Pleistocene if the method can be refined sufficiently.

Other geochemical methods.[4, 5] The long-lived uranium isotopes, ^{238}U and ^{235}U, decay through short-lived isotopes to lead. Some of these intermediate isotopes have half-lives suitable for Pleistocene dating. These are shown in table 9.5. There is also the measurement of accumulation of helium, produced by the decay of ^{238}U.

These geochemical methods of dating based on the Uranium series are under active development at present, and they will certainly contribute much to Pleistocene chronology in the future. The bases and use of some of the uranium decay isotopes is briefly as follows:

1. *^{230}Th deficiency*. In some biogenic substances (marine carbonates, peat) there is an initial deficiency of ^{230}Th, but in time there is a growth towards equilibrium with uranium.

   The $^{230}Th/U$ method has been used for dating marine organisms.[5] Assuming that the carbonate, shortly after death of the organism, contains uranium but no thorium, and that the carbonate system remains closed, then the ratio ^{230}Th/^{234}U is a function of age until equilibrium is reached at 0·5 m.y. Using this method, apparent ages of corals from Barbados, West Indies, have been obtained ranging from 80,000 to 130,000 years, and agreeing with the relative stratigraphical age of the corals. Dating of fossil molluscs has not been so successful and much work remains to be done on the geochemistry of the processes involved.

2. *^{230}Th excess.* In ocean sediments ^{230}Th decays with time. Assuming the excess ^{230}Th of recent sediments has remained constant in time at any point on the ocean floor, the amount of decay with depth gives rate of accumulation of sediment. A similar method has been devised based on the decay of ^{231}Pa.
3. *^{231}Pa–^{230}Th method.* In recent ocean sediments, it is assumed that the ratio of ^{231}Pa and ^{230}Th is constant, as it is related to the ^{238}U and ^{235}U abundance in the oceans. As decay rates of the two isotopes are different, the change in ratio of the two with depth of sediment will give the rate of accumulation of sediment. This method has been used in dating marine sediments used for isotopic temperature determinations (fig. 9.8).
4. *Tritium method.* Tritium, ^{3}H, formed by cosmic ray bombardment, has a half-life of about 12 years. On oxidation to water

Table 9.5 Decay series of uranium isotopes used in Pleistocene dating, showing half-lives of the various isotopes (after Broecker, 1965). Principal isotopes underlined.

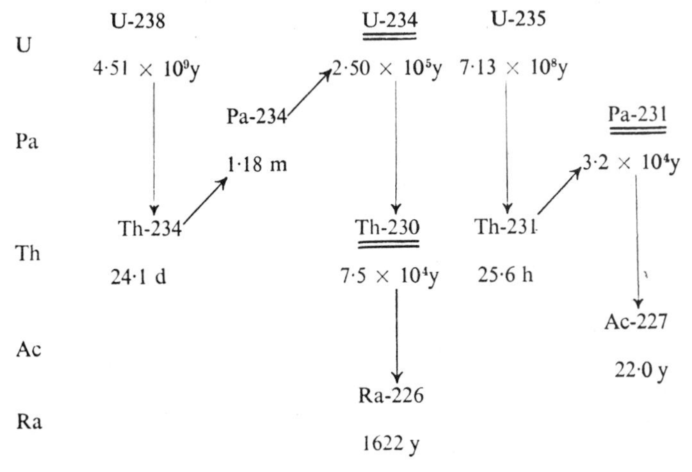

it enters into glacier ice, and its abundance in ice may be used to date ice within the last 50 to 100 years.

Amino-acid racemisation.[1] Amino-acids of the L-configuration are predominant in living organisms. In fossil bone it has been found that the amount of racemisation increases with age. The rate is dependent on temperature. If the rate of racemisation at a site can be calibrated by reference to radiocarbon dating, then age may be estimated by measuring the D to L ratio. Isoleucine and aspartic acid have been used. The latter has a half-life for racemisation, at 20°C, of about 15,000 years, so that the age limit for dating bone should be much greater than that of radiocarbon.

Fission track dating.[18] Fragments produced by fission in a mineral by a uranium nucleus produce a detectable trail or track of intense damage. These tracks can be picked out by chemical etching and can be counted on a prepared surface of mineral or glass. The sample is then subjected to bombardment by a known neutron dose and the number of tracks counted again. The ratio of the two track densities, together with the known neutron dose allows the age to be calculated. Hornblende, apatite and natural glass are considered to be the most suitable components of igneous rocks for dating in the Pleistocene. The fission track age of Bed I, Olduvai Gorge, 2·03 m.y., tallies with that obtained by the potassium-argon method.

Thermoluminescence dating.[27] This method of dating pottery may be extended to the dating of burned flint, which should give a further useful method for dating Pleistocene events.

Astronomical method. On the supposition that terrestrial climatic changes are determined by variation in elements of the earth's orbit round the sun, and because these variations can be calculated for past ages, it should then be possible to calculate the age, frequency and nature of past climatic changes. The theory of the relation between terrestrial climate and insolation changes resulting from variation in orbit is associated with the name of Milankovitch, and the theory has been developed by Zeuner (1958). In recent years our knowledge of dated climatic cycles has been greatly increased, especially through work on cores of ocean sediment, and there is a strong current of opinion that variations in the earth's orbit produces periodicities of change in the past 150,000 years matching geologically recorded climatic events.[8] It will be of great interest to see whether this correlation will be further substantiated, since a chronological basis of this type will obviously be of prime use in the Pleistocene.

CHRONOLOGY OF THE PLEISTOCENE[9]

The radiocarbon method and varve chronology provide a satisfactory chronological framework for the most recent 30,000–40,000 years of the Pleistocene, carrying us back to more or less halfway through the last glaciation. The potassium–argon method has provided dates of rocks older than 150,000 years, though many of these dates are difficult to relate to the climatic cycles of the Pleistocene.[22, 30] An important series of dates spanning 150,000 to 570,000 has been obtained from Eifel volcanics, and used to date Rhine terraces and associated climatic events. Uranium series dating, combined with palaeomagnetic chronology and oxygen isotope chronology (Ch. 10), have provided an outline chronology for marine sequences.[30]

At the upper end of the scale, in the later part of the last glaciation, correlation of the isotope chronology and radiocarbon chronology has been achieved. In the older part of the last glaciation radiocarbon dating of climatic changes in the range 50,000–70,000 years does not appear easy to reconcile with the details of the isotope stages. The oxygen isotope evidence places the warmth maximum of the last interglacial around 120,00 years (isotope stage 5e) with the amelioration at the beginning of the interglacial at around 128,000 years. The dating of the end of the last interglacial depends on the definition used. The dates vary from 116,000 to 73,000. The problem is that the period 128,000 to 73,000 shows three temperature maxima and two minima. The first maximum is thought to be the last interglacial (Eemian, Ipswichian) of the terrestrial sequence. The later two may be the early interstadials of the last glaciation known from the terrestrial sequence. If so the radiocarbon dates of the terrestrial sequence are younger than expected. The solution of these problems lies in better correlation of the marine and terrestrial sequences, more dating, and a definition of interglacial acceptable to both those working in the marine field and those working in the terrestrial field.[56]

The problem of correlation of the marine and terrestrial sequences is even more difficult in the part of the Pleistocene older than the last interglacial. The paleotemperature curve for the Brunhes (palaeomagnetic) Epoch, back to 700,000, is well established, but the correlation of the climatic stages with the terrestrial succession is as yet unsettled.[50] It is the same with the older parts of the Pleistocene. Here we are dealing with the long period 700,000 to 2 m.y. or so, with a very imperfect terrestrial record and less marked isotopic fluctuations.

Lamination of lacustrine interglacial deposits has led to the following estimates of the length of interglacials (definition based on palynology): Eemian 11,000 years, Holsteinian 16,000 years, Hoxnian 20,000–25,000 years.[61] These provide a floating chronology, which may eventually be fitted to a global chronology of the Pleistocene.

Time scale 10^3 yrs (not linear)	Polarity	N.W. Europe[28]	K/A dating		Deep-sea cores isotope stages dated by ^{14}C and extra-polation, and palae-omagnetic scale	Th ^{230}/U dates reef building stages
			Europe	Africa		
0		Flandrian t (^{14}C dating)				
10		Weichselian g / Devensian g			1 t	
30					2 g	
50					3 c	
70		? Brørup i–s			4 g	
90					5 t (a t, b 'c, c t, d c, e t)	Barbados I
110						Barbados II
130		Eemian/Ipswichian i–g				Barbados III
150		Saale g, c	Holstein i–g		6 c	
200			Mindel g (Rhine middle and old Middle Terraces) 'Cromer' I–g		7 t,c	
250					8 c	
300					9 t	
350					10 c	
400			Gűnz g		11 t	
450					12 c	
500		"Cromerian" complex t, c			13 t	
550					14 c	
600					15 t	
700					16 c, 17 t, 18 c, 19 t, 20 c, 21 t, 22 c	
800		Menapian c				
900						
1000		Waalian t	Villafranchian f			
1500		Eburonian c				
		Tiglian t		Olduvai f		
2000						
		Praetiglian c				
2500		Pliocene				

g. glacial
i–g, interglacial
i–s, interstadial
t, temperate stage
c, cold stage
f, fauna

Fig. 9.8 Outline chronology of the Pleistocene.

Figure 9.8 gives an outline of Pleistocene chronology assembled from these various dating methods, and against a background of the palaeomagnetic chronology and the marine oxygen isotope stages. The chronology of the terrestrial climatic stages of the Lower Pleistocene is indeed sketchy. There is a slight improvement in the Middle Pleistocene, and a great improvement in the Upper Pleistocene, thanks to radiocarbon dating. No doubt the further application of the dating methods described in this chapter will greatly improve the chronology in the next few years.

REFERENCES

BISHOP, W. W. and MILLER, J. A. 1972. *Calibration of hominoid evolution.* New York: Wenner-Gren Foundation/Scottish Academic Press.

OAKLEY, K. P. 1969. *Frameworks for dating fossil man.* 3rd edn. London: Weidenfeld and Nicolson.

ZEUNER, F. E. 1948. *Dating the past.* 4th edn. London: Methuen.

1 BADA, J. L. and DEEMS, L. 1975. 'Accuracy of dates beyond the ^{14}C dating limit using the aspartic acid racemisation reaction', *Nature (Lond.)*, **255,** 218–19.

2 BENDA, L. 1974. 'Die Diatomeen der niedersächsischen Kieselgur-Vorkommen, palökologische Befunde und Nachweis einer Jahresschichtung', *Geol. Jb.*, A, **21**, 171–97.

3 BISHOP, W. W. 1972. 'Post-Conference Commentary'. In *Calibration of Hominoid Evolution*, ed. W. W. Bishop and J. A. Miller, 455–77. New York: Wenner-Gren Foundation/Scottish Academic Press.

4 BROECKER, W. S. 1965. 'Isotope geochemistry and the Pleistocene climatic record'. In *The Quaternary of the United States*, eds. H. E. Wright and D. G. Frey. Princeton University Press.

5 BROECKER, W. S. and BENDER, M. L. 1972. 'Age determinations on marine strandlines'. In *Calibration of Hominoid evolution*, ed. W. W. Bishop and J. A. Miller, 19–35. New York: Wenner-Gren Foundation/Scottish Academic Press.

6 BUCHA, V. 1970. 'Influence of the earth's magnetic field on radiocarbon dating'. In *Radiocarbon variations and absolute chronology*, ed. I. U. Olsson, 501–11. Stockholm: Almqvist and Wiksell.

7 CALVERT, S. E. 1964. 'Factors affecting distribution of laminated diatomaceous sediments in Gulf of California'. In *Marine geology of the Gulf of California—a Symposium.* Memoir No. 3, Amer. Assoc. Petrol. Geologists, 311–30.

8 CHAPPELL, J. 1973. 'Astronomical theory of climatic change: status and problem', *Quaternary Research*, **3,** 221–36.

9 COOKE, H. B. S. 1973. 'Pleistocene chronology: long or short?', *Quaternary Research*, **3,** 206–20.

10 COX, A. 1969. 'Geomagnetic reversals', *Science, N.Y.* **163,** 237–45.

11 COX, A., DOELL, R. R. and DALRYMPLE, G. B. 1965. 'Quaternary palaeomagnetic stratigraphy'. In *The Quaternary of the United States*, eds. H. E. Wright and D. G. Frey. Princeton University Press.

12 CRAIG, A. J. 1972. 'Pollen influx to laminated sediments: a pollen diagram from northeastern Minnesota', *Ecology*, **53,** 46–57.

13 CREER, K. M., THOMPSON, R., MOLYNEUX, L. and MACKERETH, F. J. H. 1972. 'Geomagnetic secular variation recorded in the stable magnetic remanence of recent sediments', *Earth Planet. Sci. Lett.*, **14,** 115–27.

14 DONNER, J. J. and JUNGNER, J. 1974. 'Errors in the radiocarbon dating of deposits in Finland from the time of deglaciation', *Bull. Geol. Soc. Finland*, **46,** 139–44.

15 EMILIANI, C. and SHACKLETON, N. J. 1974. 'The Brunhes Epoch: isotopic palaeotemperatures and geochronology', *Science, N.Y.* **183,** 511–14.

16 ERICSON, D. B., EWING, M. and WOLLIN, G. 1964. 'The Pleistocene Epoch in deep-sea sediments', *Science, N.Y.* **146,** 723–32.

17 FITCH, F. J. 1972. 'Selection of suitable material for dating and the assessment of geological error in potassium-argon age determination'. In *Calibration of Hominoid evolution*, ed. W. W. Bishop and J. A. Miller, 77–91. New York: Wenner-Gren Foundation/Scottish Academic Press.

18 FLEISCHER, R. L. and HART, H. R. 1972. 'Fisson track dating: techniques and problems'. In *Calibration of Hominoid evolution*, ed. W. W. Bishop and J. A. Miller, 135–70. New York: Wenner-Gren Foundation/Scottish Academic Press.

19 FLETCHER, J., TAPPER, M. C. and WALKER, F. S. 1974. 'Dendrochronology—a reference curve for slow grown oaks, A.D. 1230 to 1546', *Archaeometry*, **16,** 31–40.

20 FRECHEN, J. and LIPPOLT, H. J. 1965. 'Kalium-Argon-daten zum alter des Laacher Vulkanismus, der Rheinterrassen und der Eiszeiten', *Eiszeitalter und Gegemwart*, **16,** 5–30.

21 FRITTS, H. C. 1965. 'Dendrochronology'. In *The Quaternary of the United States*, eds. H. E. Wright and D. G. Frey. Princeton University Press.

22 FUNNELL, B. M. 1964. 'The Tertiary period'. In *The Phanerozoic time-scale*, *Q. J. Geol. Soc. London*, **120s,** 179–91.

23 GEER, G. DE 1912. 'A geochronology of the last 12,000 years', *C.R. XI Int. Geol. Congress* (*Stockholm*), **1,** 241–53.

24 GEER, G. DE 1934. 'Geology and Geochronology', *Geogr. Annaler*, **16,** 1–52.

25 GEER, G. DE 1940. 'Geochronologia Suecica Principles', *Kungl. Svenska Vetensk. Handl.*, Ser. 3, **18,** No. 6.

26 GODWIN, H. 1969. 'The value of plant materials for radiocarbon dating', *Amer. J. Bot.*, **56,** 723–31.

27 GÖKSU, H. Y. and FREMLIN, J. H. 1972. 'Thermoluminescence from unirradiated flints: regeneration thermoluminescence', *Archaeometry*,

14, 127–32.

28 HAMMEN, T. VAN DER, MAARLEVELD, G. C., VOGEL, J. C. and ZAGWIJN, W. H. 1967. 'Stratigraphy, climatic succession and radiocarbon dating of the last glacial in the Netherlands', *Geol. en Mijnboyw*, **46,** 79–95.

29 HARING, A., DE VRIES, A. E. and DE VRIES, H. 1958. 'Radiocarbon dating up to 70,000 years by isotopic enrichment', *Science, N.Y.* **128,** 472–3.

30 HARLAND, W. B. and FRANCIS, E. H. (ed.). 1971. 'The Phanerozoic Time-scale. Supplement, Part 1', *Geol. Soc. Lond., Spec. Publ.* **5.**

31 JOHNSON, F. 1966. 'Half-life of radiocarbon', *Science, N.Y.* **149,** 1326.

32 KAY, G. F. 1931. 'Classification and duration of the Pleistocene Period', *Bull. Geol. Soc. America*, **42,** 425–66.

33 KAYE, C. A. and BARGHOORN, E. S. 1964. 'Late Quaternary sea-level change and crustal rise at Boston, Massachusetts, with notes on the autocompaction of peat', *Bull. Geol. Soc. America*, **75,** 63–80.

34 LIBBY, W. F. 1965. *Radiocarbon Dating*. 2nd edn., with additional notes. Chicago: Phoenix Science Series.

35 MANGERUD, J. and GULLIKSEN, S. 1975. 'Apparent radiocarbon ages of recent marine shells from Norway, Spitsbergen and arctic Canada', *Quaternary Research*, **5,** 263–73.

36 MILLER, J. A. 1972. 'Dating Pliocene and Pleistocene strata using the potassium-argon and argon-40/argon-39 methods'. In *Calibration of Hominoid evolution*, ed. W. W. Bishop and J. A. Miller, 63–76. New York: Wenner-Gren Foundation/Scottish Academic Press.

37 MONTFRANS, H. M. VAN, 1971. *Palaeomagnetic dating in the North Sea basin.* Rotterdam: Princo, N.V.

38 NEUSTUPNY, E. 1970. A new epoch in radiocarbon dating. *Antiquity*, **44,** 38–45.

39 NÖEL, M. and TARLING, D. H. 1975. 'The Laschamp geomagnetic event', *Nature*, (*Lond.*) **253,** 705–6.

40 OAKLEY, K. P. 1969. 'Analytical methods of dating bones'. In *Science in Archaeology*, 2nd edn., ed. D. Brothwell and E. Higgs, 35–45. London: Thames and Hudson.

41 OLSON, E. A. and BROECKER, W. S. 1958. 'Sample contamination and reliability of radiocarbon dates', *New York Acad. Sci. Trans.*, Ser. II, **20,** 593–604.

42 OLSSON, I. U. (ed.). 1970. *Radiocarbon variations and absolute chronology*. See also explanation of Plate IV, p. 625. Stockholm: Almqvist & Wiksell.

43 OLSSON, I. U. and OSADEBE, F. A. N. 1974. 'Carbon isotope variations and fractionation corrections in ^{14}C dating', *Boreas*, **3,** 139–46.

44 PENCK, A. 1908. 'Das Alter des Menschengeschlechtes', *Zeit für Ethnol.*, **40,** 390–407.

45 PERRIN, R. M. S., WILLIS, E. H. and HODGE, C. A. Ü. 1964. 'Dating of humus podzols by residual radiocarbon activity', *Nature* (*Lond.*), **202,** 165–6.

46 PRESTWICH, J. 1888. *Geology*. Vol. 2, Stratigraphical and physical. Oxford: Clarendon Press.

47 RENFREW, C. and CLARK, R. M. 1974. 'Problems of the radiocarbon calendar and its calibration', *Archaeometry*, **16,** 5–15.

48 RUDDIMAN, W. F. and MCINTYRE, A. 1973. 'Time-transgressive deglacial retreat of polar waters from the North Atlantic', *Quaternary Research*, **3,** 117–30.

49 SAURAMO, M. 1923. 'Studies on the Quaternary varve sediments in southern Finland'. *Comm. géol. de Finlande Bull.*, No. 60.

50 SHACKLETON, N. J. and OPDYKE, N. D. 1973. 'Oxygen isotope and palaeomagnetic stratigraphy of equatorial Pacific core V28-238: Oxygen isotope temperatures and ice volumes on a 10^5 year and 10^6 year scale', *Quaternary Research*, **3,** 39–55.

51 SHACKLETON, N. J. and TURNER, C. 1967. Correlation between marine and terrestrial Pleistocene successions. *Nature* (*Lond.*), **216,** 1079–82.

52 SHOTTON, F. W. 1967. 'The problems and contributions of methods of absolute dating within the Pleistocene Period', *Quart. J. Geol. Soc. Lond.*, **122,** 356–83.

53 STOKES, M. A. and SMILEY, T. L. 1968. *An introduction to tree-ring dating*. Chicago: University of Chicago Press.

54 STUIVER, M. 1971. 'Evidence for the variation of atmospheric C^{14} content in the Late Quaternary'. In *Late Cenozoic Glacial Ages*, ed. K. K. Turekian, 57–70. New Haven: Yale University Press.

54a STUIVER, M. 1975. 'Climate versus changes in ^{13}C content of the organic component of lake sediments during the Late Quaternary', *Quaternary Research*, **5,** 251–62.

55 SUESS, H. E. 1970. Bristlecone-pine calibration of the radiocarbon time scale 5200 B.C. to present. In *Radiocarbon variations and absolute chronology*. Ed. I. U. Olsson. Stockholm: Almqvist & Wiksell.

56 SUGGATE, R. P. 1974. 'When did the last Interglacial end?', *Quaternary Research*, **4,** 246–52.

57 SWITSUR, V. R. 1973. 'The radiocarbon calendar recalibrated', *Antiquity*, **47,** 131–7.

58 TAUBER, H. 1970. The Scandinavian varve chronology and ^{14}C dating. In *Radiocarbon variations and absolute chronology*. Ed. I. U. Olsson. 173–96. Stockholm: Almqvist & Wiksell.

59 THOMPSON, R. 1973. 'Palaeolimnology and palaeomagnetism', *Nature* (*Lond.*), **242,** 182–4.

59a THOMPSON, R. 1975. 'Long period European geomagnetic secular variation confirmed', *Geophys. J. R. astr. Soc.*, **43,** 847–59.

60 THOMPSON, R., AITKEN, M. J., GIBBARD, P. L. and WYMER, J. J. 1974. 'Palaeomagnetic study of Hoxnian lacustrine sediments', *Archaeometry*, **16,** 233–7.

61 TURNER, C. 1975. 'The correlation and duration of Middle Pleistocene interglacial periods in Northwest Europe', In *After the Australopithecines*, ed. K. Butzer and G. L. Isaac, 259–308. The Hague: Mouton Publishers.

62 WELTEN, M. 1944. 'Pollenanalytische, stratigraphische und geochronologische Untersuchungen aus dem Faulenseemoos bei Speiz', *Veröff. Geobot. Inst. Rübel*, **21.**

63 WILCOX, R. E. 1965. 'Volcanic ash chronology'. In *The Quaternary of the United States*, eds. H. E. Wright and D. G. Frey. Princeton University Press.

64 WILLIS, E. H., TAUBER, H. and MÜNNICH, K. O. 1960. 'Variations in the atmospheric radiocarbon concentration over the past 1300 years', *Radiocarbon*, **2,** 1–4.

65 WOLLIN, G., ERICSON, D. B. and RYAN, W. B. F. 1971. 'Magnetism of the earth and climatic changes', *Earth Planet. Sci. Lett.*, **12,** 175–83.

CHAPTER 10

CLIMATIC CHANGE

THE importance of climatic change in the Pleistocene is made evident by its use as the basis for classification of the Pleistocene. There is a contrast between the climatic fluctuations about a low mean temperature in the Pleistocene and the changes known from Tertiary time (fig. 10.1), though the transition between them must be gradual, with glacierisation starting earlier in the higher latitudes than in the lower latitudes where the great Pleistocene ice sheets developed.

The historical records of climatic change, of changes in such external factors as atmospheric circulation, and of changes in the distribution of plants and animals, give an inkling of the complex relations between climate, glacier advance or retreat and life; they allow theoretical models to be made of possible causes and of some effects of climatic change. Such models may in the future be a basis for relating environmental changes of recent time to climatic change, and from these it may then be possible to extrapolate back to explain the older Pleistocene climatic fluctuations. A more detailed knowledge of the nature, magnitude and succession of climatic changes will be the necessary prerequisite for finding the causes of climatic change.

The important climatic factors are temperature and precipitation (solid and rain), not only their annual totals but, just as important, their seasonal distribution throughout the year. It is temperature which has received the greatest attention from the palaeoclimate point of view. Correlated with these factors are changes in atmospheric circulation, storminess and windiness, which greatly affect the climate of such an oceanic area as Britain. There is the possible variation from an oceanic climate with high rainfall and low variation in temperature to a continental climate with a more seasonal distribution of rain and greater seasonal differences of temperature.

As well as these regional climatic changes, we have also to consider climatic conditions which may be prevalent on a much more local scale. It is important to distinguish between those evidences of cli-

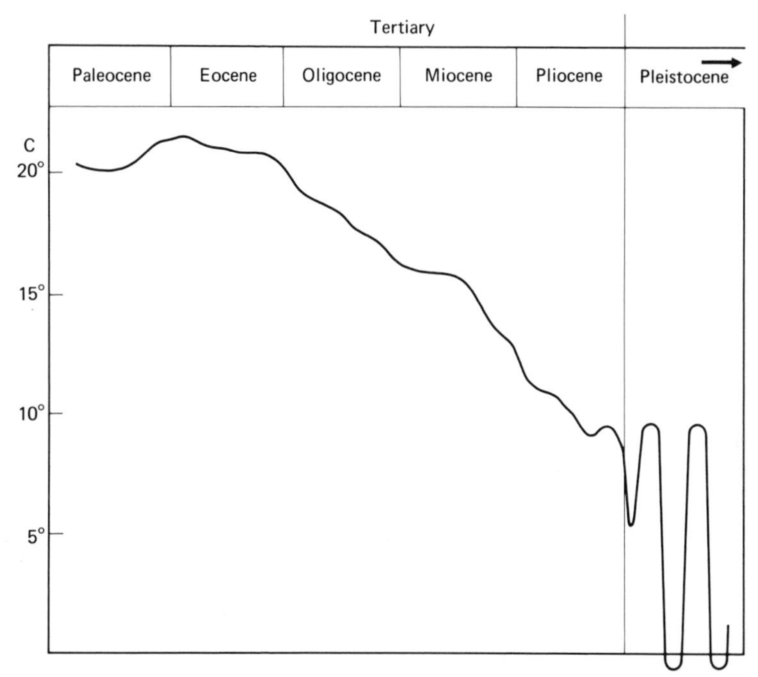

Fig. 10.1 Diagram of fall of mean annual temperature in the Lower Rhine valley in Tertiary and Pleistocene times (after Teichmuller). Only the early fluctuations of the Pleistocene are shown. Fluctuations in Tertiary time occurred, but not about the low mean temperatures seen in the Pleistocene.

mate which give regional information and those which give micro-climatic information. Day-length and angle of insolation, dependent on latitude, will also be important factors, such as where they may determine species presence, or the duration of freeze-thaw cycles.

The presence of ice sheets will have affected local and perhaps regional climates. Without doubt they affected areas near the ice margins, with lowering of mean temperatures and cooler winds, and probably this influence extended well into the periglacial areas.

The existence of colder conditions in lower latitudes than they are now present brings the certainty that periglacial climatic conditions are not repeated in the present climatic systems. This makes for difficulty in reconstructing glacial and periglacial climates. However, enough has been said of the complexities of climate and the interpretation of palaeo-climates, and we shall now examine the principal lines of evidence which enable us to estimate the nature and magnitude of past climatic change, including evidence from the changing distribution of plants and animals, both terrestrial and marine, geological evidence, such as that for changing snow-lines, and evidence from oxygen isotope studies. First to be considered are indices of climatic change of these types, and then we shall consider evidence of change in historic time.

INDICES OF PAST CLIMATIC CHANGE

The evidence to be considered is biological, geological, geomorphological and geochemical.

Biological evidence

Biological evidence of climatic history comes from the study of changing distributions, in time, of either individual species or assemblages of species. The assumption is made that the individual's ecological responses to physical conditions of the environment have remained unchanged, and therefore change in distribution means change in environmental conditions.

The fossil assemblage gives a clear indication of the general climatic type, e.g. temperate or arctic, oceanic or continental. It may appear very similar to particular assemblages seen today, and then it may be possible to infer more detailed climatic conditions. As an example of such a type of interpretation, we can refer to the interstadial deposits in the Weichselian (last) glaciation at Chelford in Cheshire,[60] where pollen spectra from biogenic muds were very similar to the pollen rain in the *Betula-Pinus-Picea* forest of northern Finland. Some approximations to the climate of the Chelford area during the deposition of the interstadial organic mud may be given by the climate of the area of distribution of this type of forest in northern Finland, with an annual rainful of 400 to 700 mm, annual temperature amplitude 20° to 27°C, mean temperature of coldest month (February) −10° to −15°C, and that for the warmest month (July) 16° to 12°C. Of course, there is the assumption here that in each of the compared areas the vegetation was in equilibrium with the climate. This interpretation on the basis of the plants is supported by deductions of similar climatic conditions based on the study of the insect fauna of the same deposits.[8]

To proceed with climatic deducations more detailed than this, it is necessary to rely on present knowledge of the distribution of species, of the factors which determine this distribution, and of ecological behaviour. Where the factors are known to some extent, then it may be possible to make more detailed deducations about climate from the fossil occurrence of species.

Two examples may be given of this type of approach, the first dealing with terrestrial plants, the second with marine planktonic assemblages. The pollen frequencies of *Viscum*, *Hedera* and *Ilex* in Denmark have been used to determine climates in the Flandrian of Denmark,[27] on the basis that their flowering and optimum growth in the recent past can be related to particular conditions of summer and winter temperature. For example, *Hedera*, whose northern and eastern limits for normal growth in northern Europe are set by mean temperatures for the coldest month being warmer than −1·5°C, or *Viscum*, which thrives where mean temperatures of the warmest month are higher than 16° to 18°C. The fall in *Hedera* pollen frequencies at the change from Atlantic to Sub-Boreal times in the

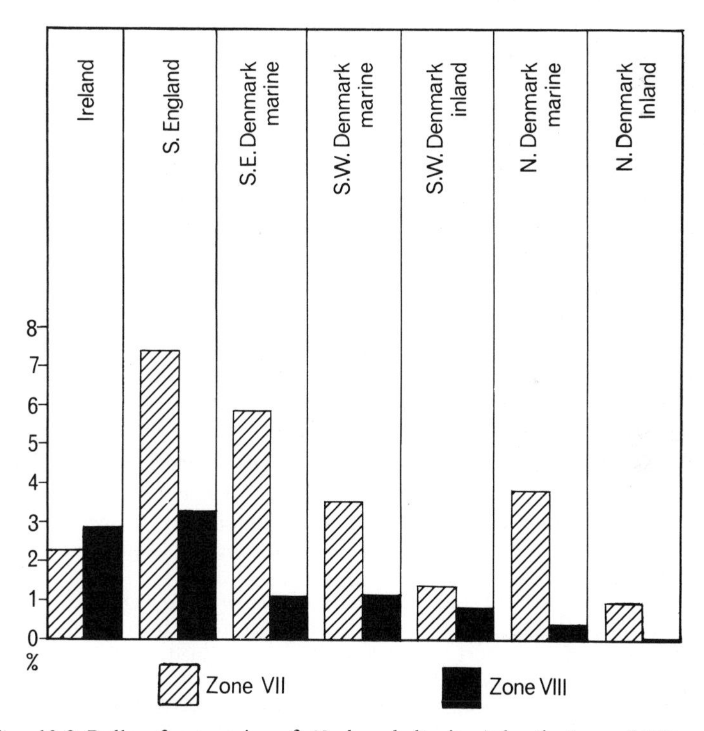

Fig. 10.2 Pollen frequencies of *Hedera helix* in Atlantic (zone VII) and Sub-Boreal (zone VIII) times in north-west Europe (Iversen 1960). The basis for calculation of the frequency is total tree pollen excluding *Corylus*. The Atlantic frequencies are greatest to the west and in the east where coastal influence is present (in marine deposits). The fall across the Atlantic–Sub-Boreal boundary is greatest in the east, least in the oceanic west.

more continental parts of northern Europe, as shown in fig. 10.2, is interpreted to indicate a fall in mean winter temperatures at this time in the more continental areas, contrasted with the smaller fall (or rise in Ireland) in the more oceanic west. The work by Iversen[28] and Andersen[1] on the Danish floras of Late Weichselian and Weichselian times respectively are also good examples of climatic determination using fossil floras.

In the study of climatic changes deduced from ocean core data, it has been found possible to relate particular foraminiferal assemblages of surface sediments to summer sea-surface temperature, winter sea-surface temperature and average sea-surface salinity. A set of transfer functions is obtained from this data which relates these physical parameters of the surface ocean waters to foraminiferal distributions at the present time, and these functions are applied to the fossil assemblages to obtain past values for these parameters.[26, 50]

Some results of this type of approach are shown in fig. 10.3. Figure 10.3a shows a comparison between winter sea-surface temperatures deduced in this way from a Caribbean core, oxygen isotope variations and the frequency of *Globorotalia menardii*, a warmth-indicating foraminifer. Figure 10.3b shows temperature and salinity values during and after the last interglacial in a north Atlantic core.

Such methods of quantitative palaeocology as these are of increasing importance in the study of past climatic changes, as they lead to classification of the ocean/atmospheric relationship. They have been applied mainly to fossil assemblages from ocean cores, including foraminiferal and radiolarian assemblages, where the uniformity of ecological conditions simplifies the situation. Thus it has been possi-

Fig. 10.3 Indices of changing environmental parameters in deep sea cores, obtained from foraminiferal assemblages by the application of transfer functions.

a. Caribbean core V12–122. Percentage of *Globorotalia menardii* (warmth-indicating), $\delta^{18}O$ ‰ in *G. ruber*, and index for winter surface water temperature (T_w). Letters on the left are zones defined on the presence or absence of *G. menardii* (Imbrie *et al.* 1973).

b. North Atlantic core V23–82. Indices for summer surface water temperature (T_s), winter surface water temperature (T_w) and salinity. Crosses indicate ^{14}C dates and circles stratigraphic dates. Horizontal lines indicate levels lacking in coccoliths; they coincide with levels of high mineral detritus content in the > 149 μm fraction (stippled pattern 10–20 per cent of sample, solid pattern > 20 per cent of sample). The detritus is thought to be ice rafted, so showing periods of polar conditions in the north Atlantic. (Sancetta *et al.*, 1973).

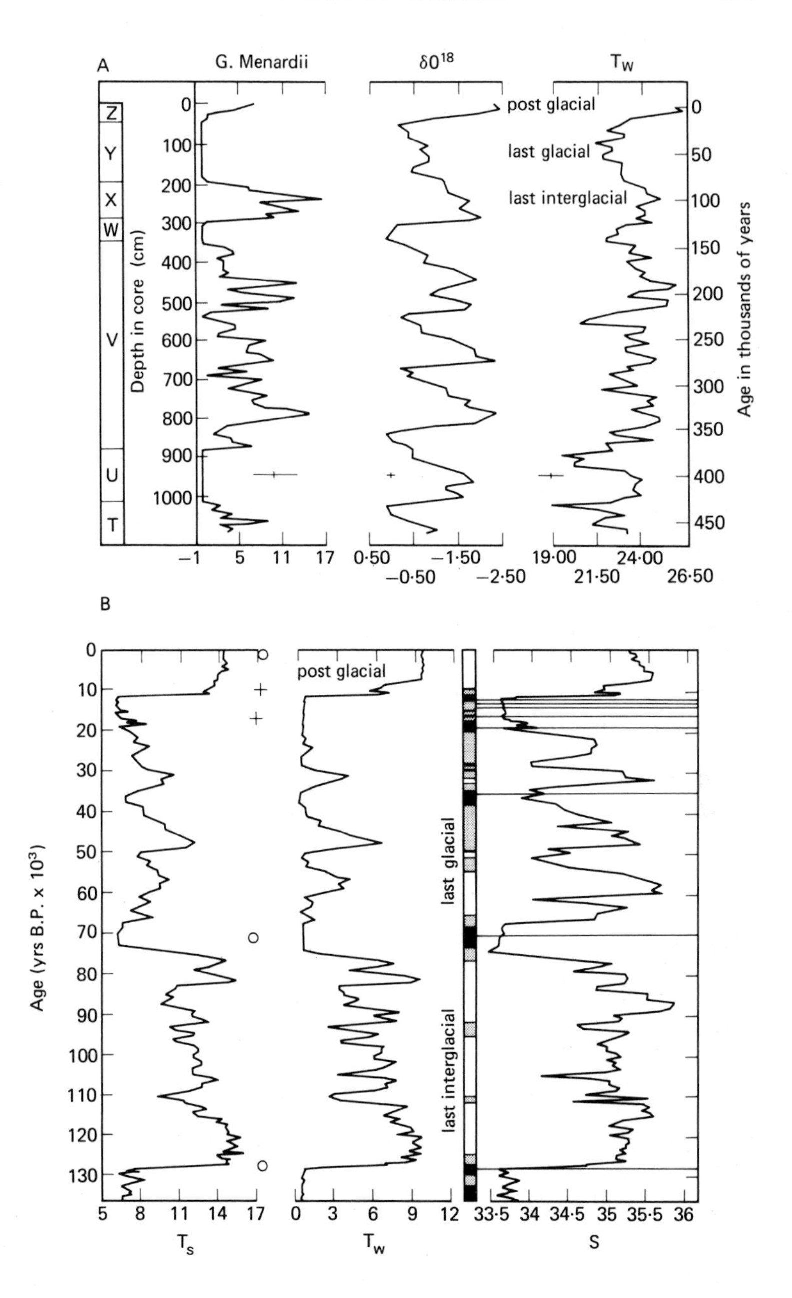
A
G. Menardii
δ0[18]
T_W
Z
Y
X
W
V
U
T
Depth in core (cm)
0
100
200
300
400
500
600
700
800
900
1000
post glacial
last glacial
last interglacial
Age in thousands of years
0
50
100
150
200
250
300
350
400
450
−1 5 11 17
0·50 −1·50
−0·50 −2·50
19·00 24·00
21·50 26·50
B
Age (yrs B.P. x 10³)
0
10
20
30
40
50
60
70
80
90
100
110
120
130
post glacial
last glacial
last interglacial
5 8 11 14 17
T_s
0 3 6 9 12
T_w
33·5 34 34·5 35 35·5 36
S

ble to draw a world map indicating climatic parameters of ocean and land at 18,000 B.P.[7a]. But such methods are also being applied to terrestrial assemblages, with, for example, the derivation of transfer functions relating modern pollen rain to climatic data, and the use of these functions to deduce past climatic conditions from pollen diagrams.[62]

Other evidence from marine sediments.[26] A quantitative approach for investigating climatic change from marine assemblages was discussed above. As the temperature of surface sea water is in equilibrium with the temperature of the air above, and the abundance and distribution of pelagic species is correlated with temperature, there is a record of temperature change in the sediments containing fossil planktonic species of Foraminifera, coccoliths and radiolarians. Both species distribution and abundance reflect temperature change.[18] Faunal variation in cores has been expressed in various ways; for example, by the changing percentages of warmth-indicating Foraminifera (e.g. *Globorotalia menardii*) and cold-indicating Foraminifera (e.g. *Globigerina pachyderma*), and the abundance of *Globorotalia menardii*. On the basis of the foraminiferal studies, a series of faunal zones allied to the climatic episodes have been proposed for the Atlantic and Caribbean cores and are shown on the left of fig. 10.3. Studies of the coccolith and foraminiferal assemblages of the North Atlantic have revealed changing conditions of water masses which are of prime importance for climatic history, especially in the British Isles, where the climate is very much subject to the North Atlantic circulation (fig. 10.4).[44]

Details of temperature and other environmental changes have also been deduced from studies of Foraminifera and molluscs of inshore deposits, particularly in southern Norway.[5, 21] In fact, the climatic optimum of the Flandrian was first postulated by Lloyd Praeger as a result of his investigations of the fauna of estuarine clays of northern Ireland.

Other faunal evidence. Vertebrates, insects, especially Coleoptera, and molluscs all give evidence of climatic change from their changing distribution in the Pleistocene. Vertebrate fossils may give evidence of climate by their skeletal adaptations and species frequency per unit of area.[32] Some wide-ranging animals with distribution governed by vegetation give regional evidence of climate; for example, the relation of horse and steppe. Each of the vegetation regions shown in fig. 10.5 will have its own characteristic fauna. Narrow-ranging organisms, for example, terrestrial molluscs, give indications of local climatic conditions, and assemblages may show a regional property

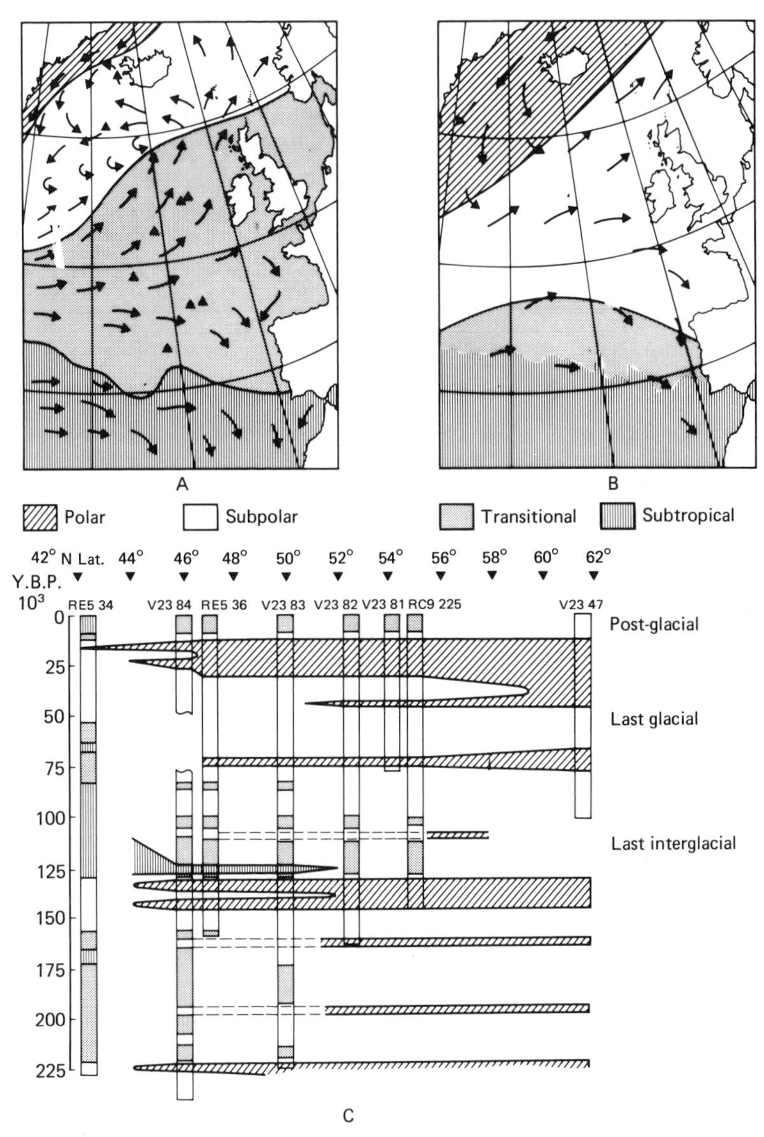

Fig. 10.4 History of northeast Atlantic (McIntyre and Ruddiman, 1972).
a. Modern surface water mass distribution. Triangles mark positions of cores in transect in C.
b. Surface water mass distribution at about 110,000 years ago, post the last interglacial climatic optimum.
c. Polar front migration revealed by distribution of floral and faunal assemblages characteristic of the four water masses.

of the climate; for example, the indication of warm summer temperatures by the presence of *Corbicula fluminalis* and *Belgrandia marginata*, both species now found in the Mediterranean region. There is here the usual difficulty of determining what are the climatic factors which really control the distribution of the species (living or extinct) found fossil. For example the climatic regime under which the extinct Siberian mammoth lived is unknown, though from its association with living cold-indicating mammals, its presence is usually thought to indicate a cold climate.

The importance of vertebrates, insects and molluscs for Pleistocene environmental history is discussed and illustrated in Chapter 13.

Evidence from fossil plants. As we have seen in Chapter 7, pollen assemblages give evidence of regional vegetation types, which, by analogy with the distribution of present-day vegetation types, indicate regional climatic type (fig. 10.5). Thus assemblages indicate a temperate forest climate, oceanic or continental according to the

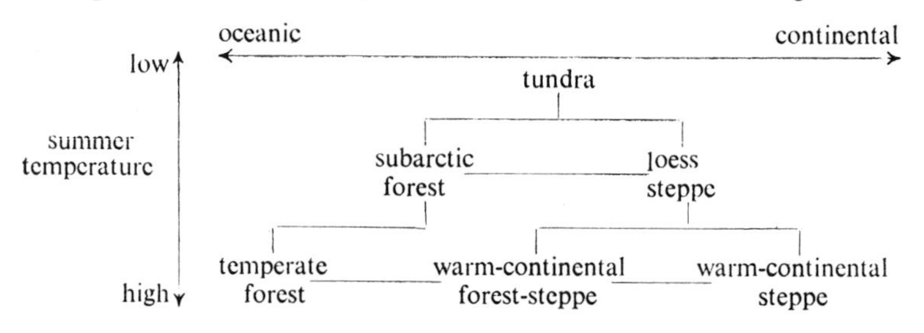

Fig. 10.5 Pleistocene biotopes in Europe (Zeuner, 1959).

genera represented, or an open vegetation climatic type, such as tundra or steppe, each with its own characteristic climate. Occasionally single genera may give such climatic indications; for example the abundance of *Ilex*, a tree of oceanic distribution type, far east of its present limits in Middle Europe in the Hoxnian interglacial, suggests that the climate of this interglacial was more oceanic than that at present. The climatic optimum of the Flandrian was early demonstrated by the fossil occurrence of thermophilous plant species north of their present range (fig. 10.6).

Macroscopic plant fossils of local derivation are of particular use as indicating local habitat conditions. Individual species may indicate particular climatic conditions, depending on whether enough is known of the present-day distribution of the species and the climatic conditions governing its distribution. Aquatic plants in particular

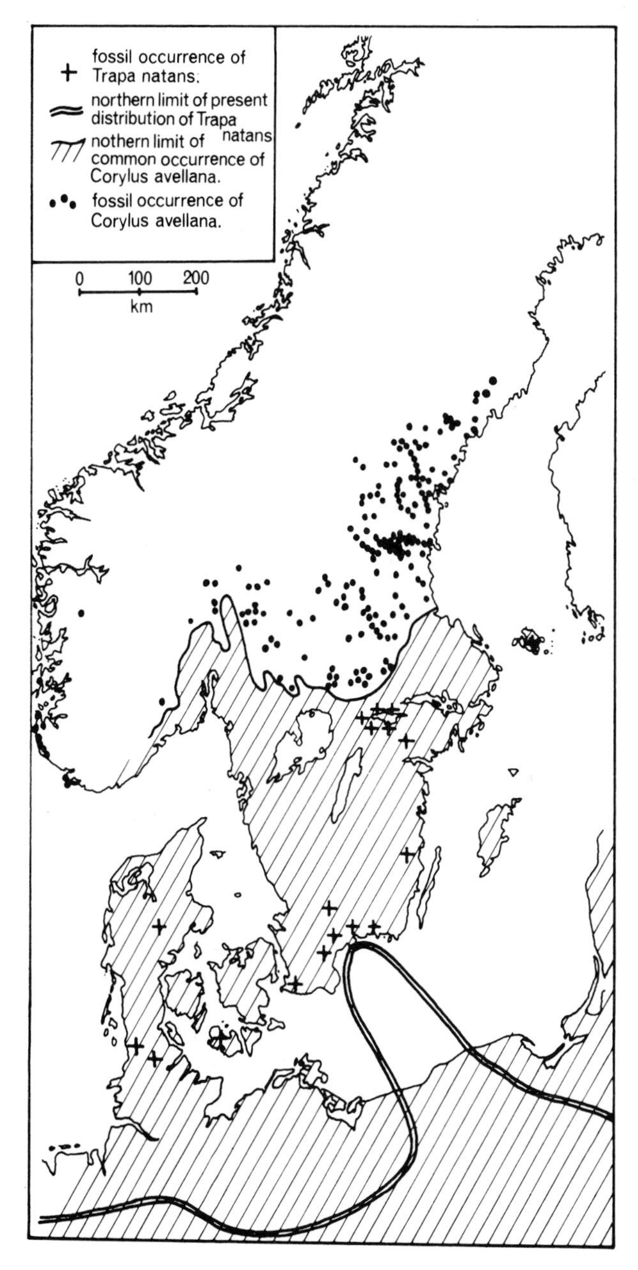

Fig. 10.6 Fossil Flandrian occurrences of *Corylus avellana* and *Trapa natans* in western Scandinavia (after Andersson, 1910).

have been considered good indicators of regional climate, because of the more uniform environment they live in.[29, 61] Seasonal temperatures have been estimated from studies of climatic conditions thought to be important in determining present distributions. For example, Iversen's detailed work on *Viscum*, *Hedera* and *Ilex* described above makes these genera particularly valuable as climatic indicators. The difficulty is that present distribution is not necessarily determined only by present climate. Historical factors, competition and edaphic factors will also affect the distribution, in ways often unknown. There is also the possibility, mentioned before, that the ecological tolerance of particular species may have changed, whilst the morphology of the part of the plant identified as a fossil remains unchanged. It would be rather surprising if some selection of biotypes had not taken place during the many migrations enforced by the repeated climatic changes of the Pleistocene.

There are some fossil assemblages which give contradictory climatic indications, for example, mixtures of cold- and warm-indicating species. Excluding the possibility that the mixture results from secondary deposition of fossils, there are a number of other possible explanations for such mixtures. It may be that under conditions of changing climate certain groups of organisms may migrate faster than others—e.g. water plants faster than land plants. We may then find a mixture of thermophilous water plants and boreal forest trees. It seems reasonable to expect that during the Pleistocene there was a lag between the changes of climate and the distributional responses of the different groups of plants and animals.

Another explanation of such anomalies is to assume that there is much microclimatic variation leading to admixture of fossilised organisms of different climatic or ecological requirements. Or there is the likelihood that past Pleistocene climates (such as those of the present temperate zone during a time of maximum glaciation) will have been different from any of the zonal climates of today. Thus the presence of thermophilous aquatic plants during past cold periods in the present temperate zone has been explained by the warming effect of the sun on water-bodies at these lower latitudes.

A last possibility would be that there has been a change in climatic tolerance of some of the species involved, and that the fossil taxon is now extinct. Such a change in tolerance, in a climatic or a wider ecological sense, might be demonstrated by the repeated finding of a particular species in association with other species which do not associate with it at the present day.

Table 10.1 shows estimations of past conditions of climate based on both plant assemblages and on individual species.

Table 10.1 Some estimates of Pleistocene temperature change

Type of evidence	*Area*	*Age*	*Parameter*	*Temperature difference from present value in area concerned*	*Reference*
Plants	Denmark	Flandrian, Atlantic, Sub-boreal	July mean	+2°C	Iversen (1944)
	Scandinavia	Late Weichselian zone III	July mean	−6°C	Iversen (1954)
	S. Germany	Late Weichselian zone III	July mean	−5·6 to −7°C	Firbas (1949–52)
	Scandinavia	Late Weichselian zone II (Allerød)		−2°C to −3°C	Iversen (1954)
	England (Fladbury)	Middle Weichselian	July mean	−5½°C	Shotton (1962)
	England (Chelford)	Early Weichselian Chelford interstadial	July mean	−2°C to −3°C	Shotton (1962)
	Denmark	Eemian, zone f	July mean	+2°C	Jessen and Milthers (1928)
	S. Poland	Günz glacial (Mizerna II/III)	annual mean	−4°C to −5°C	Szafer (1954)
Animal *Emys orbicularis*	Denmark	Flandrian, Atlantic, Sub-boreal	July mean	+2°C to +3°C	Degerbøl and Krog (1951)
Beetles	England (SE)	Ipswichian zone II	July mean	+3°C	Coope (1975)
Periglacial					
patterned ground	England (E)	Devensian	annual mean	> −12½°C	Shotton (1962)
involutions, ice wedges	England (C)	Devensian coldest part	July mean	−7°C	Williams (1975)
			January mean	−30°C	
Geochemical	Atlantic	Pleistocene	oscillation amplitude about 6°C		Emiliani (1955)

The Blytt–Sernander scheme of post-glacial climatic change in Scandinavia. The Norwegian botanist Axel Blytt[3] described arctic, sub-arctic, boreal, atlantic, sub-boreal, sub-atlantic elements in the Norwegian flora. He supposed that these elements immigrated into Norway in this order during alternating dry and wet periods. He sought to demonstrate this by a study of tree layers in peat deposits, the drier (continental) periods (boreal, sub-boreal) being marked by tree stump horizons, the wetter (oceanic) periods (atlantic, sub-atlantic) by active peat growth. R. Sernander[54] considered that Blytt's scheme could be applied with some modifications to Swedish peat deposits, and it was his work which led to the wider use of the terms Boreal, Atlantic, Sub-Boreal, Sub-Atlantic to denote climatic phases of the Flandrian. He related these phases to the stages of the Baltic and thus gave them a stricter chronological meaning. At a later date, they were related to the Scandinavian pollen zonation.

On the other hand, these ideas, formulated by the Uppsala school, were hotly contested by others, expecially G. Andersson,[2] who thought there was no good evidence for these divisions, but rather a single climatic optimum, then deterioration.

However, the Blytt–Sernander periods are often used at present as names of divisions of Flandrian time, but their reality and meaning in climatic terms is still a matter for discussion. It has recently been proposed that the names should continue to be used for Flandrian chronozones in Norden.[41]

Geological evidence

Sediments. Glacial deposits, both on land and in water, though the most obvious products of glaciation, give little evidence of the detail of climatic change. The spread of glaciers in the Alps and Scandinavia appears to be related to lower temperatures as indicated by the fact that former estimated snow-lines are lower but parallel to present snow-lines, but in other areas, for example, the Arctic, glacial advances appear to have been produced by precipitation changes. The formation and spread of glaciers and ice sheets are merely one result of climatic change. The size or duration of the change is not to be found by studying glacial deposits. Great thicknesses of till can be deposited in a short time, or none at all in a long time, such are the variables governing erosion and deposition. However, the areal extent of drift sheets does give some evidence of climate, because drift limits tend to be parallel to present summer isotherms, corresponding to the relation between existing firn limits and summer temperatures, so suggesting summer atmospheric circulation similar to that of the present. The presence of organic rather than inorganic sediments may argue for amelioration of climate, but often local environmental factors determine which will be formed, for example, the provision of inorganic sediments by solifluction, or the organic fill of abandoned meanders in outwash plains. The presence of loess sheets and other aeolian deposits again gives only general conditions of climate. At present loess accumulates under steppe climates, but is rare in wet climates, and thus loess is usually associated with dry continental climates, where silts are available for redistribution by wind. During the glacial stages the sources of silt (e.g. from outwash) were more widespread, and the presence of arctic species in loess faunas shows the loess climate of the past was not necessarily similar to that of today. The directions of prevalent winds have been estimated[40] from dune orientation and the wind circulation from dune structures.[48]

Trade-wind vigour has been assessed by sediment analyses of deep-sea cores off the coast of west Africa in the latitudes of the Sahara.[3a, 46] Such evidence is important in determining regional circulation changes associated with glaciation.

Periglacial phenomena. The climatic significance of various periglacial features has already been mentioned in Chapter 5. Structures resulting from freeze–thaw effects give evidence of alternation of temperatures across freezing-point—thus implying low winter temperatures, but not necessarily permanently frozen ground, of which the best indication appears to be deep and well-marked casts of ice-wedges, which only form where a negative mean temperature (below about −6°C) is maintained. For example, the relation between isotherms and ice-wedge formation in Alaska has already been mentioned (fig. 5.10). Solifluction is not necessarily a sign of permanently frozen ground; with permeable sediments subsurface frozen ground is required (say in the spring) to hold up the water in the thawed layer so that downslope creep of this layer develops.

In Britain, solifluction is more common than loess deposition, which is more or less confined to the south and east. This is presumably related to an oceanic influence on the climate. But increasing continentality to the east is indicated by the great frequency of ice-wedge casts in eastern England and their rarity in the southwest. Figure 10.7 gives a reconstruction of the temperature regime in central England during the coldest part of the last glaciation, based on periglacial features.[63]

An estimate of seasonal temperature in the last glaciation for the foothills of the eastern Alps has been made based on the relation between permafrost and tree growth.[47] In the west there is evidence

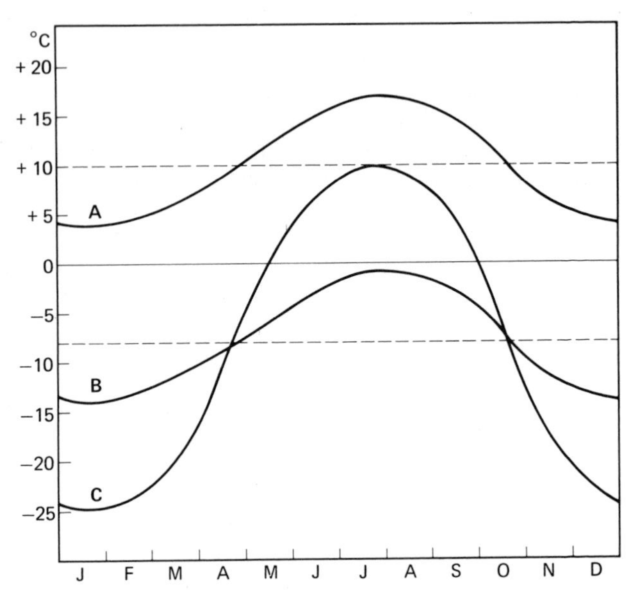

Fig. 10.7 Monthly temperatures in central England. A. Present, mean annual temperature 10°C. B. Range of temperature as now, mean annual temperature −8°C; no thawing season is possible, which does not accord with the presence of deep involutions. C. Reconstruction of a curve for the coldest part of the last glaciation, assuming mean annual temperature of −8°C, a thawing index of 900 degree days and average temperatures of the warmest month of 10°C. Such a continental temperature distribution would accord with the presence of ice-wedges and involutions (Williams, 1975).

for tundra and permafrost, to the east in Hungary for tree growth and permafrost. The northward limit of trees is said to be around the 10°C July isotherm, and the maximum annual mean temperature for permafrost around −2°C. Then, where the tree limit meets the permafrost limit in the area studied, the July temperatures would be about 10°C and the annual mean −2°C. Estimations of monthly temperatures based on these figures are shown in table 10.2. To the west, decreased summer temperatures resulted in less summer thaw and prevented tree growth, and the converse to the east.

Table 10.2 Estimations of mean monthly temperatures based on tree limit and permafrost distribution

J	F	M	A	M	J	J	A	S	O	N	D
−14	−12	−8	−2	4	8	10	8	4	−2	−8	−12° C

Geomorphological evidence. Terraces demonstrating higher lake levels than at present may be evidence of former higher precipitation but they may also indicate lower temperatures, resulting in reduced evaporation. Such a relation as the latter is apparent from the Great Salt Lake in Utah, where a drop of level was accompanied by an increased temperature in the period 1875–1937. In assessing the meaning of changed lake levels it is obviously important to know whether other factors—changed outlets, damming by lava flows or earth movements—have been important on determining lake level.

The significance of marine terraces as indicators of interglacial climates has already been discussed (Ch. 8). Conversely, deep channels in present river systems and on the continental shelf are related to glacial stages by reason of the expected custatic lowering during these stages.

Cirques are important in determining former local (orographic) snow-lines, because they form near or at the firn limit (Ch. 3). Thus the altitude of a group of cirque floors should indicate the snow-line at the time they were forming. So the altitude difference between this and the present snow-line will indicate the lowering of the snow-line during cirque formation. If we disallow changes in snowfall and apply the lapse rate (0·6°C for 100 m), a rough figure can be obtained for mean summer temperature change. The mean difference for the world as a whole has been calculated[4] at 760 m, giving a figure about 4·5°C temperature difference between glacial maxima and the present. This figure will be less if snowfall increased.

Soils. Chemical and physical weathering processes produce soil on the parent rock. The course of development of the soil depends on the parent rock type, on the climate, on vegetation growing on the soil, and on the time that has been available for soil development. The great zonal soil groups (e.g. podsol, brown earth, chernozem) are related to climate and vegetation. So the identification of a fossil soil to a particular zonal type may give information regarding the climate and vegetation of the time during which it formed. The main zonal soils and the climates relating to them are as follows:

1. Podsol. Developed under humid conditions with free drainage, with a grey bleached zone in the A* horizon caused by leaching,

* The following simplified horizon nomenclature is used:
A horizon: humus horizon, active biologically.
B horizon: enriched in iron and clay (not always present).
C horizon: parent rock.

with an accumulation of sesquioxides in the B horizon, carried down with the humic acids. The vegetation is usually heath or coniferous forest.

2. Brown-earths. These form under warmer or drier conditions than podsols, so that instead of a predominant leaching, the A horizon is well aerated and organic matter quickly oxidised. Less acid humus is produced, and sesquioxides are not leached much from the A horizon. There is loss of bases by leaching. A distinct B horizon is not usually present, and the colour of the soil is caused by the presence of oxidised and hydrated iron. The vegetation typically associated with brown-earths is temperate deciduous forest.

   The grey-brown podsolic soils are intermediate between podsols and brown-earths; they have a lower base-status than brown-earths and some podsolisation has occurred as a result of movement of sesquioxides.
3. Chernozems (black-earths). These form under conditions of continental climate with warm dry summers, with rain mainly in the spring and autumn. The main characteristic is the deep humus-rich A horizon, enriched with calcium carbonate near the base, and the absence of a B horizon. There is not sufficient rainfall for complete leaching of the bases, and the upward capillary movement of water in the summer leaves the carbonate in the basal layer of the A horizon where it concentrates to form concretions. The natural vegetation associated with the soil is short-grass prairie or steppe.
4. Chestnut soils. These occur under conditions of increased summer temperatures and decreased precipitation. The A horizon contains less humus, and calcium carbonate is well distributed throughout the profile. The vegetation is steppe.
5. 'Mediterranean' soils. Under conditions of warm dry summers and milder wet winters, the processes forming brown-earths proceed but there is a greater drying in the summer. This leads to strong oxidation, and there is less degree of hydration of the free ferric oxide, which results in red coloration of the soil.
6. Desert soils. These are often rich in alkalis or calcium carbonate, which may form encrustations or nodules (caliche) near or at the surface. Thus accumulation results from an upward movement associated with evaporation.

Such soil types as these merely show the tendency of soil development under particular climatic conditions. There are very many more types recognised, and for diagnoses and descriptions of European soil types reference should be made to Kubiena's *Soils of Europe*.[33]

The problems of interpreting fossil soils are several.[65] The first is in

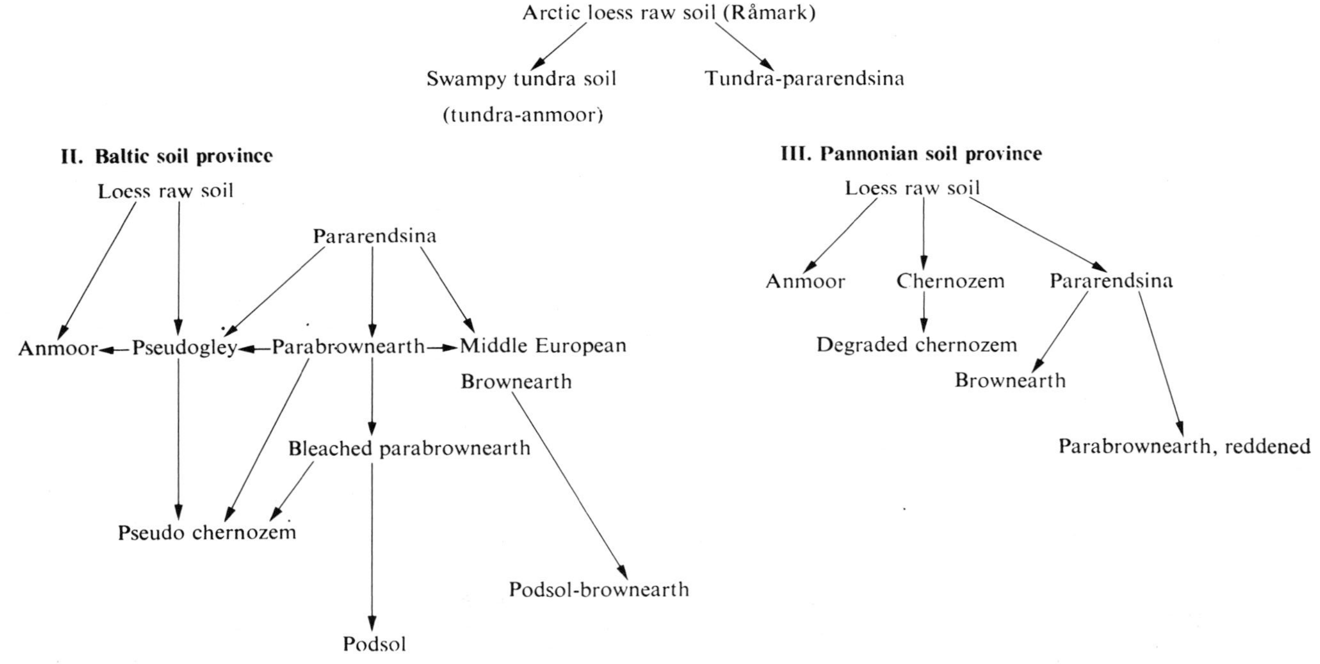

Fig. 10.8 Developmental relations of recent and fossil loess soils in the Arctic, Baltic and Pannonian soil provinces (Kubiena, 1956).

the recognition of a fossil soil. An iron rich or ferrocrete or a calcrete layer, may be the result of post-depositional changes related to former or present water levels or to impermeable layers holding up drainage. A horizon rich in organic matter may be a result of primary sedimentation.

Few extensive fossil soils have been described from the British Isles, but palaeocatenas from the Flandrian and Cromerian are known[61a] and their study has illustrated the difficulties of identifying fossil soils.

A well-developed fossil soil may be classifiable into its major type showing the general climatic regime under which it was formed. More often the profile is truncated, or the soil horizons disturbed, so that identification is made difficult. The fossil soil may be the end-product of a time of changing climate, or it may have developed

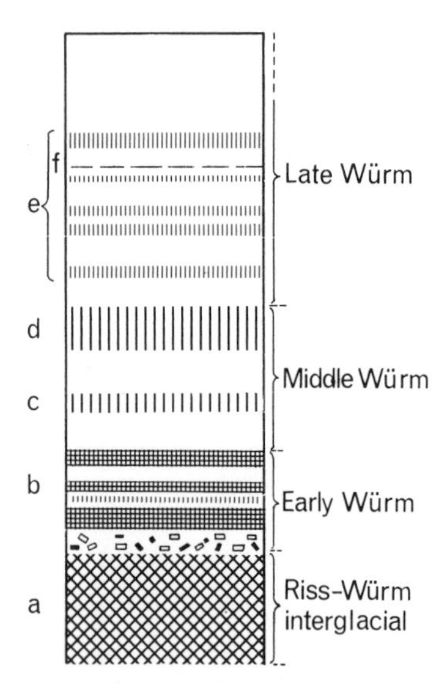

Fig. 10.9 Division of the Würm loess sequence in Hessen, Germany (Schönhals *et al.* 1964).

a. Riss/Würm interglacial soil (Homburger soil);
b. humus zones of the Early Würm (Mosbacher Humuszonen);
c, d. two soils in the loess of the Middle Würm (Gräselberger and Hahnstätter soils);
e. soil horizons in the Late Würm loess, more weakly developed than c and d (Erbenheimer soils). At f there is a tuff layer (Kärlicher tuff) important for synchronisation of the different loess sections.

during a time of no climatic change. There may be also local topographic factors influencing soil formation which are not evident in a section of sediments.

Three examples of the interpretation of fossil soils will be mentioned.

1. Fossil soils associated with loess in Europe. Much work has been done on these soils and their relation to climatic change and to chronology of the glacial and interglacial stages. The climatic interpretation of the soils has been improved by studies of thin sections of soils and by studies of the associated mollusc faunas. Figure 10.8 shows soil types associated with loess in three soil provinces, arctic, Baltic (more oceanic) and Pannonian (more continanetal). Figure 10.9 shows a typical series of soil horizons within a loess sequence. The interpretation of such fossil soils as these in environmental terms is extremely difficult,[31] and it is much aided by finds of fossil molluscs associated with the soils.[39] Thus table 10.3 shows the relation between interglacial soil development on loess and the related mollusc faunas in central Europe. Another problem is the correlation of the soil horizons from one area to another. A change of climatic province will lead to a change of soil type, and perhaps, if a continental extreme is reached the lack of appearance of a soil, as loess deposition remained continuous. However, correlation has been improved by the radiocarbon dating of charcoal associated with soils in loess sequences.
2. Gumbotil in central United States.[49] Gumbotil is a grey, leached and reduced clay overlying and grading down into till. It is thought to be a B horizon of an interglacial soil formed in upland poorly drained clay-rich areas, and is found on all of the three older till sheets in the United States (Nebraskan, Kansan and Illinoian). The presence of gumbotil has been taken to be one of the main criteria in separating the times of till-sheet formation from one another by interglacial stages. More exact climatic conditions under which the gumbotils formed are not known. Partly the difficulty is one of nomenclature. Various soil types and also possibly sediments which have accumulated in depressions—the so-called accretion gleys—have been included in the term gumbotil, and this has led to discussion on the meaning and significance of gumbotil.
3. Recurrence surfaces in ombrogenous peats.[23, 24] Some sections through ombrogenous (see p. 56) peats show bands of dark highly humified peat overlain by lighter fresh peat (plate 8). The humification and oxidation is caused by a slowing down or cessation

Table 10.3 Correlation between development of terrestrial deposits and molluscan fauna in Pleistocene warm periods in Central Europe (Lozek, 1965)

Sedimentation, soil formation		*Molluscan fauna*
Renewed sedimentation, at the beginning very slow, with concomitant formation of soils of the chernozem series in dry regions (beginning of the new cold phase)	early glacial	Disappearance of woodland species, arrival of steppe assemblages with *Chondrula tridens*, *Helicopsis striata*, *Pupilla muscorum*, persistence of *Bradybaena fruticum*, *Arianta*
Intensive soil-forming processes—climax stages of brown-earths—parabraunerde, braunerde, terra fusca, pseudochernozem, pseudogley. Cessation of the formation of calcareous tufas	climatic optimum	Fully developed woodland assemblages with slowly receding southern species, abundant in the preceding periods: *Helicodonta obvoluta*, *Helix pomatia*, *Cepaea*, *Clausiliidae*, *Zonitidae*
Sedimentation and denudation cease. Soils attain the climax stage due to intensive soil processes. A strong removal of lime into subsoil and weathering		Fauna of climatic optimum. Maximum distribution of rich moisture-requiring assemblages—*Banatica* or *Ruderatus* fauna fully developed, disappearance of steppe species
Moderate hillwash, weathering and consolidation of the ground surface more intense; soils still in initial stages (frequently calcareous up to the surface), soil sediments weakly humic		Towards the close of the interval fully developed forest faunas with a great proportion of warmth- and moisture-requiring species appear, wide distribution of molluscs on the soils, still calcareous up to the surface, enclaves of steppe assemblages pushed back by forest
Hillwash still relatively intense, start of weathering and formation of soils; on the slopes coarser waste appears; start of formation of calcareous tufas		Arrival of warmth-loving fauna, thermophilous steppe elements penetrate to steppes (e.g. *Abida*, *Cepaea vindobonensis* locally also high up into the mountains), at first prevalence of tolerant species (see below), the share of less tolerant ones gradually increasing—*Cochlodina laminata*, *Aegopinella minor*, *pura*, *Clausilia cruciata*, *Discus ruderatus*, *Gastrocopta theeli*
Intense hillwash and moderate solifluction; in cold phases yellow-brown loams with a slightly increased humus content; angular small-size rock waste. Cessation of loess sedimentation	late glacial	Tolerant, more or less moisture-requiring assemblages, faunas with *Arianta arbustorum*, *Vitrea crystallina*, *Perpolita radiatula*, *Perforatella bidentata*; in warm localities *Chondrula tridens*, *Helicopsis striata*, arrival of *Cochlicopa lubricella* and *Truncatellina cylindrica*, disappearance of *Columella columella* and *Vallonia tenuilabris*
Loess sedimentation	glacial	Loess faunas: *Columella columella*, *Vallonia tenuilabris*, *Pupilla loessica*, *Vertigo parcedentata*; only tolerant species of open environments

of peat growth caused by drying out of the surface, that is, it is incipient soil formation. Both the local hydrology of the peat bog and regional climatic changes may be responsible for the formation of such recurrence surfaces. Whether these are synchronous or not in the same or different areas of peat accumulation is important. If they are, regional climatic tendencies to drying out are suggested. If not, the hydrological balance may be locally affected only, perhaps by an increased curvature of the bog surface or by a changing relation to the surrounding water table.

The most obvious recurrence surface in north European bogs is Weber's '*Grenzhorizont*', believed to have been caused by the bog surfaces drying out in Sub-Boreal time, followed by rapid peat accumulation in Sub-Atlantic time. Many other recurrence surfaces of different age are known, but their correlation over large areas remains in some doubt. Even within one bog it is known that surfaces in different parts of the bog were not necessarily formed in the same periods,[52] suggesting local factors at work controlling the rate of growth.

Isotopic determination of palaeoclimate

Variations in the proportions of stable isotopes of hydrogen, carbon and oxygen in biogenic carbonates, organic matter and ice can be in part temperature dependent. Measurement of the isotopic composition of such materials may then lead to a knowledge of temperature at the time they were formed. The most successful investigations of this type have concerned oxygen isotopes (^{16}O, ^{18}O) in marine biogenic carbonates and in ice, but other possibilities using terrestrial materials, such as lake carbonates, peat and wood are under active investigation at present.[38] For example, the deuterium (^{2}H) concentration of peat decreases systematically with increasing carbon content in time, and since the deuterium content of living plants is related to climate, including temperature, estimations of deuterium concentration at the time peat started forming may have palaeoclimatic significance.[51]

*Oxygen isotope ratios in ice.** An important recent development has been the application of $^{18}O/^{16}O$ measurements to ice cores through the Greenland[10] and Antarctic,[17] ice sheets. Lower $\delta\ ^{18}O$ values mean a lower temperature at which the water substance composing the ice was condensed. Therefore curves of $\delta\ ^{18}O$ values show changes in

* Oxygen isotope concentrations are given as differences, δ, per mille (‰) between the concentration of the heavier isotope in the analysed sample (^{18}O) and in a standard (either Standard Mean Ocean Water (SMOW) or PD Belemnite).

temperature over the period of ice formation. The results from Greenland and the Antarctic are shown in fig. 10.10. The age–depth relations of the cores are estimated from measured accumulation rates and calculations of thinning through flow. The curves obtained are similar and indicate a similar pattern of climatic change at both poles and that the last cold stage started some 75,000 years ago and ended some 11,000 years ago, dates which are close to those obtained from other evidence.

An interesting point about the Greenland curve is that it shows both very brief and sudden fluctuations, as at 89,000 years, and longer term fluctuations. This must be of importance in the consideration of the causes of climatic change. Possible correlations of the isotope curve with continental events are also shown in fig. 10.10.

Oxygen isotope ratios in marine biogenic carbonate.[56b] The relative abundance of oxygen isotopes ^{16}O and ^{18}O in biogenic carbonates (tests of foraminifers, mollusc shells) secreted by animals living in water is related to the abundance of these isotopes in the water at the time the carbonate was formed. The ratio in the carbonate will depend on the temperature of the ocean water and on glacially-controlled changes in the isotopic composition of the ocean water. Therefore by measuring the $^{18/16}O$ ratios in calcium carbonate skeletons, say tests of foraminifers found at successive levels in deep-sea cores, it is possible to deduce past changes in temperature or isotopic composition of sea water which took place while the sediment was forming.[14,15] The ratio of the isotopes in the carbonate is measured by a mass-spectrometer. The ratio changes are very small (^{18}O enrichment in respect to ^{16}O increases 0·02 per cent per 1°C temperature fall) and the instrument used has to be very sensitive.

Ocean carbonates are the subject of measurement because of the possible variations in $^{18/16}O$ ratio in shallow marine and inland waters, though recently measurements of freshwater carbonate have indicated that the method may be applicable to deposition from inland waters (fig. 10.11).[38]

The problem of interpreting the isotope ratio curves from cores lies in assessing the contributions made by changing temperature and by glacial control of the isotopic ratio.[55] Enrichment of ^{18}O in the oceans during glacial times results from the deficiency of ^{18}O in glacial ice (as mentioned above). Variation of the surface water isotope ratio also results from evaporation ($H_2{}^{16}O$ evaporates more rapidly than $H_2{}^{18}O$), thus giving variation between the ratio in tropical and arctic waters. Evidence of the significance of these contributory factors can be gained from studies of isotope ratios of

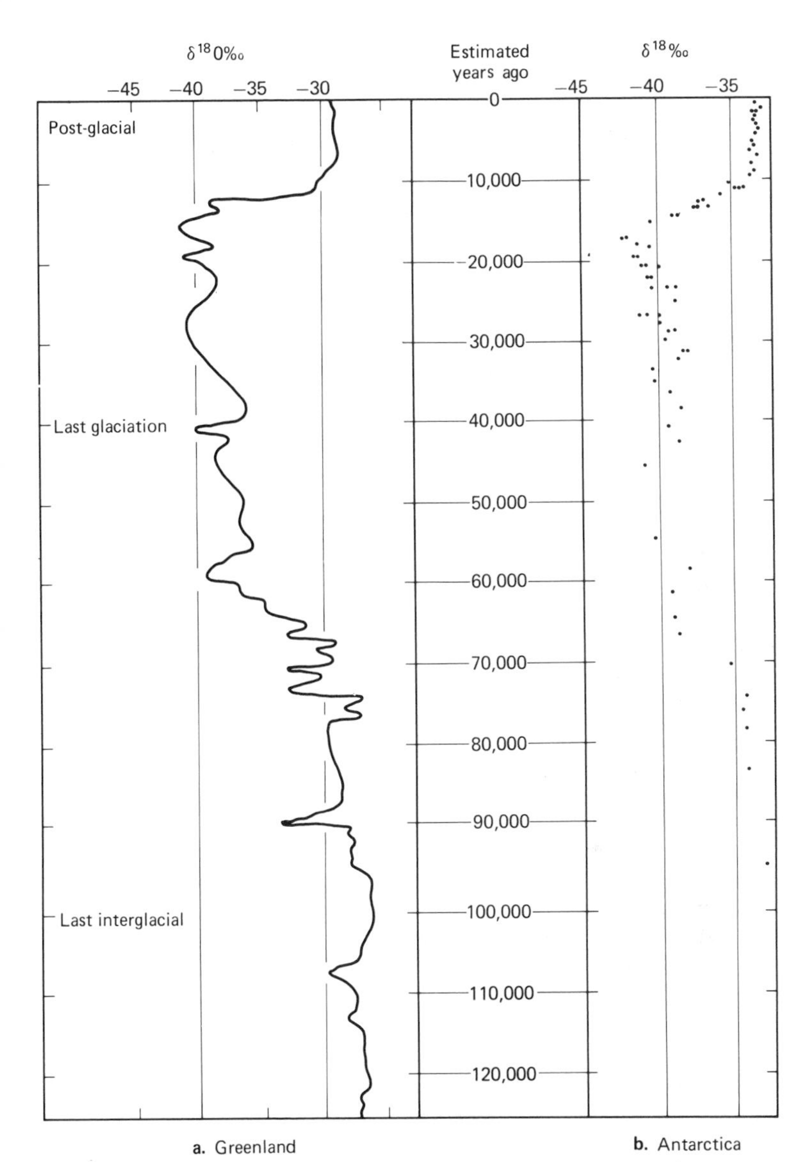

Fig. 10.10 Climatic variations in the last 100,000 years, expressed as changing $^{18}O/^{16}O$ ratios. δ is the per mille deviation of $^{18}O/^{16}O$ ratio in an ice sample from that of Standard Mean Ocean Water. Colder temperatures are indicated by move of the curves to the left (decreasing δ values).

a Variations in the Camp Century ice core, Greenland (after Dansgaard *et al.* 1971).

b Variations in the Byrd Station ice core, Antarctica (after Epstein *et al.* 1970).

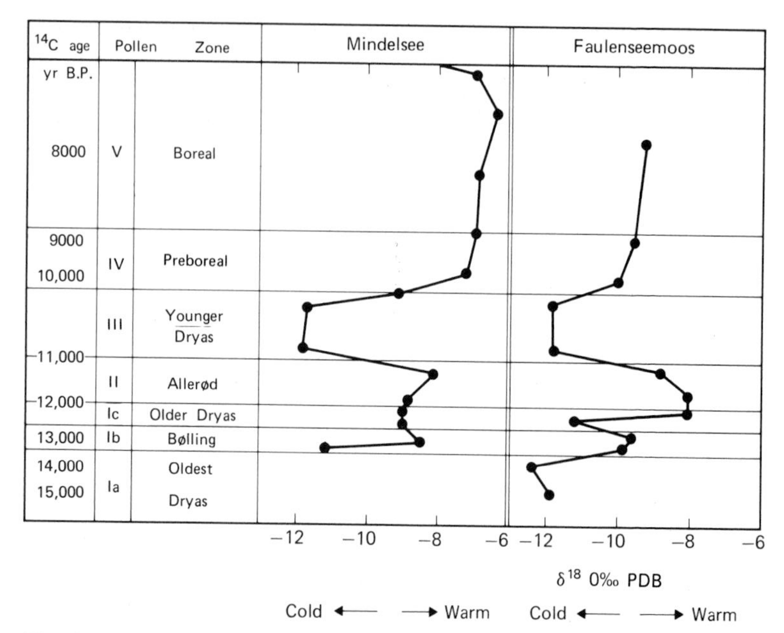

Fig. 10.11 ^{18}O content of lacustrine carbonates from two lake sites in central Europe. A correlation with pollen zones and radiocarbon years is given. The fluctuations show parallels with evidence of climatic change from sediments and vegetational history, with a climatic recession in the Younger Dryas (Lerman, 1974).

pelagic foraminifers, giving near surface ratios, and of benthic foraminifers, giving deep water ratios, associated with studies of changes in faunal composition and of isotopic ratios of water locked up in ice sheets during the glacial periods. It has been estimated that at most 30 per cent of the variation seen in the curves is due to temperature change in the oceans, with 70 per cent resulting from glacial control.[12] Whether this is right or not, it is clear that the isotope ratio curves are palaeoglaciation curves as well as palaeotemperature curves. As a palaeoglaciation curve, they will be reflecting total ice mass changes on the earth, not the growth and decay of any particular ice sheet. As a palaeotemperature curve, they suggest temperature changes in the ocean surface water of 4° to 7·5°C.

Figure 10.12 shows isotopic ratio curves of selected ocean cores. The sequences can be divided into isotope stages.

The curves obtained are important in showing that surface ocean waters fluctuated in temperature, but perhaps rather less than the estimated temperature change of the continental lowlands of present temperate regions, and that the successive fluctuations were of a

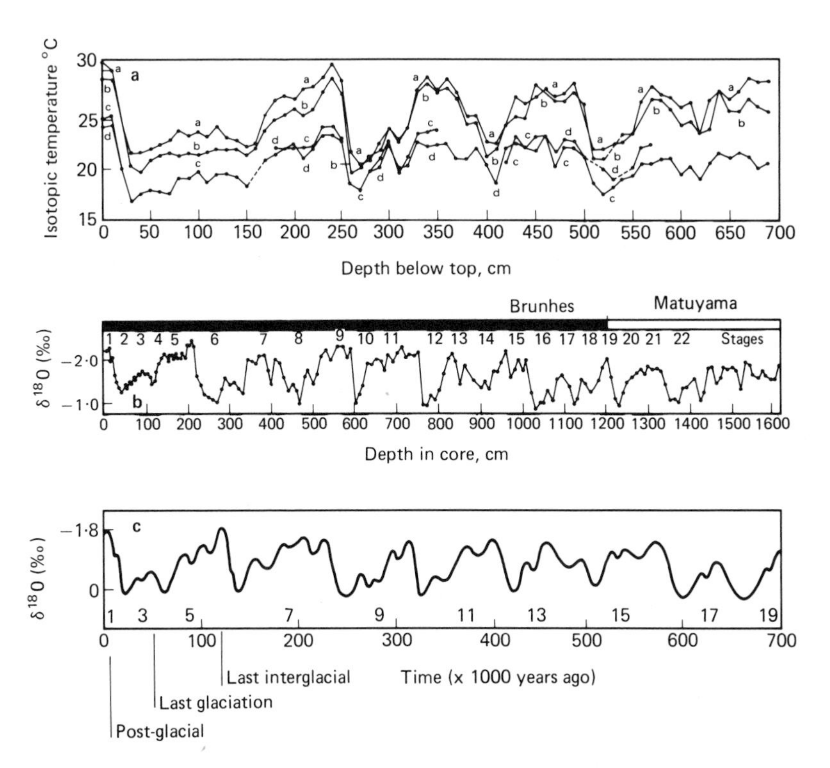

Fig. 10.12 A. Isotopic temperature curves from Caribbean core A 179–4, from tests of *Globigerinoides rubra* (a), *Globigerinoides sacculifera* (b), *Globigerina dubia* (c), and *Globorotalia menardii* (d) (Emiliani, 1955). There is good agreement between the curves from the four species.
B. Oxygen isotopic composition of *G. sacculifera* in core V28-238 complete record to 1600 cm, expressed as deviation % from Emiliani B1 standard, with magnetic stratigraphy (Shackleton and Opdyke, 1973).
C. Generalised palaeotemperature curve for the Brunhes Epoch. The numbers above the abscissa identify the isotope stages of deep-sea cores (Emiliani and Shackleton, 1974).

similar order of magnitude. There are, however, several difficulties inherent in the use of the curves as absolute indicators of change in temperature with time. First there may be difficulties in applying the method, resulting from changes in the isotope ratio because of exchange or because some organisms (e.g. coral) appear to show carbonate deposition unrelated to the environmental ratio. The curves do not show some details of temperature change which are known from studies of continental deposits. For example, an abrupt change in surface water temperature at 10,800 years ago is deduced, but no details of fluctuation around this point, as shown by the Late Weichselian zones I, II and III, are indicated. On the other hand there is a considerable degree of agreement between major climatic trends of the Weichselian in northwest Europe and the isotopic evidence from Atlantic cores[25] and the ice sheet cores from Greenland and the Antarctic.

Another difficulty results from the possibility that the record from some cores may be incomplete; parts may have been lost by slumping, etc., and so an unknown number of cycles of climatic change may be missing. However, with the abundant results now becoming available, this difficulty should recede. Even so, difficulties in the correlation of marine and terrestrial events back beyond the last interglacial still prevent convincing correlations. The isotopes curve for the whole Brunhes Epoch (fig. 10.12c), back to 700,000 years, is now well documented.[16, 56a] This period probably covers the whole of the northwest European glacial Pleistocene, and we can expect that in the not too distant future correlation of the isotope curve with the glacial/interglacial sequence will be possible.

Precipitation changes

Evidence for precipitation changes is more slender than that for temperature changes. On the one hand there are the supposed pluvial periods when lakes in the middle and equatorial lattitudes stood higher than present, and these periods have been correlated with glacial stages. Such higher levels might be caused by lower evaporation as a result of lower temperatures as well as higher precipitation. Recent radiocarbon evidence indicates that African lake levels were higher than now in the 8000- to 10,00-year period,[6] but the climatic relations of such changes with the climatic changes of higher latitudes are not yet clear. In these higher latitudes, on the other hand, there are observations on snow-line changes, for example, in the Alps, which demonstrate that snow-line depression was not accentuated in the direction of normal rain-bearing winds. Such might be expected if there was much increase in precipation. In fact, there is evidence for decreased precipitation as a whole during glacial stages. For example, it has been estimated[42,43] that in the last glacial stage the precipitation over the highlands in Britain only reached 80 per cent of today's value. This figure was suggested by the absence of cirques in southern Britain, which would have been caused by a lower firn limit related to greater precipitation, whereas much higher firn limits related to much decreased precipitation would not have produced the known ice sheets. Evidence of lower precipation in the glacial stages also comes from the presence of ice-wedge casts in the periglacial areas. Ice-wedges are only likely to form in the absence of a substantial insulating snow cover. Thus annual rain precipitation in eastern England has been estimated by Williams to be not more than 10 cm during times of ice-wedge formation in the last glaciation (60 cm at present). The presence of maritime plants in last glacial floras also indicates the development of saline conditions and a low rainfall.

The lower precipitation suggested for the glacial stages does not conflict with great expansion of ice. Lower temperatures mean less ablation, very significant for ice accumulation. Recent climatic events support this relationship. In the 1920s to 1940s storm tracks moved northwards in the northern hemisphere, bringing greater precipitation and heat to higher latitudes, with a consequent shrinkage of glaciers. Conversely, there is increasing evidence that during the glacial stages the storm tracks were forced south, with the development of an anticyclone over Europe, with accompanying continental conditions of climate—low rainfall, cold winters and warm summers.

It is thus certain that a pattern of precipitation changes accompanied each ice advance and retreat. Perhaps regional precipitation increased in our area during the early parts of the ice advances as belts of cyclonic activity moved south, and later during the glacial, at the time of maximum cold, the colder seas resulted in less moisture being available and less precipitation in the belts of cyclonic activity. Thus redistribution of precipitation accompanied ice advance and retreat.

Wind

Evidence from marine cores off the west coast of Africa for increased vigour of trade winds during the glacial stages has already been mentioned.[3a, 46] This agrees with conclusions of meteorologists that circulation was intensified during the glacial stages, with a general displacement of the zonal circulation southwards. In north-west Europe the establishment of a persistent anticyclone over Scandinavia during the glacial stages appears to have lead to cool continental conditions in eastern England. The scarcity of loess suggests no great wind intensity, but what evidence there is from windblown sands indicates westerly winds.[63] Perhaps there were impersistent westerly winds strong enough to move sand, against a background of weak easterlies.

Conclusions—past climatic change

Of all the factors involved in climatic change in the Pleistocene, it is clear that temperature changes are best documented. The magnitude of these according to various types of evidence is indicated in table 10.1. Estimates of the amount of fluctuation of temperature between the temperate and cold stages vary between 3°C and 15°C, but 5° to 8°C fluctuation of mean annual temperature is a value generally considered to be near the mark. As we have seen, there are considerable difficulties in deducing seasonal changes of temperature and precipitation values. Combination of various types of evidence, geological and biological, proves most rewarding, as illustrated by the evidence for the last glaciation climate from plant remains, insect remains and periglacial features. Each group of organisms and each type of palaeoclimate evidence provides its own insight into climates of the past. The comparison of insect and plant evidence is especially interesting in this respect, beetles appearing to be more immediately responsive in terms of distribution changes than plants, so that oscillations in climate may be more easily recognised from beetle assemblages than from plant assemblages. Another example of such an approach is shown in fig. 10.13, giving conclusions about the Late Devensian climate of the British Isles. The combination of palaeoclimate evidence from the terrestrial sequence and the marine sequence gives much detail to climates of the cold and temperate stages. The terrestrial sequence provides detail of the interglacial sequences and the bald outline of glacial advances. The marine

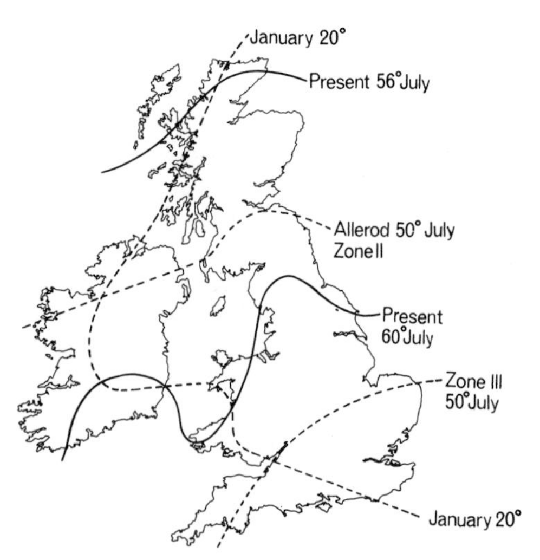

Fig. 10.13 Suggested isotherms for the Allerød interstadial (zone II) and the succeeding climatic recession of zone III, during the Devensian lateglacial of the British Isles (Manley, 1951).

record shows how short the temperate episodes were, and provides much detail of the cold stages.

We may expect in future that the marine stratigraphy will provide a basis for the Pleistocene sequence of climatic changes, but the terrestrial sequence will be needed for its biographical and evolutionary resolving power and for the expression of the consequences of the climatic changes seen in the marine cores.

RECENT CLIMATIC CHANGE[35, 37, 42]

We gain insight into climatic change by studies of recent changes and associated effects in Britain.

In brief the major changes which are believed to have occurred over the period of the last 7000 years are shown in Table 10.4.

Table 10.4 Recent climatic trends in Britain

5000–3000 B.C.	Flandrian climatic optimum. Annual mean temperatures about 2°C above present.
500 B.C.	climatic deterioration (Sub-Boreal).
1100–1300 A.D.	mediaeval warm period.
1300 A.D.	climatic deterioration set in.
1550–1700 A.D.	'little ice age'.
1850–1940 A.D.	warmth increase.
1950 A.D. to present	warmth decrease.

The detailed instrumental records over the last 100 years or so are of particular interest. There has been a period of general warming, arctic sea ice has thinned and decreased in area, deglaciation has been prevalent, and plants and animals have changed their ranges in response to the amelioration.[19, 35] These changes are also documented in the oxygen isotope curve from the Greenland ice sheet, shown in fig. 10.14 and compared with the temperature record from Iceland and England.[11] Figure 10.15 gives details of temperature, precipitation and westerly wind incidence in this period in the British Isles, together with a record of changing global temperatures. Higher temperatures were characteristic of three or four decades around 1930, and terminated a long period of temperature rise. Annual rainfall increased in Europe, including Britain, and in much of western Asia and the western parts of other continents, and there was also a general increase of snow and rain at higher latitudes. There was also an increase in the incidence of westerly winds, involving an increase in strength of the world's main wind-streams. This period also appears to have been a time of minimum precipitation in the eastern parts of temperate continents and intensification of desert

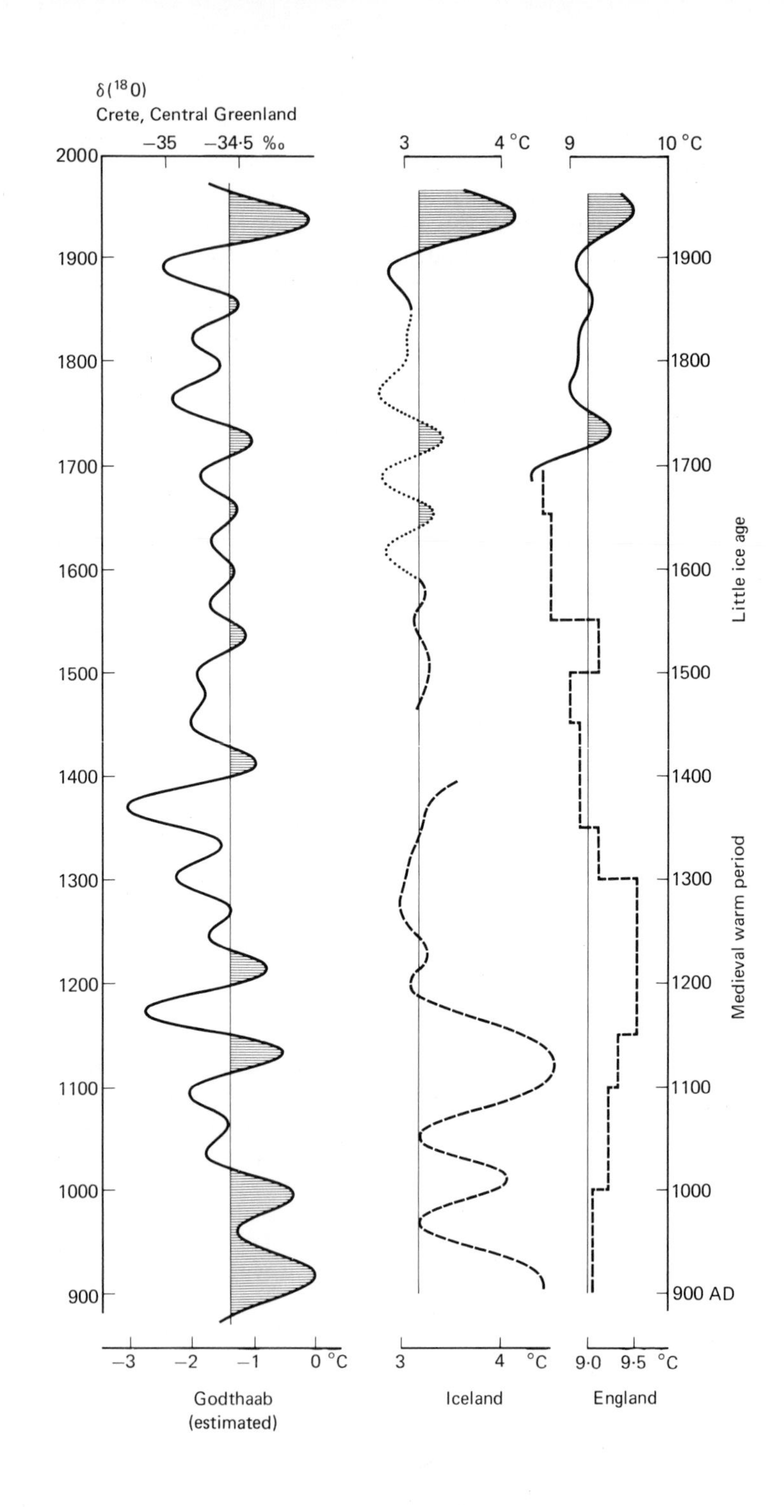
δ(18O)
Crete, Central Greenland
−35
−34·5 ‰
3
4 °C
9
10 °C
2000
1900
1800
1700
1600
1500
1400
1300
1200
1100
1000
900
900 AD
Little ice age
Medieval warm period
−3
−2
−1
0 °C
3
4
°C
9·0
9·5
°C
Godthaab
(estimated)
Iceland
England

conditions in arid regions. A weaker circulation has predominated in more recent years, especially since 1958; westerlies have decreased and so have temperatures and rainfall.

The atmospheric circulation is an important part of these changes, both in intensification, making the polar regions warmer by the penetration of warm westerly wind-streams to the north, and by encouraging warm ocean currents northwards as well. As a result storminess in middle latitudes decreases and the climate generally ameliorates. Conversely, weakening of the circulation and an increase of northerly and easterly winds at the expense of westerlies, as observed since 1940, is related to climatic deterioration, as shown by the increase of sea ice during the last decade in the Arctic and in the north Atlantic. A general correlation can be seen here of temperature, rainfall and circulation.

The question is whether there is a relation between an external factor such as intensity of solar radiation and the atmospheric circulation. There have been suggestions of a relation of sunspot activity and ultraviolet intensity to circulation changes, but it remains to be demonstrated exactly what is the relation between circulation itself, dependent on the unequal heating of the equatorial and polar areas, and external factors which might produce changes in the thermal gradient. Other factors will also affect the thermal gradient, e.g. the high albedo of snow and ice will give significance to the area covered by seasonal or perennial snow and ice; and also changes in moisture content of atmosphere and cloudiness, themselves dependent on the general atmospheric circulation.

CAUSES OF CLIMATIC CHANGE

Climatic change can take place on very different time scales: very long-term changes (10^6 to 10^9 years) seen in the geological record, medium term changes (10^4 to 10^5 years), corresponding in scale to the major Pleistocene fluctuations, short-term recurrent changes of a cyclic or quasiperiodic nature (1 to 10^3 years), which may be expressed as changes in barometric pressure, frequency of westerly-type days, etc., and short term changes which may be related to a particular cause, e.g. a solar flare.

Fig. 10.14 Comparison between the ^{18}O content of fallen snow at Crête, Central Greenland and temperatures for Iceland and England. The continuous curves are based on direct observation, the dotted curve (Iceland) on ice observations and the dashed curves on indirect evidence (Dansgaard *et al.* 1975).

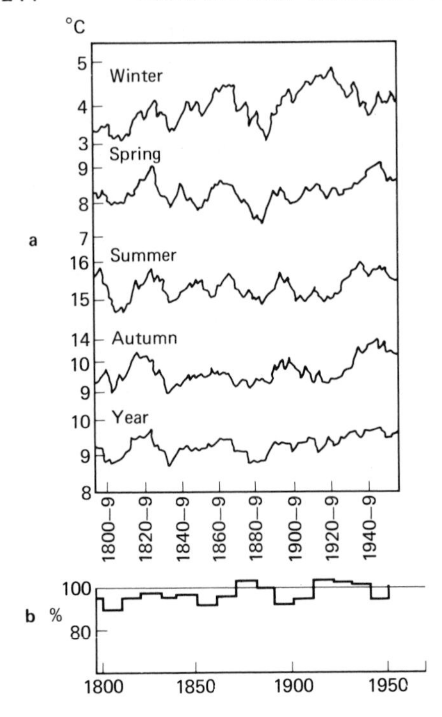

°C
a
Winter
Spring
Summer
Autumn
Year
5
4
3
9
8
7
16
15
14
10
9
10
9
8
1800–9
1820–9
1840–9
1860–9
1880–9
1900–9
1920–9
1940–9
b %
100
80
1800
1850
1900
1950

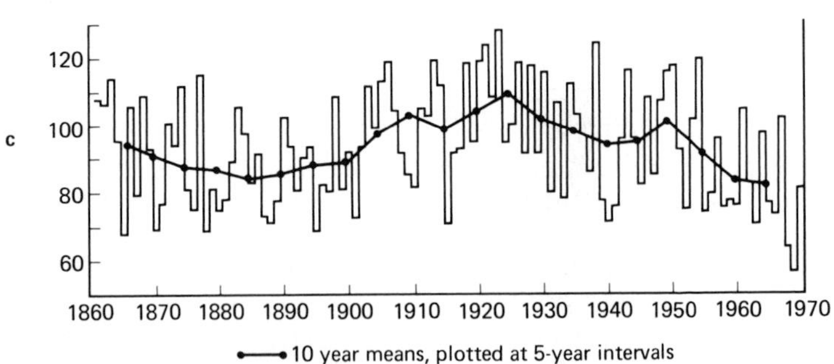

c
120
100
80
60
1860
1870
1880
1890
1900
1910
1920
1930
1940
1950
1960
1970
10 year means, plotted at 5-year intervals

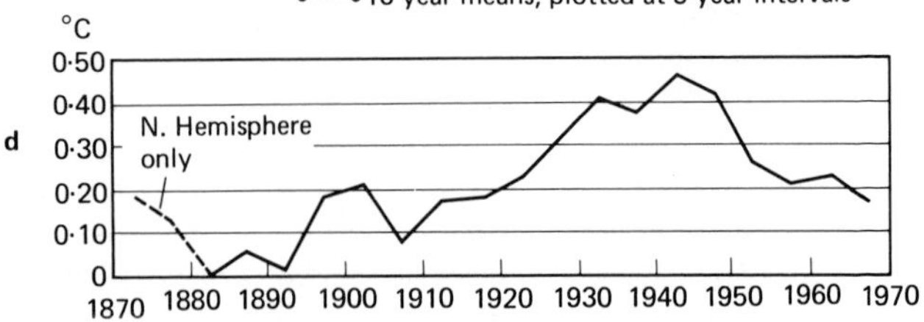

°C
d
0·50
0·40
0·30
0·20
0·10
0
N. Hemisphere only
1870
1880
1890
1900
1910
1920
1930
1940
1950
1960
1970

It is the medium-term changes of the major Pleistocene climatic events which require explanation if we are to find the cause or causes of the ice age.

A large number of hypotheses have been put forward to account for the climatic fluctuations characteristic of the Pleistocene. Any explanation should say why at the end of the Tertiary, during which there was a gradual decrease in temperature, world-wide temperatures were depressed sufficiently to ensure formation and expansion of glaciers in mountainous areas where precipitation was high enough. And it should also account for the fluctuations in temperature after this time, fluctuations of the magnitude of 5 to 8°C. Any theory of causation of ice ages should also be applicable to the pre-Pleistocene glaciations, such as those of the Permo-Carboniferous and Eo-Cambrian.

We can first briefly mention the factors which have been regarded as possible agents of climatic change. These are terrestrial and extra-terrestrial. The terrestrial factors suggested include changes in configuration of land and sea, the increased building of highlands in the late-Tertiary on which glaciers could develop, and the effect of changes of carbon dioxide concentration in the atmosphere (increase leads to inhibition of outgoing radiation from the earth to space, and more absorption of the incoming solar radiation on its passage through the atmosphere, resulting in raised temperatures at the earth's surface), and increased cloudiness (leading to depression of temperatures by loss of radiation by reflection). Extra-terrestrial factors involve variations of the solar radiation 'constant' in time and the possibility of variation in solar radiation received by the earth because of interception by inter-stellar matter. Variations in solar radiation have been linked to observed short-term climatic fluctuations, and changes in the earth's magnetic field have also been linked to climatic change.[64] A third class of factors involves changing the radiation received on various parts of the earth because of variation in the elements of the earth's orbit round the Sun.

Fig. 10.15 Recent climatic changes in Britain.

a. Average temperatures in Central England (°C), shown by 10-year running means for each season and for the year (after Manley).
b. Rainfall in England and Wales, plotted as decade averages which are shown as percentages of the average 1900–39 (932 mm) (Lamb, 1965).
c. Number of days classified as westerly type for each year since 1861 in the British Isles (Lamb, 1972).
d. The variation of global mean surface air temperature, expressed as deviation from the mean, in successive 5-year averages from 1880–84 to 1965–69, after J. M. Mitchell and the National Center for Atmospheric Research (NCAR), Boulder, Colorado (Lamb, 1974).

To illustrate hypotheses put forward we may take four examples, one dependent on change in solar radiation, one dependent on changes in solar radiation resulting from variation in the elements of the earth's orbit, one dependent on terrestrial changes, and one combining the two.

Simpson's hypothesis[58, 59]

This is interesting as an attempt to demonstrate that an increase in solar radiation results in multiple glaciation. The course of events is shown in fig. 10.16. From point A to point C solar radiation increases, temperature rises, circulation between the Equator and the Poles increases as a result and results in increased cloudiness and precipitation. This increased precipitation lowers the snow-line and glaciation ensues. From point C to point D there is a continued rise in radiation, the temperature increases further, snow-lines rise and glaciation ends. The process reverses during a decrease of radiation. Thus Simpson obtains (fig. 10.12), during an oscillation of the solar 'constant', a glaciation preceded by a cool dry interglacial and followed by a warm wet interglacial. In the extra-glacial area these

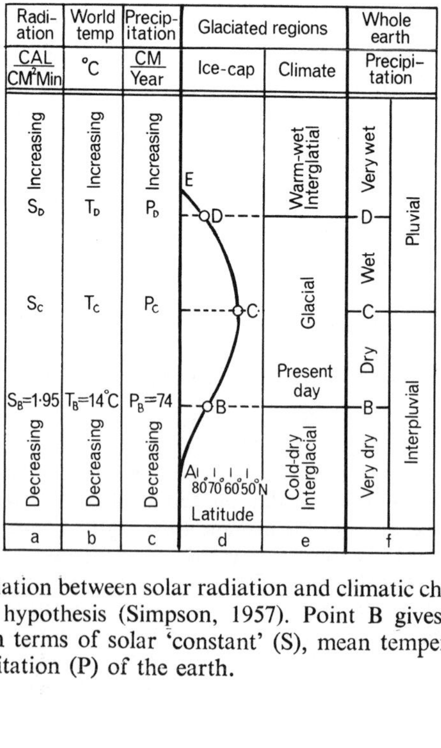

Fig. 10.16 Relation between solar radiation and climatic change according to Simpson's hypothesis (Simpson, 1957). Point B gives conditions at present day in terms of solar 'constant' (S), mean temperature (T) and annual precipitation (P) of the earth.

fluctuations would be represented by an arid interpluvial period and a pluvial period.

There is in fact no evidence from interglacial climates to support the two types of interglacial resulting from this scheme. But the ideas are valuable in demonstrating possible links between radiation changes and climatic events in the Pleistocene.

Variation in solar radiation received by the earth[67]

Even though the radiation emitted by the sun may remain constant, variation in radiation received in different latitudes and at different seasons will result from secular changes of the earth's orbital elements. The astronomical theory of climatic change says that this variation is the cause of the Pleistocene fluctuations.[67] The radiation received at the top of the atmosphere in the course of time can be calculated for the different latitudes and seasons. The result is a radiation curve of the type associated with the name of Milankovitch.

Various attempts have been made to relate radiation curves to the climatic events revealed by geological studies. For example, there is evidence that the major cold and warm events of the last 150,000 years can be matched in the curve giving variation of solar radiation resulting from the precession of the equinoxes. Also the sea-level curve for the last 230,000 years shows a correlation with radiation curves for 55°/45°N.[7] The problem is then to find the mechanism which relates radiation variation directly to a process which will end in causing glacial advance and retreat. The analysis of latitudinal change of radiation in time may indicate, since the variations are dated, a correlation with terrestrial climatic events (such as the occurrence of an interglacial), and this may be a starting point for working out a casual relation.[45]

Ewing and Donn's hypothesis[20]

This tries to explain recurrent glaciation as a result of terrestrial changes, on the premise that an ice-free Arctic ocean favours the formation of glaciers on the surrounding land masses. During a warm (interglacial) period the Arctic is ice free as a result of a flow of warm water north from the Atlantic. This may have been the cause of increased precipitation, and thus glacier growth, in the surrounding areas, with a resultant lowering of world sea-level. This causes the shallowing of the sea or the emergence of a land bridge between Iceland and the Faroes, preventing a northerly penetration of warm waters. The Arctic Ocean freezes over, precipitation to the surrounding glaciers and ice sheets is reduced and they wane. Sea-level is restored, the Arctic Ocean ice melts and the cycle starts again.

The cycles began, according to this hypothesis, when the North Pole reached a position favourable to the process, at the beginning of Pleistocene time.

There are several objections to this hypothesis. For example, there is little evidence for much polar wandering in the late Tertiary, and if the source of precipitation was the Arctic Ocean there surely would have been more glaciation north of the mountains of Alaska and Anadyr than is now recognised.

Solar-topographic hypothesis of R. F. Flint

This relies on the observed relation between late-Tertiary mountain-building and the advent of Pleistocene ice sheets and glaciers to bring about the first cold period of the Pleistocene, and on fluctuations in solar radiation (already discussed) to produce temperature variations thereafter.

The solar-topographic hypothesis seems the most reasonable general explanation for recurrent glaciation in the Pleistocene. But whether variation in the received radiation depends on variation in solar 'constant' or on changes in the elements of the earth's orbit or some other property of the sun (e.g. sunspot cycles), or some or all of these, it is not possible to say. If the correlation between a Milankovitch-type curve and the geological evidence for climatic change is substantiated then the point will be decided.

If the major climatic changes are caused in the way suggested by the solar–topographic hypothesis, it will still be possible that minor fluctuations, such as those demonstrated in the ice sheet cores, are caused by other factors, such as changes in distribution of land and sea or ice surges.

As stated at the beginning of the chapter, we need more evidence of the nature, magnitude and succession of climatic change before we can find the cause or causes, and the link between cause and effect, in this matter of Pleistocene climatic change.

REFERENCES

Understanding climatic change. 1975. Washington: National Academy of Sciences.

LAMB, H. H. 1972. *Climate: present, past and future*. Vol. 1. *Fundamentals and climate now*. London: Methuen.

MITCHELL, J. M. 1965. 'Theoretical palaeoclimatology'. In *The Quaternary of the United States*, eds. H. E. Wright and D. G. Frey. 881–901. Princeton University Press.

SCHWARZBACH, M. 1963. *Climates of the Past.* London: van Nostrand.
WRIGHT, H. E. 1961. 'Late Pleistocene climates of Europe: a review', *Bull. Geol. Soc. Amer.*, **72,** 933–84.

1 ANDERSEN, S. T. 1961. 'Vegetation and its Environment in Denmark in the Early Weichselian Glacial', *Danm. Geol. Unders.*, II Raekke, No. 75.
2 ANDERSSON, G. 1910. 'Swedish climate in the Late-Quaternary Period'. In *Die Veränderungen des Klimas seit dem maximum der letzten Eiszeit*, pp. 247–94. XI Int. Geol. Congress. Stockholm. Stockholm: Generalstabens Litografiska Anstalt.
3 BLYTT, A. 1876. *Essay on the immigration of the Norwegian flora during alternating rainy and dry periods.* Christiana: Cammermeyer.
3a BOWLES, F. A. 1975. 'Paleoclimatic significance of quartz/illite variations in cores from the eastern equatorial north Atlantic', *Quaternary Research*, **5,** 225–35.
4 BROOKS, C. E. P. 1951. 'Geological and historical aspects of climatic change'. In *Compendium of Meteorology*, ed. Malone, T. F. pp. 1004–23. Boston: American Meteorological Soc.
5 BRØGGER, W. C. 1900–01. 'Om de senglaciale og postglaciale nivåforandringer i Kristianfeltet (Molluskenfaunan)', *Norges Geol. Unders.*, No. 31.
6 BUTZER, K. W., ISAAC, G. L., RICHARDSON, J. L. and WASHBOURN-KAMAU, C. 1972. 'Radiocarbon dating of East African lake levels', *Science, N.Y.*, **175,** 1069–76.
7 CHAPPELL, J. 1973. 'Astronomical theory of climatic change: status and problem', *Quaternary Research*, **3,** 221–36.
7a Climap project members. 1976. 'The surface of the Ice-Age earth', *Science, N.Y.*, **191,** 113–7.
8 COOPE, G. R. 1959. 'A Late Pleistocene insect fauna from Chelford, Cheshire', *Proc. R. Soc. London*, B, **151,** 70–86.
9 COOPE, G. R. 1975. 'Climatic fluctuations in northwest Europe since the Last Interglacial, indicated by fossil assemblages of Coleoptera'. In *Ice-Ages: ancient and modern*, ed. A. E. Wright, and F. Moseley, 153–68. Liverpool: Seel House Press.
10 DANSGAARD, W., JOHNSEN, S. J., CLAUSEN, H. B. and LANGWAY, C. C. 1971. 'Climatic record revealed by the Camp Century ice core'. In *The Late Cenozoic Glacial Ages*, ed. K. K. Turekian, 37–56. New Haven: Yale University Press.
11 DANSGAARD, W., JOHNSEN, S. J., REEH, N., GUNDESTRUP, N., CLAUSEN, H. B. and HAMMER, C. U. 1975. 'Climatic changes, Norsemen and modern man', *Nature* (*Lond.*), **255,** 24–8.
12 DANSGAARD, W. and TAUBER, H. 1969. 'Glacier oxygen-18 content and Pleistocene ocean temperatures', *Science, N.Y.*, **166,** 499–502.
13 DEGERBØL, M. and KROG, H. 1951. 'Den europaeiske Sumpskildpadde (*Emys orbicularis L.*) i Danmark (with English summary)', *Danm. Geol. Unders.*, II Raekke, No. 78.

14 EMILIANI, C. 1955. 'Pleistocene temperatures', *J. Geol.*, **63,** 538–78.

15 EMILIANI, C. 1971. 'The amplitude of Pleistocene climatic cycles at low latitudes and the isotopic composition of glacial ice'. In *The Late Cenozoic Glacial Ages*, ed. K. K. Turekian, 183–97. New Haven: Yale University Press.

16 EMILIANI, C. and SHACKLETON, N. J. 1974. 'The Brunhes Epoch: isotopic paleotemperatures and geochronology', *Science, N.Y.*, **183,** 511–14.

17 EPSTEIN, S., SHARP, R. P. and GOW, A. J. 1970. Antarctic ice sheet: Stable isotope analyses of Byrd station cores and interhemispheric climatic implications. *Science, N.Y.*, **168,** 1570–1572.

18 ERICSON, D. B., EWING, P. M. and WOLLIN, G. 1964. 'The Pleistocene epoch in deep-sea sediments', *Science, N.Y.*, **146,** 723–32.

19 ERKAMO, V. 1956. 'Untersuchungen über die pflanzenbiologischen und einige andere folgeerscheinungen der neuzeitlichen klimaschwankung in Finnland', *Ann. Bot. Soc. 'Vanamo'*, **28,** No. 3.

20 EWING, M. and DONN, W. L. 1958. 'A theory of ice ages', *Science, N.Y.*, **127,** 1159–62.

21 FEYLING-HANSSEN, R. W. 1964. 'Foraminifera in Late Quaternary deposits from the Oslofjord area', *Norges Geol. Unders.*, 225.

22 FIRBAS, F. 1949–52. *Waldgeschichte Mitteleuropas.* Jena: G. Fischer.

23 GODWIN, H. 1954. 'Recurrence-surfaces', *Danm. Geol. Unders.*, II Raekke, No. 80, 22–30.

24 GRANLUND, E. 1932. 'De svenska högsmossarnas geologi', *Sver. Geol. Unders.*, Ser. C., No. 373.

25 HAMMEN, T. VAN DER, MAARLEVELD, G. C., VOGEL, J. C. and ZAGWIJN, W. H. 1967. 'Stratigraphy, climatic succession and radiocarbon dating of the last glacial in the Netherlands', *Geol. en Mijnbouw*, **46,** 79–95.

26 IMBRIE, J. VAN DONK, J. and KIPP, N. G. 1973. 'Palaeoclimatic investigation of a Late Pleistocene Caribbean deep-sea core: comparison of isotopic and faunal methods', *Quaternary Research*, **3,** 10–38.

27 IVERSEN, J. 1944. '*Viscum*, *Hedera* and *Ilex* as climate indicators', *Geol. Fören. Förh.*, **66,** 463–83.

28 IVERSEN, J. 1954. 'The Late-Glacial Flora of Denmark and its Relation to Climate and Soil', *Danm. Geol. Unders.*' II Raekke, No. 80, 87–119.

29 IVERSEN, J. 1960. 'Problems of the Early Post-glacial forest development in Denmark', *Danm. Geol. Unders.*, IV Raekke, **4,** No. 3.

30 JESSEN, K. and MILTHERS, V. 1928. 'Stratigraphical and palaeontological studies of interglacial freshwater deposits in Jutland and north west Germany', *Damn. Geol. Unders.*, II Raekke, No. 48.

31 KLIMA, B., KUKLA, J., LOZEK, V. and DE VRIES, H. 1962. 'Stratigraphie des Pleistozäns und alter des Paläeolithischen rastplatzes in der Ziegelei von Dolni Vestonice (unter-Wisternitz)', *Anthropozoicum*, **11,** 93–145.

32 KOWALSKI, K. 1971. 'The biostratigraphy and paleoecology of Late Cenozoic mammals of Europe and Asia'. In *The Late Cenozoic Glacial Ages*, ed. K. K. Turekian, 465–77. New Haven: Yale University Press.

33 KUBIENA, W. L. 1953. *The soils of Europe*. London: Murby.

34 KUBIENA, W. L. 1956. 'Zur Micromorphologie, Systematik und Entwicklung der rezenten und fossilen Lössböden', *Eiszeitalter u. Gegenwart*, **7**, 102–12.

35 LAMB, H. 1965. 'Britain's changing climate'. In *The biological significance of climatic changes in Britain*, ed. C. G. Johnson, 3–34. London, Institute of Biology: Academic Press.

36 LAMB, H. H. 1974. 'Reconstructing the climatic patterns of the historical past', *Endeavour*, **33**, 40–7.

37 LAMB, H. H. 1975. 'Changes of climate: the perspective of time scales and a particular examination of recent changes'. In *Ice Ages: ancient and modern*, ed. A. E. Wright and F. Moseley, 169–88. Liverpool: Seel House Press.

38 LERMAN, J. C. 1974. 'Isotope "palaeothermometers" on continental matter: assessment'. In *Colloques Internat. CNRS*, No. 219, Les méthodes quantitatives d'étude des variations du climat au cours du Pléistocène, 163–81.

39 LOZEK, V. 1965. 'The relationship between the development of soils and faunas in the warm Quaternary phases', *Sbornik geologickych Ved*, Antropozoicum, rada A. sv. 3, 7–33.

40 MAARLEVELD, G. C. 1960. 'Wind directions and cover sands in the Netherlands', *Biul. Peryglacjalny*, **8**, 49–58.

41 MANGERUD, J., ANDERSEN, S. T., BERGLUND, B. E. and DONNER, J. J. 1974. 'Quaternary stratigraphy of Norden, a proposal for terminology and classification', *Boreas*, **3**, 109–28.

42 MANLEY, G. 1951. 'The range of variation of the British climate', *Geogr. J.*, **117**, 43–68.

43 MANLEY, G. 1959. 'The Late-glacial climate of north-west England', *Liverpool and Manchester Geological Journal*, **2**, 188–215.

44 MCINTYRE, A. and RUDDIMAN, W. F. 1973. 'Northeast Atlantic post-Eemian paleooceanography: a predictive analog of the future', *Quaternary Research*, **2**, 350–4.

45 MITCHELL, J. M. 1972. 'The natural breakdown of the present interglacial and its possible intervention by human activities', *Quaternary Research*, **2**, 436–45.

46 PARKIN, D. W. 1974. 'Trade winds during the glacial cycles', *Proc. R. Soc. Lond.*, A, **337**, 73–100.

47 POSER, H. 1948. 'Boden- und Klimaverhälnisse in Mittel und Westeuropa während der Würmeiszeit', *Erdkunde*, **2**, 53–68.

48 POSER, H. 1954. 'Zur rekonstruktion der Spätglazialen Luftdruckverhältnisse in Mittel- und Westeuropa auf Grund der vorzeitlichen Dünen', *Erdkunde*, **4**, 81–8.

49 RUHE, R. V. 1965. 'Quaternary palaeopedology'. In *The Quaternary of the United States*, eds. Wright, H. E. and Frey, D G. 755–64. Prince-

ton University Press.

50 SANCETTA, C., IMBRIE, J. and KIPP, N. G. 1973. 'Climatic record of the past 130,000 years in north Atlantic deep sea core V23–82: correlation with the terrestrial record', *Quaternary Research*, **3,** 110–16.

51 SCHIEGL, W. E. 1972. 'Deuterium content of peat as a palaeoclimatic recorder', *Science, N.Y.*, **175,** 512–13.

52 SCHNEEKLOTH, H. 1965. 'Die rekurrenzfläche in Grossen Moor bei Gifhorn—eine zeitgleiche Bildung', *Geol. Jb.*, **83,** 477–96.

53 SCHÖNHALS, E., ROHDENBURG, H. and SEMMEL, A. 1964. 'Ergebnisse neuer Untersuchungen zur Würmlöss-Gliederung in Hessen', *Eiszeitalter u. Gegenwart*, **15,** 199–206.

54 SERNANDER, R. 1908. 'On the evidence of Postglacial changes of climate furnished by the peat-mosses of Northern Europe', *Geol. Fören. Förh.*, **30,** 465–78.

55 SHACKLETON, N. J. 1967. Oxygen isotope analyses and Pleistocene temperatures re-assessed. *Nature* (*Lond.*), **215,** 15–17.

56a SHACKLETON, N. J. and OPDYKE, N. D. 1973. 'Oxygen isotope and palaeomagnetic stratigraphy of equatorial Pacific core V28–238: oxygen isotope temperatures and ice volumes on a 10^5 and 10^6 year scale', *Quaternary Research*, **3,** 39–55.

56b SHACKLETON, N. J. and OPDYKE, N. D. 1976. 'Oxygen-isotope and palaeomagnetic stratigraphy of Pacific core V28–239 late Pliocene to latest Pleistocene'. *Geol. Soc. Amer. Mem.* **145,** 449–64.

57 SHOTTON, F. W. 1962. 'The physical background of Britain in the Pleistocene', *Adv. Sci.*, **19,** 1–14.

58 SIMPSON, G. C. 1957. 'Further studies in World Climate', *Quart. J. R. Meteor. Soc.*, **83,** 459–85.

59 SIMPSON, G. C. 1959. 'World temperatures during the Pleistocene', *Quart. J. R. Meteor. Soc.*, **85,** 332–49.

60 SIMPSON, I. M. and WEST, R. G. 1958. 'On the stratigraphy and palaeobotany of a late-Pleistocene organic deposit at Chelford, Cheshire', *New Phytol.*, **57,** 239–50.

61 SZAFER, W. 1954. 'Pliocene flora from the vicinity of Czorsztyn, (West Carpathians) and its relationship to the Pleistocene', *Instytut Geologiczny*, Prace Tom XI. Warsaw.

61a VALENTINE, K. W. G. and DALRYMPLE, J. B. 1975. 'The identification, lateral variation, and chronology of two buried paleocatenas at Woodhall Spa and West Runton, England', *Quaternary Research*, **5,** 551–90.

62 WEBB, T. and BRYSON, R. A. 1972. 'Late- and post-glacial climatic change in the northern Midwest, USA: quantitative estimates derived from fossil pollen spectra by multivariate statistical analysis', *Quaternary Research*, **2,** 70–115.

63 WILLIAMS, R. B. G. 1975. 'The British climate during the Last Glaciation; an interpretation based on periglacial phenomena'. In *Ice Ages: ancient and modern*, eds. A. E. Wright and F. Moseley, 95–120. Liverpool: Seel House Press.

64 WOLLIN, G., ERICSON, D. B. and RYAN, W. B. F. 1971. 'Magnetism of the earth and climatic changes', *Earth Planet. Sci. Lett.*, **12,** 175–83.

65 YAALON, D. H. (ed.) 1971. *Paleopedology: origin, nature and dating of paleosols.* Jerusalem: International Society of Soil Science and Israel Universities Press.

66 ZEUNER, F. E. 1959. *The Pleistocene period.* 2nd edn. London: Hutchinson.

67 HAYS, J. D., IMBRIE, J. and SHACKLETON, N. J. 1976. 'Variations in the earth's orbit: pacemaker of the ice ages', *Science*, N.Y., **194,** 1121–32.

CHAPTER 11

PLEISTOCENE SUCCESSIONS AND THEIR SUBDIVISION

Development of the detail of Pleistocene successions in Europe

DURING the middle of the nineteenth century the acceptance of the glacial theory for the origin of the Pleistocene drifts in northern Europe led to the recognition of the Pleistocene Ice Age. Soon after came the idea of polyglacialism, involving a number of distinct ice advances over the area. Trimmer in 1858 described two tills on the coast of East Anglia and Geikie in 1877 recognised four glaciations in the East Anglian drift sequence. Later, in 1895, Geikie[10] (table 11.1) put forward a sequence of glacials and interglacials in northern Europe. Geikie's system was not, however, widely adopted as a basis for the subdivision of the Pleistocene; the evidence at that time hardly merited a comprehensive scheme. It was not until the classic work of Penck and Brückner[19] in the Alps that a general scheme, involving fourfold glaciation (Günz, Mindel, Riss and Würm glaciations), became widely accepted in Europe (table 11.1).

In Europe Penck and Brückner's scheme remained authoritative for many years. It was used, and still is, in many areas beyond the Alps: for example, the names of the Alpine glacial stages are widely used by archaeologists in many parts of Europe. In some ways it is unfortunate that Penck and Brückner's scheme became so widely used. Unlike the clearer sequence of glacials and interglacials in northern Europe, the original work on the fourfold glaciation of the Alps was done in a much more difficult area—certain valleys of the northern foreland of the Alps (figs. 11.1 and 11.2). The sequence was mainly based on relative levels of outwash bodies with interglacial periods deduced from weathering, soil profiles, and the extent of downcutting between the deposition of the successive outwash bodies.

The succession of ice advances of the Scandinavian ice sheet became formalised in the 1920s (table 11.1). The drift borders of these different glacial stages are shown in fig. 11.4. There was a consider-

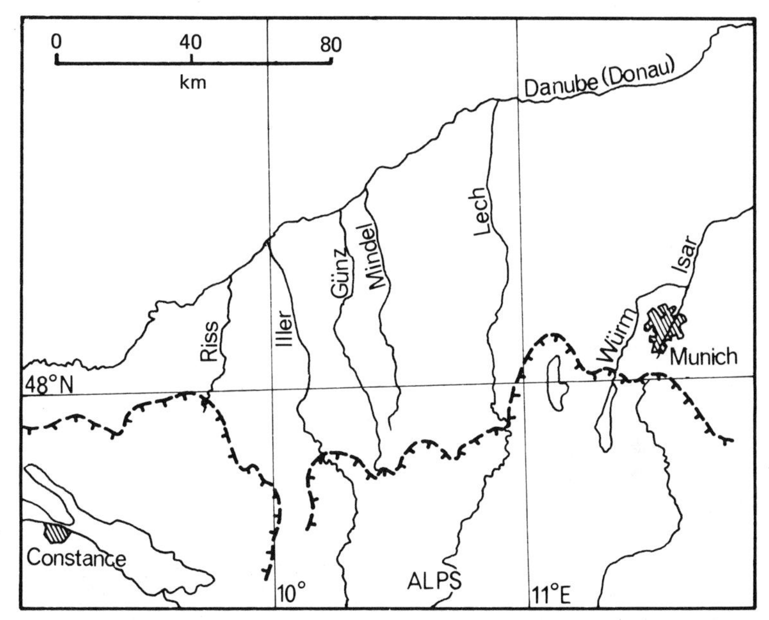

Fig. 11.1 Southern Germany, showing the approximate margin of the Würm glaciation (dashed line) and the foreland with older drifts to the north, drained by the Danube. Rivers in the foreland area give their names to the Alpine stages.

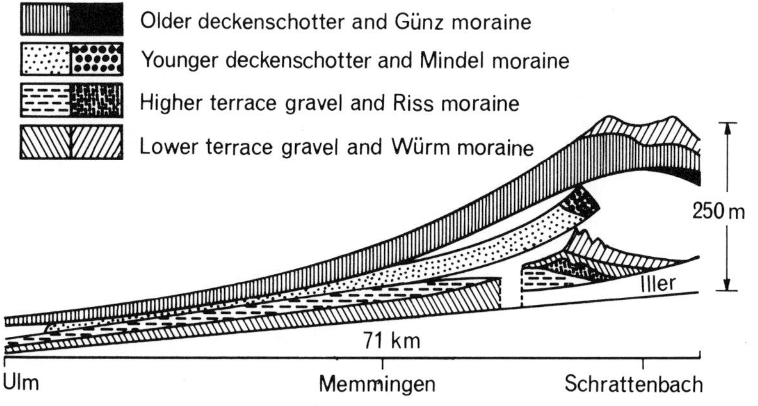

Fig. 11.2 Long profile of the gravel terraces of the river Iller (see fig. 11.1) and their relation to moraines (Wright, 1937).

Table 11.1 Classification of the Pleistocene

Names on the left of each column are cold or glacial stages, and those on the right, temperate stages. Above the double line are the classical glacial and interglacial stages of the Pleistocene; below it is the 'preglacial' part of the Pleistocene. Present conventional correlations are shown by lines connecting stage names.

N. America[9]	*N. Europe*[10]	*Alps*[26]	*N.W. Europe*[26]	*European U.S.S.R.*[15]
Wisconsin	Mecklenburgian	Würm	Weichselian	Waldai
Sangamon	Neudeckian	R/W	Eemian	Mikulino, Mgi
Illinoian	Polandian	Riss	Warthe	Moscow
Yarmouth	Helvetian	M/R	?	Odintzovo
Kansan	Saxonian	Mindel	Saale	Dnieper
Aftonian	Norfolkian	G/M	Holstein	Lichwin
Nebraskan	Scanian	Günz	Elster	Oka
			"Cromerian complex"	Belovezhsky
		Donau	Menapian	(Apsheron and
			Waalian	Akchagyl
		Biber	Eburonian	formations)
			Tiglian	
			Praetiglian	

able distance between the Alpine and Scandinavian ice sheets at the time of maximum glaciation, as seen on the map of Europe showing the distribution of Pleistocene ice (fig. 11.3), and the exact correlation of the Alpine and Scandinavian glacial advances is still a matter of some doubt.

The advances in the subdivision of the European Pleistocene really stem from the work of Jessen and Milthers[13] in 1928 on freshwater

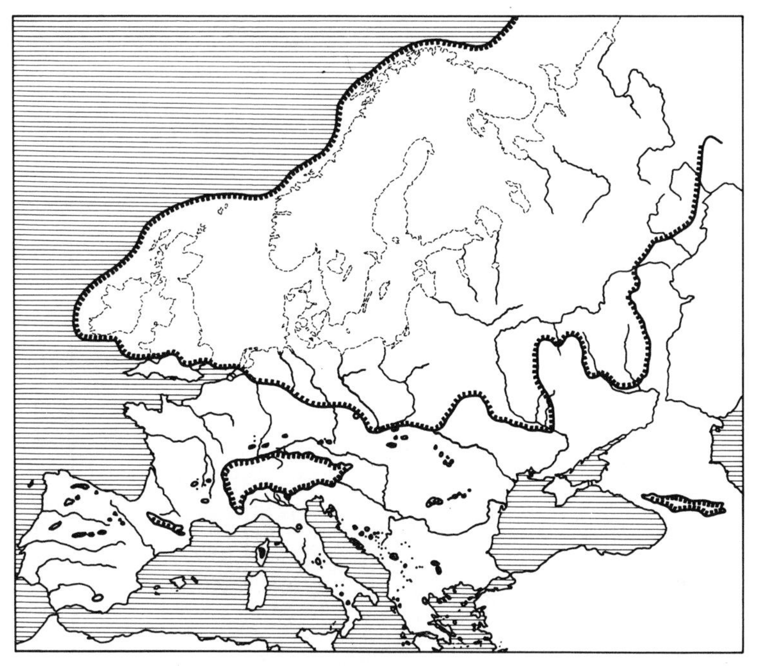

Fig. 11.3 Ice limits in Europe at time of maximum glaciation. The Scandinavian ice is shown confluent with British and Irish ice to the west and with the Urals and Novaya Zemlya ice to the northeast. Minor glaciated areas are also outlined or spotted (after Flint, 1957).

interglacial deposits. They showed that each of two interglacial periods they identified had a characteristic vegetational sequence revealed by pollen analysis. Thus for the first time was it possible to introduce definite correlations of interglacials, and hence glacials, over wide areas of northern Europe.

Moreover, as a result of Jessen and Milthers's work, it has become customary to use interglacial and interstadial as terms to define particular types of non-glacial climatic conditions in temperate north-

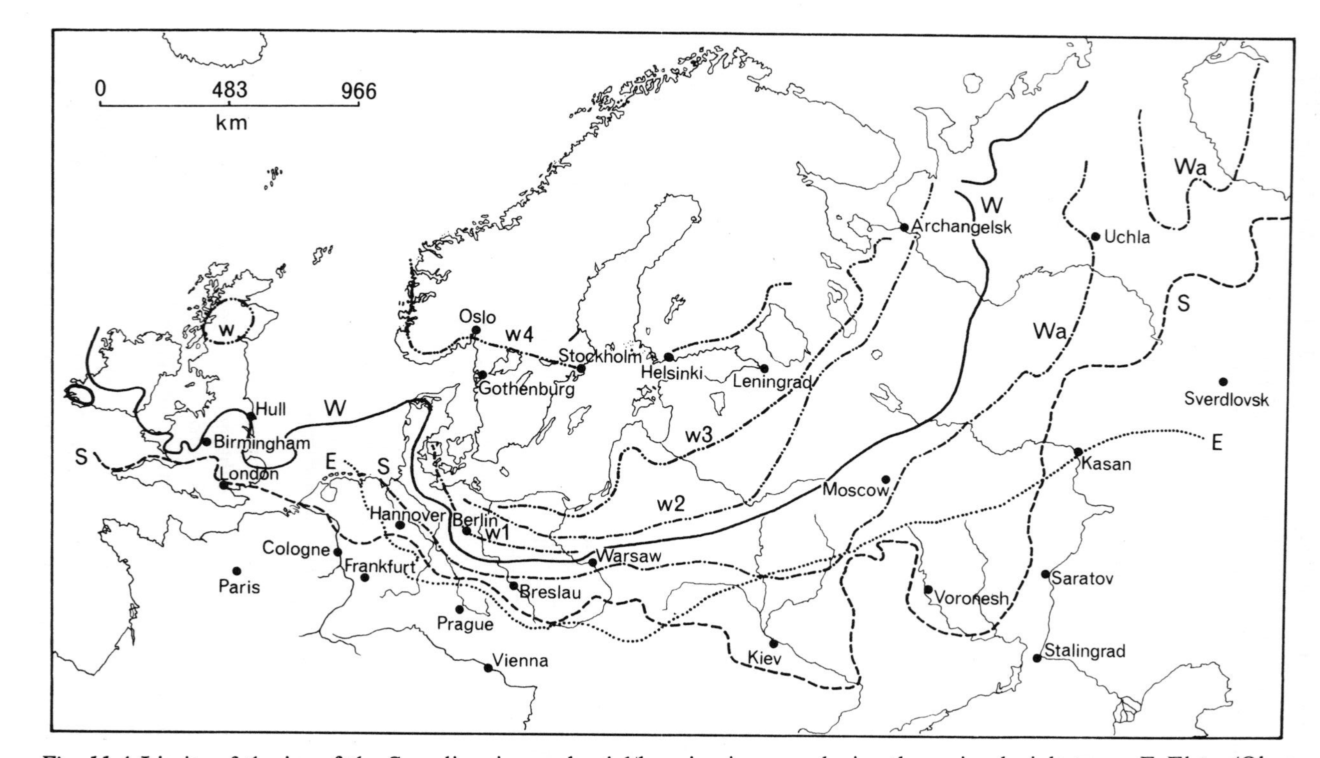

Fig. 11.4 Limits of the ice of the Scandinavian and neighbouring ice caps during the main glacial stages. E, Elster/Oka; S, Saale/Dnieper; Wa, Warthe/Moscow; W, Weichsel/Waldai maximum; w, retreat stages of the Weichselian ice; w1, Pomeranian; w2, Frankfurt/Poznan; w3, Riga; w4, Salpausselka (after Woldstedt 1958).

Table 11.2 Climatic development of interglacials and interstadials (Jessen and Milthers, 1928)

Glacial or periglacial deposits	
Interstadial climatic development	*Interglacial climatic development*
arctic	arctic
sub-arctic	sub-arctic
boreal, with summer temperature essentially lower than in the Flandrian climatic optimum of the area in question	boreal
sub-arctic	temperate, with summer temperature at least as high as during the Flandrian climatic optimum of the area in question
arctic	boreal
	sub-arctic
	arctic
glacial deposits	

west Europe, as indicated by vegetational changes (table 11.2); interglacial to describe a temperate period with a climatic optimum at least as warm as the Flandrian climatic optimum in the region concerned, and interstadial to describe a period which was either too short or too cold to permit the development of temperate deciduous forest of the interglacial type in the region concerned. This use of the terms is preferable to their use as general indicators of non-glacial conditions (fig. 11.5), e.g. the use of the word interglacial in the purely stratigraphical sense of denoting any deposit between two tills, or its use as indicating a period of erosion between two tills, or its use on the evidence of high sea-levels.

On the other hand, this definition of these terms means that they have rather a local significance, since it is based on the nature of the climatic fluctuations in the region concerned. It follows that caution must be used in correlating interglacials and interstadials across wide distances and different climatic provinces.[21]

The rapid development of pollen studies in the last twenty years has made wide correlations possible over the greater part of northern Europe. It is interesting to note, however, that few organic interglacial deposits occur in the Alps of certain date relative to the Alpine glacial stages, so correlation of glacial and interglacial stages in this area is not always clear.

The development of local Pleistocene successions which avoid the strait-jacket of the Alpine nomenclature has also resulted in advances in our knowledge of local stratigraphic sequences.

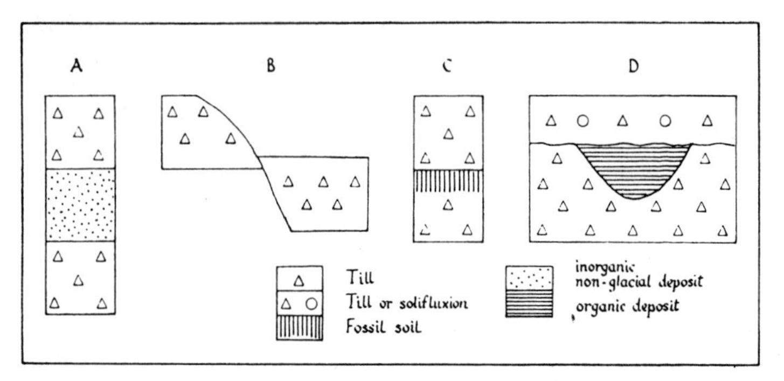

Fig. 11.5 Four situations in which the term interglacial (or interstadial) has been used (West, 1963).

a. Mid-glacial sands; no evidence of climate, only of ice recession. A 'stratigraphical' interglacial.
b. One till sheet separated from another by a period of dissection of the landscape. The erosional interval has been called interglacial.
c. Development of a soil on an older till, later sealed by deposition of a younger till. The soil type may give information on the climate of the non-glacial weathering phase. Such buried soils have been little used in British Pleistocene stratigraphy.
d. Deposition of organic deposits in a hollow in an older till during a non-glacial interval. Study of the fossils in the organic deposit gives much information on the flora, fauna and climate of the interval and can characterise it as interglacial or interstadial.

Further advances in techniques have also improved our knowledge of Pleistocene successions and hold out more promise for the future. These are dating by means of radioactive isotopes, both radiocarbon and potassium/argon dating, palaeotemperature determinations and palaeomagnetic chronology. As we have seen in Chapter 9, potassium/argon dating has given a figure for the length of time occupied by the Pleistocene succession, and radiocarbon dating has given much detail to the last glaciation and Flandrian sequences. Palaeotemperature determinations from ocean basin sediments promise to give a general scheme of climatic change during the Pleistocene which may eventually become the basis for a world-wide correlation of the Pleistocene succession.

At present, it may be said that the correlation of cold and temperate stages across Europe is far from satisfactory. The conventional correlations are shown in table 11.1. In particular the correlation of pre-Eemian interglacials across northern Europe is not clear. The nature of the interval between the Saale/Warthe and Dnieper/Moscow glaciations is not clear. There may be an additional temperate stage (Dömnitzian) after the Holsteinian but before the main

Saale glaciation,[22] and it is thought that the 'Cromerian' complex of the Netherlands contains at least three temperate stages.[29] As well as these difficulties in the correlation of stages, there are also difficulties of correlation within the last glacial. The Mid- and Early-Weichselian interstadials have not been satisfactorily correlated across Europe. The state of correlation of the pre-glacial Pleistocene is also unsatisfactory.

Some of these difficulties may be caused by the nomenclature. Thus an interval considered interglacial in continental European Russia because of its forest content may be expressed in oceanic northwest Europe as an interstadial, that is, with an absence of thermophilous forest trees.

Pliocene, Quaternary, Pleistocene, Holocene

According to present usage, the Pliocene is the final series of the Tertiary System, with the Pleistocene and Holocene Series constituting the Quaternary System. The Tertiary and Quaternary together constitute the Cainozoic Era (table 11.3). However, the boundary between the Tertiary and Quaternary Systems is not marked by such significant evolutionary changes as characterise the older system boundaries, and difficulties in definition have delayed the recognition of an isochronous and widely-applicable boundary. It may therefore be better to consider the Tertiary and Quaternary as constituting together the first system of the Cainozoic era, as advocated by R. F. Flint.[7] The boundary between the Pleistocene and Holocene is no different in character to the boundaries between the preceding glacials and interglacials, and the Holocene would therefore be better considered as the most recent stage of the Pleistocene, rather than a separate series. The name Holocene should then be discarded and replaced by a stage name (Flandrian), and the name Pleistocene would revert to its first meaning, in which sense it is used in this book. These changes from present usage are shown in table 11.3.

Originally the name Pleistocene was coined by Lyell[16] in 1839 to include rocks with a molluscan fauna containing more than 70 per cent of living species; this included the post-Pliocene strata. But the term soon came to be synonymous with the Ice Age, thus implying that the Pleistocene was characterised by glacial climates. The boundary between the Pliocene and Pleistocene was first formalised by a committee of the 18th International Geological Congress meeting in London in 1948.[14,18]

They defined the boundary in a type area of marine sedimentation in Italy, as being at the base of the Calabrian stage or its supposed continental equivalent, the Villafranchian (table 11.4). The boundary

Table 11.3 Subdivision of the Cainozoic in north-west Europe

Geological time units	*Era*	*Period*	*Epoch*	*Age*
Stratigraphical units		*System*	*Series*	*Stage*
a. *Present usage*				
		Quaternary	Holocene or Recent	Flandrian
			Pleistocene	Weichselian ↑ Pretiglian
	Cainozoic		Pliocene	Reuverian
		Tertiary	Miocene Oligocene Eocene Palaeocene	
	Mesozoic			
b. *Suggested usage*			Pleistocene	Flandrian ↑ Pretiglian
	Cainozoic	(first period of Cainozoic)	Pliocene Miocene Oligocene Eocene Palaeocene	Reuverian
	Mesozoic			

is the horizon in the Italian Neogene succession where there was thought to be good palaeontological evidence for climatic deterioration, with the appearance of a 'cold-indicating' foraminifer (*Hyalinea baltica*) and mollusc (*Artica islandica*). However, these appear at different levels. Moreover, the oxygen isotope evidence from the important section at Le Castella, Calabria, shows no great change of climatic conditions across the presumed Plio-Pleistocene boundary,[5] and the appearance of these so-called northern immigrants in the Mediterranean basin may be caused by biogeographical factors other than climatic change. In fact, climatic changes from the Neogene have been recorded,[1] so that the criterion of the first climatic deterioration is impossible to apply.

Thus the exact position of the Plio-Pleistocene boundary in relation to the Calabrian succession remains at present a matter for discussion.[8, 20] Meanwhile, it seems reasonable to maintain the 1948 recommendation that the boundary be placed at the base of the Calabrian, defined by the first appearance of *Artica islandica.*

In the Netherlands a clear climatic deterioration, evidenced by great vegetational change (Reuverian–Pretiglian), has been used to define the base of the Pleistocene (fig. 13.2).[23] This boundary coincides in northwest Europe with the spread of the genera *Elephas*, *Bos* and *Equus*.

There is pollen analytical evidence (extinction of Tertiary genera) to correlate the Netherlands Reuverian–Pretiglian boundary with the base of the Calabrian.[27] Potassium/argon dates and palaeomagnetic studies may indicate an age of *c.* 2·4 m.y. for the boundary, near to the boundary between the Matuyama and Gauss geomagnetic epochs.[27] But it has also been suggested that the boundary is near the Gilsa (Olduvai) or Reunion geomagnetic events within the Matuyama epoch, *c.* 1·8 m.y. to 2·1 m.y.[1, 2]

Several other definitions for the Plio-Pleistocene boundary have been suggested, many based on faunal zones in ocean cores, but these have not received substantial support.

Wide agreement on the position of the boundary has yet to be reached. Potassium/argon dating, palaeomagnetic studies and the oxygen isotope curve will provide the basic means for correlation of the biostratigraphical evidence in the oceans and on the continents, but even then a point of stratigraphical significance will have to be agreed as the Plio-Pleistocene boundary.

Table 11.4 Plio-Pleistocene boundary in Europe

(c, continental succession; m, marine succession)

	Italy (type area) (m)	*Netherlands* (c, m)	*East Anglia* (m)
Pleistocene	Calabrian	Pretiglian	Red Crag
Pliocene	Astian	Reuverian	Coralline Crag

Subdivision of the Pleistocene

We now have to consider the methods by which subdivisions of the Pleistocene succession can be made. There are certain differences between the bases for subdivision in the Pleistocene and in older rocks. The time covered by the Pleistocene is, geologically speaking, relatively short, so that evolutionary changes found in older parts of the geological time-scale and used as a basis for classifying the rock succession are not so evident. But although short, the record of the rocks is known in great detail, so that we have preserved over the earth's surface synchronous deposits of many different environments, again a difference from the more incomplete record of the older rocks. This variety of environments of deposition is accompanied by a variety of geomorphological processes also recorded in the earth's landforms, the study of which also gives many clues to the succession of Pleistocene events. It is all this detail which makes the classification of the Pleistocene a complicated matter.

The purpose of subdivision is twofold. First it is to split the succession into manageable units, and second it is to bring together by correlation the sequence of events and the units of the succession found in the many environments of Pleistocene time. The main difficulty of splitting the succession into units is that geological and climatic processes show continual gradual change, with few very abrupt changes. We have to decide on criteria which are at once objective and clearly expressed in the geological record, and these criteria will differ in different environments of sedimentation and landscape formation.

The ideal situation would be to find absolute age markers at all horizons in all environments, so that we could have a calendar of events divided say into 1000 year segments, giving complete correlations of sediments, processes and events over the earth's surface. The actual situation is that we have indications of absolute age (radiocarbon age) common in certain environments up to say 10,000 years ago and less common up to 60,000 years ago, and a few isotope-based dates in the middle and early part of the Pleistocene. Otherwise we have to rely for our subdivisions on the geological and geomorphological evidence for age and succession so that a series of manageable and objective units are obtained.

There are various bases for subdividing the stratigraphic succession and the time it represents. They have been much discussed recently in order to arrive at rules for obtaining uniformity of usage in subdivision and correlation.[3, 11, 12, 17]

There are various methods of subdividing the stratigraphic succession and the time it represents.

The following types of division may be mentioned:

Units based on lithostratigraphy. These units are subdivisions of the rock succession distinguished by their lithological characteristics.

Units based on biostratigraphy. These are subdivisions of the rock succession characterised by a particular content of fossils. The basic unit is the biozone. Examples are pollen biozones, mammalian faunal biozones, foraminiferal biozones in deep sea sediments. There are various kinds of biozone: e.g. assemblage biozone, characterised by a particular assemblage of fossils (e.g. a pollen assemblage biozone); peak or acme biozone, characterised by the exceptional abundance of a particular taxon.[12]

Units of the rock succession, thought to be of regional significance, which are related to type sections or areas, and characterised by a particular sequence of geological events. Both lithological and biostratigraphical evidence can contribute towards the definition of these units. For example, biostratigraphy leading to the definition of stages in the Pleistocene. These units are units of the standard stratigraphical scale.

Units based on climatic change. These are subdivisions of the succession representing inferred climatic episodes, e.g. glaciation, interglacial.

Units based partly or wholly on geomorphology. These are subdivisions of the Pleistocene resulting not wholly from the rock succession but from a consideration of the relation of constructional or erosional

Table 11.5 Stratigraphical nomenclature

	Basis for subdivision		
	Lithostratigraphical	*Biostratigraphical*	*Standard stratigraphical*
subdivisions	group	biozone	system
	formation	subbiozone	series
	member	biozonule	stage
	bed		substage

features to one another, e.g. a series of river terraces, or a series of shore-lines reflecting eustatic changes of sea-level.

There are also the means of subdividing the rock succession provided by dating methods, already discussed in Chapter 9.

Table 11.5 gives the names of various categories used in subdividing the rock or fossil succession.

Because the Pleistocene is characterised by climatic change, and because climatic fluctuation in the earth's atmosphere has governed the sequence of geological processes and events in different environments, it is usually thought that climatic change, as indicated by lithology, fossils and geomorphology, is a suitable basis for subdivision of the Pleistocene. Hence the climatic episode units (glaciation, interglacial, pluvial, interpluvial) mentioned above. But the climatic episodes are, except where extensive palaeotemperature records are found, based on interpretation of lithological, biostratigraphical and other lines of evidence, so that they are synthetic. In addition, because climatic change expresses itself at different rates and in different ways in different places, precise correlations based on climatic change are difficult over long distances.

But in the stretch of time where absolute dating by radiocarbon can be carried out reliably (the last 50,000 years or so), the climatic basis for subdivision is best subordinate to absolute dating, that is to say, subdivision and correlation may best be done on a basis of radiocarbon years.

When the geological record of the Pleistocene is complete it will be seen how the succession of lithostratigraphical, biostratigraphical and other subdivisions mentioned above are related one to another. All types of subdivision are useful in that they will indicate a different facet of the sequence of events. Meanwhile, we must attempt to build up sequences in different areas which have a local validity. The most useful method of subdivision to fulfil the purposes of subdivision will vary from place to place, depending on the criteria available.

Examples of subdivision

We can now exemplify these principles by reference to particular areas and to particular environments of deposition.

The Pleistocene sequence in northwest Europe. The glacial/interglacial sequence in northwest Europe (table 11.1) was primarily based on the rock succession, tills and other glacial deposits alternating with non-glacial sediments such as peat. Thus the units of the subdivision were of a lithostratigraphical type.

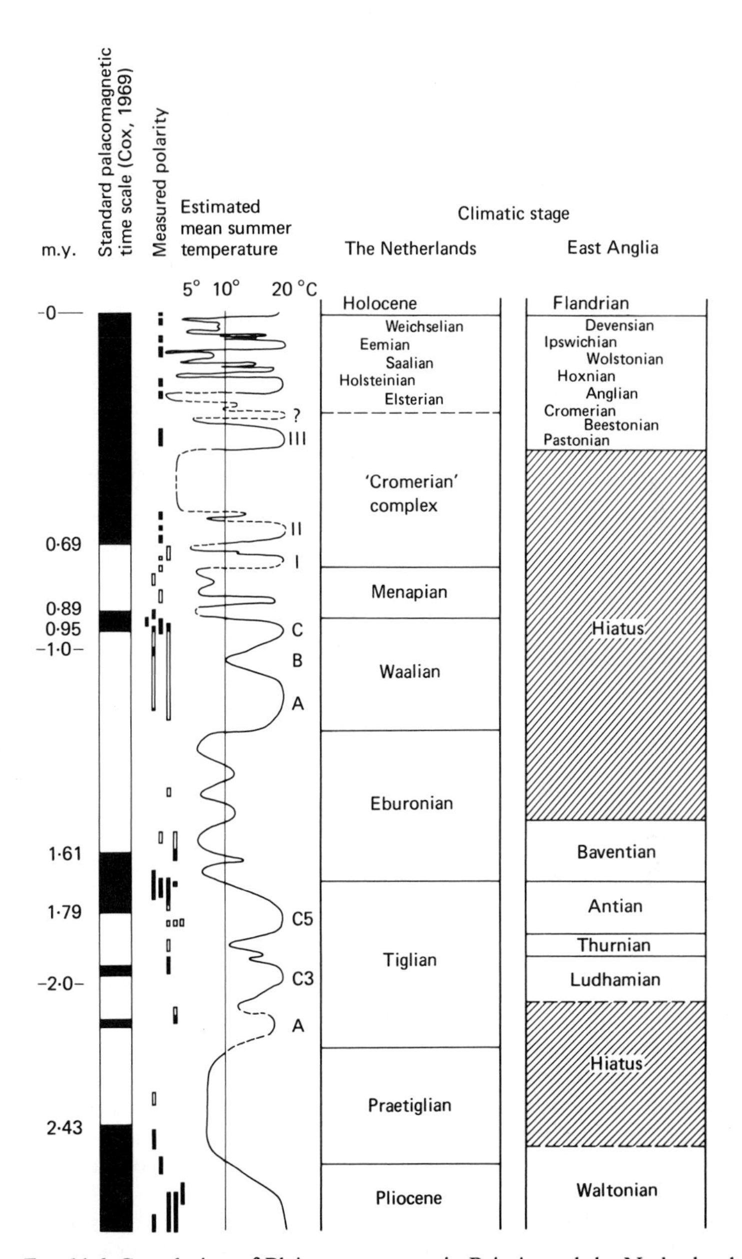

Fig. 11.6 Correlation of Pleistocene stages in Britain and the Netherlands, according to Zagwijn (1975). A climatic curve and relation to the palaeomagnetic time scale are also shown.

The increased use of palaeobotanical data, especially that derived from pollen analysis, has given much detail of biostratigraphy, and a satisfactory series of units based on biostratigraphy has been erected in the British Isles and the Netherlands. Table 11.6 summarises the British sequence and fig. 11.6 compares the Netherlands and British sequences, with estimated mean summer temperatures and correlation with the standard palaeomagnetic time scale.

Table 11.6 Stages of the British Pleistocene

			Stage	*Climate*		
(Quaternary)	(Holocene)		Flandrian	t		
	Pleistocene	Upper	Devensian	c	g	p
			Ipswichian	t		
			Wolstonian	c	g	p
		Middle	Hoxnian	t		
			Anglian	c	g	p
			Cromerian	t		
			Beestonian	c		p
			Pastonian	t		
		Lower	Baventian	c		p
			Antian	t		
			Thurnian	c		
			Ludhamian	t		
			?Pre-Ludhamian			

c, cold
p, permafrost known
g, glacial deposits known
t, temperate

Briefly, the stages resulting from biostratigraphical evidence are alternating cold (sometimes glacial) and temperate stages, as indicated by biostratigraphy, the boundary between them being placed where the vegetation changes from an open sub-arctic park landscape to a wooded landscape with temperate trees, and vice-versa. These changes are recognised in pollen diagrams by changes in the tree/non-tree pollen ratio and the frequency of pollen of thermophilous trees.

This method of subdivision leads to the formulation of stages, climatically distinct. Whether these stages be termed time-stratigraphical or biostratigraphical or climatic is not so important. What is significant is that the stages and their boundaries can be determined objectively and have a practical use for formulating the local Pleistocene succession.

Similar biostratigraphical criteria can be used to define and limit interstadial intervals in the glacial stages. Such oscillations will be accompanied by the reappearance of forest, of a boreal type, not of the temperate type indicative of interglacials (table 11.2).

Other kinds of biostratigraphical evidence, such as that based on beetles, may lead to other formulations of climatic episodes, since different groups of organisms respond to and reflect climatic change in different ways.

In the most recent part of the Pleistocene, the climatic basis for subdivision and correlation can be refined by the application of radiocarbon dating, as mentioned above.

Clearly such a scheme of subdivision as this is only of local application. If we move away much in latitude, say, further north, the same pollen changes will not be synchronous as those further south. If we move in longitude to a more continental climate, the changes in tree/non-tree pollen ratio may reflect the change to a temperate steppe vegetation. Each region must clearly use its own criteria for boundaries, having regard to the overall climatic changes, presumably synchronous, which determine the regional fluctuations.

It is thus possible with the aid of pollen analysis to devise a system of objective units for our region, based on biostratigraphy. The major units in the Pleistocene will be stages and substages. The stages will be cold or temperate, glacial and interglacial respectively if glacial deposits are present. The substages will then be the subdivisions of the stages, e.g. the stadia and interstadials of the glaciations or the major pollen zones of the cold or temperate stages.

The naming of stages and substages should be done on the type site system so that the fundamental evidence for the subdivisions is clear. Each cold, temperate and interstadial phase should have its own type site where the stratigraphical evidence is clear. There will then be the least confusion resulting from nomenclatorial difficulties. It will produce a number of new names, based on local sequences, but eventually when certain correlations emerge, regional names can replace more local ones. The application of this approach to British Pleistocene stratigraphy is discussed in the next chapter.

A division of the Pleistocene into Upper, Middle and Lower is a matter for convenience. The placings of the divisions in table 11.6 are those used in northwest Europe. The Lower/Middle boundary is placed at the beginning of the first temperate period (the Pastonian) showing an absence of the so-called Tertiary relic plants, e.g. *Tsuga*. The Middle/Upper boundary is placed at the end of the Hoxnian (temperate) stage.

The Pleistocene sequence in the north American plain.[4, 9] The lithostratigraphical basis of glacial and non-glacial deposits is used as a basis for division into stages (table 11.1). The glacials have been subdivided into substages of ice advance and periglacial loess deposition. However, not much detail is known of the sequences within non-glacial stages, and the upper and lower boundaries of the earlier stages are difficult to define. As in Europe, much more detail is known of the ice advances and retreats of the last (Wisconsin) glaciation. Ancient soil horizons provide key evidence for non-glacial conditions between ice advances and so soil-stratigraphic units form an important component of the classification of the Pleistocene succession.

Loess sequences in central Europe.[6] Loess sequences play an important part in determining the Pleistocene succession in periglacial areas. They are sensitive to climatic change, loess forming during periods of more rigorous climate and soils forming on cessation of loess formation in more temperate climates. The classification of such sequences is based on the sediments, including soils and on biostratigraphical evidence from non-marine molluscs. Both are used to formulate climatic episodes. The sequence of climatic episodes, together with radiocarbon and archaeological evidence is the basis for correlation with the glacial/interglacial sequence to the north. The respective climatic (and therefore sedimentary) provinces are very different and direct geological correlations on the basis of superposition are not possible.

Stratigraphy of cave deposits and of periglacial sequences. These are basically lithostratigraphical, as the successions concern the sequence of stalagmite, cave breccias and soils in the case of cave deposits, and sequences of solifluction, aeolian deposits, involutions, permafrost structures and soils in the case of periglacial sequences. Climatic episodes are inferred from the sequences, and it is these which form the basis for correlations. Local factors of slope, hydrology, aspect and so on may affect the sequences, so that climatic interpretation may be difficult. The definition of lithostratigraphical units will not be so difficult, and the definition of major climatic stages should be clear. Substages may be far more difficult to define.

Stratigraphy of limnic sediments. Biostratigraphical evidence from pollen analyses will provide evidence for the definition of assemblage biozones. The regional significance of such biozones will appear after the detailed regional studies which allow local variants to be isolated from the regional succession. Lithostratigraphical successions are separate, but their evidence of climatic change, together with that from biostratigraphy, should be used to formulate climatic episode units.

Sea-level sequences and terrace sequences. Because of the incomplete record of geological events given by these sequences, they will not form a series of units filling Pleistocene time, as may sequences of a lithostratigraphical type. Major episodes of sea-level standstill or of terrace aggradation or of marine or fluvial erosion will be recorded. Whether these episodes will have a climatic significance will depend on the lithology of the sediments and on the faunas and floras associated with them or on the theory that high and low sea-levels have climatic significance.

The correlation of these episodes with those of the general climatic sequence known from lithostratigraphical and biostratigraphical evidence may be very difficult, so that these geomorphologically-based episodes often form an independent sequence of events, whose exact relation with the sequence provided by other evidence is not clear.

Faunal biozones. These are based on successions of faunas, each biozone having a characteristic fauna. The zones so defined may cover several climatic stages, and so form an independent view-point of the Pleistocene succession.

Measured ages or climatic characters. A basic succession of radiocarbon years, with units of 100s or 1000s of years forms a distinct type of classification of Pleistocene time, and will be an important scale for the time covered by this method of dating. Subdivisions based on temperature changes, such as the fluctuations seen in cores of ocean sediments (fig. 10.12) may also be used to define climatic subdivisions. The evidence for the relation of these subdivisions to continental climatic subdivisions is not so far advanced as to allow correlations back beyond the last interglacial.

REFERENCES

1 BANDY, O. L. 1972. 'A review of the calibration of deep-sea cores based upon species variation, productivity, and $^{16}O/^{18}O$ ratios of planktonic foraminiferas—including sedimentation rates and climatic inferences', 37–61. In *Calibration of hominoid evolution*, ed. W. W.

Bishop and J. A. Miller. New York: Wenner-Gren Foundation.

2 BANDY, O. L. and WILCOXON, J. A. 1970. The Plio-Pleistocene boundary, Italy and California. *Geol. Soc. America Bull.*, **81,** 2939–48.

3 Code of stratigraphic nomenclature (1961). *Bull. Amer. Assoc. Petrol. Geologists*, **45,** 645–65.

4 COOKE, H. B. S. 1973. 'Pleistocene chronology: long or short?', *J. Quat. Res.*, **3,** 206–20.

5 EMILIANI, C. 1971. Palaeotemperature variations across the Plio-Pleistocene boundary. *Science, N.Y.*, **171,** 60–2.

6 FINK, J. 1969. 'Les progres de l'étude des loess en Europe', *Suppl. Bull. Assoc. franc. Quat.*, 3–12.

7 FLINT, R. F. 1957. *Glacial and Pleistocene geology.* New York: Wiley.

8 FLINT, R. F. 1965. 'The Pliocene-Pleistocene boundary', Spec. Papers Geol. Soc. America, No. 84, *International Studies on the Quaternary*, 497–533.

9 FLINT, R. F. 1971. *Glacial and Quaternary Geology.* New York: Wiley.

10 GEIKIE, J. 1895. 'The classification of European glacial deposits', *J. Geol.*, **3,** 241–70.

11 HARLAND, W. B. *et al.* 1972. 'A concise guide to stratigraphical procedure', *J. Geol. Soc. Lond.*, **128,** 295–305.

12 HEDBERG, H. D. (ed.). 1972. 'Summary of an international guide to stratigraphic classification, terminology and usage. Int. Subcom. Stratigraphic Class., Report 7b', *Boreas*, **1,** 213–39.

13 JESSEN, K. and MILTHERS, V. 1928. 'Stratigraphical and palaeontological studies of interglacial freshwater deposits in Jutland and north-west Germany', *Danm. Geol. Unders.*, II Raekke, No. 48.

14 KING, W. B. R. and OAKLEY, K. P. 1949. 'Definition of the Pliocene/Pleistocene boundary', *Nature* (*Lond.*), **163,** 186–7.

15 KRASNOV, I. I. 1965. 'Regional unified and correlative stratigraphic scheme for the Quaternary of European U.S.S.R.', Spec. Papers Geol. Soc. America, No. 84, *International Studies of the Quaternary*, 247–71.

16 LYELL, C. 1839. 'On the relative ages of the Tertiary deposits commonly called "Crag" in the counties of Norfolk and Suffolk', *Mag. Nat. Hist.*, **3** (N.S.), 313–30.

17 MANGERUD, J. *et al.* 1974. 'Quaternary stratigraphy of Norden, a proposal for terminology and classification', *Boreas*, **3,** 109–28.

18 OAKLEY, K. P. 1949. International Geological Congress XVIIIth Session, Great Britain, 1948. Proceedings of Sectional Meetings. Section H. 'Plio-Pleistocene boundary', *Geol. Mag.*, **86,** 18–21.

19 PENCK, A. and BRÜCKNER, E. 1909. *Die Alpen im Eiszeitalter.* Leipzig: Tauchnitz.

20 RUGGIERI, G. 1965. 'A contribution to the stratigraphy of the marine lower Quaternary sequence in Italy', Spec. Papers Geol. Soc. America, No. 84, *International Studies on the Quaternary*, 142–52.

21 SUGGATE, R. P. 1974. 'When did the Last Interglacial end?', *Quaternary Research*, **4,** 246–52.

22 TURNER, C. 1975. 'The correlation and duration of Middle Pleistocene interglacial periods in northwest Europe', in *After the Australopithecines*, ed. K. Butzer and G. L. Isaac, 259–308. The Hague: Mouton Publishers.

23 VOORTHUYSEN, J. H. VAN, TOERING, K. and ZAGWIJN, W. H. 1972. 'The Plio-Pleistocene boundary in the North Sea basin. Revision of its position in the marine beds', *Geol. en Mijnb.*, **51,** 627–39.

24 WEST, R. G. 1963. 'Problems of the British Quaternary', *Proc. Geol. Assoc. London*, **74,** 147–86.

25 WRIGHT, W. B. 1937. *The Quaternary Ice Age*. 2nd edn. London: Macmillan.

26 WOLDSTEDT, P. 1958. *Das Eiszeitalter*. Bd II. Stuttgart: Enke.

27 ZAGWIJN, W. H. 1974. 'The Pliocene–Pleistocene boundary in western and southern Europe', *Boreas*, **3,** 75–97.

28 ZAGWIJN, W. H. 1975. 'Variations in climate as shown by pollen analysis, especially in the Lower Pleistocene of Europe'. In *Ice Ages: ancient and modern*, ed. A. E. Wright and F. Moseley, 137–52. Liverpool: Seel House Press.

29 ZAGWIJN, W. H., MONTFRANS, H. M. VAN and ZANDSTRA, J. G. 1971. 'Subdivision of the "Cromerian" in the Netherlands; pollen-analysis, palaeomagnetism and sedimentary petrology', *Geol. en Mijnb.*, **50,** 41–58.

CHAPTER 12

THE PLEISTOCENE OF THE BRITISH ISLES

THIS chapter gives a general introduction to the Pleistocene stratigraphy of the British Isles. There is a large literature on the subject; no single up-to-date book covers the field, but Wright's classic textbook on the Quaternary Ice Age[149] gives an extensive description of the Pleistocene stratigraphy of the British Isles, and a comprehensive outline of the stratigraphy has been assembled by the Quaternary Era Sub-Committee of the Geological Society of London.[72] This contains a good bibliography. An account of a discussion on the changing environmental conditions in and around the British Isles during the Devensian cold stage has recently appeared.[72a] A useful list of names which have been used in British Pleistocene stratigraphy has been published in the International Stratigraphical Lexicon series.[64] The bibliography to this chapter is not intended to be completely comprehensive, but to give access to the recent literature on the subject concerned.

Multiple glaciation of the British Isles resulted in the formation of the great drift sheets which cover much of the lowland area (fig. 12.1) except for southern Britain and small parts of southern Ireland. From the main centres of glaciation, ice spread in the directions shown in fig. 12.2. In the periglacial areas extensive spreads of solifluction deposits were formed, and, to the south and east, loess. During permafrost times ice-wedge polygon networks were formed. Terrace sands and gravels were deposited in the drainage-ways leading from the ice fronts, and deep channels were cut in valleys as a result of low sea-levels. During the warmer temperate stages between the glaciations, rivers deposited alluvium under conditions of rising sea-level, and forest reoccupied the land. Apart from the formation of terrace deposits, the temperate stages were probably rather quiet times in terms of landscape evolution, in contrast to the erosional and depositional processes accompanying glaciation, lowered sea-levels, periglacial climates, and incomplete vegetation cover.

Glacial drifts are also found in the North Sea and Irish Sea basins,

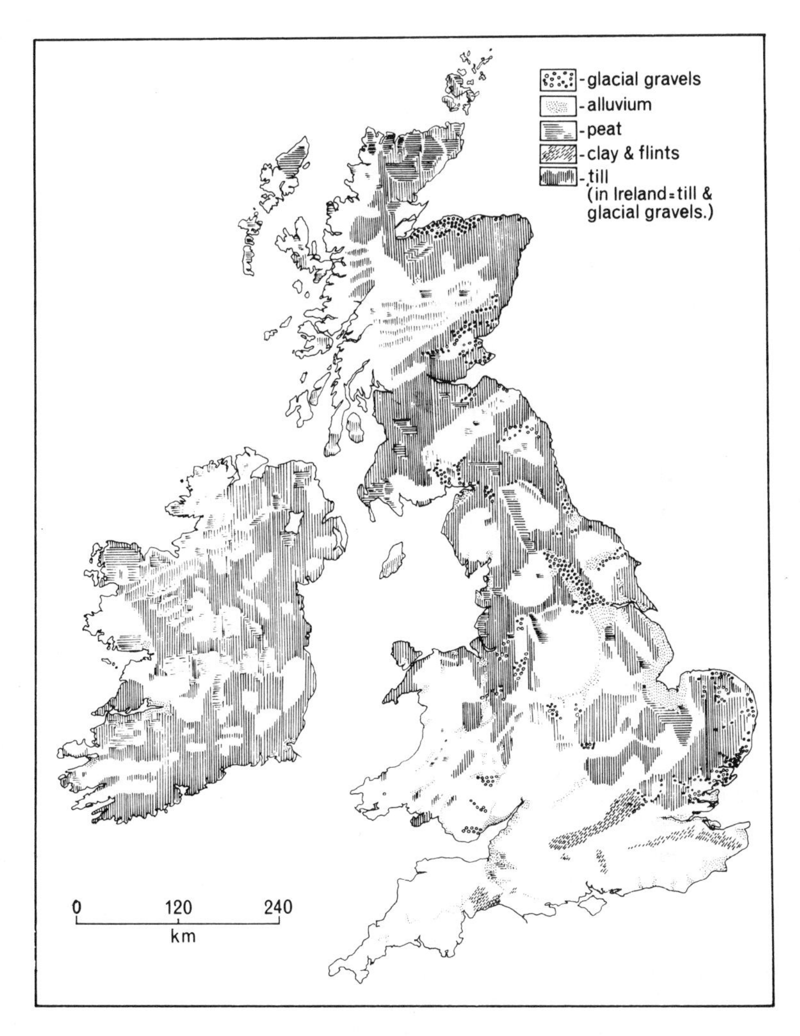

Fig. 12.1 Pleistocene drifts in the British Isles (after *Atlas of Britain* (Clarendon Press) and Geikie's Geological Map of Ireland (Bartholomew)).

often associated with marine sediments formed in a temperate climate. At the edge of the North Sea basin in southeast Britain, the changing climate is recorded in marine sediments by faunal and vegetational changes. These sediments, of Middle to Lower Pleistocene age, are often called 'preglacial', as they are found below the first true till, but their Pleistocene age is attested by their fauna and flora.

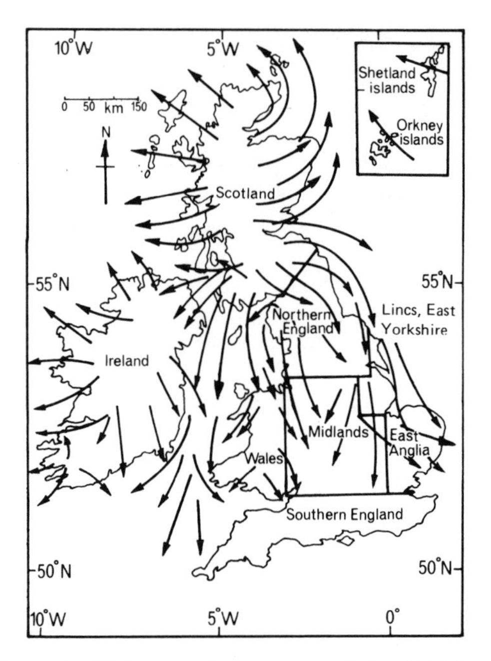

Fig. 12.2 The main Pleistocene ice streams of the British Isles, and the divisions used in the description of regional stratigraphy.

A BRIEF HISTORY OF INVESTIGATION[76]

The diluvial theory of the origin of the drift was championed by Buckland in his Reliquae Diluvianae of 1823. He was later converted to the glacial theory by Agassiz and this theory was, later in the century, accepted by most other geologists. Edward Forbes in 1846 gave detailed consideration to the effect of the northern drift ice on the distribution of plants and animals. About this time a long series of stratigraphical studies of the drift was started by officers of the Geological Survey and others. Joshua Trimmer recognised two tills in the cliff sections on the East Anglian coast, and James Geikie furthered the principle of polyglacialism by his recognition of four glaciations. By the end of the nineteenth century, much of our present knowledge of the drift sequence was obtained.

The origin of man in relation to the glacial period became an important point of discussion from the late 1850s onwards. The age of man was disputed—post-glacial or interglacial, and there was great interest in the relation of palaeolithic implements to the drift sequence. The archaeological side of the matter became rather confused towards the end of the century, with the fashion for pushing back the finding of artefacts to the older drift deposits; for example, the Kent eoliths described by Benjamin Harrison and his postulation of 'Plateau Man'.

The main difficulties of the interpretation of Pleistocene stratigraphy in the first half of this century were the lack of any reliable means of effecting correlations between distant drift sequences, and too facile correlation with the Alpine sequence. Often, at least in the south and east of Britain, correlations were based on artefacts, and a relative chronology based on them was widely mentioned.[17]

Detailed pollen-analytical studies of the Flandrian started to be made in the 1930s, and by the end of that decade the main outlines of climatic and vegetational history of this stage were known. The introduction of radiocarbon dating in the early 1950s put this climatic and vegetational history on a firm chronological basis. In the middle 1950s the main characters of the successive interglacials became clearer, largely as a result of pollen-analytical studies, and a firm basis for the correlation of till sheets was found. This also provided a scale of time to which palaeolithic cultures could be related.

A basis for a classification of the British Pleistocene on stratigraphical principles was put forward in 1963,[132] and very much extended and improved by the publication of the Geological Society's Quaternary report. It is this classification which will be used in our description of the Pleistocene of the British Isles.[72, 93] Of course, this is only a provisional classification, and new evidence will lead to complication, but its use eases the task of giving order to a multitude of facts.

CLASSIFICATION AND LOWER LIMIT OF THE PLEISTOCENE

As seen in figs. 11.3 and 12.2, the glaciation of the British Isles proceeded independently from that of continental northwest Europe, and we therefore require a separate classification of our deposits, not one based on continental stratigraphy. Believing that the major climatic changes in the whole of northwest Europe were more or less synchronous, even if differing in expression according to latitude and longitude, it follows that there should be eventually a close correlation between the sequence in the British Isles and that on the continent. But except in the Upper Pleistocene such correlation is far from being achieved, again a reason for a separate classification.

Although the Pleistocene is associated strongly with glaciation and glacial deposits, a large part of the early Pleistocene in the British Isles shows no direct traces of glaciation. This 'preglacial' Pleistocene consists of marine, estuarine and, rarely, freshwater deposits, and it includes the greater part, in terms of thickness, of the Crag deposits of East Anglia.[87] The problem is the placing of the Plio-Pleistocene boundary within the Crag sequence.

It is generally accepted that the boundary must be placed at the boundaries of or within the Red Crag. The recent view has been that the boundary is placed at the Coralline Crag/Red Crag boundary.[10] The criteria used to determine this position are the presence of a stratigraphical break between the Red and Coralline Crags, a marked increase in the proportion of northern forms of marine molluscs and the first occurrence of Lower Pleistocene molluscs of northern aspect (both indicative of a cooling in climate), and, finally, the first arrival, in the Red Crag, of elephant and horse. As described in the previous chapter, the 'type' section of the Pliocene–Pleistocene boundary in Europe is in southern Italy where the fauna of a sequence of marine deposits shows a deterioration of climate and this is taken as indicating the Pliocene-Pleistocene boundary. The Coralline–Red Crag cooling gives similar evidence for the boundary in East Anglia.

However, when we come to consider the relation of the Crags to the Netherlands sequence, where a Plio-Pleistocene boundary has been clearly distinguished on the basis of vegetational history (fig. 11.6), a position within or at the upper limit of the Red Crag appears possible. This is because the foraminifera in the Pre-Ludhamian pollen zone at the base of the Stradbroke (Suffolk) borehole resemble the assemblages from the Red Crag.[6] In the lower part of the zone they resemble those of the Red Crag at Walton-on-Naze (Waltonian); in the upper part of the zone they resemble those of the Red Crag at Butley (Butleyan). The Pre-Ludhamian pollen zone is characterised by the presence of high frequencies of *Pinus* pollen, and a comparison with the pollen zones of the Netherlands Pliocene–Lower Pleistocene suggests a correlation with the Upper Pliocene Reuverian C zone of the Netherlands. The normal polarity of the Pre-Ludhamian sediments[73] would be expected on this correlation. It means that the Plio-Pleistocene boundary should be drawn at or near the top of the Red Crag, perhaps within or at the base of the Butleyan. Harmer noted the Butleyan showed a much more northern mollusc fauna than the older parts of the Red Crag, and this fauna has been compared with that of the Pretiglian (basal Pleistocene) in the Netherlands.[150] This faunal change may be associated with the climatic deterioration which set in in the Reuverian and culminated in the Pretiglian cold stage.

In the absence of any clear evidence for a cold stage of Pretiglian magnitude, or of the presence of the characteristic basal Pleistocene foraminiferal faunas of the Netherlands in the Red Crag, it would seem that the Plio-Pleistocene boundary could be placed at the unconformity between the Red and Norwich Crags, rather than at the base of the Red Crag, even though the presence of elephant (*Archidiskodon*) and horse (*Equus*) in the Red Crag might suggest this crag is Lower Pleistocene. In the present state of knowledge it is difficult to say which of these two alternatives is correct. Dutch geologists consider the Waltonian to be Pliocene, while locally the boundary is placed at the base of the Red Crag. Since the Red Crag shows evidence of climatic deterioration, it will be convenient to include it in our account of the Pleistocene, even though it may be Upper Pliocene in part or whole, and we therefore include it in the table of stages, under the name Pre-Ludhamian.

The division of the British Pleistocene succession into stages, on the basis of principles already discussed in Chapter 11 is shown in table 11.6. It follows the usage described in the Geological Society's Quaternary report, which gives type sites for stages. Flandrian is used as a stage name in place of Holocene, following the usage of naming temperate stages after their characteristic marine transgression. The divisions into Lower (Early), Middle and Upper (Late) Pleistocene are based on current usage in northwest Europe.

The sequence of Pleistocene stages is based mainly on the East Anglian biostratigraphical and lithological evidence. In the Midlands too there is considerable biostratigraphical evidence, forming a reasonably secure basis for correlations whith the East Anglian sequence of stages. Of Ireland, the same can be said. When we extend our survey to other areas, however, there is a paucity of biostratigraphical evidence, and this leaves the lithological and geomorphological evidence as the mainstay for correlation of the succession. In Wales, north and northeast England, and Scotland, till sequences are known, but few biogenic temperate deposits are known apart from those of the Flandrian. The correlation of the till sheets is problematical, especially the question of how many are of Devensian age, for, if that were known, then tills of the earlier glacial stages might be tentatively identified and correlated.

The identification of the limits of the Devensian glacial advances, and the number of them are therefore of great importance. But there has been and still is much discussion over both these points.[124] Interstadial biogenic deposits are few and even fewer have been dated by radiocarbon assay. Figure 12.3 shows limits of the Devensian ice.

Figure 12.3 also shows the limit thought to have been reached by the older glacial advances. Again there is much discussion of the limits of these older glaciations and their correlation across the country.

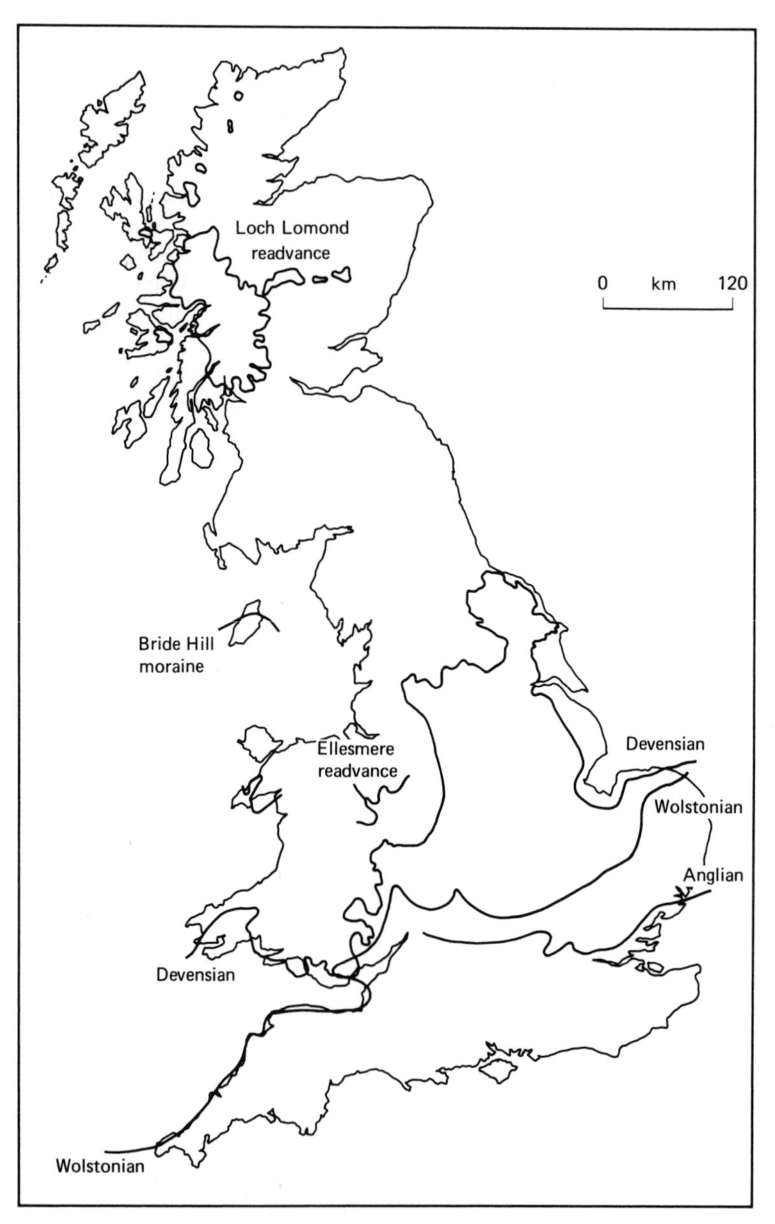

Fig. 12.3 Limits of the main ice advances in Britain.

In the following account of the Pleistocene succession in the various regions of the British Isles we will start with the classic area of East Anglia, where the Pleistocene sequence is most complete, and where there is an abundance of biostratigraphical evidence concerning Lower, Middle and Upper Pleistocene deposits. In the other glaciated areas we will be mainly dealing with the Middle and Upper Pleistocene, and, as already mentioned, correlation with East Anglia may or may not be possible. Finally, it will be necessary to consider again general topics affecting the whole region, such as periglacial phenomena, changes of relative land/sea-level, dating, and the Late Devensian and Flandrian.

In the regional accounts which follow, the emphasis is on the main stratigraphical features of the regions. Much more detail of particular sections, with references, is given by the Geological Society's Quaternary report.[72]

EAST ANGLIA

The 'preglacial' succession

The west–east section in fig. 12.4 shows the general disposition of the 'preglacial' and 'glacial' parts of the Pleistocene succession in East Anglia. We first have to consider the Lower Pleistocene Crag deposits formed in a shallow basin of deposition in the east of the area. The contours of this basin and the Crag in it are shown in fig. 12.5. The northeast trending troughs in this basin are probably scour troughs formed at a time when there was a broad connection

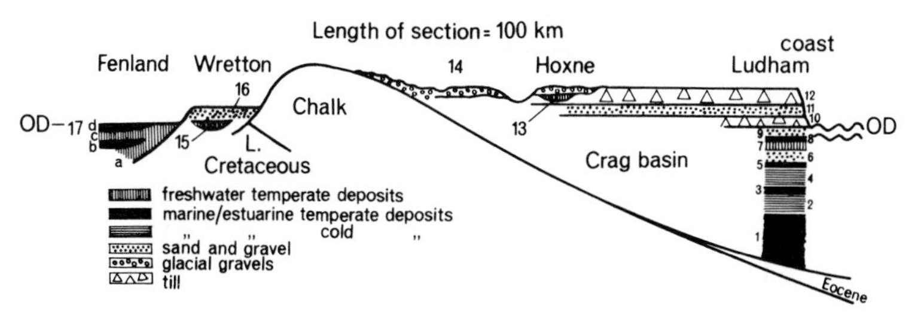

Fig. 12.4 Diagrammatic E–W composite section through the Pleistocene of East Anglia. 1, Ludhamian Crag; 2, Thurnian Crag; 3, Antian Crag; 4, Baventian sand and silt; 5, Pastonian estuarine silts; 6, Beestonian sands and gravels; 7, Cromerian freshwater muds; 8, Cromerian estuarine silts and gravels; 9, Early Anglian sands and silts; 10, Cromer till; 11, Corton sands; 12, Lowestoft till; 13, Hoxnian clay-muds; 14, Wolstonian gravels; 15, Ipswichian muds; 16, Devensian terrace sands and gravels; 17, Flandrian alluvium of Fenlands, 17a Lower peat, 17b Fen clay (estuarine), 17c Upper peat, 17d Romano-British silt.

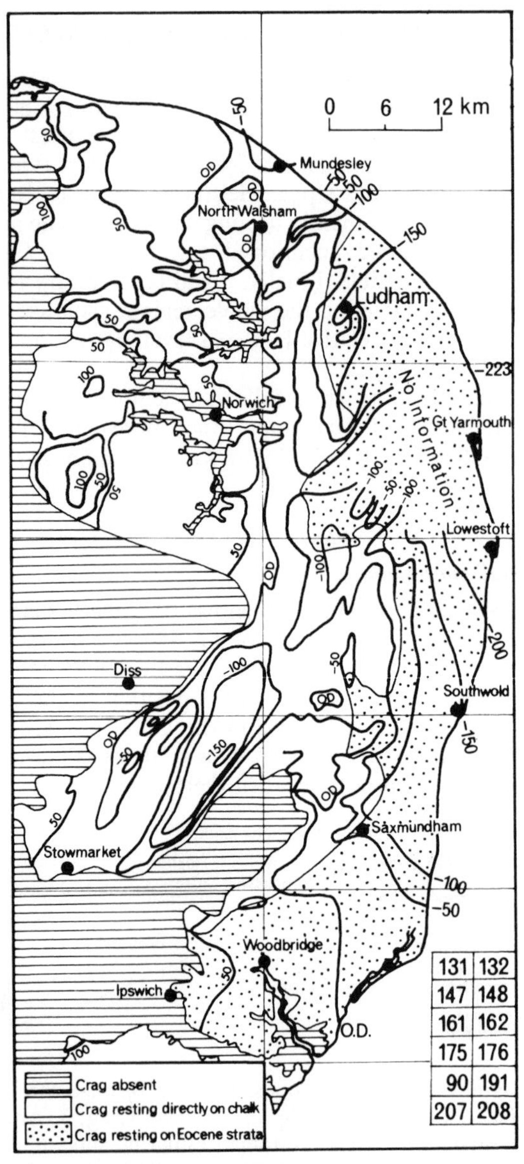

Fig. 12.5 Contours of the base of the Crag (the grid refers to the New Series 1″ geological maps). By permission of the Institute of Geological Sciences; from internal report *The hydrology of the Pleistocene deposits of East Anglia* by J. H. Price and J. Tuson, and figure 7, Wartime Pamphlet 20, Pt X, *Water supply from underground sources of Cambridge–Ipswich district*, by A. W. Woodland.

between the Atlantic and the North Sea; they are very similar in form to troughs in the Chalk in the present English Channel.[35]

The classic view of the Crag succession filling the basin was enunciated by Harmer[47] (table 12.1). This succession is largely founded on faunal rather than stratigraphical evidence, and this faunal evidence is viewed in the light of two assumptions, namely, that the proportion of northern marine molluscs increased throughout the Lower Pleistocene and the proportion of 'extinct' molluscs decreased during the same period. Both these assumptions may be questioned; the first in view of the fact that several Lower Pleistocene climatic fluctuations are known to have occurred, the second in view of the possibility that 'extinct' species may have returned with a change in climate.

The stages of the Lower Pleistocene shown in table 11.6 are based on pollen-analytical studies of a deep borehole in the Crag at Ludham,[131] and are supported by climatic evidence given by foraminiferal studies at the same locality.[34] Similar evidence from other bore-

Table 12.1 The Pleistocene Crag Succession of Harmer (1902)

Harmer's stages	*Lithological and Palaeontological Horizons*
CROMERIAN	Cromer and Kessingland Forest Bed, zone of *Elephas meridionalis*
ICENIAN	Weybourne and Belaugh Crag, zone of *Macoma balthica* Chillesford Beds, zone of *Leda oblongoides* Norwich Crag: Upper division, zone of *Astarte borealis* Lower division, zone of *Mactra subtruncata* (in other Crag classifications by Harmer the Norwich Crag remained undivided in one zone of *Astarte borealis*)
BUTLEYAN	Butley and Bawdsey Crag, zone of *Cardium (Serripes) groenlandicum*
NEWBOURNIAN	Newbourne and Sutton Crag, zone of *Mactra constricta*
WALTONIAN	Essex Crag: Oakley horizon, zone of *Mactra obtruncata* Walton horizon, zone of *Neptunea contraria*
PLIOCENE	Coralline Crag

Table 12.2 Pleistocene succession in East Anglia

Flandrian	Lake deposits, peat, alluvium, estuarine clays, blown sands
Devensian	Lake deposits of the late-glacial Intermediate and Low Terraces of the Cam Fenland river gravels Hunstanton Till
Ipswichian	Interglacial deposits at Cambridge, Bobbitshole, Stutton, Wretton. High Terrace of the Cam. March gravels
Wolstonian	Tjaele gravels, Breckland tills Cannonshot gravels of north Norfolk, kames and outwash plains
Hoxnian	Interglacial deposits at Hoxne, Marks Tey, Clacton (freshwater and marine), Nar Valley (freshwater and marine)
Anglian Lowestoft Stadial Gunton Stadial	 Lowestoft Till, Marly Drift, Springfield Till, Maldon Till Corton Sands, Gimingham Sands Cromer Tills, Norwich Brickearth Upper Arctic Freshwater Bed of Cromer Forest Bed Series
Cromerian	Estuarine clays } of Cromer Forest Bed Series Freshwater peat }
Beestonian	Sands and gravels } of Cromer Forest Bed Lower Arctic Freshwater Bed } Series
Pastonian	Estuarine clays, freshwater peat of Cromer Forest Bed Series, Norwich Crag in part, Weybourne Crag in part, Westleton Beds
Baventian	Easton Bavents clay and sand, Weybourne Crag in part
Antian	Shelly sand and gravel, Norwich Crag in part
Thurnian	Shelly sand and silt (at depth)
Ludhamian	Shelly sand and gravel (at depth)
Pre-Ludhamian	Shelly sand and clay (at depth), Red Crag

holes supports this succession of stages, with evidence of an older further, Pre-Ludhamian, stage in the deep borehole at Stradbroke, already referred to. As already discussed, this stage may be Pliocene in part.

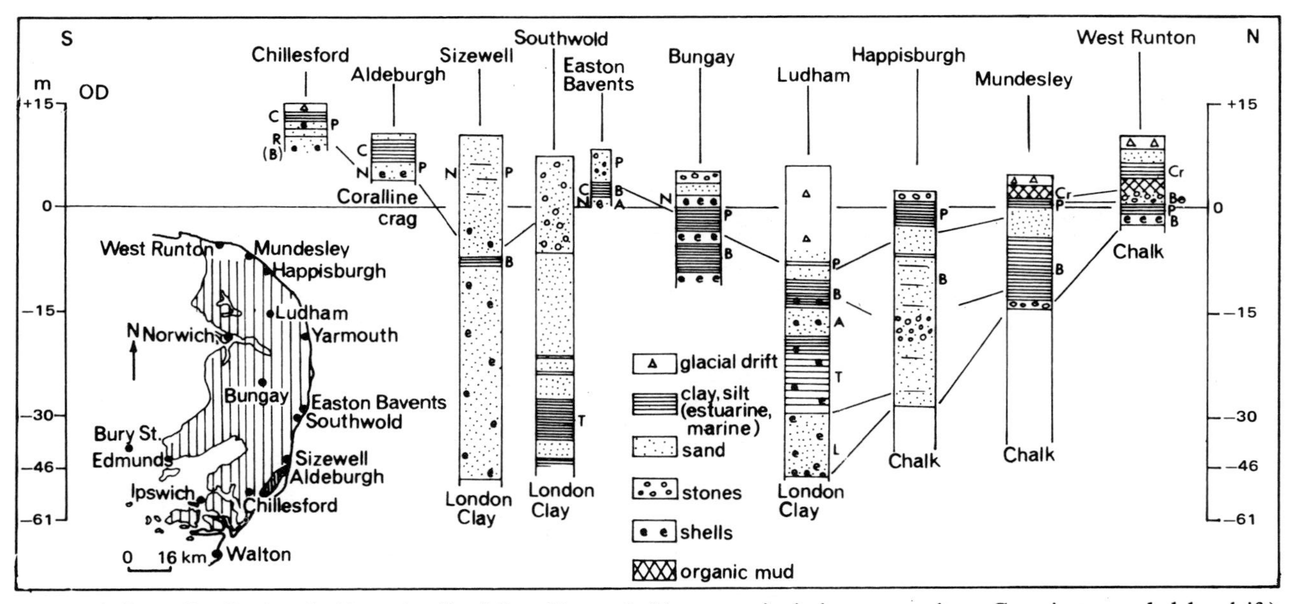

Fig. 12.6 Crag distribution in East Anglia (after Harmer) (the map includes areas where Crag is concealed by drift), and north–south section through the Crag basin in East Anglia. Letters on the left of the stratigraphical columns show Harmer's naming of the deposits: R(B), Butleyan Red Crag; N, Norwich Crag; C, Chillesford Clay and Sand; W, Weybourne Crag. Letters on the right of the stratigraphical columns show stages and correlations based on analyses of pollen and Foraminifera: L, Ludhamian; T, Thurnian; A, Antian; B, Baventian; P, Pastonian; Be, Beestonian; Cr, Cromerian.

Figure 12.6 shows a north–south section through the Crag basin. The relation between the classic Crag divisions and the stages at present known, shown in this figure, are only now becoming apparent. The Red Crag, with its shore facies a cross-bedded shelly sand, occupies an area in south Suffolk and adjoining Essex, overlapping the Pliocene Coralline Crag and the Eocene. The offshore facies, a silty clay, is found in the Stradbroke trough. The shelly sand of the Icenian Crag, with massive silty horizons, fills the centre and northern parts of the basin and overlaps the Red Crag to the south. Several stages are included within it, from the Ludhamian to the Pastonian. This last stage was apparently a time of extensive transgression, with crag and marine clays (Chillesford Crag, Chillesford Clay) being deposited in a large embayment in east Suffolk.[138]

The variation in depth of the Pastonian marine and estuarine deposits (fig. 12.6), as well as the basal Crag contours (fig. 12.5) suggest tectonic movement in East Anglia during or later than the Lower Pleistocene, with relative uplift in the Chillesford and Southwold area and downwarping in the Ludham area. However, much more study of the Crag series is required before these tectonic movements can be reconstructed in any detail.

The Pleistocene preglacial stages may be briefly described as follows:

Ludhamian. Shelly marine sands, 25 m thick at Ludham, with a temperate fauna and flora, the latter with a characteristic north American and eastern Asia genus *Tsuga*, the hemlock spruce.

Thurnian. Marine silty clays, 7 m thick at Ludham, with a foraminiferal fauna and a flora both indicating cool conditions. Deposits related to this stage are known from Ludham and Southwold.

Antian. Shelly marine sands, 3 m thick at Ludham with a temperate foraminiferal and mollusc fauna, and temperate flora, are known from Ludham and Easton Bavents. The flora again includes *Tsuga.*

Baventian. Marine silty clays, 8 m thick at Ludham, with a cool flora and fauna, are known at Ludham, Easton Bavents, South Cove, Aldeby, and on the Norfolk coast.

Pastonian. Marine silty clays and shelly sands, a few metres thick, with a temperate flora, but with only a trace of *Tsuga*, are known from the Norfolk coast, inland at Bungay, and in east Suffolk, where they include the Chillesford Crag and Clay. Beach gravels probably of this age are also known (Westleton Beds).

Beestonian. Arctic freshwater plant beds and evidence of permafrost are found in sands and gravels a few metres thick overlying the Pastonian silts in the Norfolk coast.

Cromerian. Temperate freshwater muds (plate 15a), and, upon them, estuarine and beach deposits of a marine transgression, overlie the

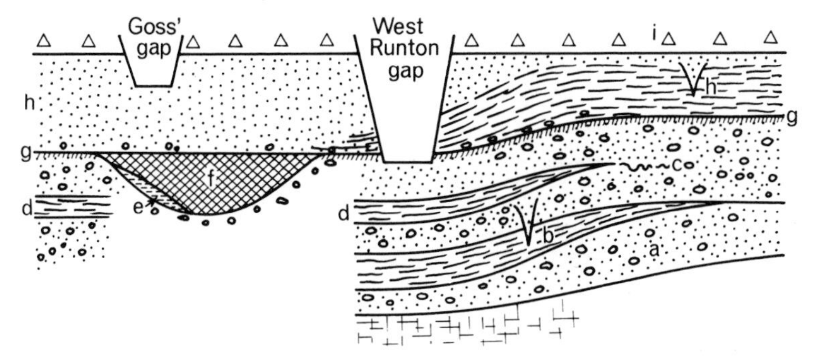

Fig. 12.7 Sketch section at West Runton, about 450 m long and 7 m depth. a, Weybourne Crag (Baventian/Pastonian); b, estuarine silts (Pastonian) pierced by ice-wedge cast of Beestonian age; c, sands and gravels, and a freshwater silt with arctic plants (d) (Beestonian); e, marl (Late Beestonian); f, muds and peats (Cromerian); g, weathered horizon (Cromerian 'soil'); h, estuarine silts and sands (Cromerian transgression deposits), pierced by ice-wedge cast of Early Anglian age; i, Cromer Till (Anglian). Chalk at base.

Beestonian sands and gravels on the Norfolk coast. They are up to 5 m thick, and are overlain by arctic plant beds and sands with evidence of permafrost conditions, both of these belonging to the earliest glacial stage of East Anglia, the Anglian.

A palaeocatena associated with this temperate stage has been described from West Runton.[124a]

The Pastonian, Beestonian and Cromerian stages are included in the Cromer Forest Bed Series;[140] fig. 12.7 shows the stratigraphical relations of the deposits of these stages in the type area of West Runton.

Table 12.2 shows the relation between these 'preglacial' stages and the formations included in them. The changes of relative land/sea-level implied by this succession is discussed later.

The glacial/interglacial succession

Deposits of three glacial and three non-glacial temperate stages are known in East Anglia, with the relations as shown in table 12.2 and fig. 12.4. The locations of important sites are shown in fig. 12.8. Good till sequences are known from the Norfolk and Suffolk coasts and from central Essex, and the relation of these to interglacial deposits is well established.[136] Figure 12.9 shows directions of ice advances into East Anglia.

Anglian. Most of the East Anglian till belongs to this stage. The sequences are complex, many different tills being recognised. The type section is in the cliffs at Corton, near Lowestoft. Here, overlying the Cromer Forest Bed Series, there is a lower grey or brown till,

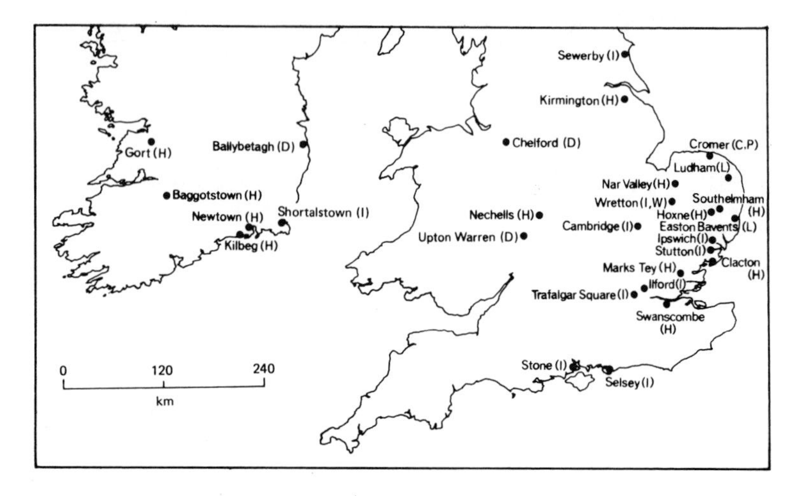

Fig. 12.8 Important Pleistocene fossiliferous sites. L, Lower Pleistocene; P. Pastonian; C, Cromerian; H, Hoxnian; I, Ipswichian; D, Devensian.

known variously as Cromer Till, Norwich Brickearth or North Sea Drift. Structural evidence indicates ice movement from the north–northeast, though stone orientation is northwest to southeast.[136] The till is overlain by the Corton Sands, several metres thick, containing a marine mollusc fauna. This fauna affords uncertain evidence of climate or of marine conditions; it has been considered secondary, or contemporary or partly both. The sands contain freshwater clays with a cold flora, and ice wedge casts have been observed in them. It is possible that the lower part of the Corton Beds have an *in situ* marine fauna, deposited on the retreat of the ice, while the upper part is related to the advance of the Lowestoft ice.

A further till, blue-grey chalky boulder clay, named the Lowestoft Till, overlies the Corton Sands.[2] This is the till which covers much of the Suffolk and south Norfolk plateau, laid down by ice which moved across East Anglia from a westerly direction. Thin clays and sands and a further chalky boulder clay in places overly the main Lowestoft Till.

The section at Corton thus covers two periods of ice advance, named the Gunton and Lowestoft Stadials, with an intervening sand, which must have been formed in a large embayment on the retreat of the first ice advance but before the advance of the Lowestoft ice.

In Norfolk, the central part of the county shows a till of Lowestoft age, but on the coast and in the north the sequence is far more complex. In the cliffs of the Happisburgh to Sidestrand area, there is a

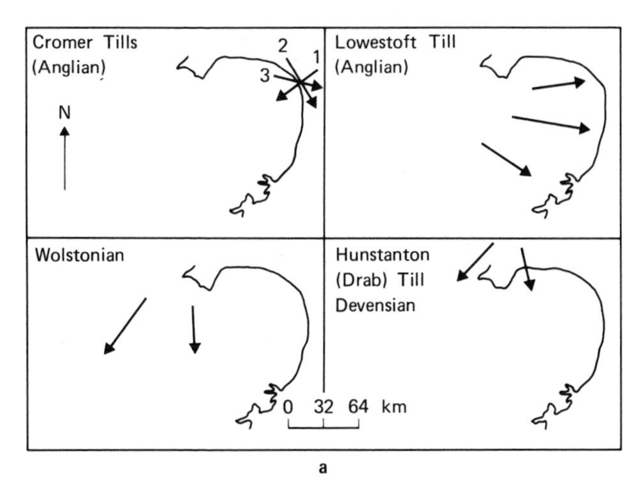

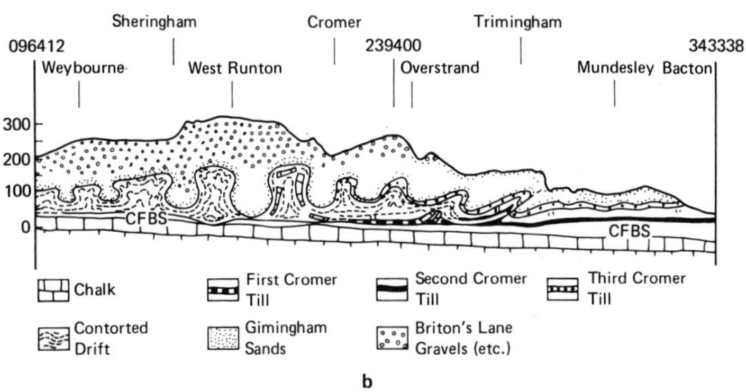

Fig. 12.9 a, Directions of ice advances in East Anglia. b. Diagrammatic interpretative section to show the structures of north Norfolk. Selected representative structures are drawn true-scale; vertical scale exaggerated. CFBS, Cromer Forest Bed Series (Banham, 1975).

complex of three tills, the First, Second and Third Cromer Tills, with laminated clays (Intermediate Beds) and sands (Mundesley Sands) between the First and Second and Second and Third Tills respectively.[4] A further series of sands (Gimingham Sands) overlies the Third Till. These may be the correlative of the Corton Sands further south. Further west, a series of sands and gravels (Briton's Lane Gravels) and a chalky till (Marly Drift) postdate the Gimingham Sands. This ice advance came from the west and is probably the same that produced the Lowestoft Till further south. Evidence from structures and fabrics associated with the three Cromer Tills indicate that the ice depositing them moved into Norfolk from the northeast, then northwest and finally west.[4]

The glacial deposits on the north and northeast Norfolk coast are heavily contorted, giving rise to the name Contorted Drift. The major structures involve the Cromer Tills and their associated sands and clays, and they have recently been shown to have resulted from load compaction and diapirism on a massive scale, with the loading probably provided by the Gimingham Sands and Briton's Lane Gravels (fig. 12.9).[4] The large Chalk rafts seen in the Norfolk cliff sections were probably emplaced by the first Cromer Till ice.

The three tills recognised in Essex are all Anglian in age.[3] They include the Springfield (with the Hanningfield) Till and the Maldon Till, separated by glacial gravels (Chelmsford Gravels).

A number of buried channels, the deepest lying at 106 m below O.D., are known beneath or in tills of Lowestoft age. They appear to be channels gouged subglacially during this ice advance.[145]

Hoxnian.[129] Many lacustrine deposits up to 20 m thick, with a temperate fauna and flora and overlying the Lowestoft Till, are known in Norfolk, Suffolk and Essex. The type site for this temperate stage is at Hoxne, Suffolk, where an extensive series of lacustrine sediments covering a large part of this non-glacial time are preserved (fig. 12.10). Marine and estuarine sediments of the latter part of this stage are known at Clacton, Essex and in the Nar Valley, Norfolk.[110]

Wolstonian. A pale chalky boulder, later than the Lowestoft Till, has been described from East Anglia. Named Upper Chalky Boulder Clay or Gipping Till (after a type area in the Gipping valley), its reality has been recently brought into question.[16] There is no doubt of a cold stage between the Hoxnian and Ipswichian temperate stages, since cold floras are found above the Hoxnian at Hoxne and elsewhere, and below the Ipswichian as typical late-glacial floras. There are certain tills which may be placed in this interval in the Breckland of Suffolk, where they are associated with interstadial deposits at Mildenhall.[94] These chalky tills contain igneous erratics of a northern provenance, suggesting ice movement from the north.[2] There are also morainic gravels found in north and west Norfolk, forming outwash plains, kames and crevasse-fillings[103] and associated with terraces in central Norfolk and the Waveney Valley. These appear to postdate the Hoxnian and so may be included within the Wolstonian. But the stratigraphical relations of these deposits require further investigation. The tjaele gravels of the Cam valley have also been placed in this cold stage.

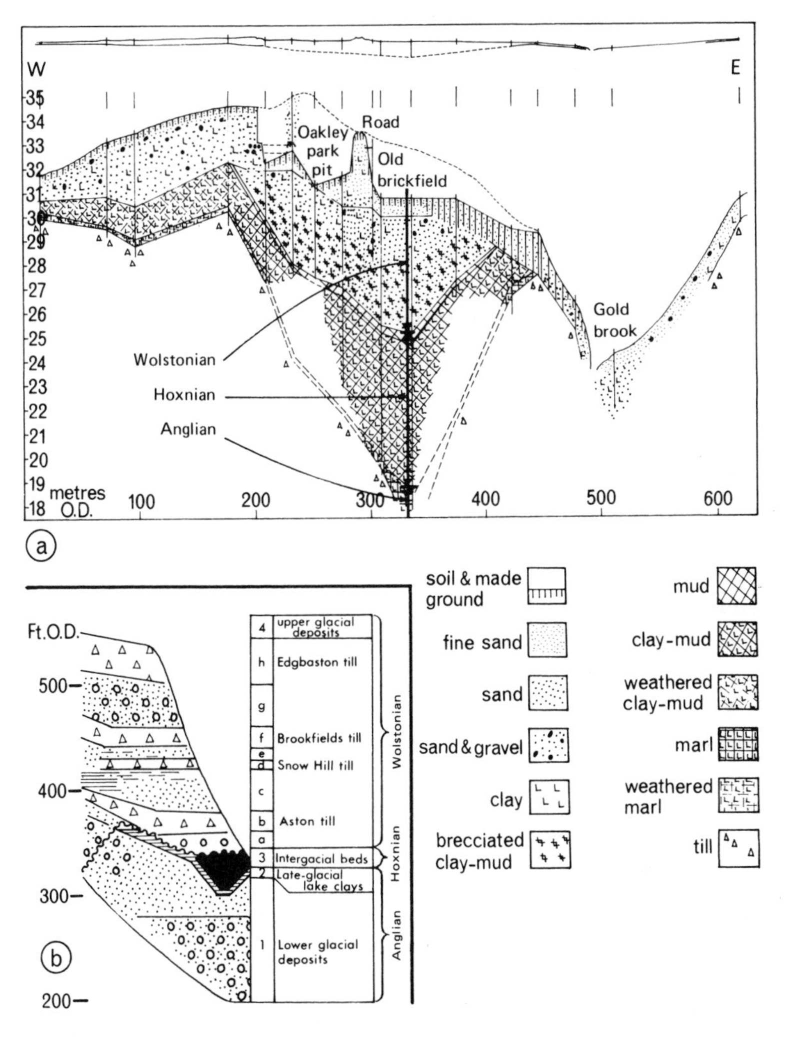

Fig. 12.10 Middle Pleistocene successions. a, at Hoxne, Suffolk (West, 1956); b, north Birmingham (Kelly, 1964).

The area and limits of ice during this stage are thus problematical, but it seems probable that an ice limit lay in the western part of East Anglia and in the north of Norfolk.

Ipswichian.[130] A further series of temperate lacustrine or fluviatile deposits, up to 5 to 10 m thick, occur in the present-day valley systems, and are younger than glacial deposits thought to be of

Wolstonian age. They belong to the Ipswichian temperate stage, with the type site near Ipswich. Brackish deposits of this stage are known at Wretton, Norfolk, and the March Gravels of the Fenland, with a good marine fauna, are thought to have been formed during an Ipswichian marine extension into the Fenland.

Devensian. The Ipswichian temperate deposits are overlain by terrace or solifluction deposits of the Devensian cold stage, during which ice reached down to the north Norfolk coast and there deposited a red brown till, the Hunstanton Till. This till is thought to be a southward extension of the Drab Till of Holderness.[66] It is associated with a few fresh landforms, such as the esker in Hunstanton Park. The low level Fen edge gravels are also associated with this final cold stage.

Flandrian. Biogenic infillings of closed depressions (meres) and the freshwater and estuarine sequences of the Fenland basin[40] and Broadland[63] are of Flandrian age. In the last two there is a lower Flandrian freshwater peat, a middle Flandrian estuarine silt or clay, and an upper Flandrian freshwater peat, overlain by estuarine silts.

Landform development

The landforms developed on the glacial deposits are generally subdued with few signs of end-moraine features. A few drainage-ways are known, such as the Little Ouse–Waveney Gap, probably of Wolstonian age. As indicated in fig. 12.4, the Hoxnian temperate deposits are found in places isolated from the present valley systems, except where sedimentation filled a buried channel of Anglian age coincident with a present drainage way. This is in contrast to the position of the Ipswichian deposits in low terraces of the present valleys. The landforms therefore took on their present aspect some time late in the Wolstonian cold stage. The terrace system of the Cam, the best known terrace system in East Anglia,[104] contains a high terrace, partly of Ipswichian deposits, and intermediate and low terraces of Devensian age. Sections in the last two often show ice-wedge casts (plate 14a).

THE MIDLANDS[120]

In the Midlands there is a substantial record of the Middle and Upper Pleistocene (table 12.3). Ice advances of the three glacial stages extended into the area, and their drainage was associated with terrace systems of the three main rivers, the Trent, Severn and Thames. The drifts of the two older glacial stages are found on the uplands, while the Devensian glacial deposits are confined to the west Midlands lowlands.

The older drifts include the deposits of two glacial stages, the earlier correlated with the Anglian stage and the later providing the type glacial sequence of the Wolstonian stage. Important biostratigraphical evidence of age is provided by interglacial deposits of Hoxnian age at Nechells[56] and Quinton in the Birmingham area.

Ice advanced into the Midlands from three main sources, each resulting drift type having its characteristic erratic suite. The Welsh drift contains Uriconian volcanic rocks, the eastern drift Mesozoic rocks, and the northern or Irish Sea drift marine molluscs derived from Irish Sea sediments and rocks from the Lake District and Southern Uplands of Scotland. Judging from the present distribution of these drift-types, during the two older glaciations the Welsh and eastern ice streams were important, while in the youngest the Irish Sea Ice was predominant.

The oldest glacial deposits are thin Welsh drifts (First Welsh ice advance) in the lower parts of the Severn river system, outwash gravels at Trysull, west of Birmingham, and northern drifts of the Coventry region, formed during ice advances from the west and north. Of the same age may be the lower boulder clay of the east Midlands and possibly the northern (plateau) drift of the Oxford region.[1] The succeeding temperate stage is recorded by peaty silts in Birmingham[56] which lie below gravels and tills of Wolstonian age (fig. 12.10). Their vegetational history can be clearly correlated with that of the Hoxnian. There is also an important erosional interval between the older glacial deposits and those of Wolstonian age.

During this second glacial stage, which includes the Second Welsh and Main Eastern ice advances, three ice streams entered the Midlands as shown in fig. 12.11. The advance of the ice impounded a large proglacial lake in a 'proto-Trent' valley wending northeast;[91] this lake, named Lake Harrison, was at its maximum size 90 km long and reached a height of 125 m O.D. The lake was eventually overridden by eastern ice, associated with a moraine at Moreton-in-the-Marsh. In this setting sequences of proglacial lake clays and tills were laid down over a wide area in the Midlands.[53, 88, 89] Overflows occurred from the lake into the upper Thames system.[9] and the link between the glacial sequence and the upper Thames terraces is given by the carriage of northern erratics (Bunter pebbles and flints), appearing in quantity in the Wolvercote Terrace of the Evenlode and Cherwell rivers. On the retreat of the eastern ice, additional stages of Lake Harrison developed, and on retreat of the western ice, the lake was

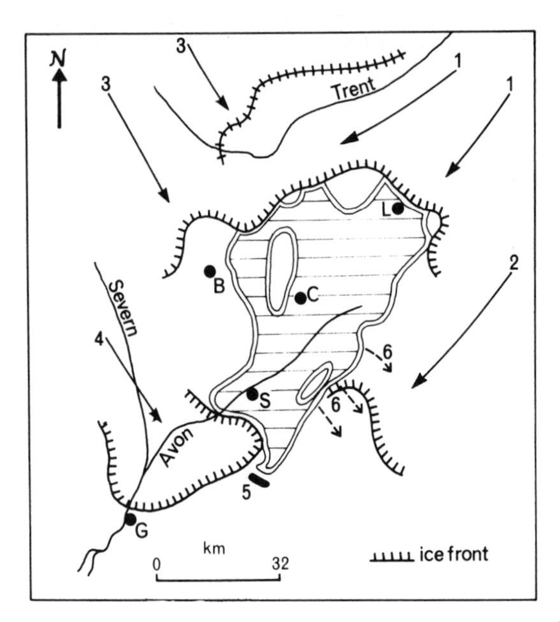

Fig. 12.11 The Wolstonian ice advance in the Midlands. Lake Harrison at its maximum stage is shaded. B, Birmingham; C, Coventry; G, Gloucester; L, Leicester; S, Stratford. 1, eastern ice margin at Lake Harrison maximum; 2, later stage of the eastern ice; 3, northern ice; 4, Severn Valley ice; 5, Moreton Moraine; 6, Lake Harrison overflows.

drained into the Severn, and rapid erosion of the lake deposits ensued. The present Warwickshire Avon was thereby initiated.

The terrace systems of the Trent,[85, 111] Severn[144] and Avon began to form on the retreat of the ice. These terrace systems in outline divide into high terraces correlated with the glacial retreat, middle terraces with a warm fauna, Ipswichian in age, and low terraces of Devensian age. Table 12.3 gives the succession of terraces in the Avon and Severn valleys.

The Midlands area is a critical one for Devensian stratigraphy and the margin for this glacial stage a subject of controversy. Meltwaters escaped east over the Pennine watershed and south through the Shropshire Hills, notably via the Ironbridge Gorge. Several proglacial lake systems, marginal drainage features and retreat stages of the Devensian ice have been described from the Cheshire Plain.[80, 84] There is much discussion over the Devensian stratigraphy of the region. Some suppose the existence of two tills separated by middle sands; from middle sands at Chelford there is a radiocarbon date of 60,800 B.P. years of wood associated with a thin peat horizon.[95]

Table 12.3 Pleistocene succession in the Midlands

andrian		Lake deposits, peat, alluvium		
evensian	Late	Lake deposits of the late-glacial		Avon Terrace 1
		Llay and Welsh Readvances	Worcester Terrace	
		Wrexham-Bar Hill Moraine		
		Wolverhampton Till (Main Irish Sea ice advance)	Main Terrace	Avon Terrace 2
	Middle	Upton Warren interstadial		Avon Terrace 3 surface
	Early	Chelford interstadial	Kidderminster Terrace	Avon Terrace 4 gravels
swichian		Gravels of Avon No. 3 terrace level, Mud at base of Four Ashes gravel		
'olstonian		Moreton (Eastern) Drift, 2nd Welsh and Older Irish Sea Ice advances,	Bushley Green Terrace	Avon Terrace 5
		Wolston Series	Severn Terraces	Avon Terraces
oxnian		Interglacial deposits at Nechells, Quinton, Trysull		
nglian		Bubbenhall Clay, 1st Welsh ice advance, gravels at Trysull Lower boulder clay of east Midlands.		

This non-glacial interval, the Chelford interstadial, is an important feature of Devensian stratigraphy in Britain. But it is likely that the separation of the middle sands as a stratigraphical unit is an over-simplification, and it is also thought that the lower boulder clay may belong in part to a Wolstonian ice advance.

On the present evidence it appears that the major Devensian ice advances are Late Devensian. At Four Ashes, north of Wolverhampton, an Irish Sea till covers gravels with organic silt lenses.[74] This site is the type site for the Devensian. Radiocarbon dates show the till is later than 30,500 years B.P., and the suite of erratics associated with this till stretches south to the 'Wolverhampton line' (fig. 12.3), which thus appears to be the limit of Devensian ice in this area. There are later readvances or standstills—the Ellesmere Readvance and the Llay Readvance. All these events must have taken place in the Late Devensian, between 30,500 and 12,000 years B.P. In the same period there were advances of the Welsh ice (Third Welsh ice advance, Welsh Readvance), but the relation of these to the movements of the Irish Sea ice is not clear. Apart from these

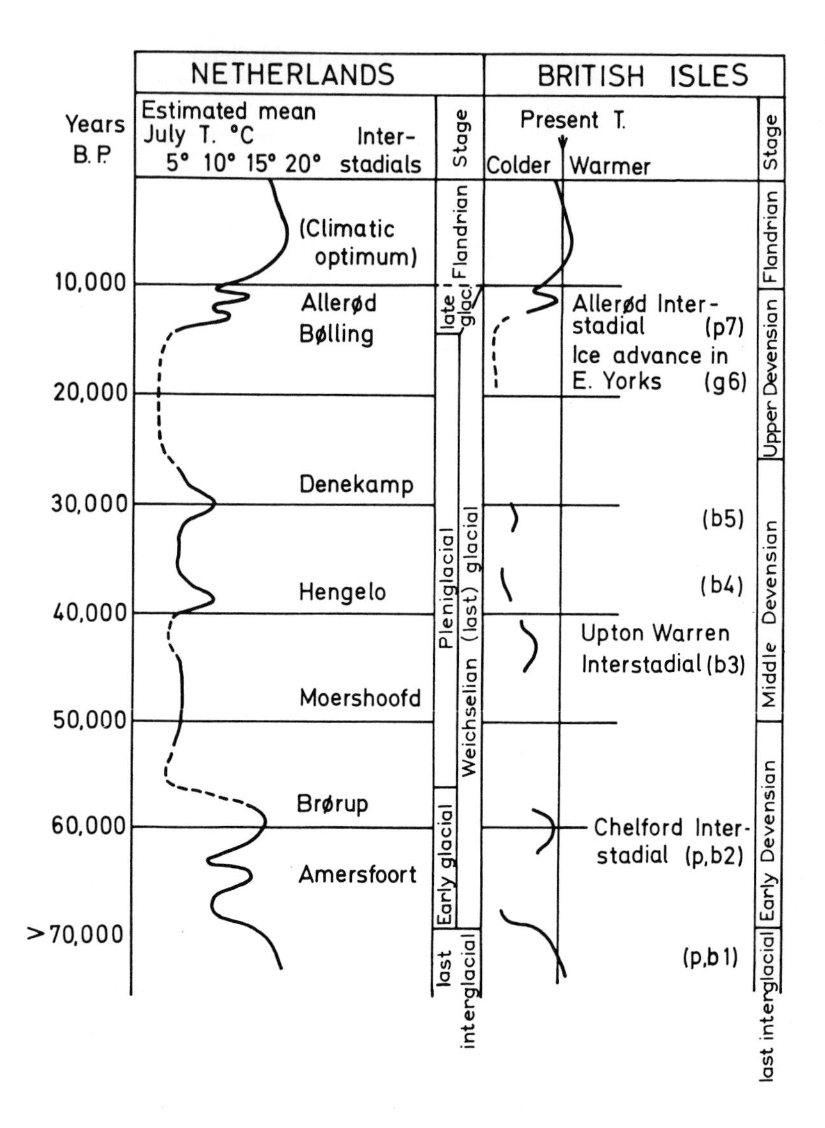

Fig. 12.12 The climatic sequence of the last glaciation in the Netherlands (after v.d. Hammen *et al.* 1967) compared with that of Britain. The evidence for climate in Britain is indicated on the right; g, based on glacial deposits; b, based on beetle evidence; p, based on palaeobotanical evidence. Numbers refer to sites: 1, several Ipswichian interglacial sites; 2, Chelford, Cheshire; 3, Upton Warren, Worcs.; 4, Fladbury, Worcs.; 5, Coleshill, Warws.; 6, Dimlington Cliffs, Yorks.; 7, many carbon-dated interstadial sites.

stratigraphical questions a number of climatic oscillations have been adduced from studies of floras and insect faunas in the terraces of the west Midlands, and there are radiocarbon dates from many of these deposits. The Upton Warren interstadial, at about 42,000 years B.P., identified on the basis of insect remains, is an important horizon.[27] Figure 12.12 shows a tentative synthesis of the course of Devensian climatic change in northwest Europe and the position of certain faunas or floras on the climatic curve.[26]

Local solifluction (tjaele) gravels and sands of the Severn and Avon valleys are also associated with the Devensian, as are fossil ice-wedge networks reported from the Wolverhampton area and the Low Terrace (No. 4) of the Avon.[92] Beyond the limit of the Devensian ice, on the limestone of the Derbyshire dome, loess is admixed with weathering residues of the limestone.[83] This may be of Devensian age, and its presence here, with the concealment of the weathered limestone surface, contrasts with the bare limestone pavements of the northern Pennines, swept by the Devensian ice.

WALES[13]

During the glacial stages Wales was overrun by ice from local ice sheets developing at various centres in the Cambrian Mountains and by ice from the Irish Sea ice sheet moving southwards and impinging on the Welsh coast. Local ice sheets developed in the north, in the mountains of Snowdonia, the Harlech Dome, Cader Idris, Arenig and Berwyn, and ice from these centres radiated west, north and east, to be confined by the Irish Sea ice at certain times. In central Wales ice moved east down the upper valleys of the Severn and Wye, and west from the Plynlimon range into Cardigan Bay. In southern Wales the principal ice centres were the Brecon Beacons and the Carmarthen Fans. Ice from these mountains filled the valleys of the Towy, Neath, Taff and Usk rivers and passed off into the Bristol Channel.

Little biostratigraphical evidence for the subdivision of the Welsh drifts has yet been obtained. Older and newer drifts have been separated on the basis of freshness of the landforms and weathering of the deposits, but there has been much argument on the stratigraphical significance of such evidence. The most important evidence for subdividing the succession comes from raised beaches in south and southwest Wales, such as the beaches at Minchin Hole, Gower.[12] These beaches lie on till or on a shore platform and are covered by head deposits, which may be associated with till of Late Devensian age. The fauna of the beaches is interglacial in character, and is believed to date from the last interglacial. At West Angle Bay, fresh-water and marine sediments show a pollen sequence which is not dissimilar to Ipswichian pollen diagrams from southeast England.

The pre-raised beach succession of older drifts is not well known. A shore platform, perhaps of Middle Pleistocene age or older, occurs at 0–15 m O.D. around the coast. Irish Sea till and Welsh drifts are found mainly on interfluves outside the limit of the Devensian ice (fig. 12.3). Erratics in the raised beach deposits are further evidence of an earlier glacial stage.

Table 12.4 Pleistocene succession in Wales

Flandrian	Lake deposits, peat, alluvium, estuarine clays
Devensian Late	Late-glacial { corrie moraines of late-glacial readvance; interstadial muds; corrie moraines } Head Irish Sea ice advance (South Wales end-moraine) Welsh ice advances Head
Middle and Early	Head
Ipswichian	Raised beaches (Minchin Hole), West Angle interglacial
Wolstonian	Irish Sea and Welsh tills of south and southwest Raised beach erratics (Gower)
pre-Wolstonian	Shore-platform

The limit of the Devensian maximum, believed to be coeval with the post-30,500 year maximum of the west Midlands, is partly that of Irish Sea ice, partly of local Welsh ice. The stratigraphy of some Devensian sites is shown in fig. 12.13. The Irish Sea ice mounted the northern slopes of Wales, moved southwest by the Lleyn Peninsula and southeast into the Cheshire Plain. An important stratigraphical point is that till was laid down in the Vale of Clwyd and sealed caves containing Palaeolithic ('Aurignacian') implements, suggesting this till was deposited in the later part of the Devensian, a dating confirmed by many radiocarbon dates. The Devensian limit in west and south Wales (fig. 12.3) has been much discussed. South Pembrokeshire and parts of the south Wales coastal region were evidently outside the limit, and much head was deposited in the Devensian. In south Wales the Devensian maximum is marked by end-moraines in the valleys. Large piedmont glaciers formed in Swansea Bay and in southeast

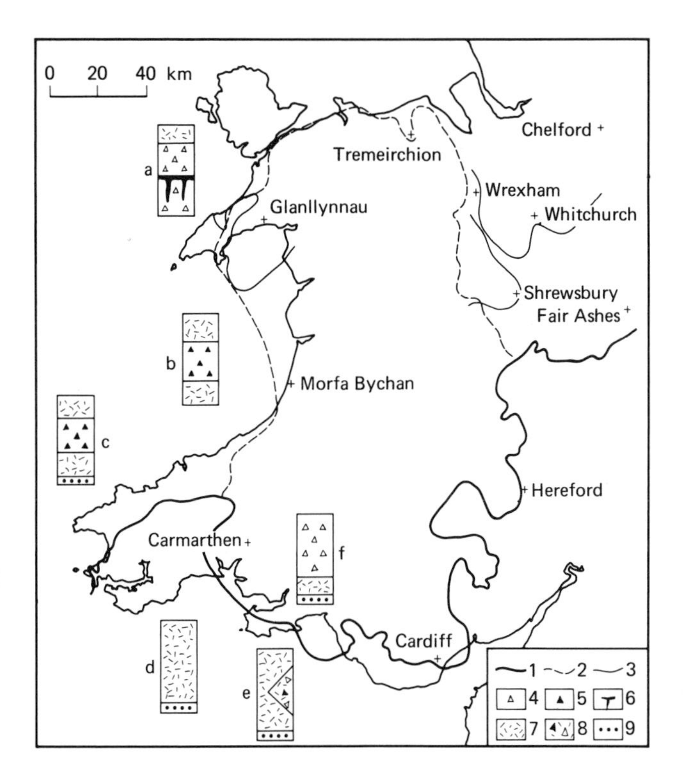

Fig. 12.13 Devensian sequences in Wales and the borderlands, from the Llyn peninsula (a), Cardigan Bay, (b), north Pembrokeshire (c), south Pembs and Carmarthen Bay (d), south Gower (e) and east Gower (f). 1, outer limit of Devensian glaciation; 2, approximate line of separation of Welsh and Irish Sea ice sheets; 3, possible readvances; 4, Welsh till/sand and gravel; 5, Irish Sea till/sand and gravel; 6, ice-wedge casts and weathering horizon; 7, head, 8, redeposited glacial drift (head); 9, raised beach (Bowen, 1973a).

Glamorgan. Substantial head deposits are also found on the west coast, inside the Devensian limit, and one of the difficulties of the Pleistocene sequence has been the differentiation of these heads from till and redeposited till. In the northwest, the Devensian sequence is more complex, with evidence for a later readvance.

Retreat phenomena of the Devensian have been recorded in many parts of Wales, e.g. the fluvioglacial terraces of the Lleyn Peninsula and marginal glacial lakes and associated overflows of the southwest.[21] The final Devensian glaciers in Wales left block-moraines, some of which have been dated to the late-glacial readvance (pollen zone III). A good Devensian record, with cold and temperate faunas, is found in cave deposits in northeast and south Wales.

The many periglacial phenomena of Devensian age in Wales are mentioned later in the chapter.

LINCOLNSHIRE AND EAST YORKSHIRE[81, 112, 115]

The margin of the Devensian ice passes north–south through this area east of the Yorkshire–Lincolnshire Wolds (fig. 12.3). The southern limit is around the Wash and on the coast of northwest Norfolk. The results of the ice advance in pounding up the drainage of the Trent and Fenland areas remain to be investigated in detail. The drifts of earlier glacial advances are represented by the chalky boulder clay of the Wolds and more clay-rich tills further west in Lincolnshire.[112]

To the east of the Wolds a sequence of tills is present (table 12.5). The uppermost till is clearly Devensian because it is associated with many fresh morainic features including kettle holes and eskers. In Holderness (fig. 12.14)[19] Drab and Purple Tills are separated from an older Basement Till by the Sewerby raised beach deposits, with a fauna indicating an Ipswichian age. A radiocarbon date of 18,240–250 years B.P. (Birm. 108) from moss silts in a basin between the Basement and the Drab Till indicates that the post-Ipswichian tills of Holderness, the Drab and Purple Tills, belong to the Late Devensian.[82] The Hessle Till, once supposed to be the final member of this sequence, is now believed to be the weathering product of the two other tills. The Upper and Lower Marsh Tills of Lincolnshire and the Hunstanton Till of northwest Norfolk are thought to belong to the same ice advance which deposited the Drab Till of Holderness.[66]

Extensive loess deposits on the Wolds are of Devensian age.[20] On the Yorkshire Wolds they are thought to date from a period before the ice reached its maximum limit.[81]

Gravels rich in marine shells occur in several areas of Holderness and east Lincolnshire. These are often associated with morainic features, and they appear to be fluvioglacial gravels containing reworked interglacial marine shells or ice-pushed interglacial gravels and silts.

Apart from these interglacial sediments, there are few biogenic temperate deposits in the region. The most important is at Kirmington in the northern part of the Lincolnshire Wolds, where, between tills, there is a thin peat horizon which is associated with estuarine and beach deposits.[14] These temperate deposits are probably Hoxnian in age. Their place in the till sequence requires renewed

Table 12.5 Pleistocene succession of Lincolnshire, east Yorkshire and northern England

	Lincolnshire	*E. Yorkshire*	*Northern England*		
Flandrian	Lake deposits, peat, alluvium, estuarine deposits, raised beaches				
Devensian Late	Late-glacial lake deposits		Corrie moraines of Lake District and Pennines		
Middle and Early	Upper Marsh Till Lower Marsh Till	Kelsey Hill shelly gravels Purple Till Drab Till Dimlington moss silts	Interstadial muds Deposits of Lake Humber (7 m and 30 m) York-Escrick moraines Silts at Oxbow, Aire Valley	Upper Till Middle Sands Lower Till	Till/sand complexes
Ipswichian		Sewerby beach deposits	Interglacial deposits at Austerfield, Armthorpe and Langham		
Wolstonian	Calcethorpe Till (chalky boulder clay) and other tills	Basement Till	Older drift of Doncaster area	Scandinavian Drift	Lower tills? Local till of Calder Valley
Hoxnian	Kirmington interglacial deposits	Bridlington Crag	Vale of York region	E. Durham	Lake District, Lancashire
Anglian	Tills below Kirmington interglacial deposits				

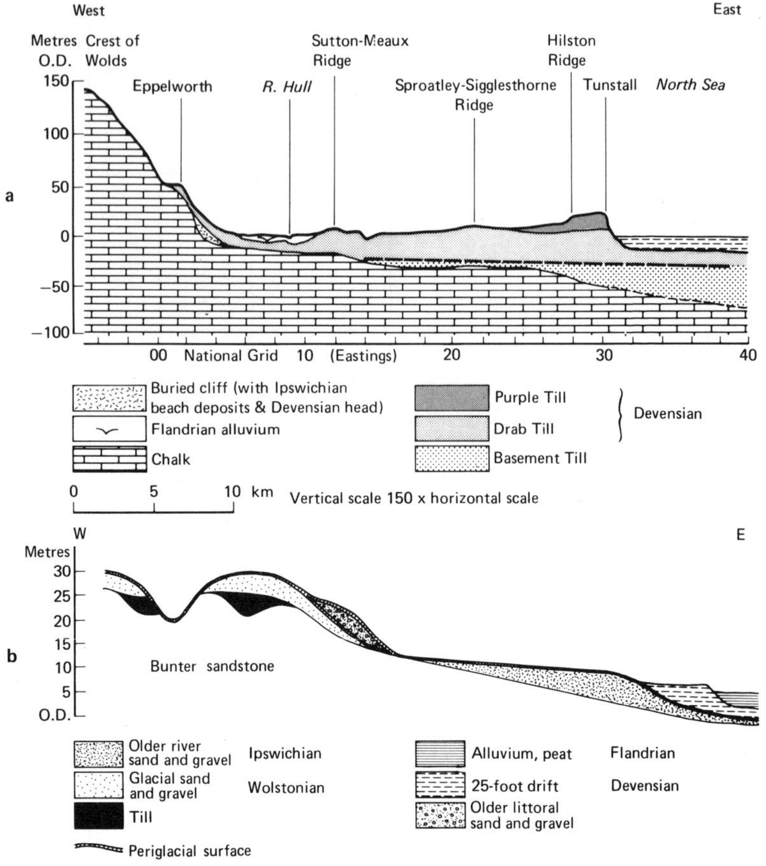

Fig. 12.14 Sections in the Yorkshire Pleistocene.

a. East–west section across Holderness (Madgett, 1975).

b. Diagrammatic section showing morphological features and stratigraphy of the Austerfield area, south Yorks. (after Gaunt *et al.* 1972).

investigation particularly since till below the interglacial deposit may be older (Anglian) than other tills of the region.

NORTHERN ENGLAND

Northern England was subject to glaciation by a local ice sheet forming in the mountains of the Lake District and by smaller ice sheets of the northern Pennines, and to glaciation by ice streams flowing from centres in the southern uplands of Scotland (fig. 12.15).

The south-flowing Scottish ice was divided by the Lake District ice, the western stream being directed towards the Irish Sea, the eastern stream across the Pennines via the Tyne Gap and Stainmore. This eastern stream was divided by the Cleveland Hills, the Vale of York receiving a western lobe, with contributions from the Dales, and an eastern lobe moving south and impinging on the coast of eastern England as far south as Norfolk.

Changes in the power and direction of these ice streams resulted in the deposition of till sequences,[101, 122] the identification of the different ice advances being facilitated by the many characteristic erratics of the drift, e.g. Norwegian erratics in the Scandinavian drift, Shap granite in tills derived from the Lake District ice. The lack of biogenic temperate deposits within the till sequences make the correlation of ice advances with the known glacial stages difficult. In northern England the problem is how many tills are Devensian. The Late Devensian margin is in the southern part of the Vale of York,[36a] with the York and Escrick moraines a later well-marked standstill phase, and a well-known system of ice-marginal features occurs in the Cleveland Hills of north Yorkshire.[46] But other moraine systems relating to ice retreat in the area are more difficult to interpret and their relation to the till sequence is problematical, as with the reality of the Scottish Readvance (fig. 12.15) within the Devensian. There are, however, late-glacial (zone III) moraines in corries of the Lake District and the Pennines.

The Upper Boulder Clay/Middle Sands/Lower Boulder Clay complex of Lancashire and the Upper and Lower Boulder Clays of Durham appear all to be of Devensian age. Tills of earlier glaciations are found on the Derbyshire Dome and on the northeast coast, where the Basement Till of east Yorks is again seen, and where the Scandinavian Drift occurs at Warren House Gill. In the Dales of the Pennines several ice advances had been recognised.[86] The Main Dales glaciation is Late Devensian, but the age of more extensive glacial deposits is problematical. Suggested correlations of these tills are given in table 12.5.

The best known sequence of the Late Pleistocene comes from the Vale of York and the area to the south of it (fig. 12.14).[37] Fluviatile deposits (Older River Sand and Gravel) containing Ipswichian interglacial deposits occur east of Doncaster,[38] lying in valleys bordered by higher ground showing older drift till and glacial sand and gravel. Incision of the fluviatile deposits was followed by periglacial conditions, then the deposition of sands and gravels associated with glacial Lake Humber, dammed between the York–Escrick ice and that at the mouth of the Humber during the Late Devensian.[36] On

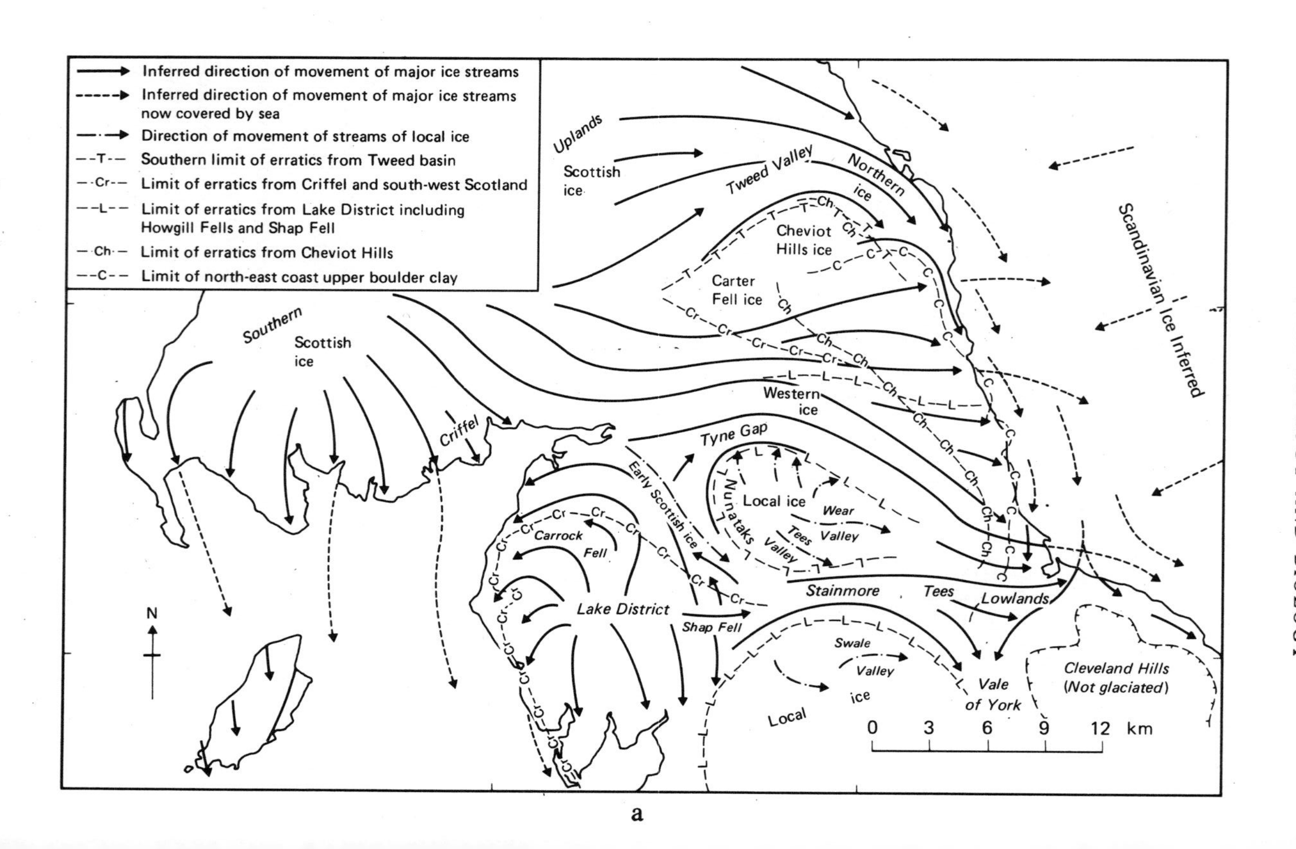

Inferred direction of movement of major ice streams
Inferred direction of movement of major ice streams now covered by sea
Direction of movement of streams of local ice
--T-- Southern limit of erratics from Tweed basin
--Cr-- Limit of erratics from Criffel and south-west Scotland
--L-- Limit of erratics from Lake District including Howgill Fells and Shap Fell
--Ch-- Limit of erratics from Cheviot Hills
--C-- Limit of north-east coast upper boulder clay
Uplands
Scottish ice
Tweed Valley
Northern ice
Cheviot Hills ice
Carter Fell ice
Scandinavian Ice Inferred
Southern Scottish ice
Western ice
Criffel
Tyne Gap
Local ice
Wear Valley
Tees Valley
Nunataks
Early Scottish ice
Carrock Fell
Lake District
Shap Fell
Stainmore
Tees Lowlands
Swale Valley ice
Local
Vale of York
Cleveland Hills (Not glaciated)
0 3 6 9 12 km
N

a

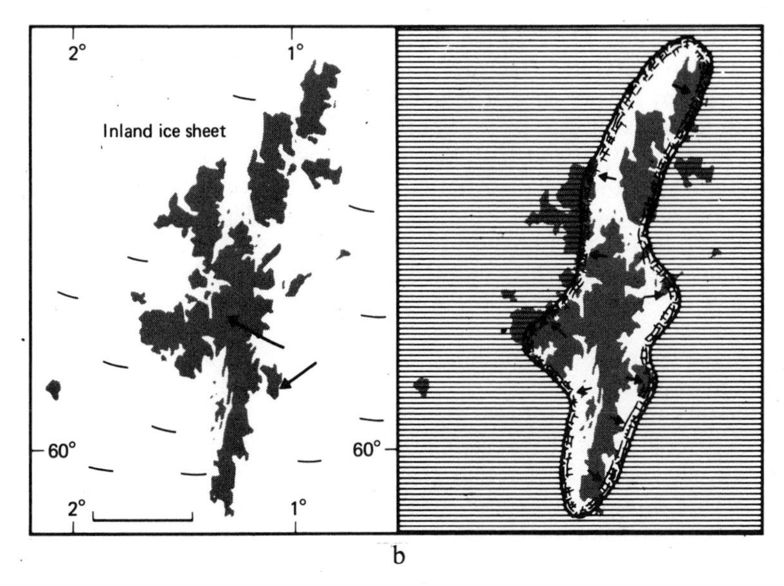

Fig. 12.15 Devensian ice flow directions in northern England (a) according to Eastwood *et al.* (1971), and Shetland (b) according to Hoppe (1974). An early glaciation and a later local ice cap are indicated.

the retreat of the Vale of York ice, the level of Lake Humber fell and it extended north of the York–Escrick moraines (25ft (7·6 m) Drift). The lake was drained by 11,000 years B.P. and sand dunes formed in the late-glacial.[67]

Ipswichian ombrogenous peats (Zones Ip II and III) have been found erratic in the Upper Till of Durham[5] and on the coast near Durham freshwater peats have been found in fissures in the Magnesian Limestone preceding the oldest till of the area, the Scandinavian Drift. The age of the freshwater deposits is problematical. They have been described as Lower Pleistocene, but this correlation, based on fossil flora, requires re-investigation.

In the Isle of Man there is again a good till sequence, with till believed to be of Anglian and Wolstonian age, separated by sands thought to be marine, and a younger Devensian till. The Bride Hills moraine of the northern part of the isle may mark a Devensian maximum or be a retreat stage. Thick head, of Devensian age, is found in the central uplands of the isle.

SCOTLAND[97]

The ice streams developing during the Scottish glaciations are shown in figs. 12.2 and 12.15. In the Highlands, ice flowed west and northwest over the Outer Hebrides and towards the east the flow was deflected by the presence of Scandinavian ice. In the south, the Highlands ice was deflected by the Southern Uplands ice, westward down the Firth of Clyde and Irish Sea, and eastwards along the Firth of Forth, where it was again deflected north and south by Scandinavian ice. The ice flowing east from the Southern Uplands was similarly deflected by the Scandinavian ice, while in the west of the Southern Uplands ice flowed partly into the Irish Sea, and partly south to the east of the Lake District ice, down the Vale of Eden and over the Pennines.

Apart from the mountainous areas, with all the normal signs of mountain glaciation,[113] the glacial scenery is of two main types: the smooth and often driftless areas of northwest Scotland, and the drift-covered areas of the east and south where the hummocky landforms, characteristic of recent glaciation, are often well developed. There are many drumlin fields in the Central Valley and in Galloway, and many areas of hummocky moraine, including the end-moraines of the late-glacial Loch Lomond Readvance, the only widely-recognised Devensian readvance. Other moraines formerly considered to be readvance end-moraines, e.g. those of the Aberdeen–Lammermuir Readvance and the Perth Readvance, are not now so regarded. In many areas, including the Outer Isles, Shetland[52] and Orkney, the drifts have been divided into a general glaciation with widespread ground-moraine, and later stages of glaciation marked by end-moraines in valleys and in corries.

Although a number of fossiliferous biogenic deposits have been described from Scotland, none show convincing evidence for dividing the Pleistocene sequence into stages, except a peat bed in Shetland,[8] tentatively correlated with the Hoxnian and lying between tills, so indicating pre- and post-Hoxnian glaciation in Shetland. Till sequences are well developed in some areas, e.g. Lothian, Aberdeen, Caithness and the Isle of Lewis, but it is uncertain how many glacial stages are concerned in the sequences. A pre-Devensian shelly till is thought to be present in northeast Scotland, but in the main the till sequences are all thought to be within the Devensian, the changes in till type being associated with changes in ice movement direction.[62] There are three radio-carbon dates of biogenic material in the range 27,000 to 28,000 from Lewis, Glasgow and Morayshire, and it is possible that the main Devensian tills all post-date this time and belong to the same period of Late Devensian ice advance that was responsible for the Devensian tills of England.

Table 12.6 Pleistocene succession in Scotland

Flandrian	Lake deposits, peat, alluvium, estuarine deposits, raised beaches
Devensian Late	Late-glacial Loch Lomond Readvance Marine interstadial muds Clyde Beds and Errol Beds
Middle and Early	Till sheets of northeast Scotland, Midland Valley, Southern Uplands Biogenic sediments at Teindland, Tolsta
Ipswichian	
Wolstonian	Shelly indigo till of Banffshire and Aberdeenshire
Hoxnian	Peat at Fugla Ness, Shetland
Anglian	Lower till at Fugla Ness, Shetland

At its maximum the Devensian ice is thought to have covered the whole mainland, including the area of Buchan termed 'moraineless'.[22] Radiocarbon dates indicate that by 13,000 years B.P. a large part of Scotland was ice-free, and indeed the ice cover may have disappeared entirely in the late-glacial interstadial. This was followed by the Loch Lomond Readvance, recorded by end-moraines at many sites from the Orkneys south to the Southern Uplands, but with the main development of ice in the western Grampians. The limits are shown in fig. 12.3. The readvance was short-lived, lasting perhaps some 500 years, with the ice disappearing at around 10,000 years ago.

The melting of the Late Devensian ice sheet was accompanied by isostatic uplift, and both marine beds with cold molluscan faunas (Errol Beds, Clyde Beds) and raised shorelines are known from the Highland periphery. The shorelines are discussed in a later section of the chapter.

IRELAND[68, 116, 118]

Except for small nunatak areas in the south, Ireland was blanketed by ice during maximum glaciation. The ice was mostly of Irish origin but near the east and northeast coasts Irish Sea ice moved inland at various times. Table 12.7 gives the sequence of temperate and glacial stages. In the table the Irish stage names are shown, together with the British correlatives. The type sites for the Irish stages are given in the Geological Society's Quaternary report.

Apart from the record of a redeposited molluscan fauna of Red Crag type in outwash gravel in Co. Wicklow, little is known of the Pleistocene record prior to a substantial temperate interglacial stage, the Gortian, correlated on the basis of vegetational history with the Hoxnian. Fossiliferous deposits of this age are found at Gort, Co. Galway, Kilbeg and Newtown, Co. Waterford, and Baggotstown and Kildromin, Co. Limerick. Indications of a glaciation earlier than this is given by stony clays with erratics under the last two of these deposits. At Newtown, the peat overlies beach material lying in the 'preglacial' shore-platform of the south and east coast of Ireland, and this beach is also correlated with the Hoxnian temperate stage, though the shore-platform itself may be older.

The succeeding glacial stage is the Munsterian, correlated with the Wolstonian. Ice movements this stage are indicated in fig. 12.16. Three distinct phases can be recognised, early and late phases of local glaciation based on upland areas, e.g. Wicklow and Kerry, which sandwich the development of a great Irish Sea ice stream which deposited shelly till (Ballycroneen Till) on the east and southeast coast. An inland ice sheet, centred on Connemara, developed after the Irish Sea ice retreated from the southeast coast.

Only one last interglacial site has so far been discovered in Ireland. It is at Shortalstown, Co. Wexford, where a marine sediment is intercalated in an irregular way between two calcareous tills.[24] The pollen diagram is unlike those from the Gortian, but has similarities with those from the early Ipswichian in southeast England. The extraordinary scarcity of last interglacial deposits parallels the situation in England, where such deposits have yet to be found in a glacigenic sequence.

The limits and ice flow directions of the last glacial stage, the Midlandian, are shown in fig. 12.16. All the ice advances are believed to belong to the Late Midlandian and post-date 26,000. A site at Derryvree, Co. Fermanagh, with organic silts dated at 30,000, shows ice-free conditions before the main glaciation. The Southern Irish (Tipperary, Ballylanders) end-moraine marks the maximum advance of the inland ice, with an Irish Sea lobe leaving and end-moraine near the southeast coast in Wexford. The main inland ice sheet was centred on an ice shed stretching from Lough Rea in the centre of Ireland through Tyrone to Lough Neagh in the northeast, while local centres of ice dispersal developed in Donegal, Connemara, and the mountains of Munster and Wicklow.[117] In the ice-free areas, corries and block-moraines are found in many mountain groups (e.g.

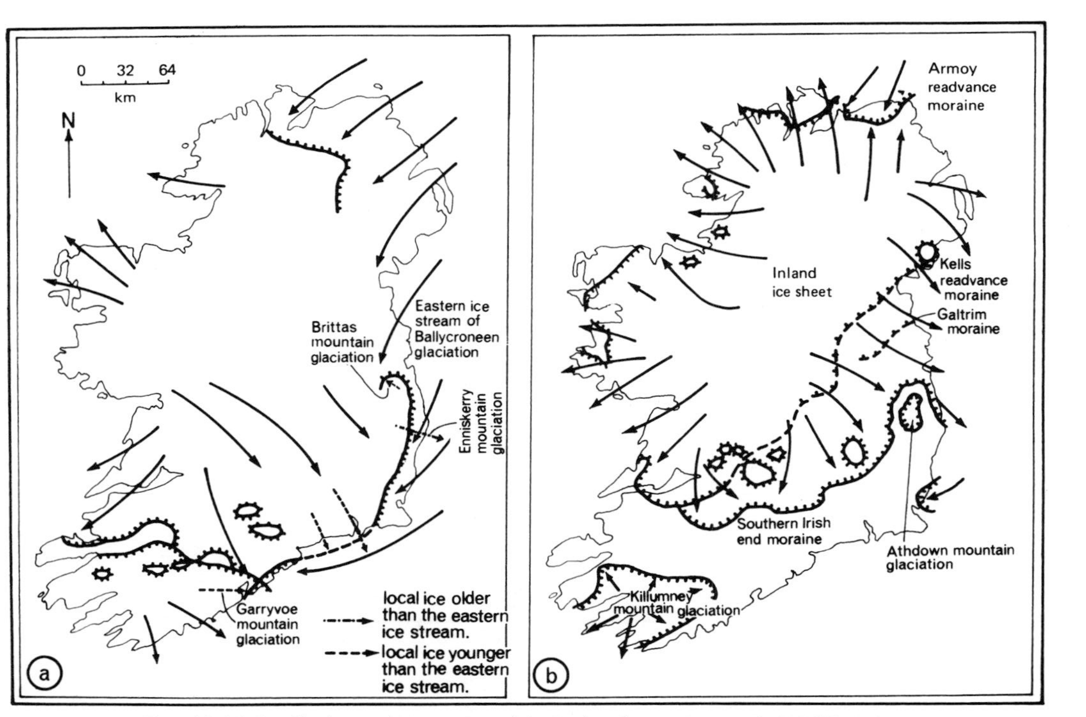

Fig. 12.16 Ice limits and movements in Ireland. a, Munsterian cold stage; b, Midlandian cold stage.

Table 12.7 Pleistocene succession in Ireland

Flandrian	Littletonian	Lake deposits, peat, alluvium, estuarine clays, raised beaches
	Midlandian Late	Late-glacial { corrie moraines interstadial muds Armoy Readvance Kells (Drumlin) moraine Galtrim moraine Southern Irish (Tipperary, Ballylanders) en-moraine, Athdown Mountain glaciation, Killumney end-moraine of Lesser Cork–Kerry glaciation
	Middle and Early	Derryvree silts
Ipswichian		Shortalstown interglacial deposit
Wolstonian	Munsterian	Brittas Till (Wicklow) Garryvoe Till (Greater Cork–Kerry glaciation) Ballyvoyle Till (inland ice) Ballycroneen Till (Irish Sea ice, Eastern General glaciation) Drogheda Till Enniskerry Till (Wicklow)
Hoxnian	Gortian	Interglacial deposits at Gort, Kilbeg, Baggotstown, Newtown, Courtmacsherry beach
	pre-Gortian	Solifluction gravel at Gort Stony clays at Kildromin, Baggotstown Erratics in Courtmacsherry beach shore-platforms Killincarrig Crag (≡ Red Crag)

the Knockmealdowns, Comeraghs, Iveragh and Dingle Peninsulas), and similar features are found in nunatak mountains within the area of the continental ice sheet, e.g. Mountains of Mourne, Slieve League. Outside the glacial limit solifluction deposits are common and ice-wedge casts, fossil ground-ice mounds and stone polygons are known.

The ice retreat from the Southern Irish end-moraine was evidently accompanied by standstills at the Galtrim moraine, a narrow ridge with feeding eskers, and at the Kells moraine further north, a belt of kame and kettle moraine marking the southern border of a drumlin and esker belt. It has been suggested that this kame and kettle moraine marks a change from orderly retreat to wholesale stagnation of the ice sheet. There appear to be no marked readvance phases except at a very late stage when the Scottish ice sheet advanced over the northeast coast of Ulster, leaving an end-moraine (Armoy Readvance). A period of marine transgression between this readvance and the main glaciation is recorded by marine clays between the tills in this area.

The late-glacial readvance is indicated by corrie moraines in the Wicklow Hills, Co. Mayo and Co. Kerry and by substantial inorganic sediments in kettle-holes of the Midlandian, covering muds of the late-glacial interstadial.

SOUTHERN ENGLAND

Most of southern England lay outside the area of glaciation, but there is much evidence of Pleistocene climatic and sea-level changes, ranging from Lower Pleistocene to the Flandrian. Periglacial climates are recorded by solifluction deposits, loess and patterned ground, and by terrace faunas and floras indicating cold conditions. Temperate stages are recorded by biogenic deposits formed in river-terrace systems or near the sea in response to eustatic sea-level changes. Changes in the relative land/sea-level are recorded by raised beaches and other marine deposits above present sea-level.

The oldest of these marine deposits are sands and gravels (*in situ*?) containing a marine fauna of Red Crag type at Netley Heath in Surrey (183 m O.D.) and Rothamsted, Hertfordshire (131 m O.D.).[28] These have been associated with a high sea-level of about 180 m, of Lower Pleistocene age. At St Erth in Cornwall marine deposits with a rich fauna occur up to 35 m O.D., indicating a transgression to at least 45 m. There has been much discussion on the age of these beds, but new evidence from the fauna, molluscs, foraminifera and ostracods indicates they are very late Pliocene in age.[69]

The northern or Plateau drift of the Oxford region[90] with its wealth of erratics mostly derived from the Bunter Beds of the Midlands antedates in part an interglacial channel filling at Sugworth, near Oxford.[15a]. The fossils of the channel filling indicate a Cromerian age for the deposit. If the Plateau drift is of glacial origin, as appears likely, then there is clear evidence of a pre-Cromerian ice advance into the upper Thames region.

The river terrace sequences of the south of England are known in some detail, e.g. those of the Solent and Bournemouth area, the Bristol area, and those of the rivers Medway, Mole and Wey.[147] The most completely known are those of the river Thames.

The terrace sequences of the Thames are shown in fig. 12.17. The

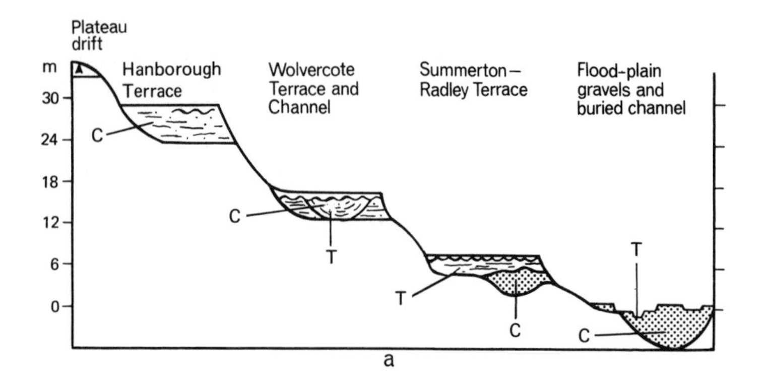

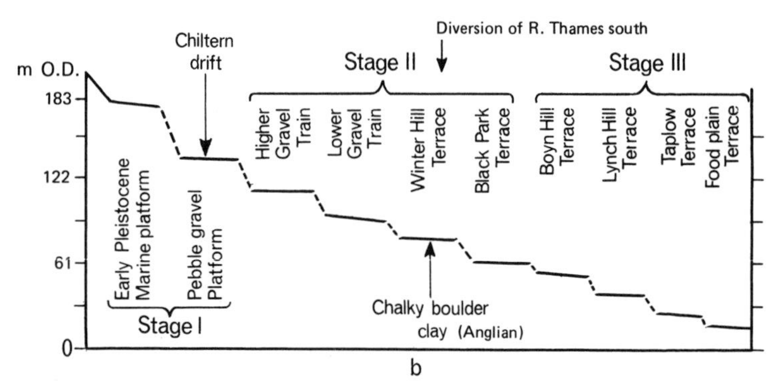

Fig. 12.17 Terrace sequences of the river Thames. a, upper Thames (after Sandford, 1954); b, middle and lower Thames (after Wooldridge, 1960). Heights Ordnance Datum (O.D.) on meridian of Slough; C, fauna and/or flora indicating a cold climate; T, fauna indicating a temperate climate.

Goring Gap, where the Thames breaks through the Chalk escarpment, separates the terrace system of the upper from that of the lower and middle Thames.

In the upper Thames,[9, 90] there are high terraces associated with the northern or Plateau drift (Combe and Freeland terraces), a lower terrace (Hanborough) with a cold fauna,[15] correlated with the early Wolstonian or Anglian, and then a terrace (Wolvercote), complicated by the presence of a channel in or near it, containing a warm fauna and, above it, a cold flora (late Hoxnian ?). The terrace itself is related to the Wolstonian glacial stage during which the Midlands Chalky Boulder Clay was laid down to the north of the area. The next lowest terrace (Summertown–Radley) is again a composite terrace, with a cold fauna and flora at the base and a temperate fauna with *Hippo-*

potamus nearer the top. Probably the cold part dates from the Late Wolstonian while the upper temperate part is Ipswichian. Then there are low-level terraces and a buried channel with a cold fauna and flora, of Devensian age.

In the middle and lower Thames, the sequence is shown in fig. 12.17.[146, 152] Though these two areas are treated together, they differ in that the succession of terraces in the lower Thames area is complicated by increased effects of aggradation, as might be expected nearer the estuary of a large river. The terrace sequence has been divided by Wooldridge into three stages. The early (Lower Pleistocene?) stage comprises marine high-level deposits and pebble gravels,[50] antedating the formation of the present valleys, and indicating an early course of the river Thames north eastward (A in fig. 12.18). The terraces of the middle stage form a lower series and often contain quartzites; they were thought to indicate an eastward course

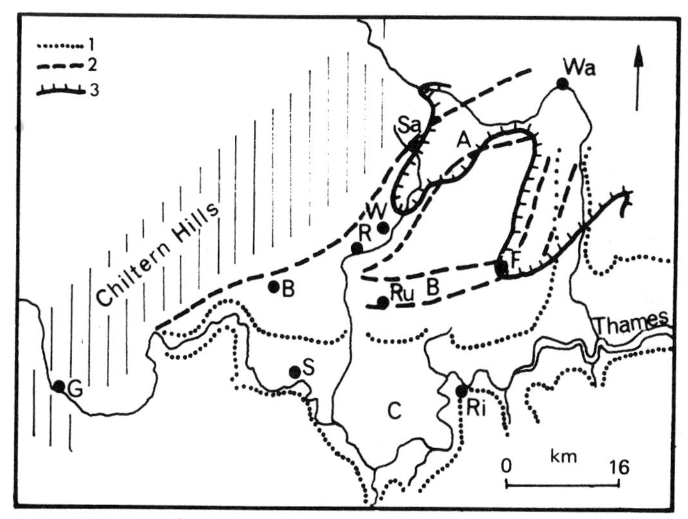

Fig. 12.18 Diversions of the river Thames envisaged by Wooldridge (1960). It is now known that diversion south from the Vale of St Albans was caused by the advance of Anglian ice from the north and north-east (Gibbard, 1977). A, Pliocene or Lower Pleistocene course; B, course after division south by local Chilterns ice; C, present-day valley following diversion from B by chalky boulder-clay ice. 1, limits of present-day valley; 2, suggested limits of earlier courses; 3, southern limit of chalky boulder-clay ice. Localities: B, Beaconsfield; F, Finchley; G, Goring Gap in the Chilterns; R, Rickmansworth; RI, Richmond; RU, Ruislip; S, Slough; SA, St Albans; W, Watford; WA, Ware.

(B in fig. 12.18) of the Thames at the time. However, a recent investigation of the stratigraphy of the Vale of St Albans indicates that the early Thames flowed along the Vale till the beginning of Winter Hill Terrace times, when Anglian ice blocked the Vale and after a period of ponding forced the river south with the formation of the Black Park Terrace.[39]

The terraces of the latest stage follow the present course of the Thames. The oldest (Boyn Hill) is considered to be Hoxnian in age, while the Taplow Terrace contains a cold fauna and is probably Wolstonian, as is probably the Main Coombe Rock of the lower Thames. The Upper Floodplain Terrace deposits probably aggraded in the Ipswichian, and associated with this interglacial are thick deposits of brickearth at Ilford[137] and Aveley in the lower Thames. The Lower Floodplain Terrace, the buried channel filling of gravels and sometimes peat, and the local coombe rocks generally hold a cold fauna and flora and are of Devensian age.

Raised beaches and erosion surfaces are known from many areas of southern England.[152] The best known are those at about 35 m O.D. at Slindon in Sussex, containing a Lower Palaeolithic industry and very probably Hoxnian in age, and a younger series at 5 to 8 m, Ipswichian in age, found on the Sussex coast (plates 8b and 13b) and also traced to southwest England. The Burtle Beds of Somerset, with a marine fauna, also probably belong to the Ipswichian transgression.

In the southwest there is some evidence for strandlines at 20 m, 7·5 m and 4 m O.D.[78] A shore-platform cut in rock at a few metres O.D. occurs in north Devon, Cornwall (fig. 8.4b, plate 13a) and the Scillies. It is thought to be of similar age to the shore-platform of southern Ireland, and if so, this would indicate a Hoxnian or older age. If it is Hoxnian, it is difficult to reconcile with the higher sea-levels noted further east for this stage, unless warping has affected the relative levels. In the southwest the shore-platform is overlain by beach gravel. There has been much discussion on the age of these beach gravels. There may be two beaches present, as in the Scillies (fig. 12.19b);[71] a lower one, with few erratics, possibly of Hoxnian age, and a younger one, rich in erratics derived from a Wolstonian glacial advance down to the Scillies, and dating from the Ipswichian The large erratics associated with the beach (fig. 12.19a) may have been emplaced at the beginning of the Wolstonian before sea-level fell. These erratics are found on the south coast as far east as Selsey, Sussex, and an alternative explanation for their presence postulates an English Channel ice lobe reaching to the Straits of Dover.[55] The evidence for such an ice advance remains to be substantiated.

However, because of the presence of till on the Scillies, the north

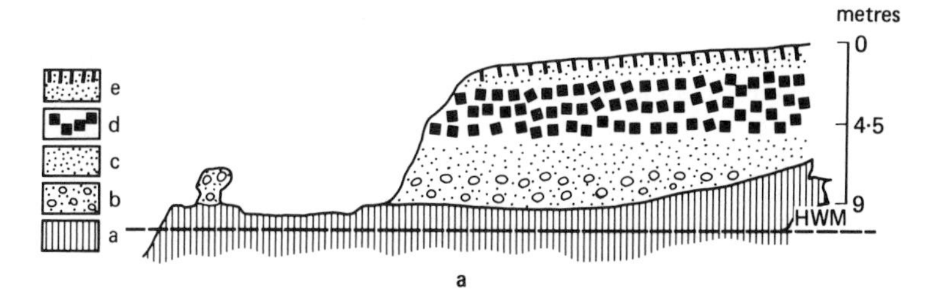

a

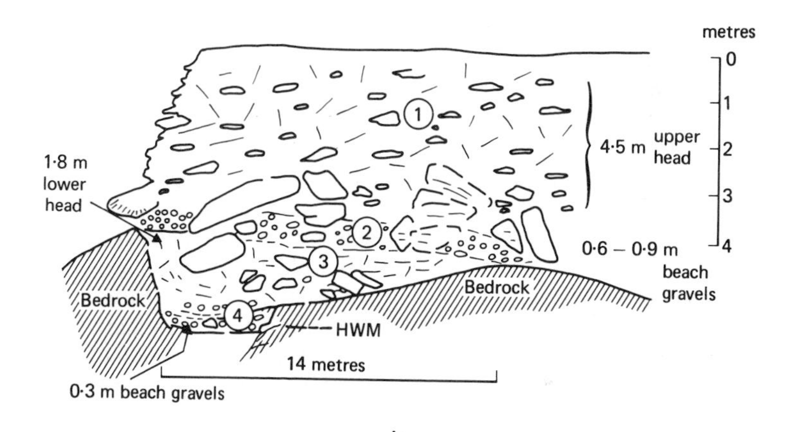

b

Fig. 12.19 a. Wave-cut shore platform at Godrevy, near St Ives, Cornwall, typifying a characteristic coastal sequence of southwest England and southern Ireland. a, rock; b, beach gravel; c, sand and pebbles; d, head; e, blown sand and soil (after Stephens and Synge, 1966).
b. Coastal section at Porth Seal, St Martin's, Scilly Isles, drawn by F. M. Synge in 1965. The section shows two heads separated by a beach and underlain by an older beach. 1, Upper Head; 2, Porth Seal beach with frequent erratics; 3, Lower Head; 4, Chad Girth beach with rare erratics (Mitchell, 1972).

coast of Cornwall (Trebetherick) and Devon (Fremington) and at Kenn, Somerset,[49] it is clear that a massive ice advance took place to the southwest, and this ice advance is correlated with the Wolstonian. This ice advance may have ponded water in the Somerset and Severn lowlands, and it has been suggested that overflow into the English Channel took place at the Chard Gap. Studies of the terrace gravels in the Chard region do not support this view.[45] Likewise, studies of the terrace gravels of the rivers draining Salisbury Plain do not support the idea of an ice advance extending into that area,[44] a thesis put forward to explain the presence of erratics in the megalithic monuments of the area.

There are few interglacial biogenic deposits in the south of England. An important one is near Abingdon in the Thames Valley and is probably of Cromerian age.[15a] Its relation to the Northern Drift is uncertain, but it is older than the Hanborough Terrace. A few biogenic deposits of Ipswichian age are known in the Solent area at about present sea-levels. They show freshwater and estuarine deposits, and record a change of sea-level which took place in the first half of the Ipswichian.[139]

Periglacial features abound in the south of England. Under the heading of solifluction deposits are included the coombe rock of the Chalk areas, the fan gravels of the limestone areas, the breccias of the shores of the Severn estuary and the block-fields of Dartmoor. Ice-wedge casts and involutions are not uncommon in sections of river gravels and large scale polygonal markings of fossil ice-wedge networks have been described from terrace surfaces. Periglacial aeolian deposits include ventifacts and also wind-blown fractions incorporated into soil horizons, and loess, particularly in Kent, Sussex and Devon. The formation of dry valleys in the Chalk may have been assisted by periglacial action, and in east Kent some coombes were cut during late-glacial readvance (zone III) times of the Late Devensian.[57]

SOME GENERAL ASPECTS OF THE PLEISTOCENE

Having described the Pleistocene of the different regions of the British Isles, there remain several topics of wider than regional significance which deserve separate mention.

Periglacial phenomena

Solifluction processes and patterned ground of recent origin are common in the mountains of the British Isles,[18] and 'fossil' periglacial phenomena have been related to many of the Pleistocene ice advances. Their presence reveals more of climate and environment than that of glacial deposits. For example, the stratigraphical position of ice-wedge casts in East Anglia, indicating former times of permafrost, is shown in table 12.8. At each of these times ground ice formed in frozen ground, and mean annual temperatures were at a maximum −4°C. Features of Devensian age are naturally the best known, and

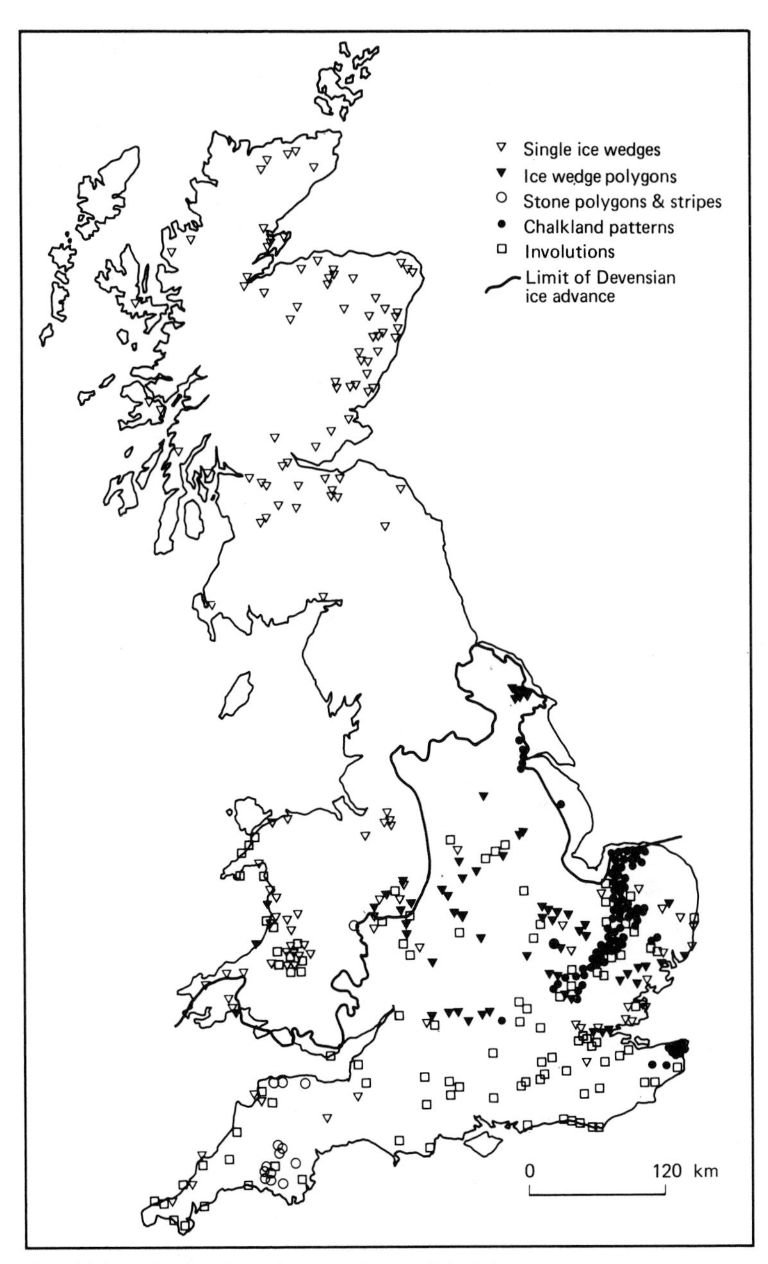

Fig. 12.20 Distribution of Devensian periglacial patterned ground features in Britain. England and Wales from Williams (1968), Worsley (1966); Scotland from Sissons (1974). Chalkland patterns refer to the non-sorted stripes and polygons of eastern England found on the Chalk (Williams, 1964).

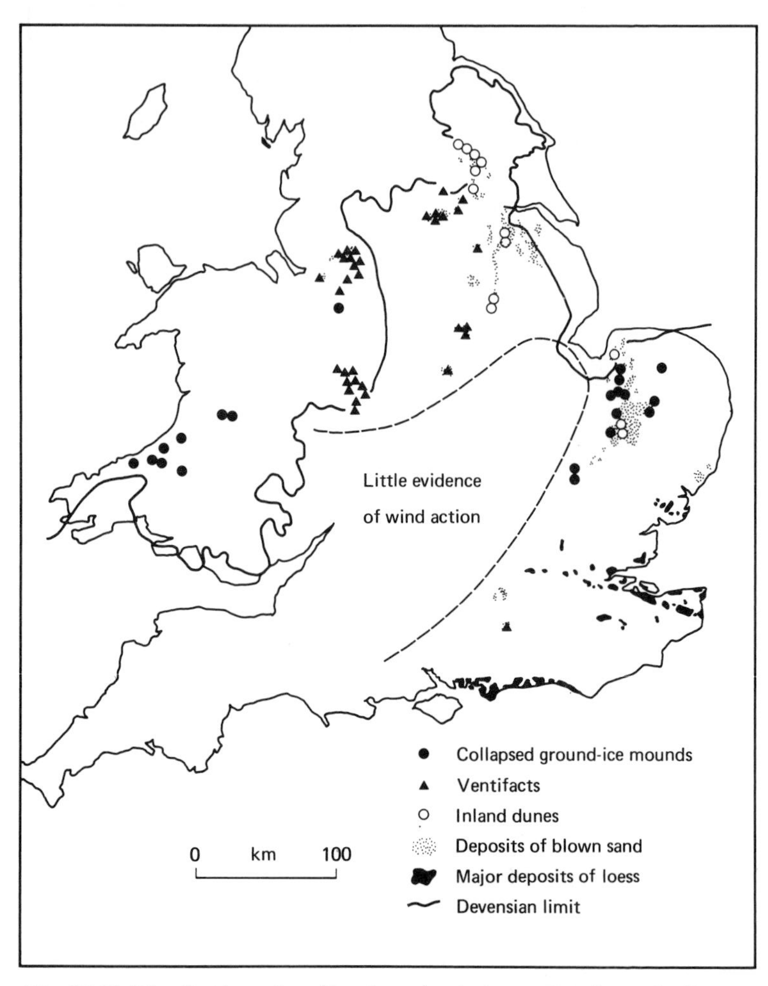

Fig. 12.21 Distribution of aeolian deposits that may date from the Devensian cold stage (Williams, 1975), and of collapsed ground-ice mounds of Devensian age. Thin loess deposits occur widely in England from Cornwall northeast to east Yorks and Durham; these are not shown on the figure.

are widely distributed in the periglacial area outside the Devensian ice limit (fig. 12.20 and fig. 12.21), but they also occur further north,[33] where they are associated with the colder climates of the Late Devensian, including the Loch Lomond Readvance.

Solifluction deposits (including head and coombe rock), patterned ground, including unsorted polygons and stripes and ice wedge networks, collapsed frost (ground-ice) mounds (pingos) and aeolian deposits have been described from many parts of southern Britain. A detailed periglacial stratigraphy for the Devensian, showing times of permafrost, loess formation and so on, has yet to be constructed but a preliminary outline can be made out (table 12.8).

Solifluction deposits. Head or coombe rock is the commonest superficial Pleistocene deposit over much of southern Britain, reaching considerable thicknesses (up to several metres) in west Britain.[127] The stratigraphy of solifluction deposits on the Chalk give a sequence of events in parts of the Devensian.[32] The deposits vary from fine solifluction earths to stone streams.[99]

Patterned ground. Ice wedge casts and polygons are known from many parts of the British Isles (plate 10b). Their distribution in Britain is shown in fig. 12.20, and indicates that permafrost covered most if not all of the periglacial parts of the British Isles at times. The severity of the Loch Lomond Readvance climate is shown by the presence of ice-wedge casts of this age down to sea-level. Unsorted stripes and polygons are well known from the Chalk (plate 10a).

Collapsed frost (ground-ice) mounds. These again have a wide distribution, being known from East Anglia,[102, 135] West Wales[128] and southern Ireland.[70] Most appear to have formed outside the Devensian limit, though some have been identified just within the Midlandian limit in Ireland. As regards age of formation, there is evidence of frost mound formation in East Anglia in the Early and Middle Devensian, as well as in the late-glacial of the Late Devensian. Those of west Wales and Ireland again appear to have formed in the Late Devensian, with sedimentation resulting from melting starting at least in the late-glacial.

Aeolian deposits. Cover sands have been identified in various periglacial sequences in eastern England, and sand dunes, of Late Devensian late-glacial age, have been found in the Vale of York[67] and in East Anglia. Ventifacts are also recorded from several sites. Loess deposits are well known from the southeast and south of England,[48] either as thick sheets as on the Kent coast, or as a loessic component in soils, as in Norfolk. The distribution of these various sediments is shown in fig. 12.21.

Landforms. Forms resulting from cambering and valley bulging, described in Chapter 5, have been widely found in the east Midlands and parts of southern England. They include surface depressions

Table 12.8 Stratigraphical positions of periglacial phenomena in south and east England

Locality	*Feature*	*Periods of freeze/thaw (F/T), permafrost (P), aeolian activity (A)*	*Outline of stratigraphy proving age*	*Age (non-periglacial intervals in brackets)*	*Reference (if none given, author's unpublished observations)*
East Walton, Norfolk	Ground-ice depressions	F/T or P	Late-glacial plant remains	Late-glacial	Sparks *et al.* (1972)
Isle of Thanet, Folkestone, Medway Valley, Kent	Chalk detritus (snow or thaw melt-water deposits) coombe cutting	F/T	Overlies ^{14}C dated soil	Late-glacial (zone III)	
	Chalk detritus (snow or thaw melt-water deposits)	F/T	Overlies ^{14}C dated soil	Late-glacial (zone I) Late Devensian	Kerney (1963) Kerney (1965) Kerney, Brown and Chandler (1964)
	Loess	A			
	snow melt-water deposits involutions, solifluction (coombe rock)	F/T	Underly ^{14}C dated soil		
Wretton, Norfolk	Ice-wedge casts penetrate involutions	P F/T	Overly organic deposits of probable Chelford age	Early Devensian	West *et al.* (1974)
				(Ipswichian)	
Beetley, Norfolk	Ice-wedge cast	P	Underlies Ipswichian, penetrates valley outwash	Late Wolstonian	
Ebbsfleet, Kent	Solifluction (coombe rock)	F/T	On Chalk, below Ipswichian	Wolstonian	Burchell in Zeuner (1959)
				(Hoxnian)	West, Lambert and Sparks (1964)
W. Runton, Mundesley, Bacton Norfolk	Ice-wedge casts, involutions	P F/T	Underly Anglian tills, penetrate Cromerian, associated with arctic plant beds	Early Anglian	Donner and West (1958) West and Wilson (1966)
				(Cromerian)	
W. Runton, Mundesley, Beeston, Norfolk	Ice-wedge casts, involutions	P F/T	Underly Cromerian, penetrate Pastonian, associated with arctic plant beds	Beestonian	West and Wilson (1966)
				(Pastonian)	
Beeston, Norfolk	Solifluction (coombe rock)	F/T	On Chalk, below Pastonian	pre-Pastonian	

resulting from gulls.[75] Dry valleys characteristic of the Chalk escarpment are another expression of periglacial erosion processes, and the stratigraphy of the valley fills has been used to determine periods of coombe cutting (table 12.8). Late-glacial landslips have also been identified in the south of England and in the Midlands.[98]

Cave deposits

The excavation of cave deposits in various parts of Britain has shown sequences covering parts of the Devensian, with thermoclastic scree sediments of the colder periods alternating with weathered finer-grained material of the less severe periods. In a few caves the record extends back further. At Minchin Hole, Gower, an Ipswichian beach is associated with a cave;[12] in south Devon temperate interglacial sediments are underlain by those of an earlier cold period (Wolstonian);[114] and at Dove Holes, Derbyshire, a small fauna of Cromerian or pre-Cromerian age has been described.[106]

Relative land/sea-level changes

There is abundant evidence in the British Isles for these changes, but the interpretation of the evidence is difficult, as we have already discussed in Chapter 8. First, there are problems of determining sea-level height from the evidence of a marine or estuarine deposit, and secondly, eustatic effects must be separated from isostatic and tectonic effects. When considering broad changes of relative land/sea-level the first of these difficulties may be gauged, but the second always remains. In general, Devensian late-glacial and Flandrian isostatic uplift formed tilted raised beaches in northeast Ireland, northern England and Scotland, particularly along the shores of the northern part of the Irish Sea. In southern England and southern Ireland there seems little evidence for tilted Late Pleistocene beaches. But isostatic changes may be possible in the southwest because ice reached the area in the Wolstonian, and, to the north, in the Devensian. Likewise, in the southeast, we are on the margin of the North Sea basin of subsidence, so tectonic factors may have affected levels in that area. This subsidence appears to be the cause of high level of the Lower Pleistocene Crag remnants in southeast England (180 m O.D.) compared with the considerable depth of Lower Pleistocene shallow-water Crag well below O.D. in East Anglia. Such depression in the southeast is related to the down-warping in the southern part of the North Sea, well known from the Netherlands. It is possible that eustatic fall may have contributed some part of the difference in levels, since evidence for a eustatic fall in the Upper Pliocene to

Table 12.9 Pleistocene marine deposits

	Northern Britain and Ireland (inside zero isobase for Flandrian beaches)	*Southern Britain and Ireland (outside zero isobase for Flandrian beaches)*
Flandrian	Post-glacial tilted shorelines Marine (Carse) clays	Marine clays in alluvial sequences
Devensian	Late-glacial tilted shorelines Marine clays (Errol and Clyde Beds)	
Ipswichian		Raised beach at Brighton to 12 m O.D. Marine clays at Selsey and Stone (*c.* O.D.) Tidal silts in S. Yorks—12 to 4 m O.D.
Hoxnian		Marine clays of Nar Valley, 6 to 20 m O.D. Marine clays at Clacton, 3 to 8 m O.D. Beach gravels in south-west England and Ireland, 3 to 15 m O.D.
Cromerian		Estuarine deposits 3 to 7 m O.D. in east Norfolk
Pastonian		Estuarine deposits—2 to 2 m O.D. in east Norfolk
Early Pleistocene		Crags—49 m O.D. to O.D. in East Anglia 183 m O.D. in southeast England

Lower Pleistocene is indicated by Pliocene deep water sediments (Lenham Beds) at the same level as shallower water Red Crag sediments.[134] In the Lower Pleistocene more exact measurements of the downwarping, and whether it was accompanied by folding, will have to await more detailed knowledge of the succession. The relative uplift of the Crag in southeast England probably caused the filling of scour troughs in the Chalk (fig. 12.5) and the increased isolation of the North Sea from the Atlantic. This isolation may then be the cause of the restricted marine fauna of the Red and Icenian Crags compared with that of the Pliocene Coralline Crag.

The shore-platforms of Ireland, Wales, west Scotland and southwest England indicate important times of sea-level standstill, but the correlation with the stratigraphical sequence is uncertain. Often they are overlain by interglacial or Flandrian beach gravels, which will give minimum ages for the platforms. In the southwest they are overlain by Ipswichian, and, some think, Hoxnian beach gravels. If the latter is correct, then the platforms may be Hoxnian or older. Or they may be features of composite origin, formed or occupied in successive interglacial still-stands of sea-levels.

Interglacial sea-levels have been identified but their correlation with sequences of wave-cut platforms and erosion surfaces in southern Britain remains obscure. In the Cromerian of East Anglia the freshwater/estuarine contact occurs near O.D. and estuarine deposits extends to a few metres above O.D. In the Hoxnian in southeast England there is evidence of change from freshwater to estuarine deposits near O.D. but marine sediments and raised beaches are known at elevations up to 40 m O.D. In the Ipswichian in southeast England the freshwater to estuarine transition again occurs near O.D. and beach gravels are known up to about 12 m O.D. In south Yorkshire mid-Ipswichian tidal sediments occur between — 12 m and 4 m O.D. Freshwater deposits of zone Ip III have been found at O.D. in west Norfolk and at 3 m O.D. in south Yorkshire.[37] Although the data are meagre, it seems that the Ipswichian transgression with its raised beach was followed by a regression in zone Ip III.

The heights mentioned for interglacial sea-levels are taken from sites where there is substantial evidence for dating. There are many isolated deposits and wave-cut features which may relate to these periods, but it is as yet impossible to make wide correlations showing sea-level heights at particular times, or even maxima in particular interglacials. Because of the possibilities of downwarping, it is also difficult to make correlations with the classic Mediterranean sequences, though there are suggestions of a correlation of the Ipswichian with the 7·5 m Monastirian II and Hoxnian with the 30 m level.[139] A very detailed discussion of sea-level changes in Britain and their relation to the Mediterranean levels can be found in Zeuner's book;[152] the difficulties of this kind of approach are discussed in Chapter 8. Table 12.9 summarises the occurrence of estuarine or marine interglacial deposits, their age and height.

The sea-levels of the glacial stages are unknown, though estimates suggest −100 m O.D. or so. Near-shore submarine benches at −107

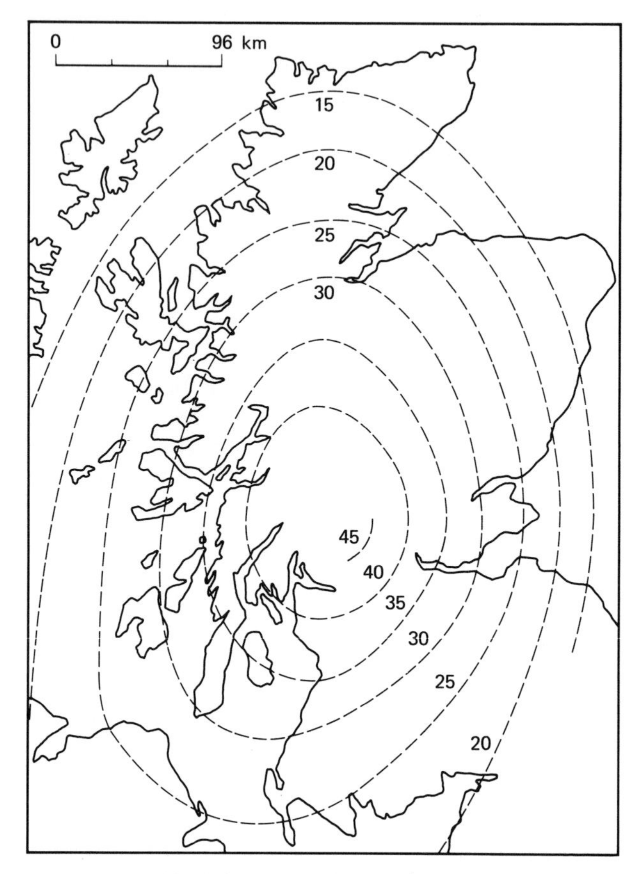

Fig. 12.22 Isobases (feet) for the Main Postglacial Shoreline in Scotland (Sissons, 1967).

and −128 m have been observed west of the Isles of Scilly, and buried channels of rivers have been traced to below −50 m O.D.[60] Whether sea-levels rose to around present heights during interstadials is also unknown, though there have been suggestions for this in the Bristol Channel for an interstadial of Devensian age.[27]

In the Devensian late-glacial and Flandrian, changes are known in greater detail. There are late-glacial tilted shore-lines in northeast Ireland[108] and Scotland (including the so-called 100ft (30 m) and 50 ft (15 m) beaches) and shore-line diagrams have been constructed (fig. 8.6);[105] the Firth of Forth raised beaches have been investigated in most detail.[97] In the same areas of Ireland and Scotland less-tilted

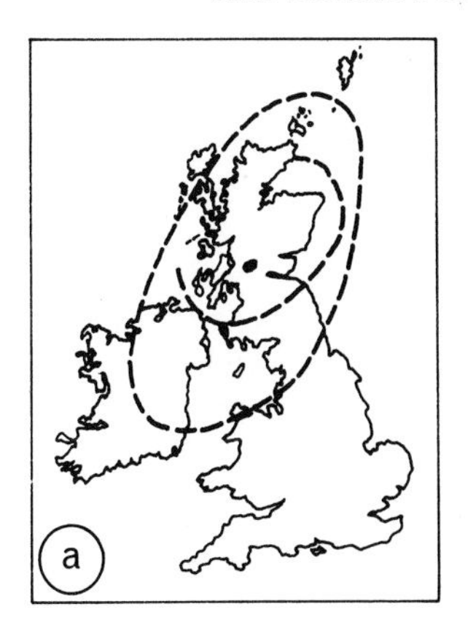

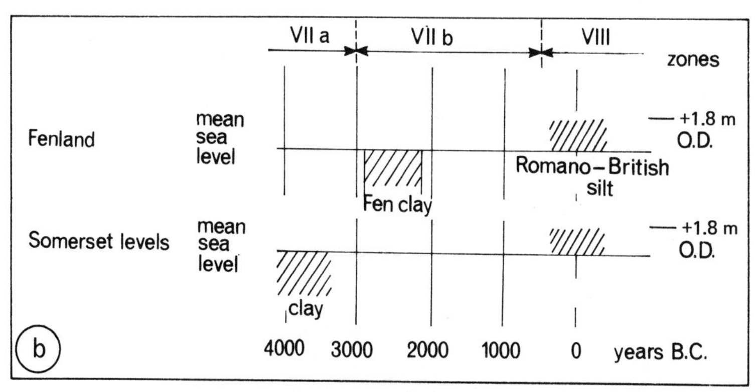

Fig. 12.23 a. Outer limits of the Devensian late-glacial ('100 ft') raised beach of Scotland (inner circle) and of the Flandrian post-glacial ('25 ft') beach (outer circle) according to Wright (1937). The dot in the Callander region in southern Scotland is the presumed centre of uplift of the late-glacial beach (Donner, 1959).
b. Age of marine transgressions as shown by estuarine clays intercalated in freshwater deposits, in two areas outside the zero isobase (Fenland and Somerset levels) (West, 1963).

shorelines of Flandrian age occur (including the so-called 25 ft (7·6 m) 'Neolithic' beach). In southeast Scotland there is a Main Postglacial Shore-line, dated at about 6000 to 7000 years ago, followed by a series of later shore-lines. Isobases for the Main Postglacial Shoreline are shown in fig. 12.22.

Outside the zero isobase for the highest postglacial shore-line (fig.

12.23), we are in the area where eustatic rise is presumably the controlling factor for sedimentation, and submerged Flandrian peats and marine silts and clays are common. In its simplest form, the alluvial Flandrian sequence consists of a lower freshwater deposit, a middle marine clay, an upper peat and a final marine clay. The middle marine clay (Carse Clay of Scotland, Buttery Clay of Fenland) reflects the main Flandrian transgression and the upper marine clay the final alluvial phase when sea-level is stationary and the lower sediments are compacting.

The sequences of the Bristol Channel,[61] Somerset Levels, the Devon coast,[23] the Fenland,[40, 100] northwest England[121] and southeast Scotland [97] are best known, and on their stratigraphy and dating of transgression and regression contacts curves of eustatic rise have been put forward (e.g. fig. 8.7a). The interpretations vary from a simple curve of sea-level rise to a curve showing numerous transgressions and regressions. Further detailed work is required on the dating of transgressions, especially in eastern and southern England to find how apparent eustatic rises have been affected by upwarping and downwarping.

Finally, we should mention that measurements of very recent sealevel change, given by tide-gauge measurements, indicate downwarping in the south of England relative to the north (Chapter 8).[123]

Devensian late-glacial and Flandrian

Sediments of this age are widespread in the British Isles, forming lake fillings, peat deposits and alluvium (fig. 12.1).[42] The subdivisions of the late-glacial and Flandrian are based on pollen zonation and radiocarbon dating. The simple numerical zone system (I to VIII) which has been applied to southern Britain and Ireland is shown in table 12.10. Many of the zone boundaries have been radiocarbon dated. With the great regional diversity of vegetation and climate within the British Isles, it is obviously not possible that a single pollen assemblage zone system will be applicable to the whole area.[133] From the point of view of vegetational history, the delineation of regional and local pollen assemblage zones and their correlation by radiocarbon dating is necessary. From the point of view of stratigraphy and chronology, radiocarbon years provide a scale for subdivision, and on this basis the Flandrian has been divided into three major chronozones, analogous to the first three zones of the interglacial zonation system (table 13.3).[51] However, the relative paucity of radiocarbon dates compared with what we would like for good chronological control means that we still use the pollen zones as a basic chronological tool in spite of the difficulties of determining the synchroneity or otherwise of the pollen zone boundaries throughout the area. In the following discussion, we refer to the numerical zone system which has been applied in the past, since this is a sufficient framework for the present purpose. In the future, perhaps, a subdivision based on millennia of radiocarbon years may also provide a satisfactory framework. We can now examine the late-glacial (zones I to III) and Flandrian (zones IV to VIII), remembering that the basis of the subdivision is pollen analytical and that its relation to climatic change is another matter of interpretation. We also note some regional differentiation of the sequence within the British Isles.

There is a tripartite division of the Devensian late-glacial, analogous to the Late-Weichselian zones I, II and III of the continent of northwest Europe.[43] Indeed, these subdivisions are often referred to the continental zones I, II and III, though exact correlations are made difficult by matters of definition. The middle of the three zones, referred to as zone II, represents the late-glacial pollen-interstadial, correlated broadly with the Allerød Interstadial of Denmark. The name Windermere Interstadial has been suggested for this interval. The pollen assemblages show higher percentages of non-herb pollen, with *Betula* and *Juniperus* prominent. In the first and third zones (the latter equivalent to the Loch Lomond Readvance) there are higher frequencies of herb pollen. The whole late-glacial sequence is typified by the preponderance of non-tree pollen and the near absence of thermophilous tree genera. The three subdivisions of zone I of the continent, with a postulated zone Ib Bølling Interstadial, have not been clearly recognised in the British Isles, though subdivisions related to changing herb pollen assemblages have been recognised. The lower boundary of the late-glacial (base of zone I) has not been clearly defined. The retreat of the Late Devensian ice leads into the late-glacial, and no doubt the herbaceous pollen assemblages of the late-glacial are close to those of the full-glacial of earlier times in the Devensian.

The boundary between the late-glacial and the Flandrian is taken at the zone III/IV boundary, dated to near 10,250 B.P., and zones IV to VIII cover Flandrian time. The main pollen analytical characters of the zones and their nomenclature is shown in table 12.10. The zone boundaries have been radiocarbon dated in many parts of the British Isles. The boundary between zones VIIa and b appears to be reasonably synchronous in Ireland and southern Britain; otherwise the diachroneity of the boundaries varies across the country, particularly from south to north (fig. 13.10). Radiocarbon dates of boundaries are shown in table 12.10 in outline.

Table 12.10 Sequences of main vegetational changes during the Devensian late-glacial and Flandrian in the British Isles

Stage	Radio-carbon years B.P.	Blytt and Sernander periods	ENGLAND AND WALES Zone	ENGLAND AND WALES Characteristics of vegetation	IRELAND Zone	IRELAND Characteristics of vegetation	SCOTLAND Zone	SCOTLAND Characteristics of vegetation
			VIII modern	Afforestation	X	*Pinus–Fagus*; Afforestation	VIII modern	Afforestation
								Fagus
	1000	Sub-Atlantic	VIII	*Alnus–Quercus –Betula* (*–Fagus –Carpinus*)	IX	*Alnus–Quercus –Betula*		
						Ulmus decline		
Fl III	2000						VIII–VIIb	*Alnus–Quercus –Betula*
	3000					*Quercus* maximum		
	4000	Sub-Boreal	VIIb	*Alnus–Quercus –Tilia*; Deforestation	VIII	*Alnus–Quercus*; Deforestation		Deforestation
	5000			*Ulmus* decline		*Ulmus* decline		*Ulmus* decline

		Years		Zone		Zone		Zone	
Flandrian	Fl II	6000 7000	Atlantic	VIIa	*Alnus–Quercus–Ulmus–Tilia*	VII	*Alnus–Quercus–Ulmus–Pinus*	VIIa	*Alnus–Quercus–Ulmus* (*Pinus* in n–e)
	Fl I	8000 9000	Boreal	VI	*Pinus–Corylus*: c *Quercus–Ulmus–Tilia*; b *Quercus–Ulmus*; a *Ulmus–Corylus*	VI	*Corylus–Pinus*: c *Pinus* max.; b *Ulmus* maximum; a *Corylus* maximum	V–VI	*Betula–Pinus–Corylus*
				V	*Corylus–Betula–Pinus*	V	*Corylus–Betula*		
		10,000	Pre-Boreal	IV	*Betula–Pinus* AP rise	IV	*Betula* AP rise	IV	*Betula* AP rise
Late Devensian late-glacial				III	Herbs	III	Herbs (Younger *Salix herbacea* period)	III	(Loch Lomond Readvance)
		11,000		II interstadial	*Betula* with park tundra	II	Park tundra with *Betula* (late-glacial birch period)	II	The interstadial is indicated by a zone with woody plants between herb zones
		12,000		I	Park tundra	I	Herbs (Older *Salix herbacea* period)	I	

Sediments in closed lakes, sediments formed in relation to changes in relative land/sea-level, and peat, constitute the most widespread deposits of the late-glacial and Flandrian. Lakes are typical of the morainic landscape of the Devensian. In many of the lakes Flandrian sedimentation has been insufficient to fill their basins. The lakes frequently have a characteristic series of late-glacial and Flandrian sediments. The late-glacial sediments are predominantly inorganic, except for the interstadial where the biogenic component may be considerable. This predominance of inorganic sediment is associated with a cold climate, when solifluction must have been common. In contrast, the Flandrian sediments are mainly biogenic, either autochthonous or allochthonous. The sediments formed during cultivation periods may have an increased proportion of inorganic sediments derived from the erosion of cleared lands. The changing chemical composition of Flandrian lake sediments has been related to rates of erosion within drainage basins, which itself may be related to climatic change.[65]

River valleys near the coast were filled with alluvial deposits during the Flandrian eustatic rise. Such alluvial areas are extensive; for example, Firth of Forth,[149] Humber, Fenland (fig. 4.5d),[40] Norfolk Broads,[63] Belfast.[149] Typically the alluvial series contain an early freshwater peat, a middle estuarine wedge of clay (Carse Clay of Scotland, Buttery Clay of the Fenland), and an upper peat. There may be an upper estuarine series of later age in some areas, e.g. Romano-British silt of the Fenland. The main clay wedge of the large alluvial areas formed as a result of the transgression of the Flandrian eustatic rise of sea level.

Coastal shingles and sand dunes of Flandrian age are widespread round the coasts.[107] A large group of inland dunes, now largely fixed, is found in Breckland in central East Anglia, where the dune sand is derived from the very sandy facies of the drift.

Raised bog and blanket bog peats are prominent in the landscapes of the western and northern parts of the British Isles.[79] Many raised bogs show the development of *Sphagnum* peats during the middle Flandrian, above earlier lake sediments and fen peats (fig. 4.5b). Bog stratigraphy and recurrence surfaces have been studied in many areas,[54, 126] and some of the latter have been dated by radiocarbon.[41]

Blanket bogs are widely developed in southwestern and northern England,[25] Wales, Scotland and Ireland.[54] Blanket bog formation started in many areas around the beginning of zone VIIa (Atlantic Period) and is associated with increased humidity of climate at that time. Blanket bog peats frequently show buried horizons of tree stumps, often pine, and these have been thought to indicate periods of cessation of growth of the bog surface related to climatic change. A complex sequence of climatic change was put forward for Scotland on such evidence, but it now seems clear that horizons with pine stumps do not have wide regional synchroneity, local factors being more important in governing the growth and death of pines.[7] Many areas of blanket bog now suffer from erosion and gullying, which may result from changed climatic conditions, unfavourable hydrological conditions following bog growth and from the prevention of their regeneration by burning or grazing.[25, 119]

Little subaerial denudation has accompanied the deposition of sediments just described. There is good preservation of features of the Devensian glaciation and of periglacial patterned ground. Solifluction in the late-glacial and hillwash in the late Flandrian may have smoothed the landforms, and in upland regions of steeper river gradients and softer bed rock, incision of rivers may have occurred. Coastal changes have resulted from marine erosion, particularly where soft drift cliffs form the coastline, as in east Yorkshire and East Anglia.

Dating

During the Devensian and Flandrian, as well as in the earlier temperate stages, each with their characteristic sequence of vegetational history, pollen analysis has been extensively used as a method of relative dating. In till and terrace sequences geomorphological criteria have often been used as a means to relative dating of a more general sort. Thus base-levels of tills have been used as a criterion of age. As already mentioned, the relative ages of biogenic interglacial deposits may often be determined from their topographical relationships. A detailed relative chronology of the Thames has been attempted based on studies of erosional benches.[152] Relative dating of bones by the fluorine method has been applied to many problems of provenance and relative age.[77]

Of the methods of absolute dating, varve analysis has hardly been used in the British Isles, though some local sequences have been studied. Radiocarbon dating has been extensively applied to problems of Devensian and Flandrian stratigraphy, vegetational history, changes in relative land/sea-levels and archaeology. Thus there is a time scale of radiocarbon years as a background for events in this period (table 12.10). But there remain very many problems of chronology, in particular the synchroneity of pollen zones, the dating of important events in the history of vegetation and the history of particular genera, and the dating of the course of relative land/sea-level changes. There are also too few dates of Devensian sediments.

Table 12.11 Some Devensian radiocarbon dates (B.P.)

Stage	*Dates*				*Events*
Flandrian					
			10,250		
Devensian Late	late-glacial	III			Loch Lomond Readvance
			10,750	dates based on a number of assays	Windermere Interstadial
		II			
			13,000		
		I			
	Colney Heath, Herts		13,560 ± 210	Q 385	Cold flora and fauna, peat erratic in Colne Valley gravesl
	Dimlington, Yorks		18,240 ± 250	BIRM 108	Moss silts deposited earlier than Drab and Purple Tills
	Barnwell, Cambridge		19,500 ± 650 / 1800	Q 590	Cold flora in Cam Valley terrace
Middle	Sandiway, Cheshire		28,000 ± 1500	I 1667	Marine molluscs, predate upper till of Cheshire Plain
	Lea Valley, Arctic Bed, Essex		28,000 ± 1500	Q 25	Cold flora, peat erratic in Lea Valley gravels
	Coleshill, Warwickshire		32,160 ± 1780 / 1450	NPL 55	Cold faunas and floras associated with Low Terrace of the Tame, Avon No. 2 Terrace, Main Terrace of Severn—(Upton Warren Interstadial Complex)
	Fladbury, Worcs		38,000 ± 700	GRO 1269	
	Upton Warren, Worcs		41,900 ± 800	GRO 1245	
Early	Chelford, Cheshire		60,800 ± 1500	GrN 1480	Boreal coniferous forest, interstadial
Ipswichian					

Apart from the difficulty of dating near the limit of the method, Devensian correlations depend to a very large extent on radiocarbon dating because the exposed sediments are at isolated sites and of little thickness. Table 12.11 gives some of these older dates related to particular Devensian episodes, but many more are required before there can be satisfactory reconstruction and correlation of the glacial and periglacial sequences.

Conclusions

It will be apparent that we have a good general knowledge of the Pleistocene succession in the British Isles. It is based on much detailed stratigraphy of sites, with perhaps too little mapping. But the former has given the outline succession. For the detail of subdivisions of the stages, especially the glacial stages, much detailed mapping is required. When we consider the doubts about the margins of the Devensian, and even more so of the older glacial stages, and the doubts about the subdivision of these stages, it is clear that only mapping in relation to the stratigraphy of biogenic deposits within and outside the drift sequence will give the required detail.

Having built up our local succession we should then turn to its correlation with the continent. Table 12.12 summarises the correlations presently suggested, but it should be remembered that even in

Table 12.12 Correlation with the continent

		British Isles	*N.W. Europe*
Pleistocene		Flandrian	Flandrian
	Upper	Devensian Ipswichian Wolstonian	Weichselian Eemian Saalian
	Middle	Hoxnian Anglian Cromerian Beestonian Pastonian	Holsteinian Elsterian 'Cromerian complex'
	Lower	Baventian Antian Thurnian Ludhamian ?Pre-Ludhamian	?Eburonian or Menapian Tiglian
Pliocene		(=Red Crag, inc. Waltonian)	Reuverian

the classic area of northern Europe, the succession, even of the glacial stages, is not yet clear. For example, the identity of temperate stages in northeast Europe in relation to those of northwest Europe is not clear, the nature of the interval between the Saale and Warthe is not clear, the number of advances within the Saale Glaciation is not clear, and Cromerian and pre-Cromerian stratigraphy is not at all settled. In fact, it has now been suggested that there are three temperate stages within the 'Cromerian complex' of the continent.[151]

All the more reason that we should not make correlations with the Alps, where the succession is equally or more difficult, and that we should use a local succession based on our own stratigraphical evidence.

REFERENCES

1 ARKELL, W. J. 1947. *The geology of Oxford.* Oxford: Clarendon Press.

2 BADEN-POWELL, D. F. W. 1948. 'The chalky boulder clays of Norfolk and Suffolk', *Geol. Mag.*, **85**, 279–96.

3 BAKER, C. A. 1971. 'A contribution to the glacial stratigraphy of West Essex', *Essex Nat.*, **32**, 317–30.

4 BANHAM, P. H. 1975. 'Glacitectonic structures: a general discussion with particular reference to the contorted drift of Norfolk'. In *Ice Ages: ancient and modern*, ed. A. E. Wright and F. Moseley, 69–94. Liverpool: Seel House Press.

5 BEAUMONT, P., TURNER, J. and WARD, P. F. 1969. 'An Ipswichian peat raft in glacial till at Hutton Henry, Co. Durham', *New Phytol.*, **68**, 779–805.

6 BECK, R. B., FUNNELL, B. M. and LORD, A. 1972. 'Correlation of Lower Pleistocene at depth in Suffolk', *Geol. Mag.*, **109**, 137–9.

7 BIRKS, H. H. 1975. 'Studies in the vegetational history of Scotland IV. Pine stumps in Scottish blanket peats', *Phil. Trans. R. Soc. Lond.*, B, **270**, 181–226.

8 BIRKS, H. J. B. and RANSOM, M. E. 1969. 'An interglacial peat at Fugla Ness Shetland', *New Phytol.*, **68**, 777–96.

9 BISHOP, W. W. 1958. 'The Pleistocene geology and geomorphology of three gaps in the Midland Jurassic escarpment', *Phil. Trans. R. Soc.*, B, **241**, 255–305.

10 BOSWELL, P. G. H. 1952. 'The Plio-Pleistocene boundary in the east of England', *Proc. Geol. Ass. London*, **63**, 301–12.

11 BOWEN, D. Q. 1973a. 'The Pleistocene history of Wales and the borderland', *Geol. J.*, **8**, 207–24.

12 BOWEN, D. Q. 1973b. 'The excavation at Minchin Hole, 1973', *Gower*, **24**, 12–18.

13 BOWEN, D. Q. 1974. 'The Quaternary of Wales'. In *The Upper Palaeozoic and post Palaeozoic rocks of Wales*, ed. T. R. Owen, 373–426. University of Wales Press.

14 BOYLAN, P. J. 1966. 'The Pleistocene deposits at Kirmington, Lincolnshire', *Mercian Geol.*, **1**, 339–49.

15 BRIGGS, D. J. and GILBERTSON, D. D. 1973. 'The age of the Hanborough Terrace of the River Evenlode, Oxfordshire', *Proc. Geol. Assoc. Lond.*, **84**, 155–73.

15a BRIGGS, D. J. *et al.* 1975. 'New interglacial site at Sugworth', *Nature* (*Lond.*), **257**, 477–9.

16 BRISTOW, C. R. and COX, F. C. 1973. 'The Gipping Till: a reappraisal of East Anglian glacial stratigraphy', *J. Geol. Soc. Lond.*, **129**, 1–37.

17 BULL, A. J. 1942. 'Pleistocene chronology', *Proc. Geol. Ass. London*, **53**, 1–45.

18 CAINE, N. 1972. 'The distribution of sorted patterned ground in the English Lake District', *Revue Géomorphol. Dynamique*, **21**, 49–56.

19 CATT, J. A. and PENNY, C. F. 1966. 'The Pleistocene deposits of Holderness, East Yorkshire', *Proc. Yorks. Geol. Soc.*, **35**, 375–420.

20 CATT, J. A., WEIR, A. H. and MADGETT, P. A. 1974. 'The loess of eastern Yorkshire and Lincolnshire', *Proc. Yorks. Geol. Soc.*, **40**, 23–39.

21 CHARLESWORTH, J. K. 1929. 'The South Wales End Moraine', *Q.J. Geol. Soc. London*, **85**, 335–58.

22 CLAPPERTON, C. M. and SUGDEN, D. E. 1975. 'The glaciation of Buchan—a reappraisal'. In *Quaternary studies in north-east Scotland*, ed. by A. M. D. Gemmell, 19–22. University of Aberdeen Geography Department.

23 CLARKE, R. H. 1970. 'Quaternary sediments off south-east Devon', *Quart. Jl Geol. Soc. Lond.*, **125**, 277–318.

24 COLHOUN, E. A. and MITCHELL, G. F. 1971. 'Interglacial marine formation and lateglacial freshwater formation in Shortalstown Townland, Co. Wexford', *Proc. R. Irish Acad.*, **B, 71**, 211–45.

25 CONWAY, V. M. 1954. 'Stratigraphy and pollen analysis of southern Pennine blanket peats', *J. Ecol.*, **42**, 117–47.

26 COOPE, G. R. 1975. 'Climatic fluctuations in northwest Europe since the Last Interglacial, indicated by fossil assemblages of Coleoptera'. In *Ice ages: ancient and modern*, ed. by A. E. Wright and F. Moseley, 153–68. Liverpool: Seel House Press.

27 COOPE, G. R., SHOTTON, F. W. and STRACHAN, I. 1961. 'A Late Pleistocene fauna and flora from Upton Warren, Worcestershire', *Phil. Trans. R. Soc. B*, **244**, 379-421.

28 DINES, H. G. and CHATWIN, C. P. 1930. 'Pliocene sandstone from Rothamsted (Hertfordshire)', *Summary of Progress, Geol. Surv. G.B. for 1929*, Pt. III, 1–7.

29 DONNER, J. J. 1959. 'The late and post-glacial raised beaches in Scotland', *Ann. Acad. Sci. Fennicae*, Series A III, No. 53.

30 DONNER, J. J. and WEST, R. G. 1958. 'A note on Pleistocene frost

structures in the cliff section at Bacton, Norfolk', *Trans. Norfolk Norwich Nat. Soc.*, **18,** 8–9.

31 EASTWOOD, T., TAYLOR, B. J., BURGESS, I. C., LAND, D. H., MILLS, D. A. C., SMITH, D. B. and WARREN, P. T. 1971. *British Regional Geology: Northern England.* 4th edn. H.M.S.O.

32 EVANS, J. G. 1968. 'Periglacial deposits of the Chalk of Wiltshire', *Wilts. Archaeol. & Nat. Hist. Mag.*, **63,** 12–26.

33 FITZPATRICK, E. A. 1956. 'An introduction to the periglacial geomorphology of Scotland', *Scottish Geogr. Mag.*, **74,** 82–36.

34 FUNNELL, B. M. 1961. 'The Palaeogene and Early Pleistocene of Norfolk', *Trans. Norfolk Norwich Nat. Soc.*, **19,** 340–64.

35 FUNNELL, B. M. 1972. 'The history of the North Sea', *Bull. Geol. Soc. Norfolk*, **21,** 2–10.

36 GAUNT, G. D. 1974. 'A radiocarbon date relating to Lake Humber', *Proc. Yorks. Geol. Soc.*, **40,** 195–7.

36a GAUNT, G. D. 1976. 'The Devensian maximum ice limit in the Vale of York', *Proc. Yorks. Geol. Soc.*, **40,** 631–7.

37 GAUNT, G. D., BARTLEY, D. D. and HARLAND, R. 1974. 'Two interglacial deposits proved in boreholes in the southern part of the Vale of York and their bearing on contemporaneous sea levels', *Bull. Geol. Survey G.B.*, **48,** 1–23.

38 GAUNT, G. D., COOPE, G. R., OSBORNE, P. J. and FRANKS, J. W. 1972. 'An interglacial deposit near Austerfield, southern Yorkshire', *Rep. No.* 72/4, *Inst. geol. Sci.* 13 pp

39 GIBBARD, P. L. 1977. 'Pleistocene history of the Vale of St Albans', *Phil. Trans. R. Soc. Lond.*, **B** (in press).

40 GODWIN, H. 1940. 'Studies on the Post-glacial history of British vegetation. III. Fenland pollen diagrams. IV. Post-glacial changes of relative land- and sea-level in the English Fenland', *Phil. Trans. R. Soc.*, B, **230,** 239–303.

41 GODWIN, H. 1960. 'Radiocarbon dating and Quaternary history in Britain', *Proc. R. Soc.*, B, **153,** 287–320.

42 GODWIN, H. 1975. *History of the British Flora.* 2nd edn. Cambridge University Press.

43 GODWIN, H. and WILLIS, E. H. 1959. 'Radiocarbon dating of the Late-glacial Period in Britain', *Proc. R. Soc.*, B, **150,** 199–215.

44 GREEN, C. P. 1973. 'Pleistocene river gravels and the Stonehenge problem', *Nature* (*Lond.*), **243,** 214–16.

45 GREEN, C. P. 1974. 'Pleistocene gravels of the River Axe in south-western England and their bearing on the southern limit of glaciation in Britain', *Geol. Mag.*, **111,** 213–20.

46 GREGORY, K. J. 1965. 'Proglacial Lake Eskdale after sixty years', *Trans. Inst. Br. Geogr.*, No. 36, 149–62.

47 HARMER, F. W. 1902. 'A sketch of the later Tertiary history of East Anglia', *Proc. Geol. Ass. London*, **13,** 416–79.

48 HARROD, T. R., CATT, J. A. and WEIR, A. H. 1973. 'Loess in Devon', *Proc. Ussher Soc.*, **2,** 554–64.

49 HAWKINS, A. B. and KELLAWAY, G. A. 1971. 'Field meeting at Bristol and Bath with special reference to new evidence of glaciation', *Proc. Geol. Assoc.*, **82,** 267–92.

50 HEY, R. W. 1965. 'Highly quartzose pebble gravels in the London Basin', *Proc. Geol. Ass. London,* **76,** 403–20.

51 HIBBERT, F. A., SWITSUR, V. R. and WEST, R. G. 1971. 'Radiocarbon dating of Flandrian pollen zones at Red Moss, Lancashire', *Proc. R. Soc. London,* B, **177,** 161–76.

52 HOPPE, G. 1974. 'The glacial history of the Shetland Islands', *Inst. British Geogr. Special Publication,* No. 7, 197–210.

53 HORTON, A. 1970. 'The drift sequence and subglacial topography in parts of the Ouse and Nene basin', *Rep. No. 70/9, Inst. Geol. Sci.,* 30 pp.

54 JESSEN, K. 1949. 'Studies in Late Quaternary deposits and flora-history of Ireland', *Proc. R. Irish Acad.*, **52,** B, 85–290.

55 KELLAWAY, G. A., REDDING, J. H., SHEPHARD-THORN, E. R. and DESTOMBES, J-P. 1975. 'The Quaternary history of the English Channel', *Phil. Trans. R. Soc. Lond.*, A, **279,** 189–218.

56 KELLY, M. R. 1964. 'The Middle Pleistocene of north Birmingham', *Phil. Trans. R. Soc.*, B, **247,** 533–92.

57 KERNEY, M. P. 1963. 'Late-glacial deposits on the Chalk of south-east England', *Phil. Trans. R. Soc.*, B, **246,** 203–54.

58 KERNEY, M. P. 1965. 'Weichselian deposits in the Isle of Thanet, East Kent', *Proc. Geol. Ass. London,* **76,** 269–74.

59 KERNEY, M. P., BROWN, E. H. and CHANDLER, T. J. 1964. 'The Late-glacial and Post-glacial history of the Chalk escarpment near Brook, Kent', *Phil. Trans. R. Soc.*, B, **248,** 135–204.

60 KIDSON, C. 1970. 'Coastline development in south-west England during the Quaternary', *Proc. Commission on coastal geomorphology,* I.G.U. Moscow 1970.

61 KIDSON, C. and HEYWORTH, A. 1973. 'The Flandrian sea-level rise in the Bristol Channel', *Proc. Ussher Soc.*, **2,** 565–84.

62 KIRBY, R. P. 1969. 'Till fabric analyses from the Lothians, central Scotland', *Geogr. Annaler,* **51**A, 48–60.

63 LAMBERT, J. M., JENNINGS, J. N., SMITH, C. T., GREEN, C. and HUTCHINSON, J. N. 1960. *The making of the Broads.* R. Geogr. Soc. Memoir No. 3.

64 LEXIQUE STRATIGRAPHIQUE INTERNATIONAL, 1963. Fascicule 3a XIII. *Neogene and Pleistocene of England, Wales and Scotland* Central National de la Recherche Scientifique.

65 MACKERETH, F. J. H. 1966. 'Some chemical observations on post-glacial lake sediments', *Phil. Trans. R. Soc.*, B, **250,** 165–213.

66 MADGETT, P. A. 1975. 'Re-interpretation of Devensian till stratigraphy of eastern England', *Nature (Lond.)*, **253,** 105–7.

67 MATTHEWS, B. 1970. 'Age and origin of aeolian sand in the Vale of York', *Nature (Lond.)*, **227,** 1234–6.

68 MITCHELL, G. F. 1972. 'The Pleistocene History of the Irish Sea:

second approximation', *Scient. Proc. R. Dubl. Soc.*, A, **4,** 181–99.

69 MITCHELL, G. F. 1973a. 'The Late Pliocene marine formation at St Erth, Cornwall', *Phil. Trans. R. Soc. Lond.*, B, **266,** 1–37.

70 MITCHELL, G.F. 1973b. 'Fossil pingos in Camaross Townland, Co. Wexford', *Proc. R. Irish Acad.*, B, **73,** 269–82.

71 MITCHELL, G. F. and ORME, A. R. 1967. Pleistocene deposits of the Isles of Scilly', *Q. Jl Geol. Soc. Lond.*, **123,** 59–92.

72 MITCHELL, G. F., PENNY, L. F., SHOTTON, F. W. and WEST, R. G. 1973. 'A correlation of Quaternary deposits in the British Isles', *Geol. Soc. Lond., Special Report*, No. 4, 99 pp.

72a MITCHELL, G. F. and WEST, R. G. 1977. 'A discussion on the changing environmental conditions in Great Britain and Ireland during the Devensian (last) cold stage', *Phil. Trans. R. Soc.* (*Lond.*), (in press).

73 MONTFRANS, H. M. VAN, 1971. *Palaeomagnetic dating in the North Sea basin.* Rotterdam: Princo N.V.

74 MORGAN, A. V. 1973. 'The Pleistocene geology of the area north and west of Wolverhampton, Staffordshire, England', *Phil. Trans. R. Soc. Lond.*, B, **265,** 233–97.

75 MUSKETT, P. J. 1971. 'Periglacial gulls in the Upper Witham valley', *Trans. Lincs. Nat. Union*, **17,** 210–16.

76 NORTH, F. J. 1943. 'Centenary of the glacial theory', *Proc. Geol. Ass. London*, **54,** 1–28.

77 OAKLEY, K. P. and HOSKINS, W. R. 1950. 'New evidence on the antiquity of Piltdown Man', *Nature* (*Lond.*), **165,** 379–82.

78 ORME, A. R. 1960. 'The raised beaches and strandlines of South Devon', *Fld Stud.*, **1,** 109–30.

79 OSVALD, H. 1949. 'Notes on the vegetation of British and Irish Mosses', *Acta Phytogeogr. Suecica*, No. 26.

80 PEAKE, D. S. 1961. 'Glacial changes in the Alyn river system and their significance in the glaciology of the north Welsh border', *Q.J. Geol. Soc. London*, **117,** 335–66.

81 PENNY, L. F. 1974. 'Quaternary'. In *The geology and mineral resources of Yorkshire*, ed. D. H. Rayner and J. E. Hemingway, 245–64. Yorkshire Geological Society.

82 PENNY, L. F., COOPE, G. R. and CATT, J. A. 1969. Age and insect fauna of the Dimlington Silts, East Yorkshire. *Nature* (*Lond.*), **224,** 65–7.

83 PIGOTT, C. D. 1962. 'Soil formation and development on the Carboniferous limestone of Derbyshire. I. Parent materials', *J. Ecol.*, **50,** 145–56.

84 POOLE, E. G. and WHITEMAN, A. J. 1961. 'The glacial drifts of the southern part of the Shropshire–Cheshire plain', *Q.J. Geol. Soc., London,* **117,** 91–130.

85 POSNANSKY, M. 1960. 'The Pleistocene succession in the middle Trent basin', *Proc. Geol. Ass. London*, **71,** 285–311.

86 RAISTRICK, A. 1934. 'The correlation of glacial retreat stages across

the Pennines', *Proc. Yorks. Geol. Soc.*, **22,** 199–214.

87 REID, C. 1890. *The Pliocene deposits of Britain.* Mem. Geol. Surv. U.K.

88 RICE, R. J. 1968a. 'The Quaternary deposits of central Leicestershire', *Phil. Trans. R. Soc. Lond.*, A, **262,** 459–508.

89 RICE, R. J. 1968b. 'The Quaternary era'. In *The geology of the east Midlands*, ed. P. C. Sylvester-Bradley and T. D. Ford, 332–55. Leicester University Press.

90 SANDFORD, K. S. 1954. 'River development and superficial deposits', in *The Oxford Region*, eds. A. F. Martin and R. W. Steel, 21–36. Oxford University Press.

91 SHOTTON, F. W. 1953. 'The Pleistocene deposits of the area between Coventry, Rugby and Leamington, and their bearing upon the topographic development of the Midlands', *Phil. Trans. R. Soc.* B, **237,** 209–60.

92 SHOTTON, F. W. 1960. 'Large scale patterned ground in the valley of the Worcestershire Avon', *Geol. Mag.* **97,** 404–8.

93 SHOTTON, F. W. 1973. 'General principles governing the subdivision of the Quaternary System'. In Mitchell, G. F. *et al.*, 1973.

94 SIEVEKING, G. DE G. 1968. 'High Lodge Palaeolithic industry', *Nature (Lond.)*, **220,** 1065–6.

95 SIMPSON, I. M. and WEST, R. G. 1958. 'On the stratigraphy and palaeobotany of a late-Pleistocene organic deposit at Chelford, Cheshire', *New Phytol.*, **57,** 239–50.

96 SISSONS, J. B. 1967. *The evolution of Scotland's scenery.* Edinburgh: Oliver & Boyd.

97 SISSONS, J. B. 1974. 'The Quaternary in Scotland: a review', *Scott. J. Geol.*, **10,** 311–37.

98 SKEMPTON, A. 1976. 'Introductory remarks', *Phil. Trans. R. Soc. Lond.*, A, **283,** 423–26.

99 SMALL, R. J., CLARK, M. J. and LEWIN, J. 1970. 'The periglacial rock-stream at Clatford Bottom, Marlborough Downs, Wiltshire', *Proc. Geol. Assoc. Lond.*, **81,** 87–98.

100 SMITH, A. G. 1970. 'The stratigraphy of the northern Fenland'. In *The Fenland in Roman times*, ed. C. W. Phillips, 147–64. Roy. Geogr. Soc. Research Series No. 5.

101 SMITH, D. B. in Duscussion in Catt and Penny, 1966.

102 SPARKS, B. W., WILLIAMS, R. B. G. and BELL, F. G. 1972. 'Presumed ground-ice depressions in East Anglia', *Proc. R. Soc. Lond.*, A, **327,** 329–43.

103 SPARKS, B. W. and WEST, R. G. 1964. 'The drift landforms around Holt, Norfolk', *Trans. Inst. Br. Geogr.*, No. 35, 27–35.

104 SPARKS, B. W. and WEST, R. G. 1965. 'The relief and drift deposits'. In *The Cambridge Region 1965*, ed. J. A. Steers. Cambridge University Press.

105 Special number on the vertical displacement of shore-lines in highland Britain. 1966. *Trans. Inst. Br. Geogr.*, No. 39.

106 SPENCER, H. E. P. and MELVILLE, R. V. 1974. 'The Pleistocene

mammalian fauna of Dove Holes, Derbyshire', *Bull. Geol. Survey G.B.*, **48,** 43–53.

107 STEERS, J. A. 1962. *The Sea Coast*. 3rd edn. London: Collins.

108 STEPHENS, N. and SYNGE, F. M. 1965. 'Late-Pleistocene shorelines and drift limits in North Donegal', *Proc. R. Irish Acad.*, **64,** B, 131–53.

109 STEPHENS, N. and SYNGE, F. M. 1966. 'Pleistocene shorelines'. In *Essays in Geomorphology*, ed. G. H. Dury, 1–51. London: Heinemann.

110 STEVENS, L. A. 1960. 'The interglacial of the Nar Valley, Norfolk', *Q.J. Geol. Soc. London*, **115,** 291–316.

111 STRAW, A. 1963. 'The Quaternary evolution of the Lower and Middle Trent', *East Midland Geogr.*, **3,** 171–89.

112 STRAW, A. 1969. 'Pleistocene events in Lincolnshire: a survey and revised nomenclature', *Trans. Lincolns. Nat. Union*, **17,** 85–98.

113 SUGDEN, D. E. 1969. 'The age and form of corries in the Cairngorms', *Scott. Geogr. Mag.*, **85,** 34–46.

114 SUTCLIFFE, A. J. and ZEUNER, F. E. 1962. 'Excavations in the Torbryan Caves, Devonshire', *Proc. Devon Archaeol. Explor. Soc.*, **5,** 127–45.

115 SWINNERTON, H. H. and KENT, P. E. 1949. *The geology of Lincolnshire*. Lincoln: Lincolnshire Naturalists' Union.

116 SYNGE, F. M. 1970. 'The Irish Quaternary: current views 1969'. In *Irish geographical studies*, ed. by N. Stephens and R. E. Glasscock, 34–48. Queen's University of Belfast Department of Geography.

117 SYNGE, F. M. 1973. 'The glaciation of south Wicklow and the adjoining parts of the neighbouring counties', *Irish Geography*, **6,** 561–9.

118 SYNGE, F. M. and STEPHENS, N. 1960. 'The Quaternary Period in Ireland—an assessment, 1960', *Irish Geogr.*, **4,** 121–30.

119 TALLIS, J. H. 1964. 'Studies on southern Pennine peats. II. The pattern of erosion', *J. Ecol.*, **53,** 333–44.

120 TOMLINSON, M. E. 1963. 'The Pleistocene chronology of the Midlands'. *Proc. Geol. Ass. London*, **74,** 187–202.

121 TOOLEY, M. J. 1974. 'Sea-level changes during the last 9000 years in north-west England', *Geogr. J.*, **140,** 18–42.

122 TROTTER, F. M. and HOLLINGWORTH, S. E. 1932. 'The glacial sequence in the north of England', *Geol. Mag.*, **69,** 374–80.

123 VALENTIN, H. 1953. 'Present vertical movements of the British Isles', *Geogr. J.*, **119,** 299–305.

124 VALENTIN, H. 1955. 'Die grenze der letzten vereisung im Nordseeraum', Deutscher Geographentag Hamburg 1955. *Tagungsbericht u. Wiss. Abh.*, 359–66.

124a VALENTINE, K. W. G. and DALRYMPLE, J. B. 1975. 'The identification, lateral variation, and chronology of two buried paleocatenas at Woodhall Spa and West Runton, England', *Quaternary Research*, **5,** 551–90.

125 VAN DER HAMMEN, T., MAARLEVELD, G. C., VOGEL, J. C. and ZAGWIJN. W. H. 1967. 'Stratigraphy, climatic succession and radio-

carbon dating of the last glacial in the Netherlands', *Geol. en Mijnbouw*, **46,** 79–95.

126 WALKER, D. and WALKER, P. M. 1961. 'Stratigraphic evidence of regeneration in some Irish bogs', *J. Ecol.*, **49,** 169–85.

127 WATSON, E. 1965. 'Periglacial structures in the Aberystwyth region of central Wales', *Proc. Geol. Ass. London*, **76,** 443–62.

128 WATSON, E. 1972. 'Pingos of Cardiganshire and the latest ice limit', *Nature (Lond.)*, **236,** 343–4.

129 WEST, R. G. 1956. 'The Quaternary deposits at Hoxne, Suffolk', *Phil. Trans. R. Soc.*, B, **239,** 265–356.

130 WEST, R. G. 1957. 'Interglacial deposits at Bobbitshole, Ipswich', *Phil. Trans. R. Soc.* B, **241,** 1–31.

131 WEST, R. G. 1961. 'Vegetational history of the Early Pleistocene of the Royal Society Borehole at Ludham, Norfolk', *Proc. R. Soc.*, B, **155,** 437–53.

132 WEST, R. G. 1963. 'Problems of the British Quaternary', *Proc. Geol. Ass. London*, **74,** 147–86.

133 WEST, R. G. 1970. 'Pollen zones in the Pleistocene of Great Britain and their correlation', *New Phytol.*, **69,** 1179–83.

134 WEST, R. G. 1972. 'Relative land/sea-level changes in southeastern England during the Pleistocene', *Phil. Trans. R. Soc. Lond.*, A, **272,** 87–98.

135 WEST, R. G., DICKSON, C. A., CATT, J. A., WEIR, A. H. and SPARKS, B. W. 1974. 'Late Pleistocene deposits at Wretton, Norfolk II. Devensian deposits', *Phil. Trans. R. Soc. Lond.*, B, **267,** 337–420.

136 WEST, R. G. and DONNER, J. J. 1956. 'The glaciations of East Anglia and the East Midlands: a differentiation based on stone orientation measurements of the tills', *Q.J. Geol. Soc. London*, **112,** 69–91.

137 WEST, R. G., LAMBERT, C. A. and SPARKS, B. W. 1964. 'Interglacial deposits at Ilford, Essex', *Phil. Trans. R. Soc.*, B, **247,** 185–212.

138 WEST, R. G. and NORTON, P. E. P. 1974. 'The Icenian Crag of southeast Suffolk', *Phil. Trans. R. Soc. Lond.*, B, **269,** 1–28.

139 WEST, R. G. and SPARKS, B. W. 1960. 'Coastal interglacial deposits of the English Channel', *Phil. Trans. R. Soc.*, B, **243,** 95–133.

140 WEST, R. G. and WILSON, D. G. 1966. 'Cromer Forest Bed Series', *Nature (Lond.)*, **209,** 497–8.

141 WILLIAMS, R. B. G. 1964. 'Fossil patterned ground in eastern England', *Biul. Peryglacjalny*, **14,** 337–49.

142 WILLIAMS, R. B. G. 1968. 'The distribution of permafrost in southern England in the last glacial period'. In *The periglacial environment: past and present*, ed. T. L. Péwé, Montreal: McGill University Press.

143 WILLIAMS, R. B. G. 1975. 'The British climate during the Last Glaciation: an interpretation based on periglacial phenomena'. In *Ice ages: ancient and modern*, ed. by A. E. Wright and F. Moseley, 95–120. Liverpool: Seel House Press.

144 WILLS, L. J. 1938. 'The Pleistocene development of the Severn from Bridgnorth to the sea', *Q.J. Geol. Soc. London*, **94,** 161–242.

145 WOODLAND, A. W. 1969. 'The buried tunnel valleys of East Anglia', *Proc. Yorks. Geol. Soc.*, **37,** 521–78.

146 WOOLDRIDGE, S. W. 1960. 'The Pleistocene succession in the London Basin', *Proc. Geol. Ass. London*, **71,** 113–29.

147 WOOLDRIDGE, S. W. and LINTON, D. L. 1955. *Structure, surface and drainage in south-east England.* London: Philip.

148 WORSLEY, P. 1966. 'Some Weichselian fossil frost wedges from east Cheshire', *Merc. Geol.*, **1,** 357–65.

149 WRIGHT, W. B. 1937. *The Quaternary Ice Age.* 2nd edn. London: Macmillan.

150 ZAGWIJN, W. H. 1975. 'Variations in climate as shown by pollen analysis, especially in the Lower Pleistocene of Europe'. In *Ice Ages: ancient and modern*, ed. A. E. Wright and F. Moseley, 137–52. Liverpool: Seel House Press.

151 ZAGWIJN, W. H., VAN MONTFRANS, H. M. and ZANDSTRA, J. G. 1971. Subdivision of the 'Cromerian' in the Netherlands; pollen analysis, palaeomagnetism and sedimentary petrology. *Geol. en Mijnbouw*, **50,** 41–58.

152 ZEUNER, F. E. 1959. *The Pleistocene Period.* 2nd edn. London: Hutchinson.

CHAPTER 13

PLEISTOCENE HISTORY OF THE FLORA AND FAUNA OF THE BRITISH ISLES

HISTORY OF THE FLORA[67, 116]

BEFORE considering in detail the Pleistocene history of the flora, it is necessary to point out the main vegetational changes which occurred during Tertiary time in northwest Europe. Floras with tropical affinities are known from the Lower Tertiary (Palaeocene, Eocene, Oligocene), and these became replaced during the Upper Tertiary (Miocene, Pliocene) by more temperate floras. The plant geographical relationships of these Upper Tertiary floras were very wide, with strong Asian and North American elements. The few Miocene–Pliocene plant-bearing deposits in the British Isles have been little studied and we have to refer to the well-known Pliocene deposits of the Rhine Valley in Germany and the Netherlands.[190] Here the Pliocene flora contains such exotic forest genera as *Carya*,

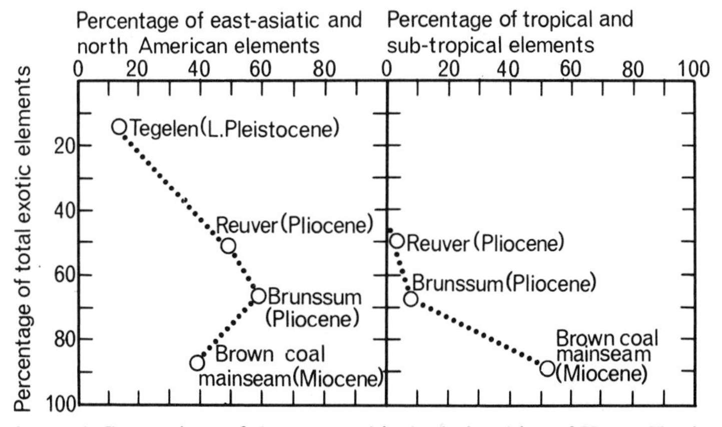

Fig. 13.1 **Comparison of the geographical relationships of Upper Tertiary** and Pleistocene macrofloras in the lower Rhine valley (Zagwijn, 1959).

Nyssa, *Tsuga*, *Sequoia* and *Taxodium*. The changes in the exotic elements from this area in Upper Tertiary and Lower Pleistocene time are shown in fig. 13.1.

The change to the Pleistocene in the Netherlands is recorded in the pollen diagram (fig. 13.2) from Meinweg. Forest is replaced by open vegetation with few thermophilous plants, the change taken to indicate the worsening of the climate marking the first cold period (Pretiglian) of the Pleistocene. On the recurrence of temperate conditions during the following interval (Tiglian), many of the exotic genera do not reappear. This reduction of the forest flora was described in detail by C. and E. Reid.[124] They explained the extinction in Europe of many exotic genera as a result of the climatic deterioration forcing temperate genera south. They pointed out that in Europe and western Asia there were barriers to this southward migration in the form of mountains, deserts and seas, while in eastern Asia and North America there were no such continuous barriers. The result was that only the more hardy of the circumboreal Tertiary forest genera remained in Europe, while in eastern Asia and North America the present forest genera survived the Pleistocene climatic deteriorations, and are much more nearly related to their Tertiary precursors.

The extinction in Europe was not entirely accomplished at the very beginning of Pleistocene time. Certain exotic genera persisted into the Early Pleistocene (e.g. *Tsuga*, *Carya*). But by the beginning of the Middle Pleistocene, the forest genera are those of modern Europe, as are most of the shrub and herbaceous species. So, for

Table 13.1 Pleistocene occurrence of some species and genera exotic to the native British flora

Species or genus with present distribution	*Stage*							
	Ludhamian	*Antian*	*Pastonian*	*Cromerian*	*Hoxnian*	*Ipswichian*	*Devensian*	*Flandrian*
Tsuga (E Asia, N. America)	+	+	+					
Pterocarya (Caucasus, E. Asia)	+	+			+			
Abies alba (C. & S. Europe)				+	+			
Picea abies (N. & C. Europe)	+	+	+	+	+		+	
Najas tenuissima (N. Europe)			+	+				
Azolla filiculoides (N., C. & S. America)			+	+	+			
Salvinia natans (C. & S. Europe)				+		+		

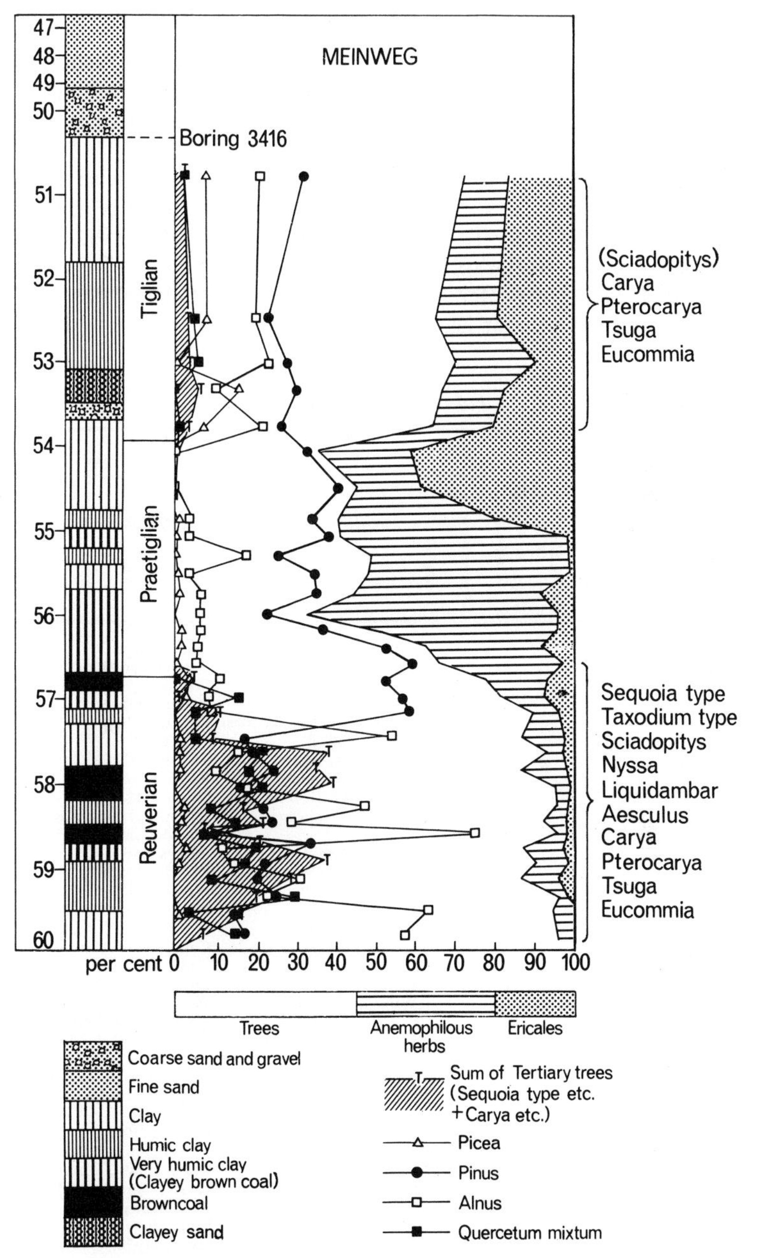

Fig. 13.2 **Pollen diagram from Meinweg, S.E. Netherlands, across the Plio/ Pleistocene (Reuverian/Pretiglian) boundary** (Zagwijn, 1960).

example, the plant species identified from the temperate Cromerian stage are nearly all found in the modern British flora. The Pleistocene occurrence of some species and genera exotic to the native British flora is shown in table 13.1.

Although the record of Miocene and Pliocene deposits with fossil flora in the British Isles is poor, in the Pleistocene there is a good record of fossil floras. Table 13.2 lists important sites with fossil floras. From the stratigraphical point of view most of the floras are well dated, as they are intercalated in the sequence of glacial deposits. The glacial advances, at least in midland England and Ireland, did

Table 13.2 Fossil floras

Early Pleistocene	*Cromerian*	*Anglian*	*Hoxnian*	*Wolstonian*	*Ipswichian*	*Early and Middle Devensian*	*Devensian late-glacial*	*Flandrian*	*Locality*	*Reference*
+									Ludham, Norfolk	West (1961a)
+									Easton Bavents, Suffolk	Funnell and West (1962)
+									Southeast Suffolk	West and Norton (1974)
+	+	+							Norfolk coast	Reid (1899)
	+								Norfolk and Suffolk coast	Duigan (1963)
		+	+	+					Hoxne, Suffolk	West (1956)
		+	+	+					Marks Tey, Essex	Turner (1970)
		+	+						Nechells, Birmingham	Kelly (1964)
			+						Nar Valley, Norfolk	Stevens (1960)
			+						Gort, Co. Galway	Jessen, Farrington and Andersen (1959)
		+	+	+					Baggotstown, Co. Limerick	Watts (1964)
				+					Dorchester, Oxfordshire	Duigan (1955)
				+	+				Ipswich, Suffolk	West (1957)
				+	+				Selsey, Sussex	West and Sparks (1960)
					+				Wretton, Norfolk	Sparks and West (1970)
					+				Shortalstown, Co. Wexford	Colhoun and Mitchell (1971)
						+			British Isles	West (1977)
						+			Wretton, Norfolk	West *et al.* (1974)
						+			Chelford, Cheshire	Simpson and West (1958)
						+			Upton Warren, Worcs	Coope, Shotton and Strachan (1961)
						+			Cambridge	Lambert, Pearson and Sparks (1962)
						+			Earith, Hunts	Bell (1970)
						+	+	+	British Isles	Godwin (1975), Pennington (1975, 1977)
							+	+	Ireland	Jessen (1949), Watts (1977)
							+	+	Scotland	H. J. B. Birks (1973), H. H. Birks (1975) Pennington *et al.* (1972)
							+	+	Wales	Moore (1970, 1972)
							+	+	Lake District	Pennington (1964)

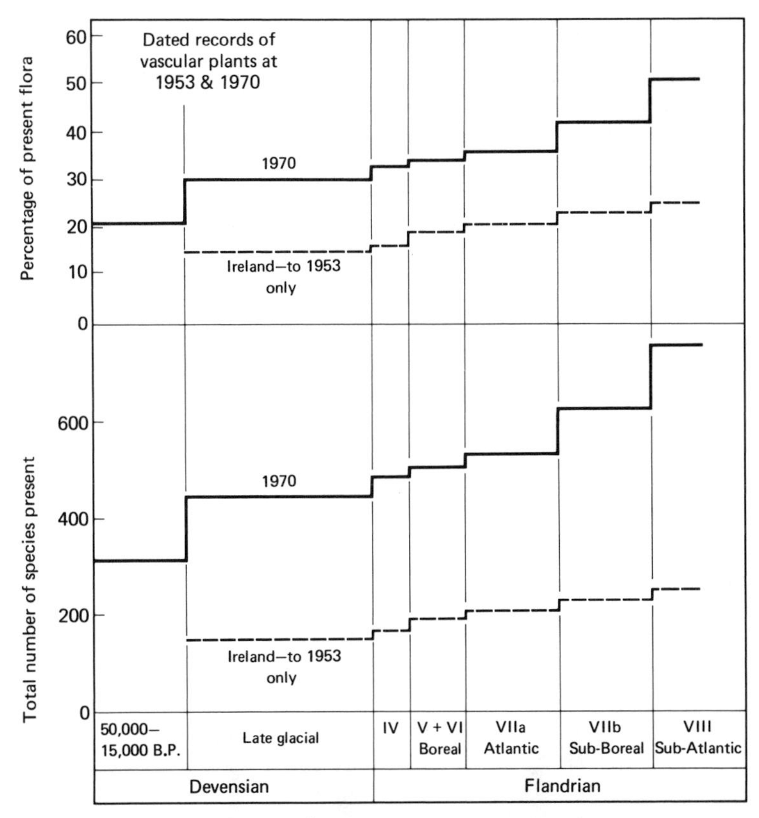

Fig. 13.3 Increment graphs for dated records of plant species excluding aliens, hybrids and crop-plants. The 1970 census comprises all identifications at species level, and genera for which no individual species has been recorded up to the time in question. The 1953 figures rest on a slightly less exacting standard of species identification. The percentages of the existing flora have been based on the assumption for this purpose of a figure of 1500 for the British Isles and 1000 for Ireland. Note the high percentage of all records already in the Devensian and by the end of the Atlantic period. The subsequent increased rate of rise is, in part at least, a consequence of the extension of disforestation and agriculture (Godwin, 1975).

not remove the older Pleistocene deposits and their contained floras. Naturally the most complete record is from the most recent stage, the Flandrian, where the history of the flora in most parts of the British Isles is known in considerable detail, and can be related to various environmental events, such as changes in sea-level, and man's clearance activities. As we go further back into the Pleistocene, the described fossil floras decrease in number. There are many from the

Devensian cold stage, fewer of the Ipswichian and Hoxnian temperate stages, fewer still of the Wolstonian and Anglian cold stages, while the few Cromerian and pre-Cromerian floras are mostly confined to coastal regions of East Anglia.

As regards completeness of the fossil record from the point of view of the native flora, up to 1970 over 750 species of vascular plants, say around 50 per cent of the present flora of the British Isles, has been recorded fossil in Devensian and Flandrian deposits (fig. 13.3). This is a very substantial proportion and means that we can reconstruct the history of the flora in great detail in this period. The fossil vascular flora is naturally the best known, but recent work on the bryophyte flora, with a large number of species identified fossil, adds much to our knowledge of the history of the British flora.[51] The records of Algae are mostly confined to the Flandrian, where sequences of diatoms have been recorded from a number of lake deposits.[130]

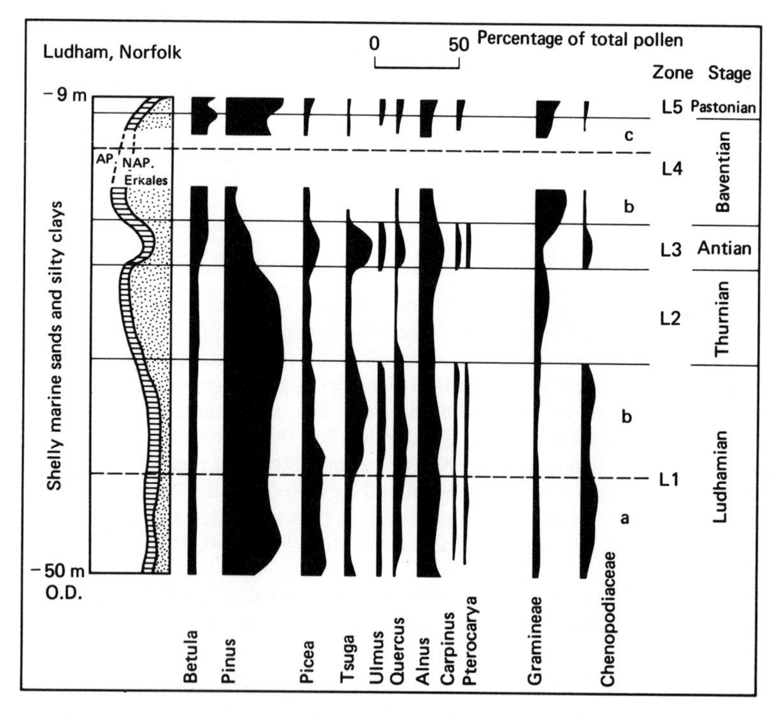

Fig. 13.4 Schematic pollen diagram through the Crag at Ludham, Norfolk (after West, 1961a).

Early Pleistocene floras

The pollen floras recorded from the Early Pleistocene Crags (fig. 13.4) are of two types, those with high non-arboreal pollen frequencies and low frequencies of thermophilous trees, representing cool periods, and those with lower non-arboreal pollen frequencies and higher pollen frequencies of thermophilous trees, representing temperate periods.[176] These temperate pollen floras show the presence of mixed coniferous and deciduous forest during two stages, the Ludhamian and the Antian. The genera *Tsuga* and *Pterocarya* are characteristic trees of these times, the former being a 'Tertiary relic', now extinct in Europe, and confined to eastern Asia and North America. The flora of the cold stages, the Thurnian and Baventian, is characterised by the high frequencies of Ericales tetrads, many of which are of *Empetrum nigrum* type. The vegetation indicated is oceanic heath, and the climate cool in comparison with the forested stages.

Possibly a flora of macroscopic plant remains from Castle Eden, Co. Durham, is of Early Pleistocene age.[125] It was originally described as being slightly younger than the Reuverian (late Pliocene) stage of the Netherlands, as it had fewer exotic species than that stage, but the plant list, which indicates temperate conditions, gives little help in age determination. The paucity of exotic species and its stratigraphical position below the oldest glacial deposits in the area, suggest an Early Pleistocene age.

Floras of the cold stages

In this category we include the fossil floras closely related to glacial or periglacial deposits and showing little or no indication of forested conditions at their time of deposition. The earliest such flora is Baventian in age, from the cold period preceding the Pastonian temperate stage, and there are floras from all the subsequent cold stages (Beestonian, Anglian, Wolstonian and Devensian). The sediments containing the floras are channel or terrace sediments, often of an organic type suggesting sedimentation in shallow transient pools (fig. 13.5b). Reworking of these organic deposits is indicated by their being found as erratics in coarse gravel, formed presumably by the undercutting of meandering channels in a flood-plain environment (fig. 13.5a).

The pollen of these deposits usually shows a small amount of tree or shrub pollen, usually *Salix*, *Betula* (including *B. nana*-type), *Juniperus* and *Pinus*, and much greater frequencies of non-tree pollen, especially sedge and grass pollen, but also varying amounts of the

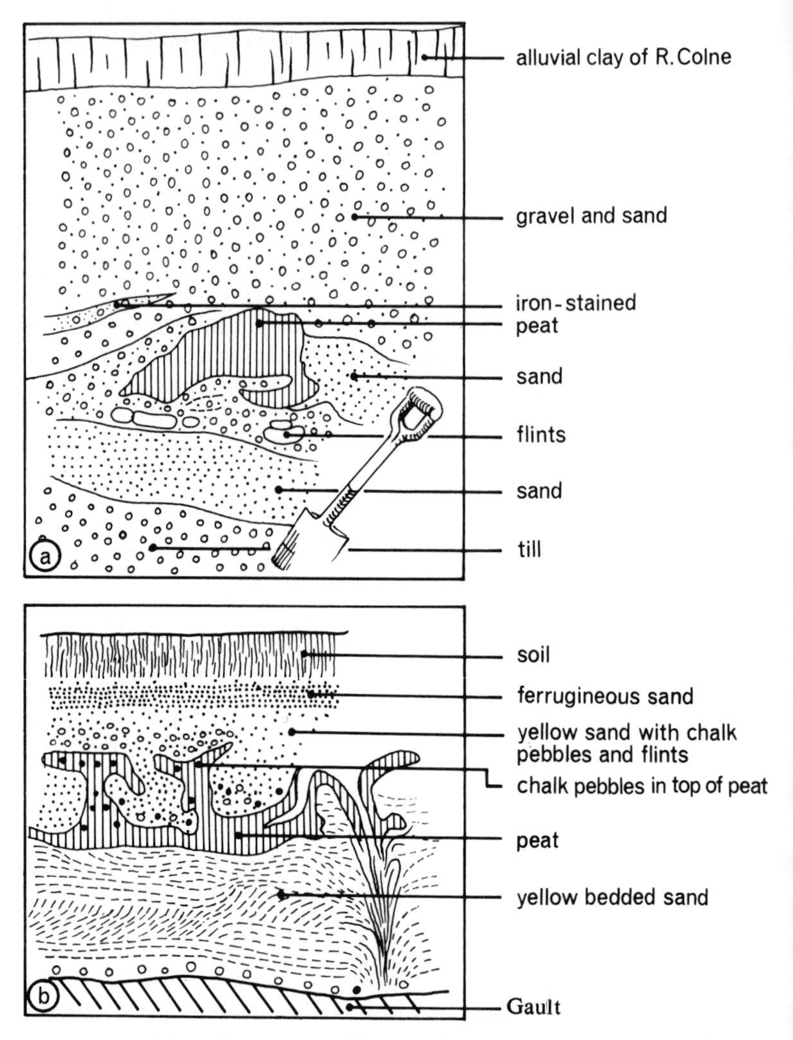

Fig. 13.5 Sections in terraces showing 'full-glacial' organic deposits. a, erratic of detritus mud in Devensian gravels in the flood plain of the river Colne, Herts. Sketch after photo by P. Evans. b, Devensian pond deposit in terrace sands and gravels at Wretton, Norfolk, penetrated by an ice-wedge cast and covered by solifluction gravels (West *et al.* 1974).

pollen of marsh and aquatic plants (*Filipendula*, *Caltha*), of ruderals and open-habitat plants (*Artemisia*, *Helianthemum*, *Plantago*), and of plants of tall-herb communities (*Polemonium*, *Urtica*, *Sanguisorba*). Amongst the macroscopic plant remains (plate 11) there is a mixture

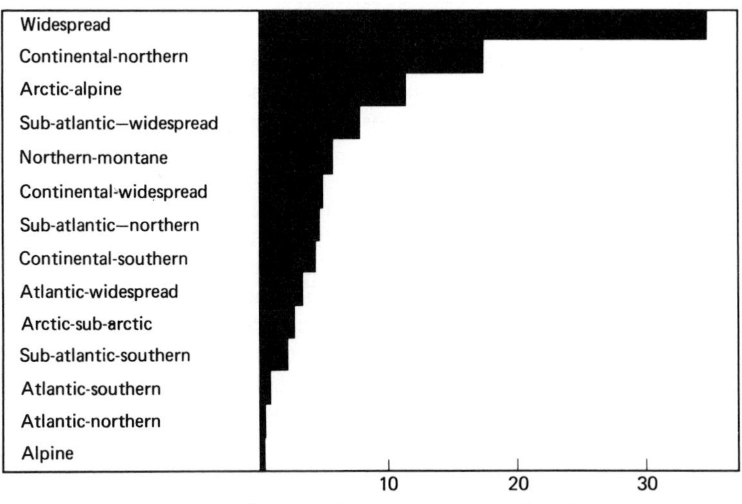

Fig. 13.6 Diagram showing the relative percentage of various geographical elements of the present British Flora (defined by Birks and Deacon, 1973) represented by Devensian late-glacial records (Godwin, 1975).

of ecological and plant-geographical categories (fig. 13.6), e.g. northern and montane plants, halophytes and other plants now commonly maritime, many widely distributed plants, often now weeds, plants indicative of open habitats and a few species of more southern distribution.[11] A few of the more characteristic species may be mentioned:

Northern and montane plants: *Draba incana, Dryas octopetala, Salix herbacea, Saxifraga oppositifolia* (plate 11a), *Thalictrum alpinum.*

Maritime plants: *Armeria maritima, Atriplex hastata, Plantago maritima, Suaeda maritima.*

Weeds: *Polygonum aviculare, Potentilla anserina, Ranunculus repens.*

Plants of a more southern distribution: *Lemna, Potamogeton crispus, Ranunculus sceleratus.*

There are indications of continentality of climate in the presence of *Hippophaë*, *Centaurea cyanus* and *Poterium sanguisorba*, and the often considerable frequencies of *Artemisia* pollen. These have been considered to give a steppe-like character to the flora. The scarcity of trees and the variety of herbs in the cold stage floras are their two characteristics. The assemblages recorded require an explanation in climatic terms. For a flora and fauna of Middle Devensian age in Worcestershire, it has been suggested that ameliorating climate and grazing by large herbivores help to explain the mixture of floristic elements and the rarity of trees.[43] At an Early Devensian site in

Cambridge a mixture of geographical elements could be related to a deteriorating climate at the beginning of the Devensian, the more southern plants and molluscs possibly being survivors from interglacial time.[87]

Assuming that there is little or no secondary deposition of plant remains, two facts require explanation: the mixture of cold- and warm-indicating geographical elements and the absence or rarity of trees. It is also relevant that similar plant assemblages are recorded from all the cold stages. The mixture of species is probably related to the great diversity of microclimates occurring in a periglacial area of lower latitude than the present arctic, occurring within a regional vegetation of grassland. Regarding the absence of trees, many periglacial environmental conditions do not favour tree growth and survival, e.g. wind, permafrost, solifluction and waterlogging of soils, as well as the climatic extremes themselves.

The key to this diversity and treelessness of the cold stage floras probably lies in the climatic effects produced by the cool Atlantic when the Atlantic polar front lay far to the south of the British Isles (see Ch. 10). Severe long winters and short, sharp summers would promote herb vegetation, such as is recorded in the high non-tree pollen biozones, and restrict the tree growth.[178a].

The repeated occurrence of such floras in the Devensian (they have been dated by the radiocarbon method at 13,560, 19,500, 28,000, 41,900 years B.P.[66] and also found in the Early Devensian), and in the earlier cold stages, indicates that this type of flora was essentially that which existed in southern Britain during 'full-glacial' and near 'full-glacial' conditions of the cold stages, and they indicate the considerable extent to which plants, of many differing ecological and geographical categories, were able to survive in this country during these cold stages.

We also consider here the late-glacial floras, which date from the times of retreat of the successive glaciations, but which precede the afforestation occurring in temperate non-glacial periods. Such late-glacial episodes are known from the Beestonian, Anglian, Wolstonian and Devensian cold stages. These floras are similar to those of the full-glacial periods except that they may show increased percentages of tree (*Betula*) and shrub pollen. For example, in the Anglian late-glacial at Hoxne, the remains of *Hippophaë rhamnoides*, the sea buckthorn, are very common. In some late-glacial sequences, e.g. those prior to the Ipswichian, the pollen spectra remain similar over a period of time, suggesting stable late-glacial conditions. These are followed by rapid vegetational changes leading to interglacial forest vegetation. In other late-glacial sequences, e.g. the Anglian and

Devensian, there are fluctuations of the climate within the sequence. The best-known late-glacial is of course that of the Devensian. The three-fold division of this late-glacial has been described in Chapter 12, and table 12.10 summarises the broad vegetational sequence. Pollen diagrams are shown in figs. 7.5, 7.6, 13.11, 13.13, 13.16 and 13.17. The pre-interstadial and post-interstadial (zones I and III) show dominant herbaceous vegetation and scattered *Betula* copses, which in the interstadial expanded, with *Juniperus*, and in the south and east may have been associated with *Pinus*.[116a, b, 172a]. In more westerly areas oceanic heath with *Empetrum* were common in the late-glacial, but the flora as a whole was still dominated by northern and montane plants.

Evidence of a minor oscillation in zone I (the Bølling oscillation of the continent?) has been reported from the north of England,[9] but it is generally thought that such an amelioration can be included in the late-glacial interstadial.[116a]

The lower boundary of the late-glacial has not been defined pollen-analytically. The lower boundary of the interstadial (Windermere Interstadial) is now defined, in pollen analytical terms, at the time pollen productivity increases rapidly, marked by sudden increase in pollen input into the sediments (measured as grains per year per unit area), a rise particularly noticeable in the frequency curves for *Betula* and *Juniperus*. The end of the interstadial is marked by the fall of *Juniperus* and *Betula* pollen frequencies and the rise of herbaceous taxa such as Gramineae, Cyperaceae, *Rumex* and *Artemisia*. The interstadial ended at around 10,800 years B.P., with the date for its beginning at around 13,000 years B.P.[116b] It will be expected that these dates will be refined when good definitions have been agreed and more profiles dated. Examples of pollen diagrams covering these late-glacial oscillations, including an absolute diagram, are shown in figs. 7.5, 7.6, 13.11 and 13.13.

Interstadial floras

This type of flora, indicating climatic oscillation towards boreal environmental conditions rather than to full interglacial temperate conditions, is rare. The best known is a flora from Chelford of Early Devensian age (^{14}C age 60,800 years B.P.), from a thin layer of sandy detritus mud within the middle sands of the Cheshire Plain.[140] Here pollen and macroscopic plant remains indicate a cool climate with a *Betula-Pinus-Picea* forest as the dominant local vegetation, similar to a forest-type now found in northern Finland. Pollen spectra of the same kind have been found in Early Devensian deposits at Wretton in Norfolk.

Temperate (interglacial) floras

During the interglacials climatic amelioration proceeded as far as in the Flandrian, and these temperate stages are characterised by the presence of deciduous forest.[76] For the purpose of vegetational comparisons we may include within the term interglacial the present temperate stage—the Flandrian—as its vegetational development so far has been of the same general nature as those of the interglacials. Interglacial deposits have been found at many sites in central and southern England and Ireland (fig. 12.8). They are usually lacustrine sediments associated with glacial deposits or with terrace deposits. They show a sequence of vegetational change, which, at its most complete, comprises four major zones, shown in table 13.3. The zones

Table 13.3 Zonation of temperate stages of the Middle and Late Pleistocene

					Vegetational aspect
Cold stage	e An	e Wo	e De		Early-glacial
Temperate stage	Cr IV	Ho IV	Ip IV		Post-temperate
	Cr III	Ho III	Ip III	Fl III (zones VIIb, VIII)	Late-temperate
	Cr II	Ho II	Ip II	Fl II (zone VIIa)	Early-temperate
	Cr II	Ho I	Ip I	Fl I (zones IV to VI)	Pre-temperate
Cold stage	l Be	l An	l Wo	l De (zones I to III)	Late-glacial

e, early; l, late; Ho, Hoxnian; An, Anglian; Cr, Cromerian; Be, Beestonian; Fl, Flandrian; De, Devensian; Ip, Ipswichian; Wo, Wolstonian

are defined on pollen assemblages, so they are strictly pollen assemblage biozones. They are outlined in subsequent tables, and their sequence is associated with the cycle of interglacial environmental change.

The detail of the vegetational history of each interglacial differs from that of the others, and these differences therefore provide possibilities of characterising each particular interglacial temperate stage by its vegetational history. But the regional differences of vegetation in any one interglacial are not yet clear, for the reason that interglacial deposits have not so far been widely found or studied.

The vegetational histories of the temperate stages of the Middle and Late Pleistocene are as follows:

Pastonian.[183] Deposits of this stage on the Norfolk coast have been insufficiently investigated. The pollen diagrams do, however, show a division into four major zones, the earliest (Pa I) with *Pinus* and *Betula*, the second (Pa II) with *Pinus*, *Quercus*, *Carpinus* and *Ulmus*, the third (Pa III) with *Pinus*, *Quercus*, *Picea* and *Carpinus*, and the fourth (Pa IV) with *Pinus*, *Betula* and *Picea*. The vegetational history leading to this forest development and the changes associated with its disappearance at the end of the stage are not yet clear.

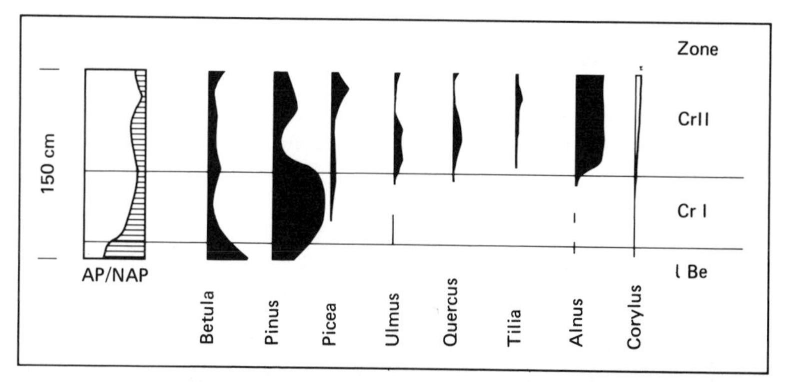

Fig. 13.7 Pollen diagram through the Upper Freshwater Bed of the Cromerian stage at West Runton, Norfolk (after Duigan, 1963). Values expressed as percentages of total tree pollen. Only the first half of the Cromerian temperate stage is recorded in this diagram.

Cromerian.[183] The vegetational history of this temperate stage has again been insufficiently studied. A pollen diagram from the Upper Freshwater Bed at West Runton is shown in fig. 13.7, but it comprises only the first half of the stage, and most of the preceding late-glacial is missing. On the basis of more recent studies of freshwater and

Table 13.4 Zone characters of the Cromerian stage and related deposits

Early-glacial	e An	High NAP*	Arctic Freshwater Bed
Post-temperate	Cr IV	*Pinus, Picea, Betula, Alnus*	Estuarine Bed
Late-temperate	Cr III	*Carpinus, Abies*	Estuarine Bed
Early-temperate	Cr II	*Pinus, Quercus, Ulmus, Tilia*	Upper Freshwater Bed
Pre-temperate	Cr I	*Pinus, Betula*	Upper Freshwater Bed
Late-glacial	l Be	High NAP	Upper Freshwater Bed

*NAP, non-arboreal pollen

estuarine deposits around the Norfolk coast, the outline pollen zonation shown in table 13.4 has been drawn up.

Apart from the pollen record, nearly 300 species are known macroscopically from the Cromer Forest Bed Series (Pastonian–Beestonian–Cromerian).[186] The plants are nearly all native to the present flora, amongst the few exotics being *Najas minor*, *Picea abies*, *Abies*, *Trapa natans*, *Azolla filiculoides* and *Salvinia natans*. There are none of the so-called Tertiary-relics. Compared with the succeeding interglacials the Cromerian vegetational history is distinct in the absence or scarcity of *Hippophaë* in the late-glacial zone, the very low frequencies of *Corylus* throughout, the high frequencies of *Ulmus* and presence of *Tilia* in zone CrII, and the late phase, zone CrIII, with *Abies* and *Carpinus*.

Hoxnian. Several lake deposits of this age are known in England. Both macroscopic and pollen remains are well known. The flora is very similar to that of the Cromerian in its paucity of exotic genera. A pollen diagram from this interglacial is shown in fig. 13.8, and the sequence of zones is shown in table 13.5.

The characters distinguishing this sequence from the others are: the high frequency of *Hippophaë* in zone e An, the abundance of *Tilia* in zone Ho II, the late rise of *Ulmus* and *Corylus* in zone Ho II, and the important *Abies* phase in zone Ho III. The Irish interglacial deposits of Gortion (≡Hoxnian) age indicate the development of a distinctive type of flora with many evergreens, e.g. *Abies*, *Taxus*, *Rhododendron*, *Ilex* and *Buxus*, after an early phase which resembles the Hoxnian interglacial in England.[75, 172] All of these, except *Rhododendron*, have been found in the English deposits. They indicate a high oceanicity of climate during this temperate stage, perhaps more so than in the Cromerian, Ipswichian and Flandrian.

An interesting feature of Hoxnian pollen diagrams from East

Table 13.5 Zone characters of the Hoxnian stage and related deposits

Early-glacial	e Wo	High NAP
Post-temperate	Ho IV	*Pinus*, *Betula*, NAP higher
Late-temperate	Ho III	*Quercus*, *Carpinus*, *Abies*
Early-temperate	Ho II	*Quercus*, *Ulmus*, *Tilia*
Pre-temperate	Ho I	*Betula*, *Pinus*
Late-glacial	l An	High NAP, *Hippophae*

Anglia is a period of deforestation towards the end of zone Ho II. At Hoxne Palaeolithic artefacts occur associated with this, but it is impossible to say whether the presence of man caused the deforestation, or the deforestation allowed the incoming of man, or whether other environmental factors were responsible.

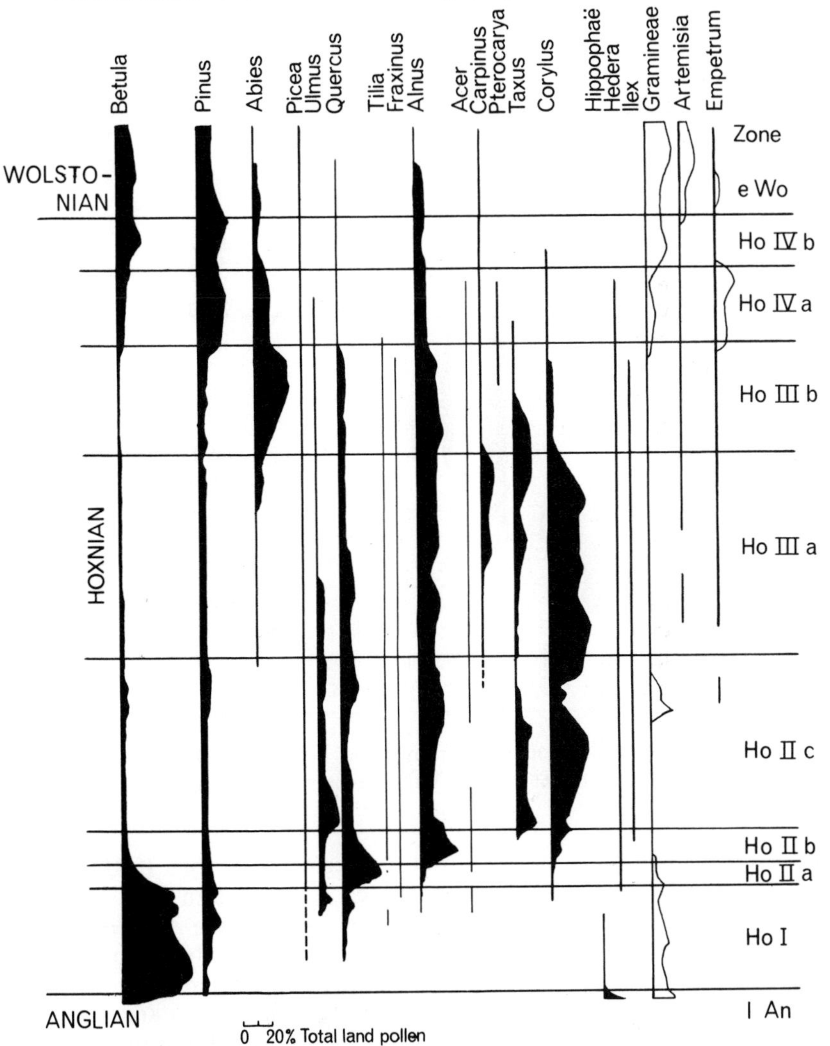

Fig. 13.8 Pollen diagram from the Hoxnian interglacial lake deposit at Marks Tey, Essex (C. Turner, 1970). Values expressed as percentages of total land pollen.

Ipswichian.[118] Floras of Ipswichian age are known from south and east England, and both macroscopic fossils and pollen have been found. The sites are nearly all in river valleys in terrace sequences. A pollen diagram from the period is shown in fig. 7.4. But there is no one site showing all the vegetational changes of the interglacial; the pollen diagram, fig. 13.9, is therefore compounded from several sites. The zone characters are given in table 13.6.

The characteristics of the Ipswichian vegetational succession are

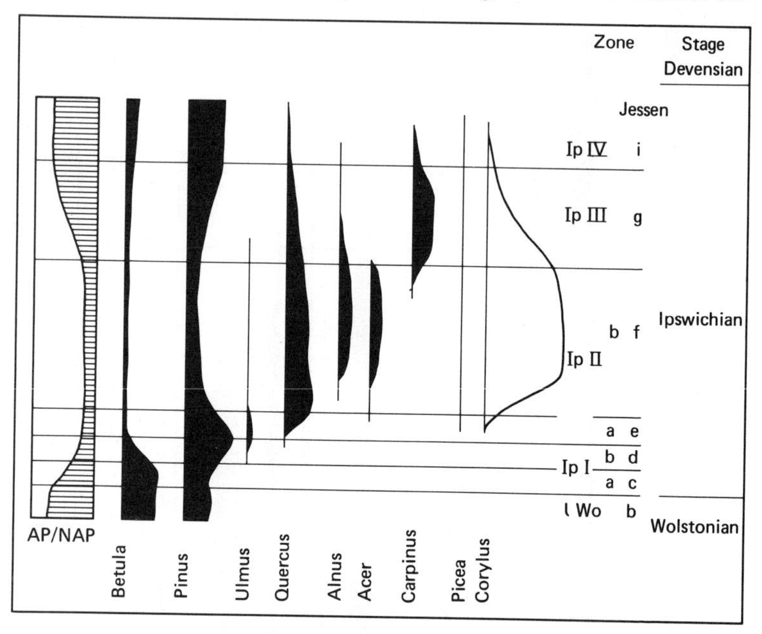

Fig. 13.9 Composite pollen diagram through the Ipswichian temperate stage, based on a number of sites. Values expressed as percentages of total tree pollen.

the high frequencies of *Corylus* at the beginning of zone Ip II; the occurrence of substantial frequencies of *Acer* pollen in zone Ip II, the scarcity of *Tilia* pollen, the varied behaviour of the *Alnus* pollen curve, the high frequencies of *Carpinus* pollen in zone Ip III, and the absence or scarcity of *Abies* and *Picea* pollen throughout the interglacial. The scarcity or absence of pollen of *Tilia*, *Picea* and *Abies* has lead to discussion of their status in the interglacial. It is probable that *Tilia* was only present in the southeast of England and that *Picea* (with a low pollen production) was locally present in the later

Table 13.6 Zone characters of the Ipswichian stage and related deposits

Early-glacial	e De	High NAP	*Equivalent zones of Jessen k Milthers (1928)*
Post-temperate	Ip IV	*Pinus*, NAP higher	i
Late-temperate	Ip III	*Carpinus*	g
Early-temperate	Ip II	*Quercus*, *Pinus*, *Acer*, *Corylus*	e, f
Pre-temperate	Ip I	*Betula*, *Pinus*	c, d
Late-glacial	l Wo	High NAP	b

part of the interglacial.[118] *Abies* appears to have been absent. A rise in NAP in the later part of the interglacial is a characteristic feature of the pollen diagrams of the Ipswichian. This feature has been related to the grazing activities of large mammals, such as *Hippopotamus*, whose remains are frequent in last interglacial deposits, and to herb communities of fluviatile and alluvial environments developing as aggradation takes place.

Most of the Ipswichian sites contain abundant macroscopic plant remains (plate 11), including only a few species not now found native, e.g. *Acer monspessulanum*, *Najas minor* (plate 11i), *Pyracantha coccinea*, *Trapa natans*, *Xanthium* and *Salvinia natans*. *Azolla filiculoides* is not found, though frequent in the earlier temperate interglacial stages. The presence of certain water plants, and of seeds of other water plants now rarely fruiting, or rarely found fossil in the Flandrian, in the British Isles, e.g. *Hydrocharis morsus-ranae*, *Lemna minor*, *Najas minor*, *Stratiotes aloides* and *Salvinia natans*, indicates warmer summers and a more continental climate than in the Flandrian.

Flandrian. The Flandrian pollen zones, equivalent to the zones I, II and III of the earlier temperate stages, are outlined in table 12.10. As with the earlier temperate stages, they show the development of temperate deciduous forests, but this development is drastically affected by man's activities after about 5000 years B.P., near the boundary of pollen zones VIIa to VIIb. After this time a widely applicable zonation has not been achieved since there are local differences of vegetational history, mainly arising from the variability of man's activities in different areas. Before this time, regional systems of zonation are widely applicable though radiocarbon dating shows that the zone boundaries are largely diachronous except at the boundary marked by the elm decline (zone VIIa/VIIb boundary).[142]

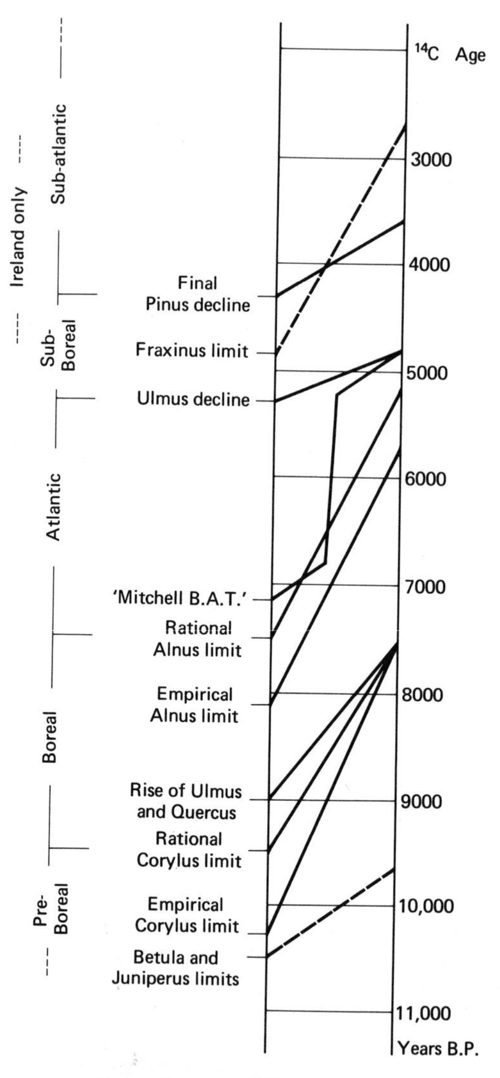

Fig. 13.10 Approximate age ranges of Flandrian pollen zone boundaries and other features (Smith and Pilcher, 1973). Empirical limit: the point at which pollen of the particular taxon becomes consistently present; rational limit: the point at which the pollen curve begins to rise to sustained high values.

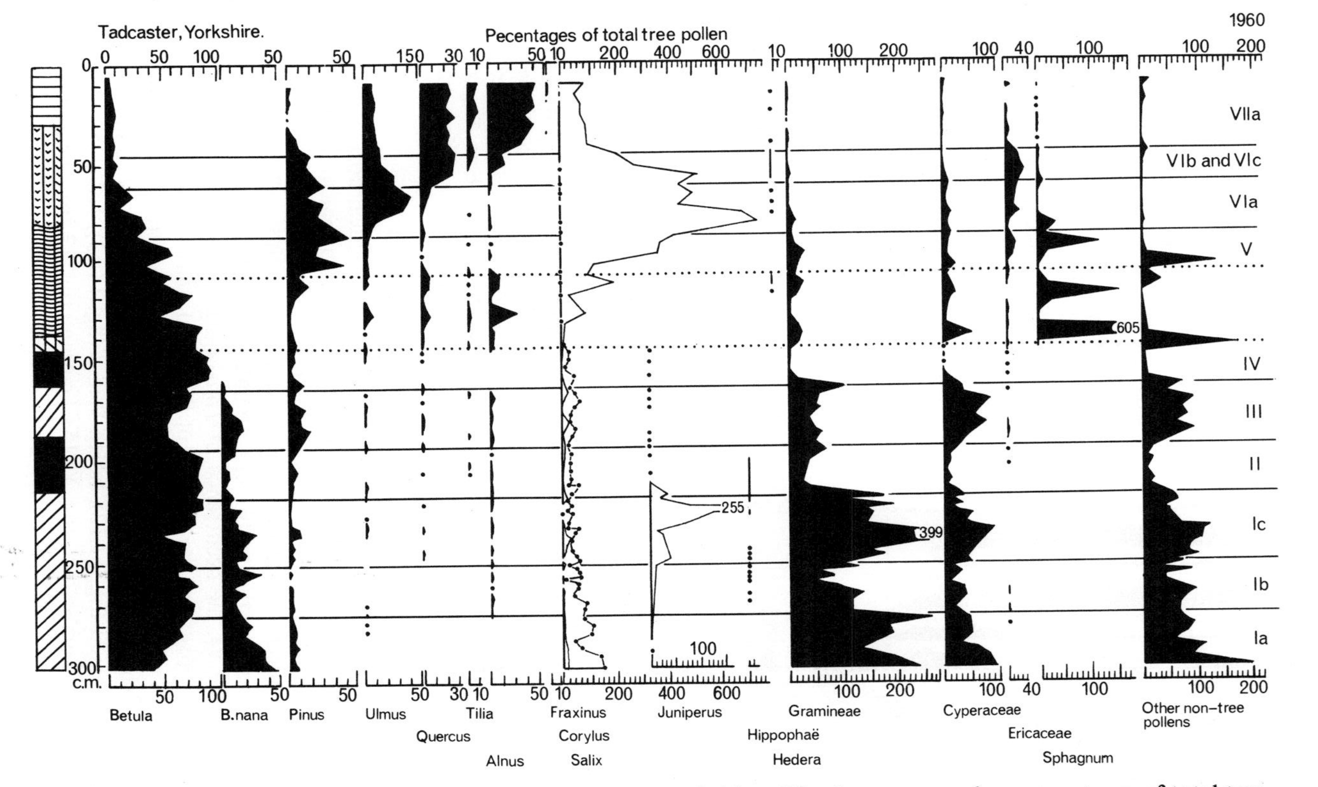

Fig. 13.11 Pollen diagram from a lake infilling at Tadcaster, Yorkshire. All values expressed as percentages of total tree pollen (Bartley, 1962).

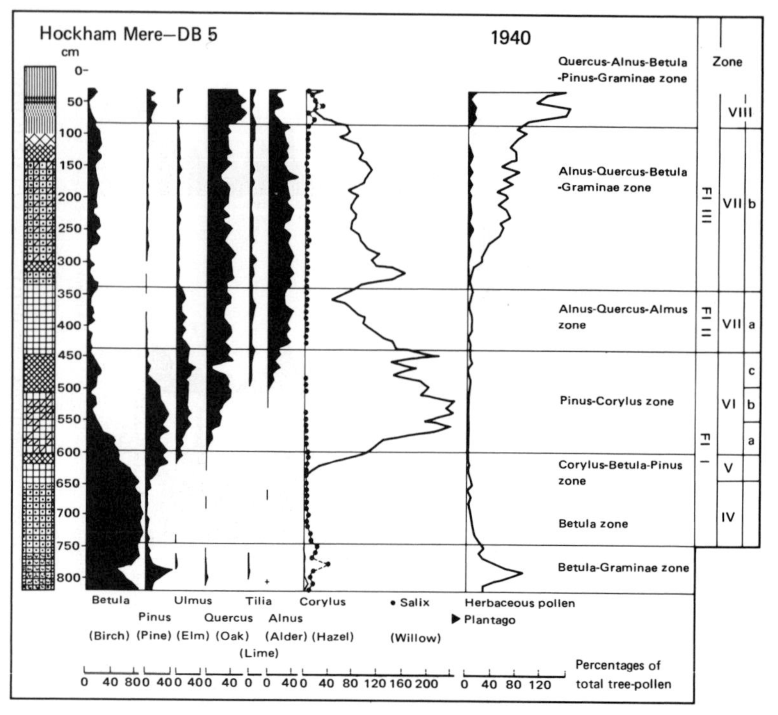

Fig. 13.12 Pollen diagram through the lake deposit at Hockham Mere in the north of the Breckland of East Anglia. The symbols on the left refer to various types of limnic mud up to 100 cm depth and to fen peat above this level. Two systems of zonation are shown. At the 340 cm level the strong decline of the *Ulmus* curve indicates the boundary between zones VIIa and VIIb. The pollen-zones are fairly characteristic for East Anglia but zone VIII is difficult to recognise in the absence of *Fagus* (after Godwin, 1975).

The extent of diachroneity for various pollen analytical horizons is shown in fig. 13.10, and is governed by both geographical location and altitude, with changes tending to lag behind in the uplands.

The Flandrian vegetation sequence differs from those of the interglacial times by the early very abundant frequencies of *Corylus* pollen (zone V), the occurrence of *Tilia*, *Carpinus* and *Fagus* in the temperate forest, and the absence of *Picea* and *Abies*. Some examples of Flandrian pollen diagrams are shown in figs. 13.11 to 13.17. These diagrams show the principal Flandrian pollen zones, as found in eastern England (fig. 13.12), north Wales (fig. 13.13), Lake District (fig. 13.14), Cumberland (fig. 13.15), Scotland (fig. 13.16) and southern Ireland (fig. 13.17). In the Cumberland diagram (fig. 13.15) a

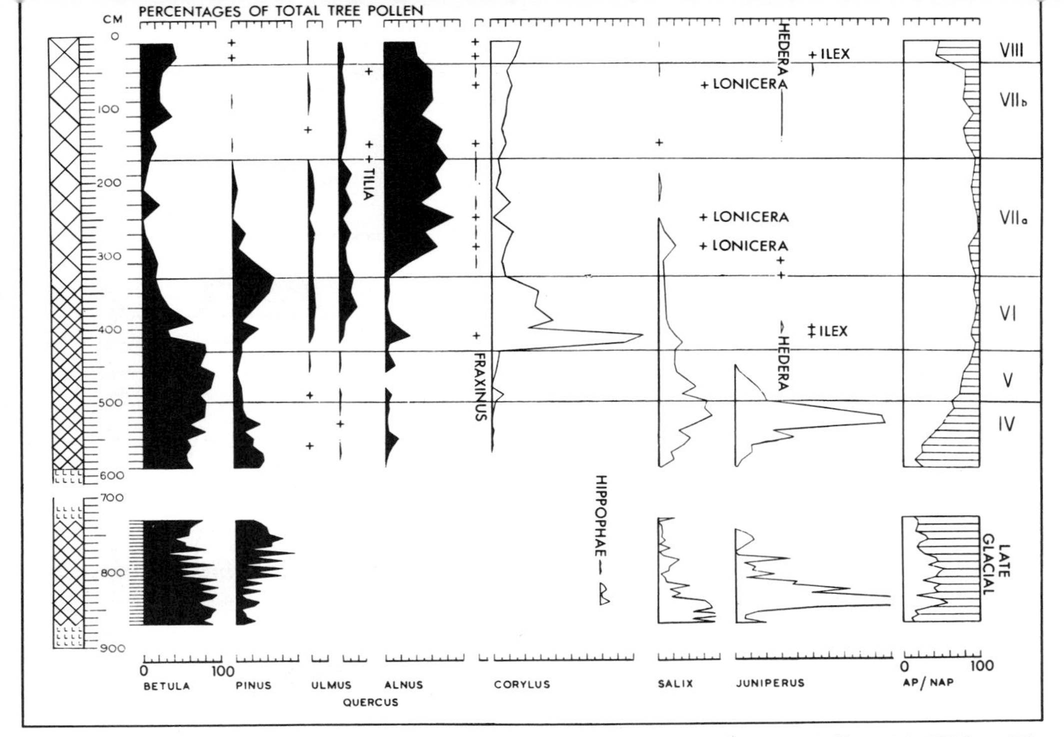

Fig. 13.13 Tree pollen diagram from lake infilling in the Nant Ffancon Valley, N. Wales. The curves shown black comprise the total on which percentages are calculated. The ratio AP/NAP shows the changing proportions of arboreal pollen (representing forest) and non-arboreal pollen (representing other vegetation) through the Devensian late-glacial and Flandrian (Seddon, 1962).

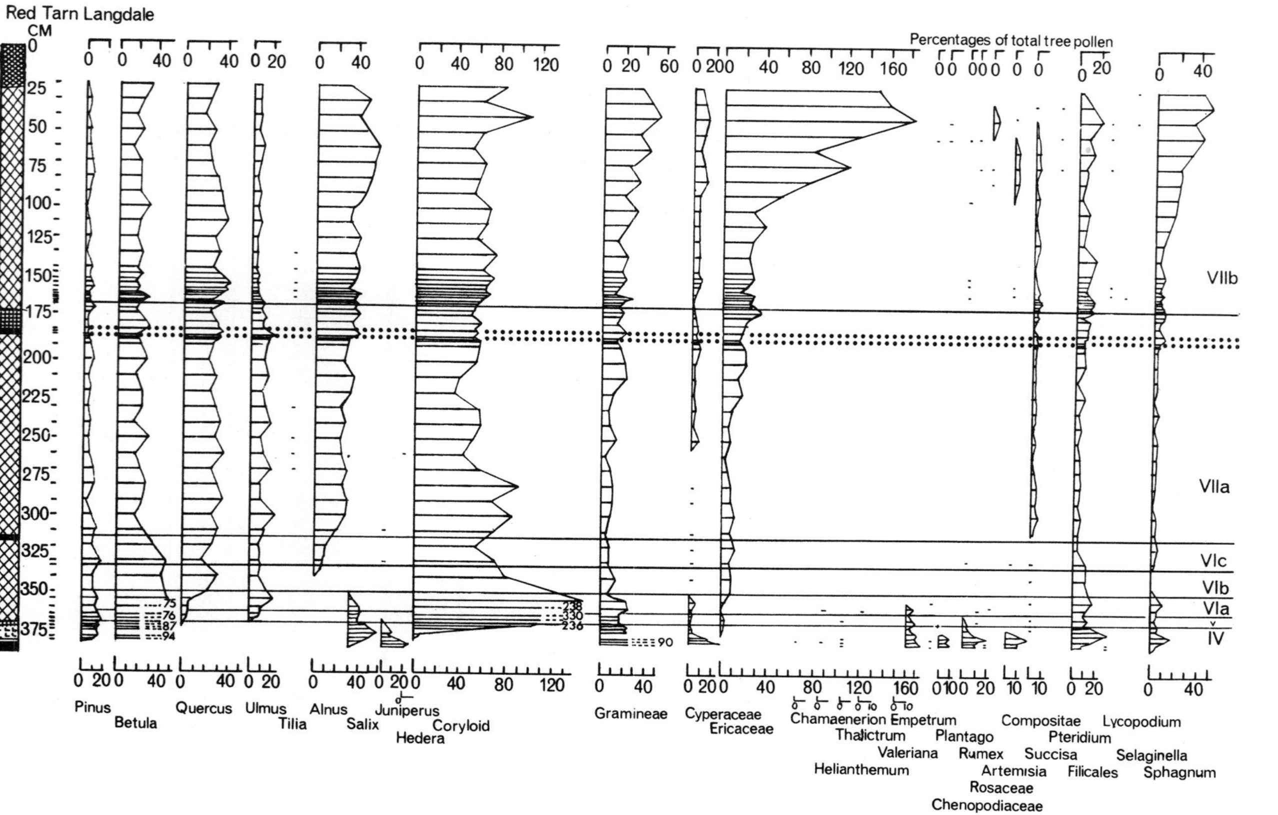

Fig. 13.14 Pollen diagram from Red Tarn, Langdale, Lake District (1700 ft O.D.). The dotted lines indicate the horizon, just below the *Ulmus* decline, where sudden fluctuations in pollen curves suggest inwash of soil containing pollen (Pennington, 1964).

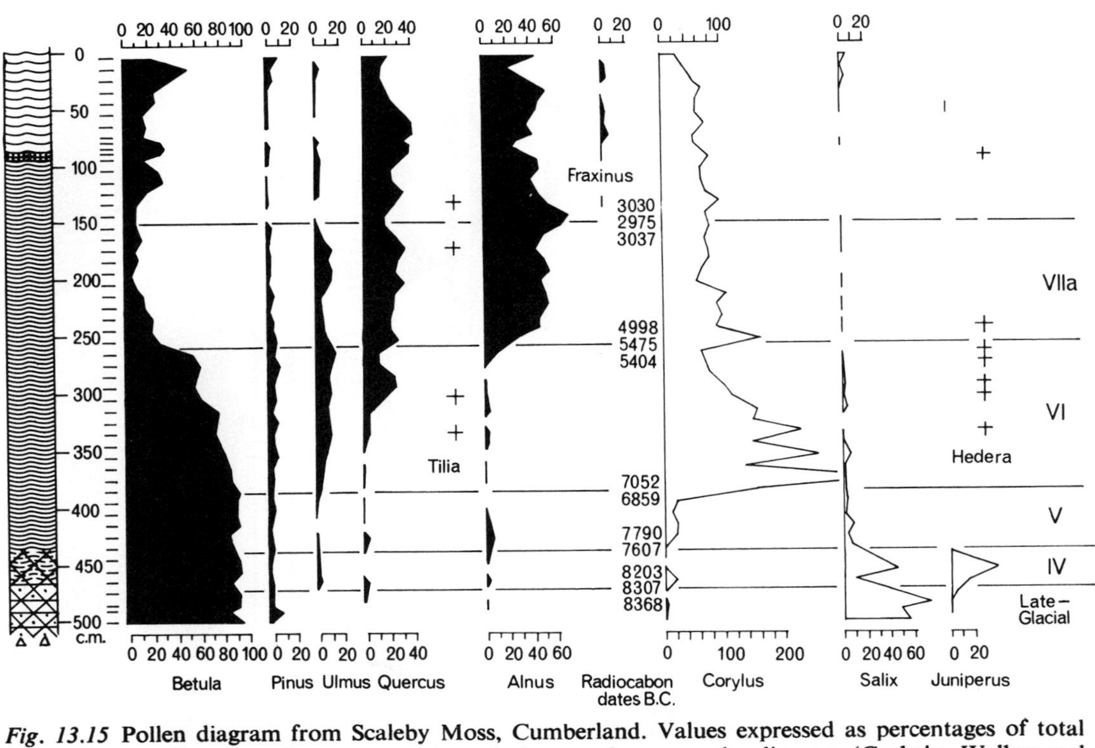

Fig. 13.15 Pollen diagram from Scaleby Moss, Cumberland. Values expressed as percentages of total tree pollen. Radiocarbon dates at zone boundaries are shown on the diagram (Godwin, Walker and Willis, 1957).

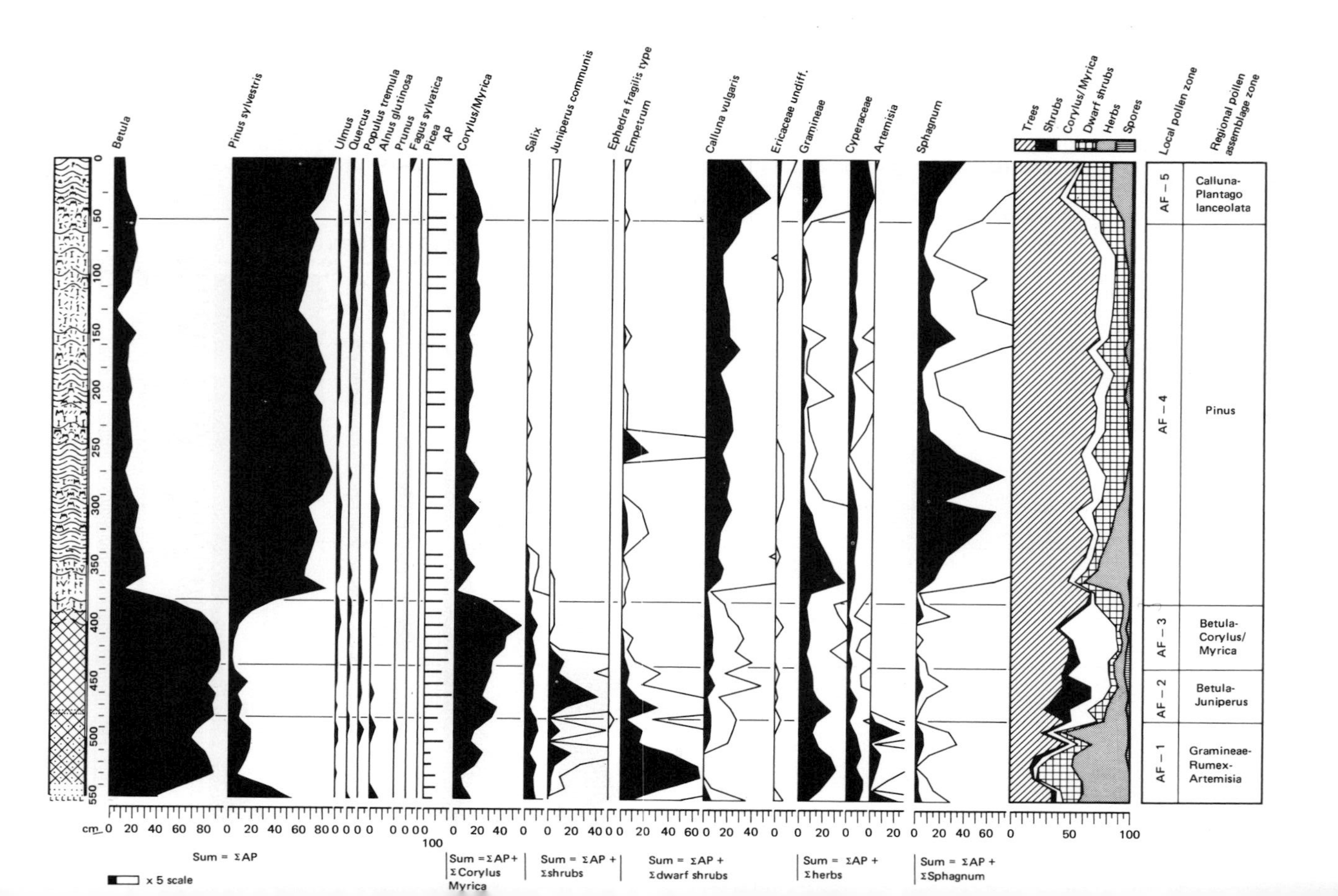
Betula
Pinus sylvestris
Ulmus
Quercus
Populus tremula
Alnus glutinosa
Prunus
Fagus sylvatica
Picea
AP
Corylus/Myrica
Salix
Juniperus communis
Ephedra fragilis type
Empetrum
Calluna vulgaris
Ericaceae undiff.
Gramineae
Cyperaceae
Artemisia
Sphagnum
Trees
Shrubs
Corylus/Myrica
Dwarf shrubs
Herbs
Spores
Local pollen zone
Regional pollen assemblage zone
AF – 5
Calluna-Plantago lanceolata
AF – 4
Pinus
AF – 3
Betula-Corylus/Myrica
AF – 2
Betula-Juniperus
AF – 1
Gramineae-Rumex-Artemisia
cm
Sum = ΣAP
Sum = ΣAP + ΣCorylus Myrica
Sum = ΣAP + Σshrubs
Sum = ΣAP + Σdwarf shrubs
Sum = ΣAP + Σherbs
Sum = ΣAP + ΣSphagnum
x 5 scale

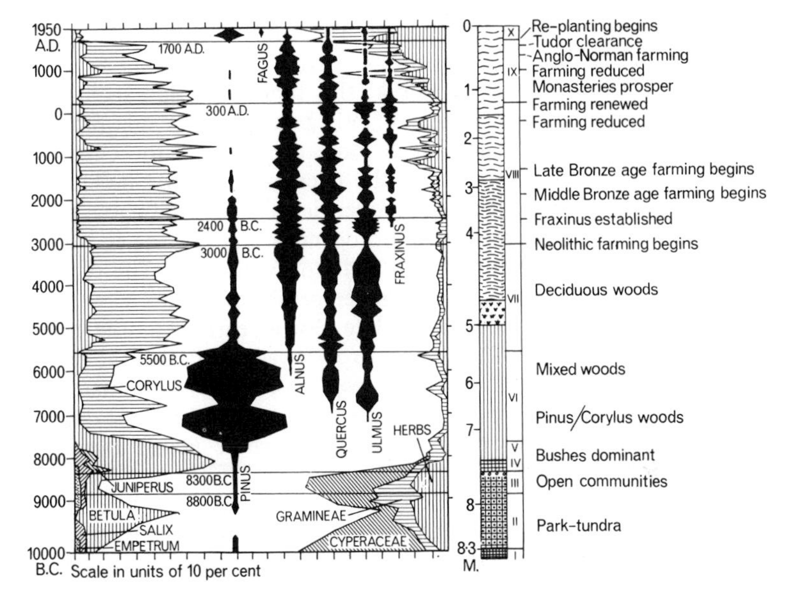

Fig. 13.17 Total pollen diagram from Littleton Bog, Co. Tipperary, Eire. Stratigraphy, pollen zonation and notes are on the right of the diagram (Mitchell, 1965).

series of radiocarbon dates gives a time scale to the sequence. These diagrams are from different regions, different elevations and different sediment types; their general uniformity demonstrates the usefulness of a system of pollen zonation for both vegetational history and relative chronology.

Table 12.10 gives the main characteristics of the pollen zones in various parts of the British Isles. Zone IV (Pre-boreal period) covers the time when birch, mostly *Betula pubescens* in Ireland, but both *B. pubescens* and *B. verrucosa* in Britain, established closed forests. In zone V, pine became an important forest tree in the south and east of the British Isles, and remained so in zone VI; in Ireland it displaced birch at a later time, in zone VI. Zone VI has been subdivided on the basis of changes in the pollen frequencies of the mixed oak forest trees. Both zones V and VI, constituting the Boreal period, show

Fig. 13.16 Pollen diagram (selected taxa) from Abernethy Forest, Inverness-shire (H. H. Birks, 1970). The white silhouettes give a ×5 exaggeration to the scale. There is little substantial change in the tree pollen frequencies following the expansion of *Pinus*.

high frequencies of *Corylus* pollen. There is evidence that towards the end of the Boreal there was a period of dryness as seen, for example, in lowered lake levels at this time. The Boreal–Atlantic transition (B.A.T.) is the zone VI/VII boundary. At this time the climate apparently became much more oceanic and the change is marked in pollen diagrams by the decrease of *Pinus* pollen frequencies and increase of *Alnus* frequencies. Corresponding with this change is the development of ombrogenous blanket bog in upland areas and ombrogenous raised bogs in lowland areas. The climatic optimum of the Flandrian (average temperatures some 2°C warmer) is thought to have occurred in late Boreal/early Atlantic times.

The time from about 7000 years to 5000 years B.P. was the period when high forest became stabilised over the country, with mixed oak forest of varying regional constitution in most of Ireland and in Britain except for the Scottish Highlands, where *Betula* and *Pinus* remained important.[18]

Ulmus (probably mostly *U. glabra*) is an important tree in the Atlantic mixed oak forest, and a sharp decline in its pollen frequencies marks the end of the Atlantic and the opening of the Sub-boreal period. This decline is taken as the boundary between zones VIIa and VIIb.[141] There has been much discussion about the cause of this decline—whether it is of climatic origin or whether it is partly the effect of the agricultural activities of Neolithic man, which start to be of widespread occurrence around this time.[72]

The approximate synchroneity of the elm decline demonstrated by radiocarbon dating,[64] 5500 to 5000 years B.P. argues for a controlling climatic cause, but coincident with the spread of Neolithic farmer cultures, active with forest clearance and the use of leaf-fodder. The elm decline is followed by a sequence of vegetational changes indicating clearance and recolonisation, and accompanied by an increase in the pollen frequencies of herbs, in particular *Plantago lanceolata*, an important indicator of clearance. Figure 13.18 shows a typical elm decline and associated clearance phenomena in northern Ireland.

The process of deforestation continued in zones VIIb and VIII or their equivalents. The boundary between these zones is more obscure than the earlier zone boundaries. In general, there are tendencies indicating climatic deterioration at the zone boundary between VIIb and VIII, as shown by the increases in *Betula* pollen and the decreases of *Tilia*, *Hedera* and *Ilex* pollen which have usually been taken to indicate these boundaries. The climate appears to have taken a turn towards wetter conditions, as bog stratigraphy indicates fresh growth of *Sphagnum* peats after earlier drier conditions in the Sub-boreal when growth was slow or had ceased. Such fresh growth gives rise to the recurrence surfaces already mentioned in Chapter 10. The radiocarbon dates of these surfaces around the VIIb/VIII boundary vary between 450 and 900 B.C., an age near the supposed age (600 B.C.) of Weber's classical *Grenzhorizont* at the opening of the Sub-atlantic period.[67] But the climatic changes at this time and later, as indicated by vegetational changes, are much confused by local anthropogenic effects. The introduction of various crop plants at Neolithic and later times becomes more and more noticeable. Both wheat and barley were cultivated from Neolithic times on, and rye and oats became important in Romano-British times.

Aspects of the history of the flora

Comparisons of the floras of temperate stages. The changes in the fossil assemblages of plants already described must now be considered from a more general environmental point of view. In the pre-Devensian temperate floras it is difficult to make out an environmental picture from the vegetational changes, partly because of the small number of sites studied, but also because the physiographic background is to a large extent unknown. When we come to consider what are the possible causes for the differing vegetational histories of the temperate stages, we are unable to distinguish the effects of the following possible factors:

1. The distances from our area of glacial refuges where species might have survived.
2. The differences in the rates of movement of different species.
3. The roles of competition, succession, and soil maturation in vegetational change.
4. Possible barriers to plant movements, e.g. high sea-levels.
5. The effects of climate.

First, factors connected with the distances of the glacial refuges of the plants from England and the rates of migration of different species. If the climatic amelioration at the beginning of a temperate stage is more rapid than the rate of migration, this rate and the distance of the glacial refuge will affect the time of appearance of, say, the mixed oak forest trees in the temperate stage. Difficulties in interpreting the differences in the pollen diagrams resulting from this

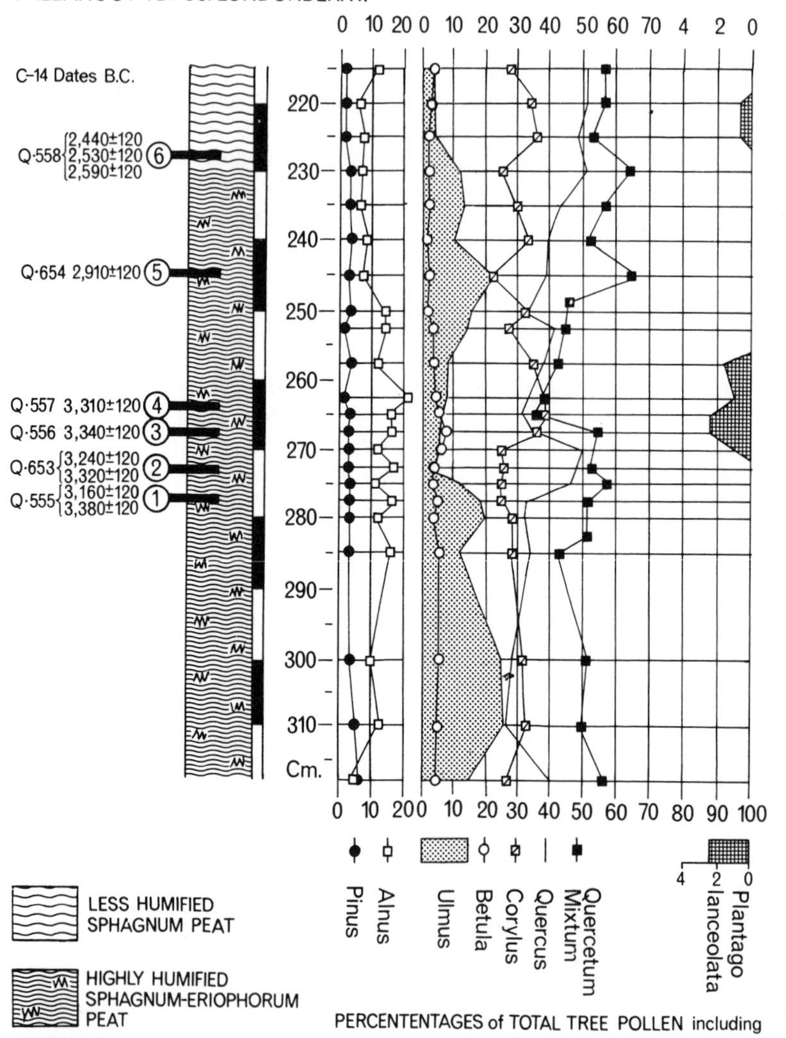

Fig. 13.18 Pollen diagram from the bog at Fallahogy, Co. Londonderry, N. Ireland. Values expressed as percentage of total tree pollen, including *Corylus*, but with totals of *Betula*, *Pinus*, *Alnus* and *Corylus* divided by

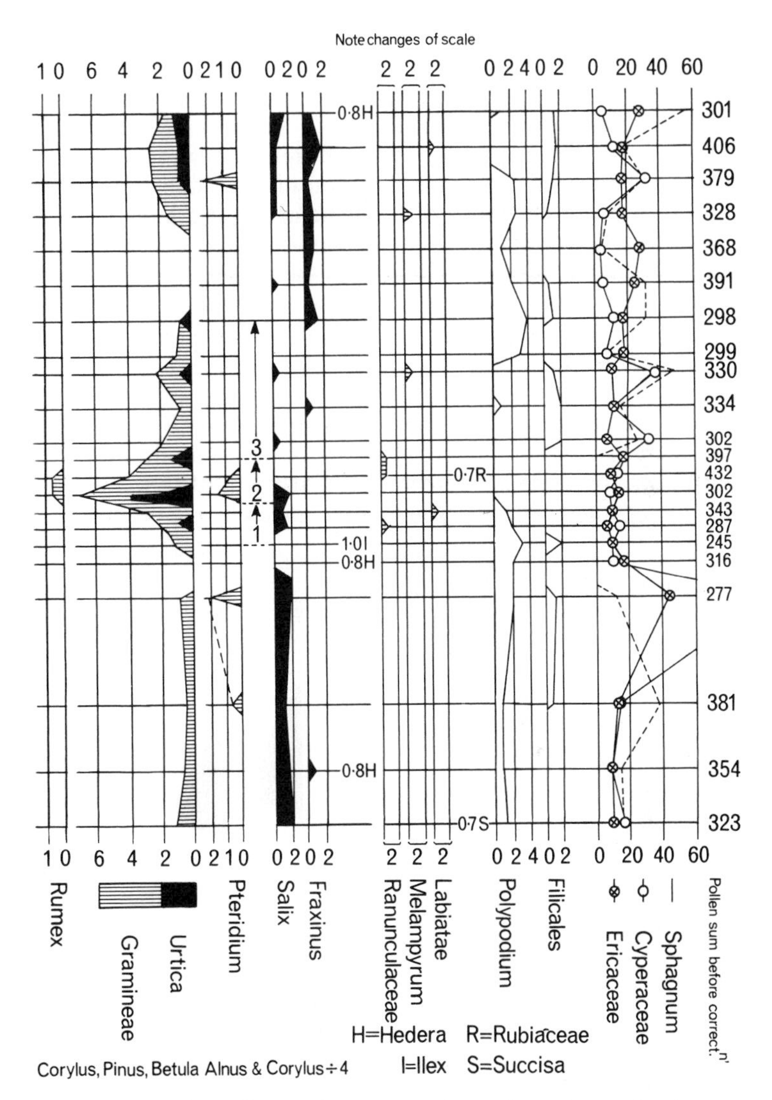

4 before calculation. The numbers 1, 2 and 3 in the centre of the diagram refer to phases of clearance (1), farming (2) and regeneration (3). Radiocarbon ages on the left (Smith and Willis, 1962).

kind of factor arise because often we do not know the species concerned in the pollen diagrams, and because we know little of the biological factors determining the rate of migration. In the absence of palaeobotanical evidence of floras of cold stages in refuge areas it is difficult to suggest the positions of these areas, though some information may be gained about the direction of migration from a comparison of pollen diagrams of the same age over a large area.

In the early-temperate substages *Ulmus* and *Quercus* are the first of the thermophilous trees to appear in the temperate stages, including the Flandrian. *Alnus* and *Corylus* appear at variable times. *Corylus* also behaves differently in the different temperate stages and the Flandrian (fig. 13.19). Nuts ascribed to *C. avellana* have been found in all the temperate stages, so it is probably with this species that we are concerned. Possibly the differing *Corylus* curves result from the behaviour of the species in competition with the mixed oak forest trees, following a different time of arrival of the species. In the earlier temperate stages, *Corylus* expands late, well into the mixed oak

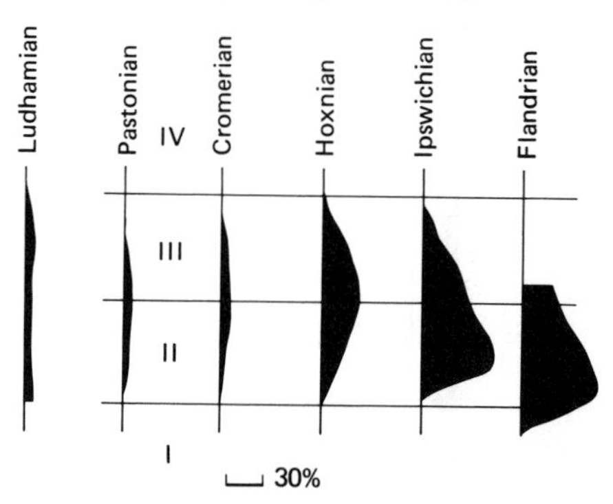

Fig. 13.19 Schematic *Corylus* pollen curves for temperate stages in East Anglia, presented as percentages of total tree pollen and related to the four-fold zone system.

forest zone, and is not very well represented in the pollen diagrams. In the Ipswichian, *Corylus* expands to high percentages with the mixed oak forest. Perhaps this early spread is related to the incomplete forest cover suggested by pollen diagrams. In the Flandrian, *Corylus* appears in quantity before the mixed oak forest and the pollen frequencies expand to high values before the mixed oak forest becomes dominant. As mentioned above, these differences might be a reflection of the differing distances of the glacial refuges of *Corylus* during the successive glaciations. But it is also possible that ecological preferences, or some other properties of the species affecting its competition with the mixed oak forest trees, have changed during the long period of time involved.

In the case of forest trees appearing later in the temperate stages, other considerations arise in the interpretation of the differences between them. The conifers of the late-temperate substages are replacing the mixed oak forest trees, either as a result of climatic change, or possibly as a successional change following soil changes as the temperate stage proceeds.

The second kind of factor to be considered is that concerning barriers to migration of plants into Britain, in particular, the effects of sea-level changes in the temperate stages. During the glacial periods, with sea-levels perhaps 100 m lower than at present, Britain must have been joined to the continent by a wide isthmus, and Ireland to Britain by at least a narrow isthmus in the north. Figure 13.20 shows the submarine contours around the British Isles. In the Flandrian, and in each temperate stage, Britain would be cut off by a

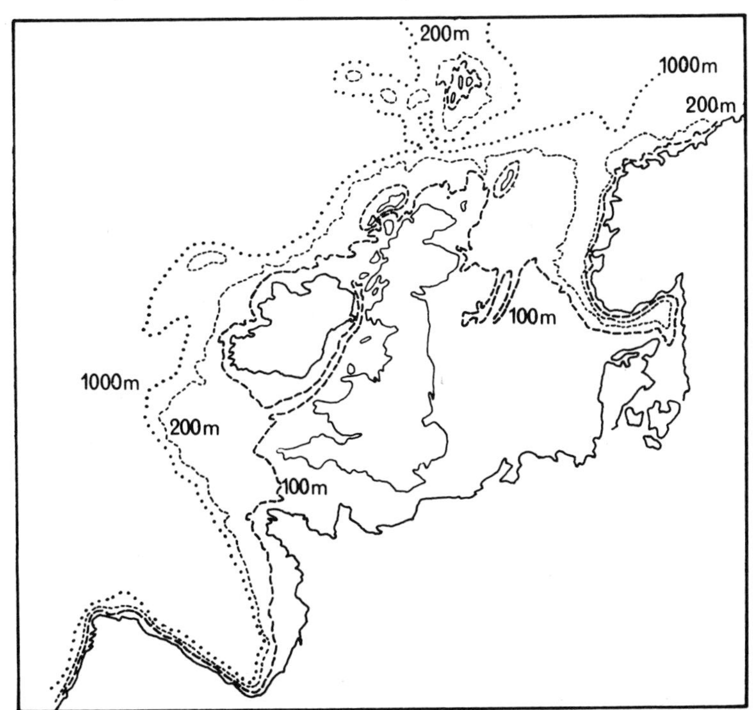

Fig. 13.20 Submarine contours around the British Isles.

eustatic rise in sea-level as the ice melted, providing that the land bridge across the Straits of Dover was flooded. The time of the rise in sea-level in relation to the development of the vegetation is obviously of importance in determining the course of its development. In the Flandrian, the rising sea-level seems to have cut off Britain from the continent at some time during the Boreal period (zone VI). As an example of the effect on plant distribution of isolation by a rising sea-level, we may mention the two trees, *Tilia* and *Fagus*, present in England but not native in Ireland. Perhaps these trees were prevented from further spread to Ireland by the rising Flandrian sea-level. In the Ipswichian there is evidence that sea-level rose above its present height in the early-temperate substage, zone I II, relatively earlier than in the Flandrian. It is possible that this early rise may account for differences between the continental and English Ipswichian interglacial pollen diagrams, e.g. the apparent lack of *Picea* in England though it is common on the continent in this interglacial. On the other hand, in the Hoxnian, there is some evidence that the rise above present sea-level was much later, late in the early-temperate substage, or in the late-temperate substage, and this may account for the similarity of the late-temperate substage vegetation in England and on the continent, with both *Abies* and *Picea* as prominent conifers. An alternative explanation for the similarity is that a land bridge with the continent existed in this interglacial, the Straits of Dover being fully cut in the subsequent glaciation.

The sea-level changes of the older interglacials, however, are difficult to reconstruct, since tectonic changes have considerably affected the North Sea area.

Lastly, there is climate as a factor determining the differences in the interglacial vegetational successions. It is difficult to distinguish the effects of climate in detail. The arrival and spread of species are not necessarily correlated with climatic change. If the rate of climatic change is faster than the rate of migration of the plants, the presence of a species gives a 'minimum' value for climate. As we have seen, the vegetational succession can obviously be determined by factors other than climatic change. Where the vegetation type shown by the pollen diagrams remains stable, general indications of climate, e.g. degree of oceanicity or continentality, may be inferred. Where the disappearance of species can be correlated with climatic amelioration or deterioration we may obtain some indication of the nature of the climatic change.

The climatic differences between the temperate stages are difficult to infer. For example, the Ipswichian shows a very different pollen diagram from the other interglacials and from the Flandrian, with high percentages of *Carpinus* pollen and much pollen of *Acer*, in part *A. monspessulanum*, fruits of which are found in the deposits concerned. Both these plants have a continental distribution and their abundance might be thought to indicate a climate more continental than that of the Flandrian or of the Hoxnian. In the latter, *Alnus*, *Hedera* and *Ilex* are all very well represented, perhaps an indication of a more oceanic climate than in the Ipswichian Interglacial where these plants are not common. But the differences between the Ipswichian and Hoxnian pollen diagrams might be due to accidents of distribution and the effect of barriers to migration. It will be seen that it is difficult to disentangle the real climatic significance of the differences between the vegetational histories reflected in the pollen diagrams.

However, the study of the behaviour of certain forest genera in the temperate stages reveals differing responses to repeated climatic change.[178] Figure 13.21 shows the distribution of some of these

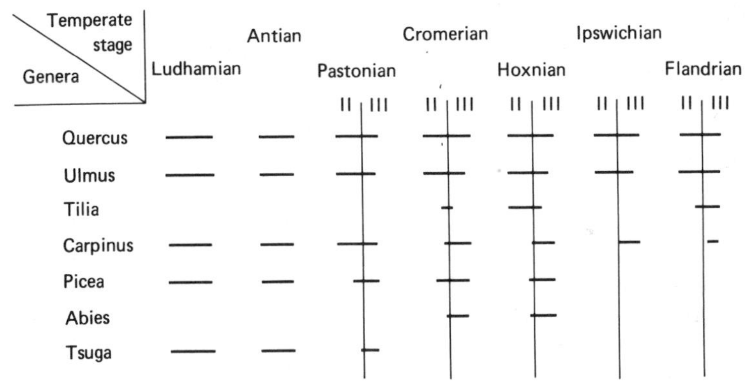

Fig. 13.21 The occurrence of some forest genera in East Anglia during zones II and III of the temperate stages of the Pleistocene.

genera in the successive temperate stages. The most successful genera, *Quercus*, *Ulmus* and *Corylus*, are those with a widespread distribution in western Europe. Others, *Carpinus*, *Picea* and *Tsuga*, have suffered restriction in the course of time, perhaps caused by loss of biotypes. Yet others, *Tilia* and *Abies*, show an inconsistent behaviour more probably related to a combination of environmental effects and change in biotype content.

Devensian and Flandrian: theories of the origin of the flora. When we come to consider the Devensian and Flandrian, however, we have a much better picture of the general physiographic conditions under which the vegetational changes took place, and we have much more detail of the history of the flora.

Here we must consider the history of biogeographical thought about the origin of the British flora, as it is in respect of the Devensian and Flandrian floras that most discussion has taken place.[49] Edward Forbes[57] in his classic essay of 1846 put forward the idea that five different geographical components in the British flora (and fauna) were the result of successive migrations into the British Isles under different climatic conditions at several distinct points of time and that gradual changes ensued as climatic changes occurred. He envisaged that the Lusitanian element was settled in Ireland in preglacial times, that the atlantic (southwest England) element and the more continental element of southeast England were probably also of preglacial 'migratory' origin, that the glacial (artic–alpine) element appeared during the glacial epoch, and that the widespread Germanic element was received after the glacial epoch.

At a later date there was much discussion over the problem of 'perglacial' survival. Clement Reid in 1899 suggested that during the maximum glaciation the flora was nearly exterminated, only arctic plants survived in the south, and temperate species were entirely lost.[123] Wilmott maintained that species of local distribution survived in their local areas during glaciation.[185] Thus the peculiar Teesdale flora[119] would represent a nunatak flora which survived glaciation in that area. However, with our now increased knowledge of the history of the flora, it is clear that no generalisations on these lines can be made. Each species has its own particular history. Some survived glaciation in the British Isles, some are restricted and local because their habitat area was reduced by the forest spread in the Flandrian, while others may be now restricted by reason of narrow climatic tolerances or by man's interference with the balance of environment.[120]

We should now briefly consider the course of Devensian late-glacial and Flandrian vegetation change in relation to the present British flora. Godwin calculated that about half of the British flora was present in the British Isles in the Devensian.[67] We know most about the late-glacial and at this time the flora, as we have seen, was a very varied one, mainly herbaceous, in which occurred many species now known as weeds and ruderals. During the Flandrian, forest was established, open habitats became rare, and there was an expansion in range of the woodland ground flora and thermophilous species. In zone VII, the extension of mires, soil changes towards podsolisation, and cultivation effects reduced forest.[52] In upland areas 'montane' species (e.g. *Armeria*, *Juniperus*) disappeared as a result of peat growth.[165] Many late-glacial relics survived these changes in steep and hilly areas, particularly where the rocks were calcareous, where trees found difficulty in growing, where grazing was an unimportant factor and where mire growth could not take place. The Flandrian vegetational changes produced a greater extinction of late-glacial species in Ireland (e.g. *Betula nana*) than in Britain, perhaps because such environments were much rarer, slopes gentler, and mires more widespread.

In Neolithic and post-Neolithic times, zone VIIb and onwards, there developed much diversity of vegetation. Deforestation was accompanied by a rapid spread of weeds and ruderals, a diminution of the woodland flora and development of scrub, and widespread cultivation resulted in the introduction of both alien crop plants and alien weeds. Peat erosion started in the Pennines at the zone VIIa/b boundary, and continued into historic times.[166] Temporary and extensive clearances have been distinguished by respectively low and high pollen frequencies of grasses. Pastoral activities have been

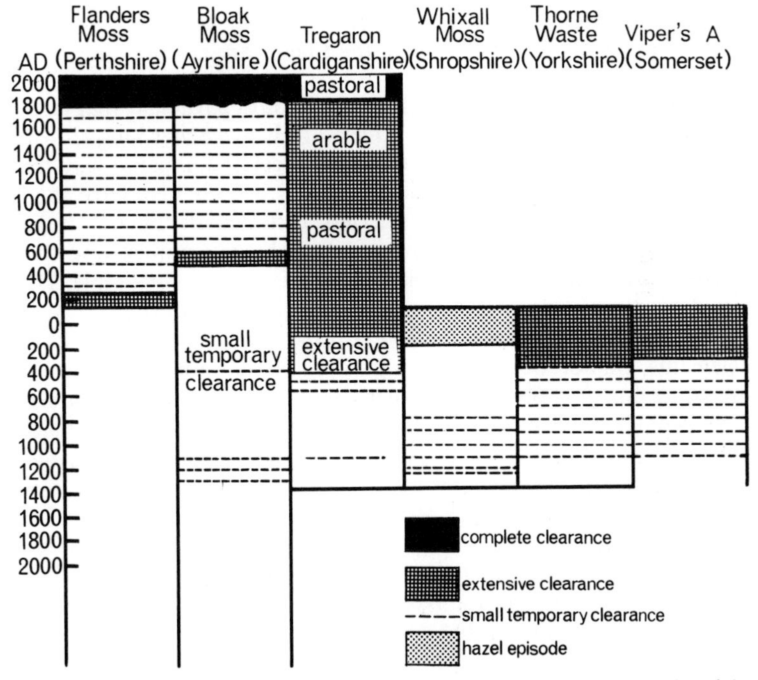

Fig. 13.22 Clearance types at six bog sites. The sequences are dated by radiocarbon (J. Turner, 1965).

distinguished from arable, the former showing high frequencies of *Plantago* and *Pteridium* pollen, and the latter with marked frequencies of *Artemisia*, Compositae, *Rumex*, Chenopodiaceae and Ranunculaceae pollen.[169] Figure 13.22 shows the distribution of these agricultural activities, recognised by pollen frequency changes, in respect of six sites in Britain.

The course of floristic change in the late-glacial and Flandrian is succinctly portrayed by changes in the representation of different phytogeographical elements in the different regions of the British Isles.[17] Such an analysis of the phytogeographical elements shows that in the late-glacial continental and northern elements were important, with little regional differentiation. By the middle Flandrian (zones VI and VIIa) the northern elements had greatly decreased especially in the south, where the atlantic and southern elements show a concomitant increase, so that by this time a north–south gradient of distribution was developing. This gradient increased in the late Flandrian. These changes are probably primarily associated with major changes in the weather systems over the British Isles, with cold, continental anticyclonic conditions in the Devensian replaced by atlantic cyclonic conditions in the Flandrian.

The Irish flora. The history of the Irish flora, in particular that of the so-called Lusitanian and North American elements, has long aroused interest. The flora is small (around 1260 species) compared with that of Britain (around 2200 species), and the area is of course much smaller than Britain, and has a less continental climate. It has already been pointed out that many British species must owe their absence in Ireland to the barrier of the Irish sea, for example, *Tilia cordata*, *Fagus sylvatica*, *Carpinus betulus*, *Paris quadrifolia*.

The species in the Lusitanian (L) and North American (N.A.) elements cannot be treated in groups. Rather they each have a separate history. Some frost-tolerant species, e.g. *Eriocaulon septangulare* (N.A.), *Erica mackaiana* (L), may have survived in Ireland or in the northwest of the British Isles during the glacials. *Eriocaulon septangulare* has been found fossil in the Hoxnian in Ireland. Both the fossil and living Irish species appear similar to each other, but distinct from the North American species, which suggests a disjunction of a former wider circumboreal distribution and slight speciation. Other species, frost-sensitive, e.g. *Arbutus unedo* (L), probably immigrated in the early Flandrian, while there is a possibility that some whose distribution is now seen to be expanding, may be introductions, e.g. *Juncus tenuis* (N.A.). There are a few Irish calcicolous species, of wide continental distribution, but not in

Britain, e.g. *Inula salicina* and *Euphrasia salisburgensis*, and here the peculiar distribution type may result from extinction in Britain as a result of the later Flandrian environmental changes and destruction of habitats.

Conclusions. This broad survey indicates how much information we have on the history of the British flora, especially in the Flandrian, but, keeping a geological perspective, we must remember that the forested episodes of the Pleistocene appear to cover a small proportion of Pleistocene time, and that the 'full-glacial' type of flora must be the normal type of British flora from the time point of view.

Even with the huge amount of data we have on the history of British flora, there are still very many gaps in our knowledge. There is yet no synthesis possible of the history of particular plant communities in the Devensian or Flandrian, of the history of particular vegetation types on particular soil types, or much knowledge of how the plant lists can be segregated into their component communities. Neither are we able to carry interpretations of pollen diagrams far enough because of absence of data on recent pollen rain originating from vegetation types similar to those we believe were present in the Pleistocene. Rates of immigration and expansion of particular genera are also largely unknown, but here refined use of radiocarbon dating may give a lot of information. Because of our detailed knowledge which has already accumulated, and the comparatively small size of the British Isles and its isolated position in relation to the continent, there is every promise that progress can be made on these problems.

HISTORY OF THE FAUNA

Forbes's essay on the history of the British flora and fauna has already been mentioned; his interpretation of the history of the faunal migrations was identical to that of the flora migrations already described. Scharff[133] continued along the lines mapped out by Forbes. As a contrast, Beirne's study of the history of the British fauna, dealing with the living fauna, is largely based on ecological, taxonomic and zoogeographical evidence.[13] Whereas we have a very rich record of plant fossils in the Pleistocene, which gives a substantial basis for reconstructing the history of the flora, it is not yet possible to reconstruct the faunal history in the same detail; the difficulties have been that many of the old records of fossils require re-study and placing in a stratigraphical framework. Nevertheless, recent work has greatly improved our knowledge of the history of the fauna, especially of molluscs, insects and vertebrates.

Foraminifera, Crustacea, Bryozoa

The most extensive records of Foraminifera are from the East Anglian Crag. A study of foraminiferal populations from boreholes and sections in the Crag has given indications of climatic and habitat change during Crag deposition and of the time of disappearance of certain Pliocene-relic forms.[60] In general two faunas are found; a Red Crag type with Pliocene-relic genera (*Pararotalia*, and the Lituolid genera *Dorothia* and *Gaudryina*), and a much poorer Icenian fauna. In the Red Crag of the Neutral Farm pit, Butley, there is some evidence for cold conditions in the presence of *Elphidiella orbiculare* and there is a similar fauna in the Pre-Ludhamian horizon of the Stradbroke borehole (see p. 283). The Red Crag type of fauna found in the Crag (Ludham Crag) at the base of the Ludham borehole has warmer indications, based on a rich fauna including genera of the Anomalininae, Lituolidae, Rotaliidae, and Elphidiidae.

The Icenian Crags, represented at outcrop and above the Ludham Crag in the Ludham borehole, have a much more restricted fauna, in the same way that the Icenian mollusc fauna is also restricted compared with the older faunas, and the tests of the Foraminifera are generally thinner than those of the older Crags. Pliocene-relic forms are no longer found, *Elphidiella* is abundant, the Anomalininae are only represented by *Cibicides lobatulus*, and '*Rotalia*' *beccarii* is common at certain levels. The frequency of these three genera varies much in the Icenian Crag and it appears that in the warmer episodes *Cibicides* and *Rotalia* are abundant, with *Elphidiella* less conspicuous, while in the cold episodes *Elphidiella* is most abundant and the other two genera are much reduced in frequency. The species of *Elphidiella* are cold-tolerant, *E. orbiculare* and *E. cf. bartletti* in particular indicating cold conditions. The colder faunas are associated with the Thurnian, Baventian (both in the Ludham borehole), and are found in the Weybourne Crag of Sidestrand (Baventian?). The warmer faunas occur in the Antian of the Ludham borehole and in the Norwich Crag of Bramerton and the Chillesford Clay.

Foraminifera have also been recorded from the marine Cromerian (*Leda-myalis* bed), and also from the Corton Sands, where they are believed to be largely derived from earlier formations.[90, 177] Faunas have been recorded from the Hoxnian,[8] Ipswichian[5] and Flandrian,[91] but no detailed studies have been made of faunal successions in the Middle and Late Pleistocene, except in the Flandrian.[88]

Few studies of Pleistocene Ostracoda have been made, though these fossils are widespread in both the cold and temperate stages. The faunas appear particularly useful for determining salinity conditions of water bodies, but there is not yet enough information on distribution in time to say whether the faunas will be useful for chronological purposes. A spinose species of *Ilyocypris* may be characteristic of the Middle Pleistocene.[163] Interglacial marine and estuarine faunas have been described from Selsey[184] and faunas of Middle Devensian age have been described from Upton Warren and Fladbury (Worcs.) and from Isleworth (Middlesex).[138] The Upton Warren fauna includes a brackish species, with oligohaline conditions resulting from permafrost or local brine springs. Otherwise the faunas are freshwater, indicating standing water or slowly-flowing conditions. A freshwater water notostracan, *Lepidurus arcticus*, with a circumpolar distribution in northern latitudes, has been found in Devensian and Middle Pleistocene deposits in association with floras of late- or full-glacial type. Studies of Cladocera in Late Devensian and Flandrian lake sediments in the Lake District have shown that these animals undergo frequency changes associated with climatic history and limnological conditions.[69] Possible crustacean and arthropod trails have been found as trace fossils in proglacial lake sediments of Anglian age.[63]

The Bryozoa of the Crag have been discussed in relation to the Plio-Pleistocene Bryozoa of the Netherlands, but no detailed evidence of faunal changes is available.[86]

Table 13.7 Localities with fossil Foraminifera, Crustacea, Bryozoa

	Icenian Crag											Locality	Reference
Red Crag and Ludhamian	*Thurnian*	*Antian*	*Baventian*	*Pastonian*	*Beestonian*	*Cromerian*	*Anglian*	*Hoxnian*	*Ipswichian*	*Devensian*	*Flandrian*		F, Foraminifera C, Cladocera O, Ostracoda B, Bryozoa
+	+	+	+	+	+							Norfolk and Suffolk	F. Funnell (1961), Beck *et al.* (1972)
+												Suffolk	B, Lagaaij (1952)
						+	+					East Anglia	F, Macfadyen (1932)
								+				Clacton-on-Sea, Essex	F, Baden-Powell (1955)
							+	+				Trysull, Staffs	O, Morgan (1973)
									+			Selsey, Sussex	O, Whatley and Kaye (1971)
									+			Fenland	F, Baden-Powell (1934)
										+		Fladbury, Worcs	O, Siddiqui (1971)
										+		Isleworth, Mdx	O, Siddiqui (1971)
										+		Upton Warren, Worcs	O, Coope, Shotton and Strachan (1961)
											+	Fenland	F, Macfadyen (1933)
										+	+	Lake District	C, Goulden (1964)
											+	Start Bay, Devon	F, Lees (1975)
										+	+	Eastern Scotland	O, F, Peacock (1974)

Insects[37, 136]

The study of insect faunas has been one of the most spectacular recent developments in Pleistocene research. Many faunas of different age have been studied (table 13.8) in relation to stratigraphy, and conclusions have been drawn about climatic conditions and the relation of the faunas to the contemporary vegetation. Of the insect skeletal fragments found fossil, most are coleopteran, both aquatic and terrestrial species, but larval heads and jaws of Diptera (Chironomids in particular) and Trichoptera are also sometimes common. Most remains are identifiable to living species, at least in the Late Pleistocene faunas which have been most frequently studied, and some even in Cromerian faunas. There is no palaeontological evidence so far for morphological evolution of insects in the Middle and Late Pleistocene.[37]

Distributions of living species are known quite well, especially in Scandinavia, so it has proved possible to interpret clmatic and vegetational conditions on the basis of the constitution of insect faunas. In the Devensian, it often appears that both northern European and southern European species of beetle are present in the same fauna, a parallel to the situation found with floras, and probably related to the effects of shifting climates and to the probability that Devensian environmental conditions cannot today be paralleled. In fact, when the Eurasian distributions are considered, rather than the European distributions, the ranges are found to be compatible in some areas, implying both that climates were other than types found now in western Europe and that vast changes in range have occurred in the Pleistocene. As an example of the latter we may take a species of *Aphodius* (dung beetle) found in the Devensian but now of Tibetan distribution.[38] Many species identified are not British today, and these exotic species occur both in cold and temperate stages; in the former, northern European species occur, in the latter, central or southern European species. Thus it is now possible to envisage changing insect populations related to the Pleistocene climatic changes in the same way, and with similar biogeographical significance, as with the plant record. The relation of the two is of course close, because of the food relations existing between the two groups.

We can now briefly mention some of the sites with insect faunas which have been studied.

Hoxnian. A very good sequence of insect remains, mainly beetles, has been isolated from the interglacial deposit at Nechells, Birmingham, covering Anglian late-glacial and much of Hoxnian times.[137] The fauna of about seventy species and many more genera is essentially a British one with only three non-British species named, two of which have disjunct distributions in southern Europe (a weevil and a species of *Platypus*). Other weevil species were abundant enough to allow their frequency to be correlated with the pollen frequency of their food plant, e.g. *Rhynchaenus quercus* and *Quercus*. The fauna indicates that at the optimum of the interglacial, climatic conditions were not greatly different from the present climate in the same region.

Ipswichian. Insect faunas from Ipswichian sites have been studied.[36, 39, 114] Several species have a distinctly southern or continental distribution in Europe, some exotic to the present British fauna, e.g. *Oodes gracilis*, a ground beetle of reedswamp habitat; *Onthophagus opacicollis*, a dung beetle of Mediterranean distribution. The Ipswichian assemblage at Bobbitshole (zones Ip I and II) indicates July temperatures of about 20°C, about 3° above that of southern England now, a warming tallying with the climatic conclusions deduced from vegetational studies.

Devensian. Many Devensian faunas have been studied, and a range of assemblages, many of them radiocarbon-dated, have been analysed.[40,42] They include forest assemblages of the Early Devensian Chelford Interstadial to restricted faunas associated with very cold times. A climatic curve for the Devensian, based on insect faunas, is shown in fig. 12.12. A small Early Devensian fauna from terrace deposits at Cambridge, associated with herb vegetation, gives no clear climatic indication towards warmth or cold, but contains halophilous species, as does the flora. At Wretton, Norfolk, a fauna of similar age indicates cool conditions.

The interstadial organic deposit at Chelford, Cheshire, contains a large fauna of about a hundred species and genera.[34] A distinct thermophilous element is absent, and there is a collection of species of a more northern type including three exotic species now found in Scandinavia and the mountains of central Europe. Many species are associated nowadays with coniferous or birch forest in south Finland (fig. 13.23), e.g. the phytophagous pine weevil *Hylobius abietis*, and these indications of the interstadial vegetation compare closely with the flora list from the same site. The climate prevalent at the time has been compared with the cool continental climate of Finland between latitude 60° to 64° north.

Of the Middle Devensian faunas, we may take the fauna of Upton Warren, Worcs. (41,900 years B.P.), type site for the Upton Warren interstadial complex, and the cold faunas of tundra type, such as that from Fladbury, Worcs.[35] This latter fauna indicates open almost

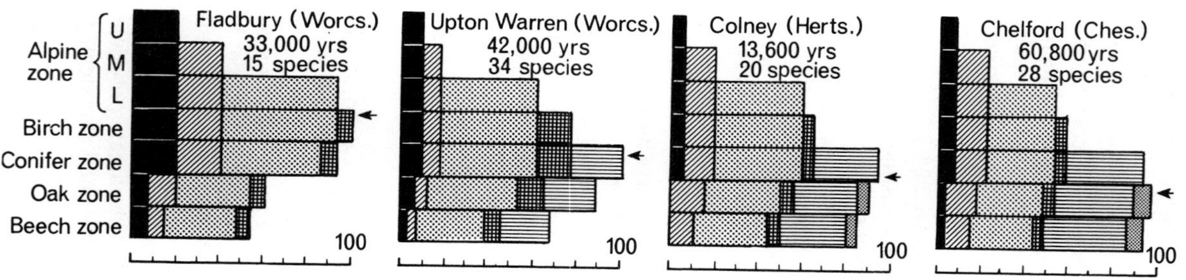

Fig. 13.23 The distribution of carabid beetles from four Devensian sites in England. The histograms show the percentage of the list of fossil carabids occurring in each of the present vegetational zones of Fennoscandia. The arrows mark the inferred vegetation equivalent at the fossil site. Chelford is seen to have the least severe environment (interstadial) and Fladbury the most severe "full-glacial" (Shotton, 1962).

treeless vegetation in the region, and contains about fifty species. Of these, eleven are exotic to the present British fauna, e.g. the beetles *Diachila artica*, *Amara torrida*, and eight in Britain are only found in the north. All except one are found today north of the Arctic circle. Figure 13.23 shows how the carabid species from Fladbury are at present distributed in the vegetation zones of Fennoscandia, and from this a climate with July mean temperature of about 11°C, 5·5°C lower than present, has been suggested. A similar comparison is made in fig. 13.23 between the Upton Warren fauna (^{14}C age 41,900 B.P.) and the vegetation zones of Fennoscandia indicating slightly more temperate conditions than the Fladbury fauna. This fauna is very large, about 190 species and genera.[43] Seventy-eight per cent of the species are living in the area today, 9 per cent are restricted to north Britain or the mountains of Wales and 13 per cent are exotic to the present British fauna and mostly occur in northern Europe. Other species are more southern and there is a small continental element. Here we have an example of the very different types of present geographical distribution present in one fauna. Another interesting point is that although the carabid species are mostly common in the conifer zone of Fennoscandia, few remains of trees were found at the site. It is not clear whether these properties of the fauna (and flora) result from rapid amelioration of climate and resulting overlapping of distributions or from other climatic factors, or from biotic factors, such as grazing by bison reducing tree cover. Possibly continentality of climate was the cause. A temperate fauna associated with a tree-less environment found at Isleworth, Middlesex, has also been dated to the Upton Warren interstadial (43,140 years B.P.).[40b]

Another fauna, of Late Devensian age, comes from Colney Heath, Herts (^{14}C age 13,560 B.P.);[113] the Fennoscandian distribution of the carabid element of this fauna is shown in fig. 13.23. Forty-seven species were recorded, including both northern and southern species, but again the flora from this site shows the vegetation to have been open, though the Fennoscandian correlation is with the conifer zone. Distributions mapped in fig. 13.24 of three Devensian late-glacial species from Corstorphine, near Edinburgh, show how ranges have changed since late-glacial time. It has been suggested that late-glacial climatic indications of cold continental type permitted the existence of *Chlaenius*, and of *Diachila* where the habitat was more exposed, but the western species of *Barynotus* was nocturnal and avoided the summer's day heat of a continental climate.[36]

The late-glacial interstadial is clearly shown by changing beetle faunas.[41] In a succession from north Wales, there is an abrupt change

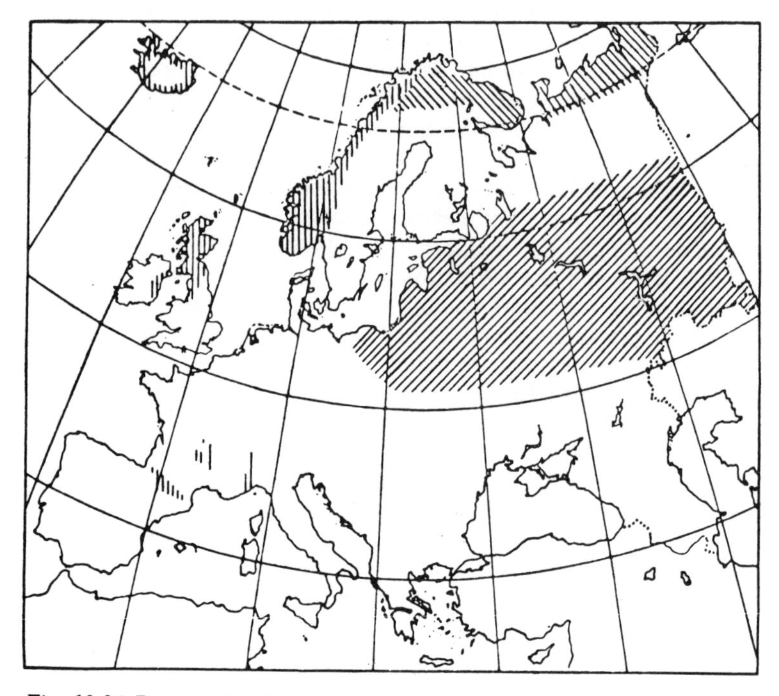

Fig. 13.24 Present distributions of three beetle species in Devensian late-glacial deposits at Corstorphine, Edinburgh (Coope, 1965). *Barynotus squamosus* |||, *Diachila arctica* \\\, *Chlaenius* costulatus ///.

from a fauna indicating an intensely cold continental climate to a climate at least as warm in the summers as at the present day at 13,000 years B.P. After 12,000 years B.P. a deterioration set in, leading to the cold phase associated with the Loch Lomond Re-advance. The beetle-interstadial thus covers part of the late-glacial zone I as well as zone II, and from the non-correspondence with the pollen-interstadial (zone II) it may be concluded that the beetle faunas are more sensitive indicators of climatic change than the floras.

Flandrian. Early Flandrian faunas show a rapid climatic change at the close of the late-glacial, with thermophilous assemblages rapidly establishing themselves.[110] A study of insect faunas from two horizons of alluvium at Shustoke, Warwickshire,[78] one near the zone VIIa/b boundary (C^{14} age 4830 B.P.), the other about 400 years old, showed the predominance of a fauna characteristic of deciduous woodland at the first level, and a great increase in the latter fauna of insects living in open habitats. Weeds of cultivated ground are the food plants of some of the phytophagous beetles of this latter horizon. In the same way that some species of the full- and late-glacial flora again become widespread on deforestation in the Flandrian, so it appears that some beetle species (e.g. the carabid beetle *Agonum mulleri*) which are found in full- or late-glacial deposits are now associated with man's clearance activities.[114] Assemblages associated with settlement sites of Bronze and Roman age include present day pest species of wood-borers and of stored food infestations.[109]

Table 13.8 Localities with fossil insect faunas

Hoxnian	*Ipswichian*	*Early and Middle Devensian*	*Devensian late-glacial*	*Flandrian*	*Locality*	*Reference*
+					Nechells, Birmingham	Shotton and Osborne (1965)
	+				Selsey, Sussex	Pearson (1963)
	+				Trafalgar Sq., London	Coope (1965)
	+				Bobbitshole, Ipswich	Coope (1974)
		+	+		General	Coope *et al.* (1971), Coope (1977)
		+			Wretton, Norfolk	West *et al.* (1974)
		+			Four Ashes, Staffs	Anne Morgan (1973)
		+			Sidgwick Ave., Cambridge	Lambert, Pearson and Sparks (1963)
		+			Chelford, Cheshire	Coope (1959)
		+			Upton Warren, Worcs	Coope, Shotton and Strachan (1961)
		+			Fladbury, Worcs	Coope (1962)
			+		Colney Heath, Herts	Pearson (1962)
			+		Glanllynnau, Wales	Coope and Brophy (1972)
				+	General	Osborne (1965)
				+	Shustoke, Warwickshire	Kelly and Osborne (1964)
				+	Lea Marston, Warwicks	Osborne (1974)

Non-marine molluscs

These have been found very commonly in Pleistocene deposits in the British Isles. Much early descriptive work was done in the period up to 1950;[79] since that time more detailed quantitative studies have been made, especially in south and east England, and these have led to important conclusions regarding the environmental significance of the faunas.[55, 149]

To judge by numbers of extinct species, it appears that assemblages have evolved in the Pleistocene. In the Cromer Forest Bed Series about 21 per cent of the species recorded appear to be extinct, in the Hoxnian about 17 per cent, Ipswichian about 15 per cent, and Early Flandrian about 5 per cent.[152] But these figures are taken from total faunal lists for each temperate stage, and there are great variations from site to site, so that the figures are less significant than they appear. A few species appear to be confined to particular stages, e.g. *Nematurella runtoniana* to the Cromerian, and *Valvata antiqua* and *Viviparus diluvianus* to the Hoxnian, but such conclusions are based on studies of few sites. In the Flandrian, presence of certain species indiciates Roman and later times, e.g. *Helix pomatia* and *H. aspersa*, while an abundance of others e.g. *Lauria anglica*, *Vertigo angustior*, indicates an early Flandrian age.

Assemblages have been found to be indicative of particular types of climate. Cold faunas show a dominance of a few species in great numbers, temperate faunas a larger number of species, many of which occur abundantly. The cold 'loess-type' fauna contains abundant *Pupilla muscorum*, and also *Columella columella* (plate 16a), *Vertigo parcedentata*, *Pisidium obtusale lapponicum* and *P. vincentianum*. Species characteristic of temperate stages include *Corbicula fluminalis*, now in the Nile area and Asia west of India, and *Belgrandia marginata* (plate 16b), a southern European species, and also several extinct species, e.g. *Theodoxus serratiliniformis* and *Helicella crayfordensis*.

A list of localities with fossil faunas is shown in table 13.9. Faunas older than the Anglian have not been systematically studied recently, except for a brief account of a Cromerian fauna from the Upper Freshwater Bed at West Runton. The Anglian late-glacial fauna at Hoxne, nearly all fresh water species, is a nondescript one with no positive indications of climate.[147] The Hoxnian fauna at Swanscombe is a larger one, of nearly a hundred species and genera;[81] here a number of distinct assemblages occur. They indicate a fluviatile environment with changing flow régimes, with terrestrial components indicating fen and marsh, woodland and dry and damp grassland. The faunas from the Lower Gravels and the Lower Loam and from the base of the Lower Middle Gravel are temperate and have been correlated with the middle and late Hoxnian, while the fauna from the Upper Middle Gravel is impoverished and indicative of a cold climate; it may be Early Wolstonian. The freshwater fauna changes between the Lower and Middle Gravels with the arrival of species described as 'Rhenish', including *Corbicula fluminalis*, *Valvata antiqua* and *Belgrandia marginata*.

A fauna of Wolstonian age has been found in tjaele gravels at Thriplow, Cambs.[148] It is very similar to the loess-type fauna, with a combination of dry and wet-loving species and abundant *Pupilla muscorum*. *Columella columella*, an indicator of cold conditions, is also present. This type of fauna is also common in Devensian terrace deposits, e.g. at the well-known sites of Barnwell Station, Cambridge and Ponders End, Herts.

Many Ipswichian faunas have been described in detail.[151] Over 120 species are known from this temperate stage. There is a steep rise in the number of species in zone Ip I, a maximum in zone Ip II, and a slow decline after this (fig. 7.2). Many southern species occur in zones Ip II and Ip III, e.g. *Corbicula fluminalis*, *Belgrandia marginata*, *Potamida littoralis* (plate 16c); their entry and expansion in number have been correlated with vegetational changes. On the whole the land molluscs first occur at a later time than the freshwater species, probably because dispersal of freshwater species is more easily accomplished by natural agents. Many of the southern species appear to survive well into the later part of the stage and even into the early part of the Devensian.[179] The Ipswichian fauna as a whole is not fundamentally different from the Flandrian, but it includes a number of species of central and southeast Europe (*Ena montana*, *Helicella striata*), a southern group (*Belgrandia marginata*, *Corbicula fluminalis*) and a southwest European group (*Potamida littoralis*). On the other hand, the Flandrian contains some central and south European species, but more species with oceanic, Mediterranean-oceanic, central and southwest European distribution types. Thus there is an indication, as with the plants, of a more continental type of climate in the Ipswichian than in the Flandrian.

The Devensian faunas include those from terrace deposits and those from chalky sludge deposits in southeast England. The former are usually of a tolerant type, with no clear climatic indications, or of the 'loess-type' already mentioned, such as those of the Flood Plain terrace at Ponder's End in the Lea valley. Freshwater faunas of Middle Devensian age at Upton Warren, Worcs, have *Pisidium vincentianum* and *P. obtusale lapponicum*, indicating cold conditions.[43] At this site there is some suggestion of water salinity dwarfing *Lymnaea peregra* and *Planorbis laevis*. A Middle Devensian loess-type faunal sequence in brickearth at Halling, Kent, shows evidence for an amelioration of climate by a temporary increase in the absolute numbers of molluscs and a rise in the percentage frequency of the most thermally demanding species of the assemblage, *Hygromia hispida*.[82]

Devensian late-glacial faunas, predominantly of land molluscs, are well known from Kent.[80] Twenty-one species have been recorded. The affinities of the fauna are alpine rather than arctic. Several geographical elements are present; a very tolerant group, two arctic-alpine species (*Columella columella* and *Vertigo genesii*), southern Scandinavia and European species, west European species not in the Scandinavian mainland (*Abide secale*), and a species now known only in the Scandinavian mountains and from maritime habitats in north-west Europe (*Catinella arenaria*). Here we have similar variance of distribution types as with plants and insects, and presence of species whose distribution scarcely overlaps today, e.g. *Columella columella* and *Hygromia hispida*. A similar mixture of xerophilous and hygrophilous species is seen in the earlier Devensian faunas. It has been suggested that with colder average temperatures, too rapid evaporation of ground moisture is prevented, and this allows the growth of species which would normally avoid such places. In zone I of the late-glacial the fauna is mainly of hardy Holarctic species, dry land species of open habitats; in zone II (the late-glacial interstadial) the assemblages increase in variety, indicating diversification of habitats, and thermophilous species spread. In zone III, the same fauna survives in spite of the climatic deterioration manifested by frost-heaving and frost-shattering of the Chalk. Some evidence, e.g. the frequency of hygrophilous species, points to an increased dampness in this zone compared with zone I, and perhaps the temperatures did not return to zone I severity.

The early Flandrian faunas show the continuance of the open

Table 13.9 Localities with fossil non-marine mollusc faunas

Cromerian	*Anglian*	*Hoxnian*	*Wolstonian*	*Ipswichian*	*Early and Middle Devensian*	*Devensian late-glacial*	*Flandrian*	*Locality*	*Reference*
+								W. Runton, Norfolk	Sparks (1963)
	+	+						Hoxne, Suffolk	Sparks (1956)
		+	+					Swanscombe, Kent	Kerney (1971a)
			+					Thriplow, Cambs	Sparks (1957)
				+	+			Wretton, Norfolk	Sparks and West (1970), West *et al.* (1974)
				+	+			Summary of many sites	Sparks (1964)
					+			Upton Warren, Worcs	Coope, Shotton and Strachan (1961)
					+	+	+	Kent	Kerney (1963), Kerney (1971b)
							+	Apethorpe, Northants	Sparks and Lambert (1961)
							+	General	Sparks (1962), Evans (1972)
						+	+	Co. Down	Stelfox *et al.* (1972)

grassland element, and a number of common European species appear at this time (*Helix nemoralis*, *Discus rotundatus*). In later Flandrian times *Pomatias elegans* becomes abundant on the Chalk and is characteristic of the warm part of the Flandrian. Many species appear about Romano-British times and later, e.g. *Helicella gigaxii*, *Monacha cantiana*. The effects of cultivation on the mollusc habitats must have diversified them and is presumably a major cause of the late spread of many species, especially the xerophilous land snails. The analysis of snail assemblages found in association with settlement sites is a major source of information on environmental history at such sites.[56]

Marine molluscs

There are abundant records of marine molluscs throughout the Pleistocene (table 13.10), in particular from the marine Early Pleistocene of East Anglia, but also from marine transgression deposits of the temperate stages and from sands closely associated with till sequences. There has been much discussion with faunas of the last type over the question of whether the faunas are *in situ* or derived. There was much early work on fossil marine molluscs, especially the systematics, in the nineteenth[187] and early twentieth centuries.[70] A

Table 13.10 Localities with fossil marine molluscs

	Icenian Crag										
Red Crag and Ludhamian	*Antian*	*Pastonian*	*Beestonian*	*Cromerian*	*Anglian*	*Hoxnian*	*Ipswichian*	*Devensian*	*Flandrian*	*Locality*	*Reference*
+	+	+	+	+						General	Reid (1890); Wood (1848–1882); Harmer (1914–25)
+	+	+								Norfolk	Norton (1967) Norton and Spaink (1973)
	+									Easton Bavents, Suffolk	Funnell and West (1962), Norton and Beck (1972)
	+									Aldeby, Norfolk	Norton and Beck (1972)
		+								Suffolk	West and Norton (1974)
					+					Corton, Suffolk	Baden-Powell (1950)
						+				Clacton-on-Sea, Essex	Baden-Powell (1955)
							+			Fenland	Baden-Powell (1934)
								+		Cheshire Plain	Thompson and Worsley (1966)
								+	+	Scotland	Jamieson (1865), Peacock (1974)
								+	+	Northeast Ireland	Praeger (1896)
									+	General	Baden-Powell (1927)
									+	Northwest Scotland	Baden-Powell (1937)

more quantiative and ecological approach has more recently been applied to the faunas.[105]

Three properties of the changing mollusc faunas of the Pleistocene have been studied, times of extinction of species, times of arrival of new species, and the changes in climate suggested by percentages of northern and southern forms. By the time of the Middle Pleistocene interglacials there are only a few species which have since become extinct e.g. *Macoma obliqua* in the Ipswichian fauna. Thus a major time of extinction was after Icenian Crag times, and before the Hoxnian temperate stage. Several species have been recognised as first appearing at particular times, e.g. *Chlamys varia* in the Corton Sands, *Tellina tenuis* in the Ipswichian. Table 13.11 shows some of the trends which have been revealed by analysis of gasteropod faunas of Early and Middle Pleistocene age.

Several hundred species have been recorded from the Crags. The sequence of Crags determined by Harmer has already been referred to in Chapter 12. As with the Foraminifera, it appears that the Pleistocene Crag sequence is conveniently divisible into Red Crag, older or partly coeval with the Ludhamian, and the Icenian Crag, containing faunas of ages varying between the Antian and Cromerian. The Red Crag fauna is a rich one of several hundred species, with a marked percentage of northern species, the Crag itself being a shallow water shore deposit of landlocked bays or inlets. In contrast, the Icenian faunas appear to have lived in a more open and shallow sea, and are much impoverished, with about 150 species, only 40 of which are common. The shells are commonly thinner and more fragile than those of the Red Crag, perhaps an indication of decreased salinity, and the assemblages are developed in various facies, including infralittoral, wadden and open coast facies.[181] Characteristic species include *Littorina littorea*, *Rissoa semicostata*, *Leda oblongoides*, *Cardium edule*, *Astarte borealis*, *Scrobicularia piperata*, and *Mactra subtruncata*. The Icenian type of fauna is found at Easton Bavents[61] in the Antian stage, and at Bramerton and Chillesford in later temperate parts of the Early Pleistocene and lower Middle Pleistocene. The fauna with *Leda myalis*, *Mya truncata* and *Astarte borealis*, at West Runton, Norfolk, belongs to the marine transgression deposits of the Cromerian stage. Where the climatic conditions are more severe, e.g. in the basal stone bed at Sheringham, Norfolk, the Icenian fauna is much reduced and a prominent member is *Macoma balthica*, which characterises the so-called Weybourne Crag, once thought to be a specific horizon in Crag stratigraphy, but more probably of wider age limits.

A rich fauna from the Corton Sands in the Lowestoft area has been the subject of much controversy over whether it is wholly, partly or not derived.[177] The subject is still an open one, but the fauna is interesting in being intermediate in type between that of the Crag and that of the present day, with a small number of extinct species, some cold indicators and some warm indicators.

A fauna from the Hoxnian deposits at Clacton-on-Sea, Essex, is of a modern intertidal type,[8] and those from gravels thought to be associated with marine transgression of the Ipswichian, e.g. at Selsey, Sussex and the March Gravels of Fenland, are again modern in aspect.[5] Faunas also occur in Devensian sands and gravels, as in the Cheshire Plain[167] and east Linconshire, but these shells are most likely to be derived from Ipswichian interglacial marine sediments or even perhaps partly from interstadial marine sediments, as suggested by a radiocarbon date of 28,000 B.P. from two shells of *Nucella lapillus* from fluvioglacial sands in the Cheshire Plain. Devensian late-glacial marine clays of the Clyde area and eastern Scotland show an early restricted fauna of arctic type (*Portlandia arctica, Pecten groenlandica, Astarte borealis*) in the Errol Beds, and a later more diverse fauna in the interstadial Clyde Beds.[112]

Flandrian faunas are of modern type, but there is evidence that distributions have changed during this time; in particular, that ranges were extended during the warmer part of the Flandrian. For example, in northeast Ireland several species absent in the existing fauna are present in the mid-Flandrian estuarine clays and raised beaches, including *Tapes decussatus, Thracia pubescens* and *Rissoa albella*. In fact this difference between modern and mid-Flandrian faunas in northeast Ireland was the basis for the postulation of the climatic optimum by R. L. Praeger and it is graphically represented in fig. 13.25.[121]

Table 13.11 Distribution of marine Gastropoda in the Crags (Boswell, 1952)

	Northern %	*Southern* %	*Not known living* %	*Total no. of species*
Icenian	26	0	28	98
Butleyan	21	2	40	163
Red Crag Newbournian	12	7	53	226
Waltonian	16	11	48	352
Coralline Crag (Pliocene)	4	12	57	303

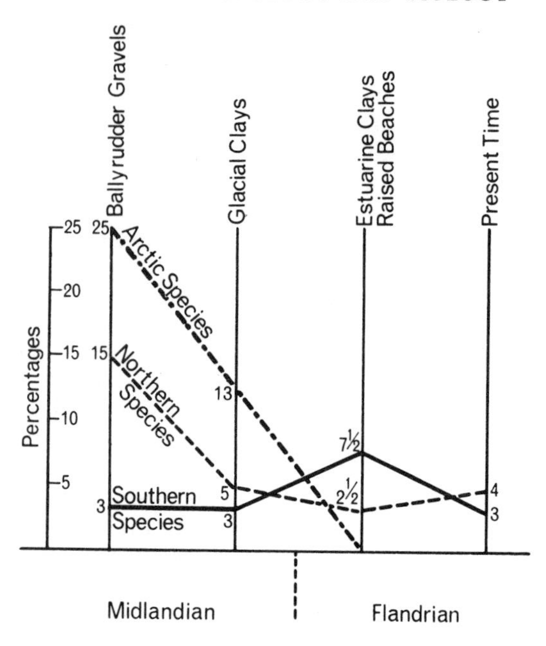

Fig. 13.25 Changes in the percentage composition of marine mollusc faunas in northeast Ireland, showing the evidence for the climatic optimum in raised beach ('Neolithic') times (after Praeger, 1896).

Vertebrates[85, 158]

Although there is a large body of evidence concerning the Pleistocene vertebrate fauna, rarely are fossils found abundantly enough or in sufficiently continuous stratigraphical series to make possible detailed reconstruction of faunal history. Many of the past finds are difficult to relate to the geological sequence and many genera (*Elephas* in particular, *Equus*, and the voles) are in need of revision.

There are records of fishes, reptiles, amphibians and birds throughout the Pleistocene. Several species of bird have been discovered in Devensian and Flandrian cave deposits and from Flandrian peats.[23] But most attention has been paid to the mammal faunas, particularly the land mammals. British Pleistocene mammals have been monographed in considerable detail,[1, 47, 48, 127] but in many instances the finds have been recorded with insufficient stratigraphical data; consequently their provenance, age and ecological relations are in doubt. An example is the long list of species from the Cromer Forest Bed Series.[104] This sequence covers a number of climatic changes, and the stratigraphical context of each find is not often recorded. The process of faunal change and the nature of the faunal assemblages are therefore not as clear as they might be. A recent development has been the careful correlation of vertebrate records with vegetational history, resulting in better reconstructions of vertebrate faunal history.

While Pleistocene taxa of Pisces, Reptilia, Amphibia and Aves have been referred to living species, mammals exhibit a considerable degree of evolution and extinction, together with change in distribution patterns. This has led to some species or genera having a limited range within the Pleistocene, giving a basis for the use of faunal stage names in Pleistocene stratigraphy; for example, the Villafranchian faunal stage of the Early Pleistocene, which contains a number of species common to the late Tertiary.[84] In British faunal stratigraphy such stages have not been defined, except that the pre-Cromerian faunas have a Villafranchian aspect. Superimposed on the gradual faunal change resulting from evolution and extinction there are the changes related to climatic and vegetational changes. Thus in the Middle and Upper Pleistocene occur faunas characteristic of the interglacials (*Elephas* (*Palaeoloxodon*) *antiquus*, *Rhinoceros* (*Dicerorhinus*) *kirchbergensis* (*merckii*), *Hippopotamus*), and of the glacials (*Elephas* (*Mammuthus*) *primigenius*, *Rhinoceros tichorhinus* (*Tichorhinus antiquitatis*), *Rangifer tarandus*).

Of the temperate species, some are thought to be characteristic of woodland, others of steppe, or more open habitats,[158] and there is some support for this in their association with plant remains indicating such environmental types.

Table 13.12 lists mammal faunas from various Pleistocene stages, and it is possible to outline a history of the mammal fauna from this evidence. A much more comprehensive account has been given by Stuart.[158]

The pre-Cromerian (including Crag) faunas[155] are characterised by a Villafranchian element, more strongly represented in the Red and Icenian Crags, including *Elephas* (*Archidiskodon*) *meridionalis*, *Equus stenonis*, *Rhinoceros* (*Dicerorhinus*) *etruscus*, *Cervus* (*Euctenoceros*) *ctenoides*, *Cervus* (*E.*) *falconeri*, *Libralces gallicus* and *Mimomys pliocaenicus*. There are also a large number of extinct cetacean species associated with this period. There is a diversification of species within genera, including at least four types of elephant (*E. meridionalis*, *E. antiquus*, *E.* cf. *trogontherii* and *E.* cf. *primigenius*), fifteen species of deer,[3] and a number of species of the extinct vole *Mimomys*.[71] Unfortunately it is not yet possible to relate faunal changes in the Early Pleistocene with the known stages, but it appears that the

Villafranchian element is well represented in the Pastonian, because the rich fauna of the shelly crag of east Runton is thought to be of this age. But this faunal element is greatly reduced in the Cromerian.[159] Two interesting records from this part of the Middle Pleistocene in Norfolk are a glutton (*Gulo luscus*) probably from the Pastonian, and a suslik (*Citellus*), a mainly steppe genus, from the early Anglian (Arctic Freshwater Bed).

By the time of the Hoxnian, the Villafranchian element has been very much reduced or lost. Thus *Elephas meridionalis* and *Rhinoceros etruscus* have become extinct, and *Arvicola* has replaced *Mimomys*. Faunas of Hoxnian age are well known from the Thames valley, particularly from Swanscombe.[161] Here a fauna of many woodland species (*Elephas antiquus, Rhinoceros kirchbergensis, Sus, Cervus clactoniana*) is found in the Lower Gravels, together with a Clactonian industry, and above this in the Middle Gravels there are indications of increasing openness in the vegetation, with *Bos primigenius* and *Equus caballus* showing an increase in abundance.

In the Ipswichian similar woodland faunas are found, but *Hippopotamus*, thought to be absent in the Hoxnian, is also present.[161] In the Thames valley such Ipswichian faunas are found in London (Trafalgar Square), Ilford and Aveley. The first named site is apparently earlier than the last two and contains a different fauna with no *Equus* or *Elephas primigenius*. At Ilford, in the late part of the interglacial, the palaeobotanical evidence shows open vegetational conditions; two species of elephant are present, *E. antiquus* and an early form of *E. primigenius*, together with *Bison*, *Equus* and *Rhinoceros hemitoechus*. At Aveley, Essex, *E. primigenius* appears at the time when the forest opens out in the latter half of the interglacial. This fossil is a form of *Elephas primigenius*, related to the closely similar true *E. primigenius* of the Devensian. The development of the Ipswichian fauna in relation to vegetational history, shows an earlier forest fauna and later faunas associated with more open vegetational conditions.[160]

Of the faunas from the cold stages,[160a] several examples are known from the Wolstonian, and many more from the Devensian.[83] They are characterised by the presence of *Elephas primigenius, Rhinoceros tichorhinus, Rangifer tarandus*, with, less commonly, *Megaceros giganteus*,[96] *Ovibos moschatus*,[139] *Microtus nivalis, Lemmus lemmus*, and *Dicrostonyx henseli*. Sometimes *Equus caballus* and *Bison* are present, perhaps indicating conditions towards steppe rather than tundra. The mixture of steppe and tundra animals is characteristic of cold stage faunas. But a tundra species of *Bison* has been recorded in the Late Weichselian (≡Devensian late-glacial) of the continent,[50] and studies of such remains from this country are necessary before closer ecological interpretation can be made.

The characteristic large mammals of the interglacials are not found in the Flandrian and the fauna is reduced to the fauna we know today, apart from recent extinctions from the British Isles, such as the wolf *Canis lnpus* and the beaver *Castor fiber*.[45, 58, 146] Other species, such as *Cervus elaphus*, have undergone reduction in size since late-glacial time.[89]

Regarding the smaller British land mammals, present distribution patterns have been used to interpret the history of the fauna in Devensian and Flandrian time. Species now common to Britain and Ireland, e.g. *Sorex minutus* and *Sciurus vulgaris*, are thought to have spread before Ireland was isolated by the rising Flandrian sea-level, while others found only in Britain, e.g. *Talpa europea*, *Sorex araneus*, were prevented from further spread to Ireland by the rising sea-level.[132] Particular problems are the presence of the common European vole *Microtus arvalis* in Orkney and Guernsey, though absent in the rest of the British Isles and the subspecies of *Microtus agrestis* in the Hebrides and mainland of Scotland. There are conflicting views about whether the peculiar geographical distributions of these species and subspecies are a result of speciation after introduction or after isolation following natural migration. There is insufficient fossil evidence to decide the matter.[44]

Table 13.12 Localities with fossil vertebrate faunas

	Icenian Crag										
Red Crag and Ludhamian	*Antian*	*Pastonian*	*Beestonian*	*Cromerian*	*Hoxnian*	*Wolstonian*	*Ipswichian*	*Devensian*	*Flandrian*	*Locality*	*Reference*
+	+	+	+	+						Norfolk and Suffolk	Newton (1891), Spencer (1964)
		+	+	+						Norfolk	Azzaroli (1953) (deer), Stuart (1975)
				+						Westbury, Somerset	Bishop (1974)
					+		+			Swanscombe, Kent and Lower Thames	Sutcliffe (1964)
						+	+	+		Upper Thames	Sandford (1925) (elephants)
					+	+	+	+		Lower Thames	King and Oakley (1936)
						+	+	+	+	Torbryan Caves, Devon	Sutcliffe and Zeuner (1962)
								+		British Isles	Stuart (1977)
								+		Pin Hole Cave, Derbys	Armstrong (1931)
									+	Star Carr, Yorks	Fraser and King (1954)

Man. Apart from finds of *Homo sapiens* in the Flandrian, two occurrences of *Homo* remains in older Pleistocene deposits deserve mention. One is the skull from the terrace gravels of the 30 m Boyn Hill Terrace of the Lower Thames at Swanscombe, Kent,[111] associated with an Acheulian industry. The occipital was found in 1935, the left parietal in 1936 and the right parietal in 1955. The skull is believed to be similiar to the Steinheim skull, and comes within the range of variation of *Homo sapiens.*

The other is the find published in 1912 of a jaw and cranial fragments at Piltdown, Sussex, in terrace gravels either Middle or Late Pleistocene in age. The discovery resulted in much discussion on their significance in the evolution of man, and the remains were ascribed to *Eoanthropus dawsoni.* In 1953, following fluorine tests on these remains, it was demonstrated that the Piltdown skull was a forgery.[173]

Much more is known concerning the relation of man's artefacts to the Pleistocene succession,[129] and a brief relation of the main industries and cultures to the Pleistocene sequence is given in table 13.13. The table emphasises the richness of the area and the completeness of the record. Eoliths, regarded by some as artefacts, have been recorded from the Early Pleistocene Crags. The earliest artefacts appear to be present in a small flint assemblage associated with a Cromerian fauna at Westbury-sub-Mendip, Somerset.[20] It is not until later in the Middle Pleistocene that the first well-defined Palaeolithic industries occur. Artefacts of the Clactonian flake industry [81] and the Acheulian hand-axe industry occur in deposits of Hoxnian age, proved by pollen analysis of freshwater sediments. Acheulian artefacts are more frequently found in river terrace gravels, and may be reworked into later terraces. Thus Acheulian artefacts have been found in gravels correlated with the Hoxnian, Wolstonian and Ipswichian stages. Levalloisian artefacts occur in the Wolstonian stage, in the Lower Thames area beyond the ice margin of the time. The Lower Palaeolithic sites appear to be settlements by lakes or rivers and they are mainly known from the southeast of England. The first cave industries occur in the Upper Palaeolithic, in the Pennines and in north Wales. These industries appear to be Devensian in age but their exact correlation with the sub-stages of the Devensian remains in doubt.

The Mesolithic and later stages spread over Britain and into Ireland. The sites are usually open, often by lakes in the Mesolithic, and on raised beaches, e.g. the so-called Neolithic raised beach of northeast Ireland and southwest Scotland. Artefacts of most of the Flandrian industries occur in biogenic deposits so that their dating by the radiocarbon method and by pollen analysis has been possible in considerable detail.

CONCLUSIONS

We have been concerned mainly with the fossil evidence for the history of the flora and fauna of the British Isles, and it will now be of interest to make comparisons between, and general observations on, the histories of different groups as deduced from the fossil record. The total change we see in any group in the Pleistocene is the net result of environmental changes acting in conjunction with evolution of species. Our knowledge of evolution of species is based on the harder parts of the organisms only. Concerning the history of the flora, a few 'Tertiary relic' living genera are known in the Early Pleistocene, but by Cromerian times the flora appears to be largely of modern European species. With the land vertebrates, the Early Pleistocene Villafranchian element just survives into the Cromerian, but even then the faunas younger than this are not of modern species, but characteristically contain modern genera with extinct species. An exception is *Cervus elaphus*, one of the oldest modern species and known in the lower Middle Pleistocene. With the marine molluscs, the number of extinct species is reduced at the Plio–Pleistocene boundary and again at the Red Crag–Icenian Crag boundary. Following Cromerian times very few extinct species are found. A similar change in the foraminiferal fauna takes place at the Red Crag–Icenian Crag boundary, and both these changes are probably related not only to climatic change but also to the isolation of the North Sea from the Atlantic by uplift of the southern part of the Red Crag area, as indicated by Red Crag at high levels (at 120 to 182 metres O.D.) in southeast England (Netley Heath, Rothamsted). There is little or no evidence concerning the history of the insects or non-marine molluscs in the Early Pleistocene. In the Middle and Late Pleistocene, the floras and faunas chiefly contain modern species, except that the land vertebrates show high percentages of extinct species of modern genera or genera closely related to modern genera. Considering the gross climatic and other environmental changes of these times, it is surprising that so little evidence of evolution or extinction is seen in the lists of fossil plants and animals, apart from the vertebrates.

But changes of distribution which modern species have undergone are evident. Such changes in distribution have presumably in the first place resulted from climatic change, and secondly from geographical barriers to species movement. In the British Isles relative land/sea-level changes must have played an important part in permitting or hindering movements. In the case of the Flandrian, where the rise in sea-level is well known, it is possible to explain certain plant and

Table 13.13 The archaeological sequence related to stratigraphy

Stage (time scale not linear)	*Industry/culture*	*Local example (with radiocarbon dates B.P.)*	*Reference*	*Archaeological stage*	*General reference*
	Romano-British			Romano-British	Frere (1967)
VIII	Iron Age A, B (la Tène), C			Iron Age	Cunliffe (1974)
	Late Bronze Age	Trackway, Shapwick, Somerset (2642)	Godwin (1960b)	Late	
	Urn			Middle Bronze Age	Burgess (1974)
	Food vessel			Early	
	Beaker				
VIIb	Chambered-tomb				
	Windmill Hill	Avebury, Wilts (4530)		Neolithic	Clark and Godwin (1962) Watts (1960) Smith (1974)
	Larnian-Obanian	Oronsay (5450)	Mellars and Payne (1971)		
		Co. Dublin (4260)	Mitchell (1971)		
Flandrian					
VIIa	Horsham	Oakhanger, Hants (6300)	Rankine *et al.* (1960)		
	Sauveterrian-type	Shippea Hill, Cambs (7230)	Clark (1955)	Mesolithic	Clark (1936) Mellars (1974)
VI	Maglemosian	Brandesburton, Yorks	Clark and Godwin (1956)		
	Larnian	Co. Londonderry (7680)	Mitchell (1971)		Mitchell (1971)
IV and V	Protomaglemosian	Thatcham, Berks (9840)	Wymer and Churchill (1962)		
		Seamer, Yorks (9557)	Clark (1954)		

Devensian	L	Creswellian	Aveline's Hole, Somerset	Garrod (1926)	Upper Palaeolithic	
			Kent's Cavern, Devon (12180)	Campbell and Sampson (1971)		
	M	'Proto-Solutrean/ Aurignacian'	Kent's Cavern, Devon (28720)			Garrod (1926)
		Mousterian	Oldbury, Kent	Collins and Collins (1970)		Mellars (1974)
	E		Kent's Cavern, Devon	Campbell and Sampson (1971)		
Ipswichian		'Levalloisian'	Brundon, Suffolk	Moir and Hopwood (1939)		
Wolstonian		Baker's Hole Levalloisian	Northfleet, Kent	Smith (1911)		Bordes (1968) Burkitt (1955)
Hoxnian		Acheulian	Hoxne, Suffolk (zone IIc)	West and McBurney (1955)	Lower Palaeolithic	Roe (1968)
			Swancombe, Kent	Ovey (1964)		Wymer (1968)
		Clactonian	Clacton-on-Sea, Essex (zone IIb)	Warren (1951), Kerney (1971)		
			Swancombe, Kent	Ovey (1964)		
Anglian		derived Abbevillian	Sidestrand, Norfolk	Moir (1927)		
Cromerian			Westbury-Sub-Mendip, Somerset	Bishop (1975)		
Early Pleistocene		Crag eoliths		Moir (1927)		

animal distributions as being a result of isolation by rising sea-level, for example, the absence of *Tilia*, *Carpinus*, and *Talpa* from Ireland. The relative numbers of species of land mammals in west Europe (167), Britain (41) and Ireland (21) is a result of this isolation, and the flora shows a similar trend. During the glaciations at times of low sea-level, Britain was certainly connected with the continent, and indeed the Devensian and Wolstonian faunas of Britain and the continent are similar. Ireland too has a limited Devensian fauna with *Megaceros* and *Rangifer tarandus*, so land connections are indicated in that glacial stage, but earlier cold faunas in Ireland are not certainly known,[164] and there is some doubt about earlier land connections. The history of the Irish fauna and flora previous to the Devensian poses interesting biogeographical questions. If isolation was the rule, then the survival of the Hoxnian floras through earlier glaciations may be possible, and this would suggest the Irish flora in Hoxnian times was a survival from an early western European oceanic flora, possibly of late Tertiary age.

In Britain also, there is the question of isolation during the interglacials. In Cromerian and Hoxnian times, it is highly probable that land connections existed, because of the similarity of faunas and floras across the channel. Perhaps this land connection was brought into existence by the Early Pleistocene uplift already referred to, for the Pliocene shell beds of Kent (Lenham Beds) and East Anglia (Coralline Crag) have strong southern affinities, suggesting junction of North Sea and the Atlantic. In the Ipswichian, faunal comparisons[135] (e.g. *Hippopotamus* on both sides of the Channel) indicate the possibilities of migration from the continent, though the presence of southern marine molluscs in the Eemian (last) interglacial fauna of the Netherlands shows that migration was possible from the Atlantic to the North Sea, and the vegetational history is different from that on the continent, particularly in the absence of the late zone f(Ip IIb) expansion of *Tilia* and the absence of the *Picea* zone (h) of Jessen and Milthers. But perhaps there are climatic or edaphic or other reasons for such differences.

Such geological problems as these are presented by a consideration of the fossil evidence, as well as the problems of a biological nature. Of prime interest is the nature of the response to climatic change by species: questions of speciation, biotype formation and extinction, and rates of change of distribution area. It may be possible in future to relate fossil evidence to other lines of evidence concerning the history of the flora and fauna, such as the study of present distributions, and cytogenetic and ecological studies of biotypes and the ecological amplitudes of biotypes. It is only in this way that there will be more understanding of the nature of the changes in the fauna and flora which are so evident in the Pleistocene.

REFERENCES

1 ADAMS, A. L. 1877–81. 'Monograph on the British fossil elephants', *Mon. Palaeontogr. Soc.*

2 ARMSTRONG, A. L. 1931. 'Excavations in the Pin Hole Cave, Cresswell Crags, Derbyshire', *Proc. Prehist. Soc. E. Anglia*, **6,** 330–4.

3 AZZAROLI, A. 1953. 'The deer of the Weybourne Crag and Forest Bed of Norfolk', *Bull. Br. Mus. Geology*, **2,** 3–96.

4 BADEN-POWELL, D. F. W. 1927. 'On the present climatic equivalent of British raised beach Mollusca', *Geol. Mag.*, **64,** 433–8.

5 BADEN-POWELL, D. F. W. 1934. 'On the marine gravels at March, Cambridgeshire', *Geol. Mag.*, **71,** 193–219.

6 BADEN-POWELL, D. F. W. 1937. 'On a marine Holocene fauna in north-western Scotland', *J. Anim. Ecol.*, **6,** 273–83.

7 BADEN-POWELL, D. F. W. 1950. 'Field meeting in the Lowestoft district', *Proc. Geol. Ass. London*, **61,** 191–7.

8 BADEN-POWELL, D. F. W. 1955. 'Report on the marine fauna of the Clacton channels', *Q.J. Geol. Soc. London*, **111,** 301–4.

9 BARTLEY, D. D. 1962. 'The stratigraphy and pollen analysis of lake deposits near Tadcaster, Yorkshire', *New Phytol.*, **61,** 277–87.

10 BECK, R. B., FUNNELL, B. M. and LORD, A. 1972. 'Correlation of Lower Pleistocene at depth in Suffolk', *Geol. Mag.*, **109,** 137–9.

11 BELL, F. G. 1969. 'The occurrence of southern, steppe and halophyte elements in Weichselian (last-glacial) floras from southern Britain', *New Phytol.*, **68,** 913–22.

12 BELL, F. G. 1970. 'Late Pleistocene floras from Earith, Huntingdonshire', *Phil. Trans. R. Soc.*, B, **258,** 347–78.

13 BEIRNE, B. P. 1952. *The origin and history of the British fauna.* London: Methuen.

14 BIRKS, H. H. 1970. 'Studies in the vegetational history of Scotland I. A pollen diagram from Abernethy Forest, Inverness-shire', *J. Ecol.*, **58,** 827–46.

15 BIRKS, H. H. 1975. 'Studies in the vegetational history of Scotland IV. Pine stumps in Scottish blankets peats', *Phil. Trans. R. Soc. Lond.*, B, **270,** 181–226.

16 BIRKS, H. J. B. 1973. *Past and present vegetation of the Isle of Skye.* Cambridge University Press.

17 BIRKS, H. J. B. and DEACON, J. 1973. 'A numerical analysis of the past and present flora of the British Isles', *New Phytol.*, **72,** 877–902.

18 BIRKS, H. J. B., DEACON, J. and PEGLAR, S. 1975. 'Pollen maps for the British Isles 5000 years ago', *Proc. R. Soc. Lond.*, B, **189,** 87–105.

19 BISHOP, M. J. 1974. 'A preliminary report on the Middle Pleistocene

mammal bearing deposits of Westbury-sub-Mendip, Somerset', *Proc. Univ. Bristol. Spelaeol. Soc.*, **13,** 301–18.

20 BISHOP, M. J. 1975. 'Earliest record of man's presence in Britian', *Nature* (*Lond.*), **253,** 95–7.

21 BORDES, F. 1968. *The Old Stone Age*. London: World University Library.

22 BOSWELL, P. G. H. 1952. 'The Plio-Pleistocene boundary in the east of England', *Proc. Geol. Ass. London*, **63,** 301–12.

23 BRAMWELL, D. 1960. 'Some research into bird distribution in Britain during the Late-glacial and Post-glacial periods', *Bird Rept. Merseyside Nat. Ass.*, 1959–60, 51–8.

24 BURGESS, C. 1974. 'The bronze age'. In Renfrew 1974, 165–232.

25 BURKITT, M. C. 1955. *The Old Stone Age*. Cambridge: Bowes and Bowes.

26 CAMPBELL, J. B. and SAMPSON, C. G. 1971. 'A new analysis of Kent's Cavern, Devonshire, England', *Univ. Oregon* (*Eugene*) *Anthropol. Papers*, **3,** 1–40.

27 CLARK, J. G. D. 1936. *The Mesolithic settlement of Northern Europe*. Cambridge University Press.

28 CLARK, J. G. D. 1954. *Excavations at Star Carr*. Cambridge University Press.

29 CLARK, J. G. D. 1955. 'A Microlithic Industry from the Cambridgeshire Fenland and other industries of Sauveterrian affinities from Britain', *Proc. Prehist. Soc.*, **21,** 3–20.

30 CLARK, J. G. D. and GODWIN, H. 1956. 'A Maglemosian site at Brandesburton, Holderness, Yorkshire', *Proc. Prehist. Soc.*, **22,** 6–22.

31 CLARK, J. G. D. and GODWIN, H. 1962. 'The Neolithic in the Cambridgeshire Fens', *Antiquity*, **36,** 10–23.

32 COLHOUN, E. A. and MITCHELL, G. F. 1971. 'Interglacial marine formation and late glacial freshwater formation in Shortalstown Townland, Co. Wexford', *Proc. R. Irish Acad.*, B, **71,** 211–45.

33 COLLINS, D. and COLLINS, A. 1970. 'Excavations at Oldbury in Kent: cultural evidence for Last Glacial occupation in Britain', *Bull. Inst. Archaeol. Lond.*, **8–9,** 151–76.

34 COOPE, G. R. 1959. 'A late-Pleistocene insect fauna from Chelford, Cheshire', *Proc. R. Soc.*, B, **151,** 70–86.

35 COOPE, G. R. 1962. 'A Pleistocene Coleopterous fauna with arctic affinities from Fladbury, Worcestershire', *Q.J. Geol. Soc. Lond.*, **118,** 103–23.

36 COOPE, G. R. 1965. 'Fossil insect faunas from Late Quaternary deposits in Britain', *Adv. Sci.*, **21,** 564–75.

37 COOPE, G. R. 1970. 'Interpretations of Quaternary insect fossils', *Ann. Rev. Entomology*, **15,** 97–120.

38 COOPE, G. R. 1974a. 'Tibetan species of dung beetle from Late Pleistocene deposits in England', *Nature* (*Lond.*), **245,** 335–6.

39 COOPE, G. R. 1974b. 'Interglacial coleoptera from Bobbitshole, Ipswich, Suffolk', *J. Geol. Soc. Lond.*, **130,** 333–40.

40 COOPE, G. R. 1975. 'Climatic fluctuations in northwest Europe since the Last Interglacial, indicated by fossil assemblages of Coleoptera'. In *Ice ages: ancient and modern*, ed. by A. E. Wright and F. Moseley, 153–68. Liverpool: Seel House Press.

40a COOPE, G. R. 1977. 'Fossil coleopteran assemblages as sensitive indicators of past climatic changes'. In Mitchell and West, 1977.

40b COOPE, G. R. and ANGUS, R. B. 1975. 'An ecological study of a temperate interlude in the middle of the Last Glaciation, based on fossil Coleoptera from Isleworth, Middlesex', *J. Anim. Ecol.*, **44,** 365–91.

41 COOPE, G. R. and BROPHY, J. A. 1972. 'Late glacial environmental changes indicated by a coleopteran succession from north Wales', *Boreas*, **1,** 97–142.

42 COOPE, G. R., MORGAN, A. and OSBORNE, P. J. 1971. 'Fossil Coleoptera as indicators of climatic fluctuations during the last glaciation in Britain', *Palaeogeogr., Palaeoclimatol., Palaeoecol.*, **10,** 87–101.

43 COOPE, G. R., SHOTTON, F. W. and STRACHAN, I. 1961. 'A Late Pleistocene fauna and flora from Upton Warren, Worcestershire', *Phil. Trans. R. Soc.*, B, **244,** 379–421.

44 CORBET, G. B. 1961. 'Origin of the British insular races of small mammals and of the "Lusitanian" fauna', *Nature (Lond.)*, **191,** 1037–40.

45 CORBET, G. B. 1964. *The identification of British Mammals.* London: British Museum (Nat. Hist.).

46 CUNLIFFE, B. 1974. 'The iron age'. In Renfrew, 1974, 233–62.

47 DAWKINS, W. B. 1869. 'On the distribution of the British Post-glacial Mammals', *Q.J. Geol. Soc. London*, **25,** 192–217.

48 DAWKINS, W. B. and SANFORD, W. A. 1866–87. 'The British Pleistocene Mammalia', *Mon. Palaeontogr. Soc.*

49 DEEVEY, E. S. 1949. 'Biogeography of the Pleistocene', *Bull. Geol. Soc. Am.*, **60,** 1315–1416.

50 DEGERBØL, M. 1964. 'Some remarks on late- and post-glacial vertebrate fauna and its ecological relations in northern Europe', *J. Anim. Ecol.*, **33** (Supplement), 71–85.

51 DICKSON, J. H. 1973. *Bryophytes of the Pleistocene.* Cambridge University Press.

52 Discussion on the development of habitats in the Post-Glacial. 1965. *Proc. R. Soc.*, B, **161,** 293–375.

53 DUIGAN, S. L. 1955. 'Plant remains from the gravels of the Summertown-Radley Terrace near Dorchester, Oxfordshire', *Q.J. Geol. Soc. London*, **111,** 225–38.

54 DUIGAN, S. L. 1963. 'Pollen analyses of the Cromer Forest Bed Series in East Anglia', *Phil. Trans. R. Soc.*, B, **246,** 149–202.

55 EVANS, J. G. 1970. 'Interpretation of land snail faunas', *Univ. London Inst. Archaeol. Bull.*, **8–9,** 109–16.

56 EVANS, J. G. 1972. *Land snails in archaeology.* London: Seminar Press.

57 FORBES, E. 1846. 'On the connexion between the distribution of the existing fauna and flora of the British Isles, and the geological changes which have affected their area, especially during the epoch of the Northern Drift', *Mem. Geol. Surv. G.B.*, **1,** 336–432.

58 FRASER, F. C. and KING, J. E. 1954. 'Faunal remains'. In J. G. D. Clark. *Excavations at Star Carr*, 70–95. Cambridge University Press.

59 FRERE, S. S. 1967. *Britannia: a history of Roman Britain.* London: Routledge & Kegan Paul.

60 FUNNELL, B. M. 1961. 'The Palaeogene and Early Pleistocene of Norfolk', *Trans. Norfolk Norwich Nat. Soc.*, **19,** 340–64.

61 FUNNELL, B. M. and WEST, R. G. 1962. 'The Early Pleistocene of Easton Bavents', *Q.J. Geol. Soc. London*, **118,** 125–41.

62 GARROD, D. 1926. *The Upper Palaeolithic Age in Britain.* Oxford: Clarendon Press.

63 GIBBARD, P. L. and STUART, A. J. 1974. 'Trace fossils from proglacial lake sediments', *Boreas*, **3,** 69–74.

64 GODWIN, H. 1960a. 'Radiocarbon dating and Quaternary history in Britain', *Proc. R. Soc.*, B, **153,** 287–320.

65 GODWIN, H. 1960b. 'Prehistoric wooden trackways of the Somerset levels: their construction, age and relation to climatic change', *Proc. Prehist. Soc.*, **26,** 1–36.

66 GODWIN, H. 1964. 'Late-Weichselian conditions in south-eastern Britain: organic deposits at Colney Heath, Herts', *Proc. R. Soc.*, B, **160,** 258–75.

67 GODWIN, H. 1975. *History of the British Flora.* 2nd edn. Cambridge University Press.

68 GODWIN, H., WALKER, D. and WILLIS, E. H. 1957. 'Radiocarbon dating and post-glacial vegetational history: Scaleby Moss', *Proc. R. Soc.*, B, **147,** 352–66.

69 GOULDEN, C. E. 1964. 'The history of the Cladoceran fauna of Esthwaite Water (England) and its limnological significance', *Arch. Hydrobiol.*, **60,** 1–52.

70 HARMER, F. W. 1914–1925. 'The Pliocene Mollusca of Great Britain', *Mon. Palaeontogr. Soc.*

71 HINTON, M. A. C. 1926. *Monograph of the voles and lemmings (Microtinae).* London: British Museum (Natural History).

72 HOVE, H. A. 1968. 'The *Ulmus* fall at the transition Atlanticum-Subboreal in pollen diagrams', *Palaeogeogr., Palaeoclimatol., Palaeoecol.*, **5,** 359–69.

73 JAMIESON, T. F. 1865. 'On the history of the last geological changes in Scotland', *Q.J. Geol. Soc. London*, **21,** 161–203.

74 JESSEN, K. 1949. 'Studies in Late Quaternary deposits and flora-history of Ireland', *Proc. R. Ir. Acad.*, **52** B, 85–290.

75 JESSEN, K., FARRINGTON, A. and ANDERSEN, S. T. 1959. 'The interglacial deposit near Gort, Co. Galway, Ireland', *Proc. R. Ir. Acad.*, **60** B, 1–77.

76 JESSEN, K. and MILTHERS, V. 1928. 'Stratigraphical and palaeontological studies of interglacial freshwater deposits in Jutland and north-west Germany', *Danm. Geol. Unders.*, II Raekke, **48.**

77 KELLY, M. R. 1964. 'The Middle Pleistocene of north Birmingham', *Phil. Trans. R. Soc.*, B, **247,** 533–92.

78 KELLY, M. R. and OSBORNE, P. J. 1964. 'Two faunas and floras from the alluvium at Shustoke, Warwickshire', *Proc. Linn. Soc. London*, **176,** 37–65.

79 KENNARD, A. S. and WOODWARD, B. B. 1922. 'The post-Pliocene non-marine Mollusca of the east of England', *Proc. Geol. Ass. London*, **33,** 104–42.

80 KERNEY, M. P. 1963. 'Late-glacial deposits on the Chalk of south-east England', *Phil. Trans. R. Soc.*, B, **246,** 203–54.

81 KERNEY, M. P. 1971a. Interglacial deposits in Barnfield Pit, Swanscombe, and their molluscan fauna. *Q.J. Geol. Soc. London*, **127,** 69–93.

82 KERNEY, M. P. 1971b. 'A Middle Weichselian deposit at Halling, Kent', *Proc. Geol. Assoc. Lond.*, **82,** 1–11.

83 KING, W. B. R. and OAKLEY, K. P. 1936. 'The Pleistocene succession in the lower parts of the Thames Valley', *Proc. Prehist. Soc.*, 1936, 52–76.

84 KURTEN, B. 1963. 'Villafranchian faunal evolution', *Soc. Sci. Fenn. Comment. Biol.*, **26,** No. 3.

85 KURTEN, B. 1968. *Pleistocene Mammals of Europe*. London: Weidenfeld and Nicolson.

86 LAGAAIJ, R. 1952. 'The Pliocene Bryozoa of the Low Countries', *Med. Geol. Sticht.*, Ser. C, V, No. 5.

87 LAMBERT, C. A., PEARSON, R. G. and SPARKS, B. W. 1962. 'Pleistocene deposits from Sidgwick Avenue, Cambridge', *Proc. Linn. Soc. London*, **174,** 13–29.

88 LEES, B. J. 1975. 'Foraminiferida from Holocene sediments in Start Bay', *J. Geol. Soc. Lond.*, **131,** 37–49.

89 LOWE, V. P. W. 1961. 'A discussion on the history, present status and future conservation of the Red Deer (*Cervus elaphus* L.) in Scotland', *Terre et la vie*, **108,** 9–40.

90 MACFADYEN, W. A. 1932. 'Foraminifera from some late Pliocene and glacial deposits of East Anglia', *Geol. Mag.*, **69,** 481–97.

91 MACFADYEN, W. A. 1933. 'The Foraminifera of the Fenland clays at St. Germans, near King's Lynn', *Geol. Mag.*, **70,** 182–91.

92 MELLARS, P. A. 1974. 'The palaeolithic and mesolithic'. In Renfrew, 1974, 41–99.

93 MELLARS, P. A. and PAYNE, S. 1971. 'Excavation of two Mesolithic shell middens on the island of Oronsay (Inner Hebrides)', *Nature (Lond.)*, **231,** 397–8.

94 MITCHELL, G. F. 1965. 'Littleton Bog, Tipperary: an Irish vegetational record', *Geol. Soc. Am.* Special Paper No. 84, 1–16.

95 MITCHELL, G. F. 1971. 'The Larnian culture: a minimal view', *Proc. Prehist. Soc.*, **37,** 274–83.

96 MITCHELL, G. F. and PARKES, H. M. 1949. 'The Giant Deer in Ireland', *Proc. R. Ir. Acad.*, **52** B, 291–314.

97 MITCHELL, G. F. and WEST, R. G. 1977. 'Discussion meeting: Devensian (last) glaciation', *Phil. Trans. R. Soc. Lond.*, B (in press).

98 MOIR, J. R. 1927. *The antiquity of man in East Anglia.* Cambridge University Press.

99 MOIR, J. R. and HOPWOOD, A. T. 1939. 'Excavations at Brundon, Suffolk (1935–37)', *Proc. Prehist. Soc.*, **5**, 1–32.

100 MOORE, P. D. 1970. 'Studies in the vegetational history of Mid-Wales II. The Late-glacial period in Cardiganshire', *New Phytol.*, **69**, 363–75.

101 MOORE, P. D. 1972. 'Studies in the vegetational history of Mid-Wales III. Early Flandrian pollen data from west Cardiganshire', *New Phytol.*, **71**, 947–59.

102 MORGAN, ANNE, 1973. 'Late Pleistocene environmental changes indicated by fossil insect faunas of the English Midlands', *Boreas*, **2**, 173–212.

103 MORGAN, A. V. 1973. 'The Pleistocene geology of the area north and west of Wolverhampton, Staffordshire, England', *Phil. Trans. R. Soc. Lond.*, B, **265**, 233–97.

104 NEWTON, E. T. 1891. *The Vertebrata of the Pliocene deposits of Britain.* Mem. Geol. Surv. U.K.

105 NORTON, P. E. P. 1967. 'Marine Mollusca in the Early Pleistocene of Sidestrand, Bramerton, and the Royal Society borehole at Ludham, Norfolk', *Phil. Trans. R. Soc.*, B, **253**, 161–200.

106 NORTON, P. E. P. and BECK, R. B. 1972. 'Lower Pleistocene molluscan assemblages and pollen from the Crag of Aldeby (Norfolk), and Easton Bavents (Suffolk)', *Bull. Geol. Soc. Norfolk*, **22**, 11–31.

107 NORTON, P. E. P. and SPAINK, G. 1973. 'The earliest occurrence of *Macoma balthica* (L.) as a fossil in the North Sea deposits', *Malacologia*, **14**, 33–7.

108 OSBORNE, P. J. 1965. 'The effect of forest clearance on the distribution of the British insect fauna', *Proc. 12th Int. Congress Ent.* (*London*), 454–5.

109 OSBORNE, P. J. 1971. 'An insect fauna from the Roman site at Alcester, Warwickshire', *Britannia*, **2**, 156–65.

110 OSBORNE, P. J. 1974. 'An insect fauna of Early Flandrian age from Lea Marston, Warwickshire and its bearing on the contemporary climate and ecology', *Quaternary Res.*, **4**, 471–86.

111 OVEY, C. D. 1964. *The Swancombe Skull. A survey of research on a Pleistocene site*, ed. C. D. Ovey. Royal Anthropological Institute, Occasional Memoir No. 20.

112 PEACOCK, J. D. 1974. 'Borehole evidence for Late- and Post-glacial events in the Cromarty Firth, Scotland', *Bull. Geol. Survey G.B.*, **48**, 55–67.

113 PEARSON, R. G. 1962. 'The Coleoptera from a detritus deposit of full-glacial age at Colney Heath, near St Albans', *Proc. Linn. Soc. London*, **173**, 37–55.

114 PEARSON, R. G. 1963. 'Coleopteran associations in the British Isles during the Late Quaternary Period', *Biol. Rev.*, **38,** 334–63.

115 PENNINGTON, W. 1964. 'Pollen analyses from the deposits of six upland tarns in the Lake District', *Phil. Trans. R. Soc.*, B, **248,** 205–44.

116 PENNINGTON, W. 1974. *The history of British vegetation.* 2nd edn. London: English Universities Press.

116a PENNINGTON, W. 1975. 'A chronostratigraphic comparison of Late-Weichselian and Late-Devensian subdivisions, illustrated by two radiocarbon-dated profiles from western Britain'. *Boreas*, **4,** 157–71.

116b PENNINGTON, W. 1977. 'The Late Devensian flora and vegetation of Britain'. In Mitchell and West, 1977.

117 PENNINGTON, W., HAWORTH, E. Y., BONNY, A. P. and LISHMAN, J. P. 1972. 'Lake sediments in northern Scotland', *Phil. Trans. R. Soc. Lond.*, B, **264,** 191–294.

118 PHILLIPS, L. 1974. 'Vegetational history of the Ipswichian/Eemian interglacial in Britain and continental Europe', *New Phytol.*, **73,** 589–604.

119 PIGOTT, C. D. 1956. 'The vegetation of upper Teesdale in the north Pennines', *J. Ecol.*, **44,** 545–86.

120 PIGOTT, C. D. and WALTERS, S. M. 1954. 'On the interpretation of the discontinuous distributions shown by certain British species of open habitats', *J. Ecol.*, **42,** 95–116.

121 PRAEGER, R. L. 1896. 'Report upon the raised beaches of the north-east of Ireland, with special reference to their fauna', *Proc. R. Ir. Acad.*, **4** (3rd series), 30–54.

122 RANKINE, W. F., RANKINE, W. M. and DIMBLEBY, G. W. 1960. 'Further excavations at a Mesolithic site at Oakhanger, Selbourne, Hants', *Proc. Prehist. Soc.*, **26,** 246–62.

123 REID, C. 1899. *The origin of the British Flora.* London: Dulau.

124 REID, C. and E. M. 1915. 'The Pliocene floras of the Dutch-Prussian border', *Med. Rijksopsporing v. Delfstoffen*, No. 6.

125 REID, E. M. 1920. 'Two preglacial floras from Castle Eden', *Q.J. Geol. Soc. London*, **76,** 104–44.

126 RENFREW, C. (ed.). 1974. *British Prehistory.* London: Duckworth.

127 REYNOLDS, S. H. 1902–12. 'A monograph of the British Pleistocene Mammalia', *Mon. Palaeontogr. Soc.*

128 ROE, D. 1968. 'British Lower and Middle Palaeolithic handaxe groups', *Proc. Prehist. Soc.*, **34,** 1–82.

129 ROE, D. 1971. *Prehistory.* London: Paladin.

130 ROUND, F. E. 1961. 'The diatoms of a core from Esthwaite Water', *New Phytol.*, **60,** 43–59.

131 SANDFORD, K. S. 1925. 'The fossil elephants of the upper Thames basin', *Q.J. Geol. Soc. London*, **81,** 62–86.

132 SAVAGE, R. J. G. 1966. 'Irish Pleistocene mammals', *Ir. Nat. J.*, **15,** 117–30.

133 SCHARFF, R. F. 1907. *European animals: their geological history and*

geographical distribution. London: Constable.

134 SEDDON, B. 1962. 'Late-glacial deposits at Llyn Dwythwch and Nant Ffrancon, Caernarvonshire', *Phil. Trans. R. Soc. Lond.*, B, **244**, 459–81.

135 SHOTTON, F. W. 1962. 'The physical background of Britain in the Pleistocene', *Adv. Sci.*, **19**, 1–14.

136 SHOTTON, F. W. 1965. 'Movements of insect populations in the British Pleistocene', *Geol. Soc. Am.* Special Paper No. 84, 17–33.

137 SHOTTON, F. W. and OSBORNE, P. J. 1965. 'The fauna of the Hoxnian interglacial deposits of Nechells, Birmingham', *Phil. Trans. R. Soc.*, B, **248**, 353–78.

138 SIDDIQUI, Q. A. 1971. 'The palaeoecology of non-marine Pleistocene ostracoda from Fladbury, Worcestershire and Isleworth, Middlesex'. In *Colloquium on the paleoecology of ostracodes* (Pau, 1970), ed. H. Oertli, 331–9. Pau; Société Nationale des Pétroles d'Aquitaine.

139 SIMONS, J. W. 1962. 'New records of Musk Ox from Plumstead, Kent and Cosgrove, Northants', *Lond. Nat.*, No. 41, 42–53.

140 SIMPSON, I. M. and WEST, R. G. 1958. 'On the stratigraphy and palaeobotany of a late-Pleistocene organic deposit at Chelford, Cheshire', *New Phytol.*, **57**, 239–50.

141 SMITH, A. G. 1961. 'The Atlantic—Sub-boreal transition', *Proc. Linn. Soc. London*, **172**, 38–49.

142 SMITH, A. G. and PILCHER, J. R. 1973. 'Radiocarbon dates and vegetational history of the British Isles', *New Phytol.*, **72**, 903–14.

143 SMITH, A. G. and WILLIS, E. H. 1962. 'Radiocarbon dating of the Fallahogy Landnam phase', *Ulster J. Archaeol.*, **24–5**, 16–24.

144 SMITH, L. F. 1974. 'The neolithic'. In Renfrew, 1974, 100–36.

145 SMITH, R. A. 1911. 'A Palaeolithic industry at Northfleet, Kent', *Archaeologia*, **62**, 515–32.

146 SOUTHERN, H. N. 1964. *The handbook of British Mammals*. Oxford, Blackwell.

147 SPARKS, B. W. 1956. 'The non-marine Mollusca of the Hoxne Interglacial', Appendix, in West, 1956.

148 SPARKS, B. W. 1957. 'The tjaele gravel near Thriplow, Cambridgeshire', *Geol. Mag.*, **94**, 194–200.

149 SPARKS, B. W. 1961. 'The ecological interpretation of Quaternary non-marine Mollusca', *Proc. Linn. Soc. London*, **172**, 71–80.

150 SPARKS, B. W. 1963. 'Non-marine Mollusca from the Cromer Forest Bed at West Runton'. In Duigan, 1963, 197–9.

151 SPARKS, B. W. 1964. 'The distribution of non-marine Mollusca in the last Interglacial in south-east England', *Proc. Malac. Soc. London*, **36**, 7–25.

152 SPARKS, B. W. 1969. 'Non-marine Mollusca and archaeology'. In *Science and Archaeology*, ed. D. R. Brothwell and E. S. Higgs. 2nd edn. London: Thames and Hudson.

153 SPARKS, B. W. and LAMBERT, C. A. 1961. 'The post-glacial deposits at Apethorpe, Northamptonshire', *Proc. Malac. Soc. London*, **34**, 302–15.

154 SPARKS, B. W. and WEST, R. G. 1970. 'Late Pleistocene deposits at Wretton, Norfolk I. Ipswichian interglacial deposits', *Phil. Trans. R. Soc. Lond.*, B, **258,** 1–30.

155 SPENCER, H. E. P. 1964. 'The contemporary Mammalian fossils of the Crags', *Trans. Suffolk. Nat. Soc.*, **12,** 333–44.

156 STELFOX, A. W., KUIPER, J. G. J., MCMILLAN, N. F. and MITCHELL, G. F. 1972. 'The Late-glacial and Post-glacial Mollusca of the White Bog, Co. Down', *Proc. R. Irish Acad.*, **72,** B, 185–207.

157 STEVENS, L. A. 1960. 'The interglacial of the Nar Valley', *Q.J. Geol. Soc. London*, **115,** 291–316.

158 STUART, A. J. 1974. 'Pleistocene history of the British vertebrate fauna', *Biol. Rev.*, **49,** 225–66.

159 STUART, A. J. 1975. 'The vertebrate fauna of the type Cromerian', *Boreas*, **4,** 63–76.

160 STUART, A. J. 1976. 'The history of the mammal fauna during the Ipswichian/last interglacial in Britain', *Phil. Trans. R. Soc. Lond.*, B, **276,** 221–50.

160a STUART, A. J. 1977. 'The vertebrates of the Last Cold Stage in Britain and Ireland'. In Mitchell and West, 1977.

161 SUTCLIFFE, A. J. 1964. 'The mammalian fauna'. Ch. 9 in Ovey, 1964.

162 SUTCLIFFE, A. J. and ZEUNER, F. E. 1962. 'Excavations in the Torbryan Caves, Devonshire. 1. Tornewton Cave', *Proc. Devon Arch. Expl. Soc.*, **5,** 127–45.

163 SYLVESTER-BRADLEY, P. C. 1973. 'On *Ilyocypris quinculminata* Sylvester-Bradley sp. nov.', *Stereo-Atlas of Ostracod shells*, **1,** 85–88.

164 Symposium on the insularity of British Pleistocene Mammals. 1969. *Bull. Mammal Soc. Br. Isles*. **31,** 2–25.

165 TALLIS, J. H. 1964a. 'The pre-peat vegetation of the southern Pennines', *New Phytol.*, **63,** 363–73.

166 TALLIS, J. H. 1964b. 'Studies on southern Pennine peats. II. The pattern of erosion', *J. Ecol.*, **53,** 333–44.

167 THOMPSON, D. B. and WORSLEY, P. 1966. 'A late Pleistocene marine molluscan fauna from the drifts of the Cheshire Plain', *Geol. J.*, **5,** 197–207.

168 TURNER, C. 1970. 'The Middle Pleistocene deposits at Marks Tey, Essex', *Phil. Trans. R. Soc. Lond.*, B, **257,** 373–440.

169 TURNER, J. 1965. 'A contribution to the history of forest clearance', *Proc. R. Soc.*, B, **161,** 343–53.

170 WARREN, S. H. 1951. 'The Clacton flint industry; a new interpretation', *Proc. Geol. Ass. London*, **62,** 107–35.

171 WATTS, W. A. 1960. 'C-14 dating and the Neolithic in Ireland', *Antiquity*, **34,** 111–16.

172 WATTS, W. A. 1964. 'Interglacial deposits at Baggotstown, near Bruff. Co. Limerick', *Proc. R. Ir. Acad.*, **63,** B, 167–89.

172a WATTS, W. A. 1977. 'The Late Devensian flora of Ireland'. In Mitchell and West, 1977.

173 WEINER, J. S., OAKLEY, K. P. and LE GROS CLARK, W. E. 1953. 'The solution of the Piltdown problem', *Bull. Br. Mus. Geology*, **2,** 139–46.

174 WEST, R. G. 1956. 'The Quaternary deposits at Hoxne, Suffolk', *Phil. Trans. R. Soc.*, B, **239,** 265–356.

175 WEST, R. G. 1957. 'Interglacial deposits at Bobbitshole, Ipswich', *Phil. Trans. R. Soc.*, B, **241,** 1–31.

176 WEST, R. G. 1961a. 'Vegetational history of the Early Pleistocene of the Royal Society Borehole at Ludham, Norfolk', *Proc. R. Soc.*, B, **155,** 437–53.

177 WEST, R. G. 1961b. 'The glacial and interglacial deposits of Norfolk', *Trans. Norfolk Norwich Nat. Soc.*, **19,** 365–75.

178 WEST, R. G. 1970. 'Pleistocene history of the British Flora'. In *Studies in the vegetational history of the British Isles*, ed. D. Walker and R. G. West, 1–11. Cambridge University Press.

178a WEST, R. G. 1977. 'Early and Middle Devensian flora and vegetation'. In Mitchell and West, 1977.

179 WEST, R. G., DICKSON, C. A., CATT, J. A., WEIR, A. H. and SPARKS, B. W. 1974. 'Late Pleistocene deposits at Wretton, Norfolk II. Devensian deposits', *Phil. Trans. R. Soc. Lond.*, B, **267,** 337–420.

180 WEST, R. G. and MCBURNEY, C. M. B. 1955. 'The Quaternary deposits at Hoxne, Suffolk, and their archaeology', *Proc. Prehist. Soc.*, **20,** 131–54.

181 WEST, R. G. and NORTON, P. E. P. 1974. 'The Icenian Crag of southeast Suffolk', *Phil. Trans. R. Soc. Lond.*, B, **269,** 1–28.

182 WEST, R. G. and SPARKS, B. W. 1960. 'Coastal interglacial deposits of the English Channel', *Phil. Trans. R. Soc.*, B, **243,** 95–133.

183 WEST, R. G. and WILSON, D. G. 1966. 'Cromer Forest Bed Series', Nature (*Lond.*), **209,** 497–8.

184 WHATLEY, R. C. and KAYE, P. 1971. 'The palaeoecology of Eemian (last interglacial) ostracoda from Selsey, Sussex'. In *Colloquium on the paleoecology of ostracodes* (Pau, 1970), ed. H. Oertli, 311–30. Pau: Société Nationale des Pétroles d'Aquitaine.

185 WILMOTT, A. J. 1935. 'Evidence in favour of survival of the British Flora in glacial times'. In Discussion on the origin and relationship of the British Flora. *Proc. R. Soc.*, B, **118,** 197–241.

186 WILSON, D. G. 1973. 'Notable plant records from the Cromer Forest Bed Series', *New Phytol.*, **72,** 1207–34.

187 WOOD, S. V. 1848–1882. 'A monograph of the Crag Mollusca', *Mon. Palaeontogr. Soc.*

188 WYMER, J. J. 1968. *Lower Palaeolithic archaeology in Britain as represented by the Thames Valley*. London: John Baker.

189 WYMER, J. and CHURCHILL, D. M. 1962. 'Excavations at the Maglemosian sites at Thatcham, Berkshire, England. The stratigraphy of the Mesolithic sites III and IV at Thatcham, Berkshire, England', *Proc. Prehist. Soc.*, **28,** 329–70.

190 ZAGWIJN, W. H. 1959. 'Zur stratigraphischen und pollenanalytischen Gliederung der Pliozänen ablagerungen im Roertal-Graben und Venloer Graben der Niederlande', *Fortschr. Geol. Rheinld. u. Westf.*, **4,** 5–26.

191 ZAGWIJN, W. H. 1960. 'Aspects of the Pliocene and Early Pleistocene vegetation in the Netherlands', *Med. Geol. Sticht.*, series C III, **1,** No. 5.

Appendix 1

METHODS OF ISOLATING AND COUNTING FOSSILS

The following is a brief account of the simpler methods for extraction and counting of fossils. An extended account is available in the *Handbook of palaeontological techniques*.[9]

EXTRACTION OF THE SMALLER MACROSCOPIC FOSSILS (SEEDS, FRUITS, MOLLUSCA, SMALL VERTEBRATES)

For quantitative analyses, start the extraction with a known volume of sediment.

Method 1

Break up the lumps of material along the bedding planes and pick out any leaves or large moss stems. Put small lumps of material into a pudding basin and cover with 10 per cent HNO_3, or 5 per cent NaOH; the NaOH tends to soften plant tissue, but is better for breaking down some muds. Leave in a fume chamber for at least 24 hours, and stir occasionally until all the lumps have broken down.

Remove any seeds or other fossils which have floated to the top and put successive small amounts of the broken-down sediment, say 20 ml, on a sieve (100 mesh to the inch), and wash with a spray of water, stirring or agitating the sediment until the washing water becomes clear. If there is coarse detritus it may be necessary to use a coarser sieve (about 35 mesh to the inch) above the fine one, and then to examine the material in two fractions.

Tip the coarse fraction on to a soup plate, add water and use a fine paint brush to lift out seeds and other identifiable remains. Put the finer fraction into a smaller dish and extract the seeds under a low-power binocular microscope (about $\times 6$).

Collect the identifiable remains on damp filter paper in a Petri dish or partially sort them into 4-part paint palettes, or similar dishes, in water and cover with a Petri dish. Surface patterns of the remains are often best seen while they are drying out after wetting with alcohol. But the remains should be kept either dry or wet, as alternate wetting and drying tends to damage them.

When the remains have been identified, they should be stored in small tubes in a glycerine, alcohol and formalin or thymol mixture, and labelled with glass writing ink.

Method 2

This method for the extraction of shells, fruits, seeds and small bones is abridged from the detailed account by Sparks;[11] see also Evans.[5]

To break down sands and silts only gentle washing may be necessary, but a compacted sediment may need soaking in dilute NaOH, KOH or Calgon. When the matrix is broken down the residue is dried and sifted dry into fractions of over 2 mm, 1–2 mm, $\frac{1}{2}$–1 mm, and $\frac{1}{4}$–$\frac{1}{2}$ mm. The largest grade is sorted on a black glass plate, the next two by sprinkling the material thinly on a lined porcelain plate and systematically scanned with a low-power binocular microscope. All identifiable remains can be picked out with a moistened brush. The $\frac{1}{4}$–$\frac{1}{2}$ mm grade is examined for very small seeds, *Chara* oospores and the tests of Foraminifera.

The fruits and seeds should be kept dry, since alternate wetting and drying tends to damage them. They can be stored for short periods dry in glass tubes. The shells and small vertebrate remains can be stored dry permanently in glass tubes.

EXTRACTION OF INSECTS[3]

In predominantly inorganic sediments such as sands and silts, the material is first broken down (as above) then washed through a series of sieves, and the residues sorted under a low-power binocular microscope. It is found that most beetle remains are in the fractions retained by sieves of 50 mesh to the inch or less. In biogenic sediments such as felted peats, it is best to split the material along the bedding planes and pick out insect remains. Associated fragments of the same individual can be found in this way, but it has also been demonstrated that there may be a subjective bias to the advantage of the more obvious and larger insects.

EXTRACTION OF FORAMINIFERA (AND OSTRACODS)

Numerous methods of extraction of Foraminifera and ostracods from different types of sediment have been devised, details of which can be found in text books of micropalaeontology. For unconsolidated clastic sediments commonly encountered in the Pleistocene simple methods are usually sufficient. A useful summary of the factors to be considered in the sampling, preparation and investigation of Pleistocene Foraminifera is given by Feyling-Hanssen.[7] The following account has been very kindly provided by Professor B. M. Funnell.

The quantity of original sediment required to yield a few hundred Foraminifera is very variable. In sparsely fossiliferous sediments up to 2 kg may be desirable, whereas in highly fossiliferous sediments 10 g will suffice. In general, however, 100 g is often found to be a convenient amount to prepare, although larger samples may be collected to provide a reserve of material in case of need.

Dispersion and cleaning. The sediment to be prepared should be thoroughly dried and if it is clayey broken up into fragments of not more than 1 ml in size. It is then dropped into a beaker containing a hot solution of sodium carbonate, and gently boiled for ten minutes—longer if the lumps of clay have not dispersed by then. It can be stirred with a glass rod to assist dispersion but care should be taken not to crush the Foraminifera against the bottom of the beaker with the end of the rod. If the clay is tenacious and does not disperse easily it is probably more straightforward to decant the clay already dispersed, dry out the residue and start again rather than to continue boiling and stirring. An alternative method of dispersion for dense clays is to immerse the broken up dry clay in a dilute (5 per cent) solution of hydrogen peroxide. The effervescence produced is rather variable, depending on the state of oxidation and organic content of the clay, and delicate tests may be damaged by the slightly acid reaction.

Sieving. The dispersed sediment is sieved wet through a 240 mesh (65μm opening) sieve, flushing out gently with running water. This sieve retains all sand-size particles including Foraminifera, allowing silt and clay to pass through. Some Foraminifera, particularly immature individuals, are smaller than 65μm, but their systematic study is impracticable. Indeed, it is customary to make routine observations only on the Formanifera retained by a 120 mesh (125μm opening) sieve, the 125 to 65μm fraction being reserved for special study of rare forms, immature individuals, and so on. The maximum size of Foraminifera varies between different environments but for the Pleistocene of northwest Europe only rare individuals exceed 1000μm (1 mm).

Picking. After sieving, the sediment retained on the sieves is dried and spread on a small (10 cm square or less) tray with a black ground divided into a grid. It is examined systematically under incident illumination with a low-power binocular microscope at magnifications of ×20 or ×40, and the Foraminifera removed to a standard slide on the end of a '00' or '000' sable brush moistened with water (the amount of moistening required can only be learned with practice —it should be remembered that the brush may dry out relatively quickly under the heat of the incident illuminator). Some investigators prefer to divide the residue into separate fractions, e.g. 1000–500μm, 500–250μm and 250–125μm for picking: this avoids the necessity for frequent magnification changes, and scanning is made easier when all the particles are more or less the same size. If the original sediment was a relatively pure clay a high proportion of the residue being examined will consist of microfossils; if it was sandy they will be scattered amongst mineral grains. A concentration of the chambered tests of Foraminifera and of paired ostracod values can be effected by completely drying out the residue (using alcohol if necessary) and floating on the heavy liquid, carbon tetrachloride. In general mineral grains sink and the air-filled Foraminifera float and can be decanted. However the proportion of shell to chamber space is high in some Foraminifera and the chambers may be infilled with authigenic minerals, notably pyrites and glauconite, after deposition, causing some Foraminifera to sink; the method cannot therefore be relied upon to give a quantitative yield.

Mounting. The standard slide used for Foraminifera consists of a recessed black card cell, with a glass cover of the same size as a standard microscope slide. The cell is either single and circular (*c*.1 cm diameter) for types or single species, or rectangular (*c*. 4 × 2 cm) and divided into 60 or 100 subdivisions for assemblages. The black base of the cell (with its subdivisions) is frequently photographic paper with an emulsion surface. Foraminifera will adhere to this satisfactorily for most laboratory purposes by placing them in position wet and allowing them to dry without disturbance. If the surface is not adhesive a *small* amount of gum tragacanth, preferably incorporating some mould preventative (e.g. thymol), may be applied. In either case it should be possible to remove the Foraminifera easily at a later date by application of a little water at the end of a sable brush. Difficulties in removal may be experienced if the specimen is allowed to adhere by a flat or slightly concave surface. Examination for purposes of identification requires magnifications up to the limit of the low-power binocular microscope, i.e. around ×200.

EXTRACTION OF MARINE MOLLUSCS

Take a sample of about 300 g. If there is much clay, wet with water and cover with 100 volumes hydrogen peroxide solution in a pudding basin. The mixture may boil if too little water is added at the beginning. Organic matter is destroyed, so pollen or plant macroscopic samples required should be removed previously.

The sample is then washed through sieves of the following mesh numbers: 3½, 7, 20, 40, 80. This can conveniently be done by placing the sample in the coarsest sieve, agitating it in a bucket of water until separation is as complete as possible, then washing the material remaining on the sieve with a jet of water into the bucket. The sieve and its contents are then dried in an oven. This procedure is repeated with the remaining four sieves, by pouring the contents of the bucket through the second sieve into another bucket, and so on. The dried sieve contents are then assembled in a column and shaken to complete the separation.

The contents of each sieve are tipped into a 25 cm Petri dish and examined systematically under a low-power binocular microscope and fragments and whole shells are picked out with a sable brush, identified and recorded.

In counting, a gasteropod shell with an apex, or one lacking so little of the apex that the remainder cannot be identified alone, is counted as one individual. A lamellibranch fragment with a hinge, or so much of hinge that the remainder cannot be identified alone, is counted as half an individual.

Specimens can be stored in cavity slides with a black background or in tubes.

POLLEN ANALYSIS

The following procedures are used in the Sub-department of Quaternary Research at Cambridge. A much fuller discussion of methods and problems is given by Faegri and Iversen,[6] Erdtman,[4] and Brown.[2]

Samples for pollen analysis are chemically treated to remove the organic and inorganic sediment which contains the pollen. Since not all samples need every treatment, it is useful to prepare temporary mounts in glycerine after each process for each different type of sediment, to check how the preparation is proceeding. Samples are washed and centrifuged between each treatment. 50 ml Pyrex or polypropylene centrifuge tubes are most commonly used, but for fresh pollen or samples which do not settle well after centrifuging, 10 or 15 ml size tubes are better. The hydrofluoric acid (HF) treatment slightly dissolves Pyrex glass.

Samples in 50 ml tubes should be centrifuged for about 4 minutes at c. 2000 r.p.m.; those in 10–15 ml tubes for 1–2 minutes. Balance alternate pairs of tubes before each centrifugation. After each chemical treatment add water and stir well. Balance the tube, centrifuge, and then carefully pour off in one movement the supernatant liquid.

It is advisable to wear rubber gloves while handling HF. HF burns should be treated as soon as possible at hospital by an injection of 10 per cent calcium gluconate under the coagulum of the burn. An ampoule of calcium gluconate should be kept in the laboratory for taking to the hospital.

Gravity separation using heavy liquids such as bromoform has also been used to separate pollen from its sediment matrix, especially with inorganic sediments with a low pollen content. The extraction of pollen from loess by this type of method has been described by Frenzel.[8]

Quantitative analyses. Various methods have been proposed for the measurement of pollen content/cm^{-3} sediment.[10] A simple method involves the addition of a known volume of a suspension containing a known concentration of an exotic pollen grain (*Eucalyptus*, *Nyssa*) to a known volume of sediment, this addition being made before the treatment processes outline below.

A sample of known volume of sediment can be removed from a core or monolith by using a 1 cm square section brass tube with a bevelled cutting edge containing a transverse slot about 1 cm from the end into which a cutting edge can be pushed to isolate a block of sediment. The tube is pushed into the sediment and then removed. The exposed end of the sample is cut flush with the end of the sampler, and the sample is then isolated in the tube by cutting it flush with the outer end of the slot. A brass square can be inserted in the slot and the sample extruded by pushing with a rod on the square. The volume of the sampling device will be measured beforehand by vernier and displacement of water by blocks of silt cut with the sampler.

The concentration of exotic pollen grains in the suspension to be added to the sample is determined by haemacytometer. A known volume (the size of which should be related to the fossil pollen concentration in such a way that the proportion of exotic pollen in the final preparation is not vastly different from that of the fossil pollen) is added to the sample and the combined sample, after thorough mixing, is subjected to the preparation techniques outlined below, as necessary. In the final pollen count, the proportion of exotic to fossil pollen is measured. Then,

$$\frac{\text{Exotic pollen counted}}{\text{Fossil pollen counted}} = \frac{\text{Amount of exotic pollen added}}{\text{Fossil pollen concentration in original sample}}$$

The fossil pollen concentration/cm^{-3} can then be calculated, and further, if the sediment column has been dated at intervals, rates of pollen influx/year/cm^{-2} can be calculated.

Breaking up sediments

Muds and clays. Put a small amount of the sample on a watch glass and add dilute HCl. If effervescence occurs put about 2 g in a beaker with 7 per cent HCl. Stir and add more HCl until effervescence ceases. Froth may be dispersed with acetone or alcohol. Filter as for peats.

Peats. Put 1 to 2 g of the peaty sample in a boiling tube with about 20 ml of 10 per cent NaOH or 5 per cent KOH. (This removes unsaturated soil colloids.) Break up with a glass rod. Leave in a bath with boiling water until the sample is thoroughly broken down and no lumps remain (10 minutes to 1 hour).

Filter through a fine steel gauze (100 mesh to the inch/, 250μm mesh) into a centrifuge tube.

Wash the sediment towards the middle of the gauze with running water. Invert and wash the sediment into a Petri dish. Keep this residue in water for subsequent examination.

Centrifuge and decant. The supernatant liquid will be very dark if much humic material is present.

Wash, centrifuge and decant. Wash and centrifuge again if necessary, i.e. if the supernatant is not fairly clear.

Oxidation (for the removal of lignin)

Pollen is easily destroyed by oxidation so the reaction should be stopped with water as soon as the sample is bleached.

Add 8 ml glacial acetic acid and 4·5 ml sodium chlorate (a saturated solution is *c.* 100 g to 200 ml water). Stir with a glass rod and add 1 ml concentrated H_2SO_4 drop by drop. Cover the tube with a glass ball or watch glass and leave it in a fume chamber until the supernatant becomes yellow or orange. Although muds and clays may be safe when left for 12 to 16 hours, peats may oxidise in a few minutes. Slow oxidation can be accelerated by adding one or more drops of concentrated hydrochloric acid.

Centrifuge and decant.

If the sample contains coarse sand, add water, stir, and decant into a small beaker. Rinse out the centrifuge tube and decant the sample back again. Centrifuge and decant.

Treatment of siliceous samples

A chemical method is given here, but flotation and gravity separation methods are also used.

Transfer about 0·5 ml of the material into a 10 ml copper, nickel or platinum crucible. Use a little 7 per cent HCl if necessary.

Add HF (40 to 60 per cent) until the crucible is about two-thirds full.

Gently boil for 2 or 3 minutes, on a sand bath or clay triangle, in a fume chamber.

When the mixture has cooled a little, pour it (using nickel tongs) into a small beaker containing about 20 ml of 7 per cent HCl, then tip this mixture into a centrifuge tube and cover it.

Centrifuge and decant.

Wash, centrifuge and decant. If sample contains clay or silt omit this washing, and carry on with the next.

Without boiling, heat with 7 per cent HCl to remove colloidal silicon dioxide and silicofluorides.

Centrifuge while hot and decant.

Add a few drops of dilute NaOH to the next washing water, centrifuge and decant.

Examine with the aid of a microscope; if clumps of inorganic material remain, repeat with further hot HCl.

Acetolysis (Acid hydrolysis of the cellulose)

This stage may be omitted if there is very little cellulose in the sample.

Add 10 ml glacial acetic acid, stir, and carefully add $\frac{1}{2}$ ml concentrated H_2SO_4.

Leave in a bath of boiling water for half an hour and stir the sample occasionally.

Centrifuge, and wash twice. To the first washing water add a few drops of dilute NaOH.

Erdtman's acetolysis

Erdtman's acetolysis is an effective method of removing organic material which cannot otherwise be destroyed. If rapid oxidation is used, it should not precede this acetolysis or the grains may become very swollen. Badly-preserved pollen may be destroyed or swollen beyond recognition. Well-preserved grains are slightly enlarged and become yellow or brown. Staining may be omitted if preferred.

Use 10 or 15 ml tubes and dehydrate the sample by centrifuging with glacial acetic acid, and then decant.

Prepare the acetolysis mixture by adding 1 ml of concentrated H_2SO_4 to 9 ml of acetic anhydride. Thoroughly mix in a measuring cylinder with a glass rod.

Add the acetolysis fluid, put a glass rod in each tube and place the tubes in a water bath at a temperature below 70°C.

Heat the water bath to 100°C and leave tubes there for 1 to 2 minutes. Stir the contents two or three times while heating, but take care that the water does not boil into the tubes.

Balance with the acetolysis fluid or with glacial acetic acid, centrifuge for 1 or 2 minutes and decant.

Wash with glacial acetic acid, centrifuge and decant.

Wash and centrifuge twice with water. Add a few drops of dilute sodium hydroxide to the first washing.

Preparation of pollen mounts (washing, staining and mounting)

Glycerine jelly. The mounts are permanent, and the mounting medium gels. Grains cannot be moved about on the slide unless it is heated first.

Pour off all supernatant liquid, carefully invert tube and drain over filter paper.

Add twice the bulk of glycerine jelly stained with safranin, stir well and warm for half a minute.

Prepare up to 5 slides for each sample, with large thin coverslips e.g. 2×250 mm, No. 0.

Use a pipette or glass rod to streak out the material on to a slide, and gently lower the coverslip. If necessary leave the slide on a hot plate to spread the jelly evenly.

Label with glass-writing ink.

N.B. Examine with a microscope one slide prepared from each sample to see if further dilution with glycerine jelly is needed. If organic material clumps and obscures the pollen, gentle movement of the coverslip may help to disperse it.

Silicone Oil.[1] A liquid mounting medium of high viscosity, thus allowing grains to be rotated on the slide. The slides are permanent. Dehydration is necessary before staining and mounting.

Wash with a few drops of water and 96 per cent alcohol. Centrifuge and decant.

Wash with 99 per cent alcohol and stain with fuchsin if desired. Centrifuge and decant.

Wash with benzene. Centrifuge and decant.

Add *c.* 1 ml benzene, transfer to small tube, add silicone oil (viscosity 2000 centistokes or more), leave to evaporate for about 24 hours. The tubes may be stored in this state.

Add silicone oil to give suitable concentration of pollen grains, but as little as possible. Place the contents of the tube on a slide, cover with 16 mm square No. 0 coverslip.

Note that for identification of fossil grains mounted in silicone oil, reference material mounted in the same way is necessary.

Preparation of pollen reference slides

Take the stamens of the flower concerned, grind in a mortar, wash through a fine sieve with a jet of alcohol into a 10 or 15 ml centrifuge tube, and centrifuge. Then treat with either a 10 minute acetolysis or by Erdtman's acetolysis. Wash twice and prepare slides with 16 mm square No. 0 coverslips. Mount as above.

Pollen counting

A high-power binocular microscope with a mechanical stage is used for the analyses. Magnifications of at least $\times 300$ are required for routine counting, but for closer examination of grains, $\times 1000$ is also essential. This entails the use of an oil immersion lens with a high numerical aperture (1.3). The most convenient immersion oil is anisol, which can be easily removed from slides and lenses. Phase-contrast equipment is also desirable, as this type of microscopy gives greater detail of pollen grain morphology, and keys to modern pollen types may be based on phase-contrast observations.

The procedure for counting is to traverse the width of the slide at standard intervals, not less than 1½ field diameters apart. As the pollen may be unevenly distributed on the slide, it is important to traverse the whole width of the slide and space the traverses at intervals over the length of the slide. A note is kept of the number of traverses, as this gives a rough idea of pollen frequency on the slide.

Each grain encountered during the traverses is identified and recorded on a form with the main pollen types already printed on it. The traverses are continued till the total number required is achieved. The positions of unidentified grains are carefully recorded by ringing them with ink or recording the mechanical stage reading. These grains should be identified as nearly as possible after the count has been made.

DIATOM ANALYSIS

The following account has been very kindly provided by Dr G. H. Evans.

Extraction and mounting

Place a small quantity (e.g. about 0·1 g dry weight) of the sample in a small conical flask.

Add a few drops of 30 per cent hydrogen peroxide to break up the sample.

Add 10 ml normal potassium dichromate and mix thoroughly.

Add slowly 20 ml of concentrated H_2SO_4 and allow to stand for half an hour. Oxidation of organic matter occurs.

Wash and centrifuge the flask contents till the supernatant is clear.

Place the contents of the centrifuge tube in a 50 ml measuring cylinder and add distilled water to 50 ml. Shake thoroughly.

Transfer a drop of the suspension to a coverslip and allow to dry at room temperature.

Prepare a slide by placing a crystal of Mikrops 163 mountant (Refractive Index 1.63) on it and melting in an oven at *c.* 110°C.

Invert the slide on to the coverslip, place in the oven at *c.* 110°C. and allow the mountant to spread across the slide.

Counting

The procedure is similar to that used in pollen analysis. A count of 600 to 1000 valves per horizon has been used as a basis for percentage frequency calculations.

REFERENCES

1 ANDERSEN, S. T. 1960. 'Silicone oil as a mounting medium for pollen grains', *Danm. Geol. Unders.*, IV Raekke, **4,** No. 1.
2 BROWN, C. A. 1960. *Palynological techniques.* Baton Rouge.
3 COOPE, G. R. 1961. 'On the study of glacial and interglacial insect faunas', *Proc. Linn. Soc. London,* **172,** 62–5.
4 ERDTMAN, G. 1943. *An introduction to pollen analysis.* Waltham, Mass.: Chronica Botanica.
5 EVANS, J. G. 1972. *Land snails in archaeology.* London: Seminar Press.
6 FAEGRI, K. and IVERSEN, J. 1975. *Textbook of pollen analysis.* 3rd edn. Copenhagen: Munksgaard.
7 FEYLING-HANSSEN, R. W. 1964. 'Foraminifera in late Quaternary deposits from the Oslofjord area', *Norges Geol. Unders.*, No. 225.
8 FRENZEL, B. 1964. 'Zur pollenanalyse von Lössen. Untersuchungen

der lössprofile von Oberfellabrunn und Stillfried (Niederösterreich)', *Eiszeitalter* u. *Gegenwart*, **15,** 5–39.

9 KUMMEL, B. and RAUP, D. (eds). 1965. *Handbook of palaeontological techniques*. London: W. H. Freeman and Co.

10 PECK, R. M. 1974. 'A comparison of four absolute pollen preparation techniques', *New Phytol.*, **73,** 567–87.

11 SPARKS, B. W. 1961. 'The ecological interpretation of non-marine mollusca', *Proc. Linn. Soc. London.*, **172,** 71–80.

Appendix 2

LACQUER METHOD OF TREATING SECTIONS

The following account includes notes which have been very kindly provided by Dr W. H. Zagwijn of the Geological Survey of the Netherlands.

The exposure should be cleaned very thoroughly and left at a near vertical angle, at about 70° to 80°. The surface should be neither completely dry nor wet, so that the right weather should be chosen for the purpose. If it is too wet the surface may be dried by a blow-lamp. A horizontal gully is made along the top of the section near its edge. A mixture of lacquer (polyvinyl acetate) and thinner (acetone) is then made in proportions about 3 : 1 and poured into the gully. It overflows down the section. Add more mixture in the gully till the whole cleaned surface has been covered. About 2 kg of the mixture will cover 1 m^2 of the section. A less dilute mixture is required with sands than with clays.

An alternative method, suitable for greater heights of section, is to spray the mixture on to the surface, then paint less dilute mixture over this.

The lacquer should be left to dry for a day, then, in order to strengthen the lacquer film, gauze is laid on the lacquer surface and incorporated by painting on undiluted lacquer. This should then be left to dry for another day, when the whole may be peeled off the section. The lacquer sheet with the adherent sediment may then be glued to hardboard and permanently framed if necessary. A thin film of the sediments is thus permanently preserved.

REFERENCES

AARIO, R. and VIITANEN, P. 1970. 'A lacquer film technique', *Geologi* (Helsinki), **21,** 9–11.

VOIGT, E. 1949. 'Die Anwendung der Lackfilm-Methode bei der Bergung geologischer und bodenkunlicher Profile', *Mitt. Geol. Staatsinst. Hamburg*, **19,** 111–29.

INDEX

(Page numbers in Bold type refer to text-figures)